Pure and Functionalized Carbon Based Nanomaterials

Analytical, Biomedical, Civil and Environmental Engineering Applications

Editor

Paweł K. Zarzycki

Department of Environmental
Technologies and Bioanalytics
Faculty of Civil Engineering
Environmental and Geodetic Sciences
Koszalin University of Technology
Koszalin, Poland

CRC Press
Taylor & Francis Group
Boca Raton   London   New York

CRC Press is an imprint of the
Taylor & Francis Group, an **informa** business

A SCIENCE PUBLISHERS BOOK

Cover credit: Images provided by the editor, Paweł K. Zarzycki.

CRC Press
Taylor & Francis Group
6000 Broken Sound Parkway NW, Suite 300
Boca Raton, FL 33487-2742

© 2020 by Taylor & Francis Group, LLC
CRC Press is an imprint of Taylor & Francis Group, an Informa business

No claim to original U.S. Government works

Version Date: 20200226

International Standard Book Number-13: 978-1-138-49169-4 (Hardback)
International Standard Book Number-13: 978-0-367-53214-7 (Paperback)

Library of Congress Cataloging-in-Publication Data

Names: Zarzycki, Pawel K., 1966- editor.
Title: Pure and functionalized carbon based nanomaterials : analytical, biomedical, civil and environmental engineering applications / editor, Pawel K. Zarzycki.
Description: Boca Raton : CRC Press, Taylor and Francis Group, [2020] | "CRC Press is an imprint of the Taylor & Francis Group, an informa business." | Includes bibliographical references and index.
Identifiers: LCCN 2019051105 | ISBN 9781138491694 (hardback)
Subjects: LCSH: Nanostructured materials. | Carbon.
Classification: LCC TA418.9.N35 P87 2020 | DDC 620.1/93--dc23
LC record available at https://lccn.loc.gov/2019051105

Visit the Taylor & Francis Web site at
http://www.taylorandfrancis.com

and the CRC Press Web site at
http://www.crcpress.com

Preface

The general concept of this book was predominantly inspired by my hobby, which was recreational scuba diving and underwater photography. I started this activity in the diving club "Rekin" (Gdańsk, Poland) in 1995. After a few years of extensive training, mainly under cold and dark Polish lakes and Baltic Sea conditions, I had an opportunity to discover the beauty of the underwater world and transparent seawater surrounding Elba Island at the Tyrrhenian Sea. Then, I moved with my family to Newcastle (NSW, Australia), where I worked as a visiting scientist at John Hunter Hospital, focusing on separation science and analysis of biological samples. During that time I did several diving trips, enabling macro underwater photography at night, mainly with diving staff from Charlestown Diving Academy, over the great diving spots of Port Stephens and Newcastle areas. The picture on this book cover presents the dark background, which is a fossil coal, and fragile marine snails, called bubble shell (*Cephalaspidea*), moving on it. Such rather unique underwater scenery is typical for the Newcastle seashore, where several shallow water areas are covered by natural organic carbon layers. In my opinion, this macro picture can be a good illustration of various carbon forms available on the Earth, starting from raw carbon transformed from living organisms during long-term geological processes, and the beauty of carbon-based biopolymers forming a variety of living bodies. Presently, I changed my activity more to sailing, which still enables me to combine my hobby and work, most recently resulting in non-commercial, non-hypothesis driven studies and self-funded RoSSY (research on small sailing yacht) project. This research relates to bioanalytical studies, involving elaboration of new micro separation analytical devices and testing various biopolymers and carbon materials as the extraction and samples collection media, mainly for quantification of low-molecular mass micropollutants and biomarkers in surface water ecosystems.

There is an increasing interest in research related to various carbon-based classical and new nanomaterials, and they may be considered the most versatile objects, which are presently studied by various research groups worldwide. This book consists of 14 chapters dealing with a number of key research topics and applications of pure and functionalized carbon nanomaterials and their hybrid nanocomposites. In general, individual chapters belong to three groups describing: (i) carbon nanoparticles and layers synthesis, (ii) analytical aspects of carbon nanomaterials and their characterisation under different conditions, as well as (iii) various applications of carbon nanoparticles. Particularly, the first chapter prepared by George Z. Kyzas and co-authors describes several simple methods for the synthesis of nanographene molecules in different forms that may be used as high performance electrode materials for supercapacitors, ultra-sensitive strain sensors, and Hole Transporting Material for efficient and stable Perovskite Solar cells. The method to synthesize graphene molecules on TiO_2 are described in a comprehensive manner, and the new materials using several orthogonal physicochemical approaches including SEM-EDX, TEM-SAED, XRD, BET, UV-Vis, cycling voltammetry, AFM and XPS techniques are extensively characterized. Moreover, this research group confirmed that a nanocrystalline TiO_2 film injected with polyphenylene compounds, via the Scholl reaction, can be oxidized to graphene molecules. Snigdha Gupta discusses the modern approaches for synthesis of diamond-like carbon and amorphous carbon thin films by liquid electrodeposition techniques from different organic solvents. This chapter also focuses on optoelectronics, biomedical, tribological applications of these materials, including potential material science and civil engineering applications of carbon protective layers for large objects, e.g., solar cells used in space. Manoj B. reports advances in research related to production of nanocarbons, which can be derived from various precursors, such as fullerene, pristine graphite or graphene oxides, lignite, coal, carbon fibers, and humic organic substances by chemical, hydrothermal, or solvothermal synthesis. Biomass-derived carbon compounds were of interest for Darminto research group, with particular emphasis on potential applications as electronic and magnetic materials. Chapter 4 provided by this team reviews a green synthesis method of carbon-based materials using waste biomass products, for example, coconut shell and coconut water. This group examined structures and magnetic properties of resulting nanomaterials, especially as candidates for radar absorbing materials and supercapacitors. Sining Yun and co-workers have reported simple and cost-effective methods, including hydrothermal carbonization and microwave pyrolysis, which may be used for effective preparation of bio-based carbon materials derived from various biomass, and also to utilize the waste biomass to produce green energy.

Two chapters provided by Madan L. Verma and co-authors focused on extensive literature reviews covering applications of carbon-based nanobiocatalysis for sustainable biofuels production (Chapter 6) and various types of carbon particles which may be used as biosensors, specially in agriculture and food industry, allowing efficient detection and quantification of micropollutants at the nano to picograms range (Chapter 8). Chapter 7 prepared by Nabisab M. Mubarak and co-authors deals with the whole range of new functionalized carbon nanomaterials combined with other nanoparticles. They describe several examples of such complex nanocomposites with enhanced properties and high potential to utilize these materials for energy storage and sensing applications in the electronics industry. Marcela C. Rodríguez's research group prepared Chapter 9 focusing on analytical applications of carbon nanotubes. The authors have extensively discussed several electrochemical detection strategies and approaches for selective quantification of environmental hazards agents, such as arsenic contaminations, as well as lead, cadmium, mercury, chromium, and other priority detrimental heavy metals. Contribution provided by Clara Marquina and M. Ricardo Ibarra has summarised the latest literature and their own experimental data focusing on application of carbon coated magnetic nanoparticles in life sciences. This chapter gives a broad overview of the interdisciplinary investigations involving Fe@C nanoparticles. Experimental data provided by the authors confirmed that carbon nanoparticles may have an excellent biocompatibility, both in animals and plants, and can be used for detecting diseases and for delivering active molecules driven by the magnetic field. Adam Voelkel and Beata Strzemiecka used inverse liquid chromatography for characterization of carbon materials. They reported interesting studies related to surface properties of carbon materials. Particularly, this group has examined activity of carbon blacks, carbon fibers, activated carbon, and carbon nanotubes by means of inverse gas chromatography, both at infinite dilution and finite concentration. The use of Hansen Solubility Parameters leads to interesting results in the examination of carbon nanotubes and carbon fibers/epoxy resin composites. Yoshihiro Saito's research group highlighted the problem of separation of fullerenes and polycyclic aromatic hydrocarbons using chromatography, mainly high-performance liquid chromatography, working in different modes. They provided experimental results focusing on characterization of classical and novel chromatographic stationary phases on the basis of the systematic retention behavior studies for fullerenes and PAHs. Such data are necessary for designing novel, non-expensive, and highly effective miniaturised separation systems. The chapter by Kathryn A. Mumford and Jack G. Churchill describes the applications of granular activated carbon materials for water treatment. They discuss the basic principles of adsorption kinetics and equilibrium, and introduced predictive modelling. These principles have then been used in outlining two models for determining process parameters for the design of full-scale adsorption systems. Moreover, the authors predicted the performance of existing adsorbents for newly identified contaminants of concern, and discussed the role of biological activity and development of biological activated carbon filters. Additionally, they presented a case study of granular activated carbon that has been used for designing permeable reactive barriers in Antarctica, which can be considered as an interesting application in environmental remediation. The last chapter presents a summary of an extensive literature search, where the authors discuss the latest trends in applications of various carbon related nanomaterials as active media for analytical, pharmaceutical, biomedical, and wastewater technological processes. This chapter also consists of a detailed protocol for preparation of graphene oxide particles from graphite powder using modified Hummers and Offeman procedure. Moreover, they demonstrated an ability of GO of tuning the physicochemical properties of cellulose strips, which were used as a stationary phase for planar electrochromatographic separation of selected dyes. This modification, particularly using GO layers formed as horizontal walls, can increase the separation efficiency of target analytes. Reported phenomenon may be a starting point for elaboration of new extraction and/or separation systems based on hybrid carbon/biopolymers supports.

I would like to express many thanks to all of the authors for their feedback and interesting contributions, all the referees for their critical comments and time devoted to evaluation of the manuscripts. I really appreciate the helpful cooperation of the Editorial Department staff, particularly, Raju Primlani and Danielle Zarfati for helping me solve all technical problems with book cover preparation, manuscripts handling, and for their generous cooperation in preparation of this book. I truly appreciate fruitful and long-term research cooperation with Katarzyna A. Mitura concerning detonation nanodiamonds. I would also like to express many thanks to Robert Nowak for helping me with an extensive search for the potential contributors to this book as well as to my research team, namely: Renata Świderska-Dąbrowska, Krzysztof Piaskowski, Lucyna Lewandowska, Bożena Fenert, and Michał J. Baran for literature searches, references selections, and experimental work involvement. Furthermore, their enthusiastic approach for all of the experimental work mistakes allowing seeing the new separation phenomenon based on graphene oxide, which was not visible using initial hypothesis driven experiment, is particularly appreciated. Finally, I would like to thank Mr. Ryszard Gritzman, a professional scientific photographer at Koszalin University of Technology for his help with SEM and EDS analysis.

Aug 1, 2019   **Paweł K. Zarzycki**

Department of Environmental Technologies and Bioanalytics
Faculty of Civil Engineering, Environmental and Geodetic Sciences
Koszalin University of Technology
Śniadeckich 2, 75-453 Koszalin, Poland

Contents

CHAPTER 1

Nanographenes

Efstathios V. Liakos, Ramonna I. Kosheleva,
Athanasios C. Mitropoulos and *George Z. Kyzas**

1. Introduction

Graphene is considered to be the most "hot" material of recent years due to its numerous properties (mechanical, optical, environmental, etc.). In 1986, Boehm et al. (Boehm et al. 1986) described the structure of graphite having a single atomic sheet in detail (Boehm et al. 1986). The major turn in graphene history was done during 2000s, where it was proved that 2-D crystals (like graphene) did not have thermodynamic stability, suggesting their non-existence at room temperature (conditions) (Nemes-Incze et al. 2008). Especially, graphene sheet is thermodynamically unstable if its size is less than about 20 nm (graphene is the least stable structure until about 6000 atoms), and becomes the most stable fullerene (as within graphite) only for molecules larger than 24,000 atoms (Shenderova et al. 2002).

The leader of graphene science is Konstantin Novoselov, who successfully isolated and characterized an exfoliated graphene mono-layer with various techniques (Novoselov et al. 2004); A.K. Geim and K.S. Novoselov were awarded with the highest honor (Nobel Prize) in 2010 for their impact on graphene science. But what is graphene? The reply was clear and given by IUPAC: graphene is a carbon layer (single) of graphite, having a structure/nature similar or analogous with an aromatic hydrocarbon (polycyclic) of quasi infinite size (Iupac 1995). That means that graphene is a flat mono-layer of hybridized sp^2 atoms of carbon, which are densely packed with each other into an ordered two-dimension honeycomb network (Ivanovskii 2012). A hexagonal unit cell of graphene comprises of two equivalent sub-lattices of carbon atoms, joined together by sigma (σ) bonds with a carbon-carbon bond length of 0.142 nm (Avouris and Dimitrakopoulos 2012). Each carbon atom in the lattice has a π-orbital that contributes to a delocalized network of electrons, making graphene sufficiently stable compared to other nanosystems (Zhu et al. 2010). The applicability of graphene is based on an advantageous network provided by this material: combination of high three-dimensional aspect ratio and large specific surface area, superior mechanical stiffness and flexibility, remarkable optical transmittance, exceptionally high electronic and thermal conductivities, impermeability to gases, as well as many other supreme properties. Due to all of the above, Novoselov characterized it as a miracle material (Novoselov et al. 2012).

Hephaestus Advanced Laboratory, Eastern Macedonia and Thrace Institute of Technology, Kavala, GR 65404, Greece.
* Corresponding author: kyzas@teiemt.gr

One of the most important derivatives of graphene is the oxidized form, graphene oxide. The large scale production of functionalized graphene at low cost should result in good adsorbents for water purification (Kemp et al. 2013). So, anyone can imagine the advantages of the use of graphene in nanoscience. Nanomaterials are a special topic of recent research, and are a milestone of nanoscience and nanotechnology. These materials have a broad area of development which is rapidly growing day by day. Their impact to commercial applications is huge, as well with respect to academia and education.

At this point, it is mandatory to give a brief definition of nanomaterials. So, nanoscale materials are a series of substances/compounds, in which at least one dimension has a smaller size than 100 nm. One nm is equal to 10^{-6} mm, meaning 10^{5} times smaller than a possible size of a human hair (obviously measuring the diameter). Another question which must be replied is why there is such a big interest in nanomaterials. The answer is not difficult. At this scale, interesting optical, attractive, electrical, and some other different properties rise. These innovative properties have the potential for extraordinary effects in various gadgets, especially in biomedicine, biochemistry, or other relatively similar fields. So, in the present chapter, special attention is given to the synthesis procedures of various forms of nanographenes and their respective characterizations.

2. Synthesis Procedures

2.1 Synthesis of nanographene oxide

In a study by Kim et al. (Kim et al. 2011), preparation of nanographene oxides (NGOs or nGOs) by a two-step oxidation was achieved. The first step of this process involves preparation of graphite oxide from graphite powder following Hummers method. After that, the yielded graphite oxide was oxidized again, now with potassium permanganate $KMnO_4$ as the oxidant medium. For the final step of NGOs synthesis, a mild sonication process was implemented. The detailed experimental process is as follows:

0.05 g graphite oxide was inserted into 50 mL solution of concentrated H_2SO_4, while $KMnO_4$ was added slowly to the above mixture under vigorous stirring. Regarding the concentration of $KMnO_4$, different sizes of NGOs were finally obtained, namely NGO-100 (0.05 g (100 wt% of graphite oxide)), NGO-200 (0.10 g (200 wt% of graphite oxide)), and NGO-300 (0.15 g (300 wt% of graphite oxide)), where NGO-X indicates X wt% of $KMnO_4$ compared to the weight of graphite oxide. For NGO-100, the duration of vigorous stirring process was 30 minutes at 35°C, while for NGO-200 and NGO-300, the same process lasted for 90 and 150 minutes at 45°C, respectively. The resultant mixture was cooled in an ice bath. After reaching the desired temperature, 100 mL of H_2O_2 solution (95 mL of H_2O + 5 mL of 30 wt% H_2O_2) was added very slowly under vigorous stirring for 60 minutes. In order to obtain exfoliated NGOs from nanosized graphite oxides, the mixture was further ultra-sonicated for 30 minutes. The NGO-TiO$_2$ composites of nanographene oxide-coated TiO$_2$ (NGOTs) were prepared from this resulting product (Kim et al. 2011).

2.1.1 Preparation of r-NGOT, r-LGOT, Pt/r-NGOT, and Pt/r-LGOT

In order to prepare a series of NGOT composites with different contents of NGO, TiO$_2$ (Degussa, P25) was mixed with various amounts of NGO-300. The process continued under ultra-sonication in order to disperse 0.5 g of TiO$_2$ in H_2O. After that, a calculated amount of NGO (0.4, 0.7, 2, 4, and 6 wt% of TiO$_2$) was introduced to the above solution. The aging of the mixture was achieved by an overnight vigorous stirring.

Afterwards, the mixture was filtered, and an aqueous solution of 1 M HCl containing deionized H_2O and 1% H_2O_2 was used to wash the mixture in order to remove impurities. To prepare LGOT, a 0.7 wt% of LGO compared to TiO$_2$ was exfoliated from graphite oxide with ultra-sonication in

H_2O. Then, the obtained LGO was mixed with a dispersed (ultra-sonication in H_2O) TiO_2 suspension. Before the filtering and washing process, the mixture was aged overnight.

The yielded NGOT (or LGOT) was a brownish powder, which was then suspended in a 1 M methanol solution as an electron donor, in order to reduce photocatalytically NGOT (LGOT) to r-NGOT (r-LGOT). The photocatalytic process was achieved under UV irradiation for 30 minutes with the use of a mercury lamp of 200 W. In order to synthesize platinized r-LGOT (Pt/r-LGOT) and platinized r-NGOT (Pt/r-NGOT), the photocatalytic deposition process was achieved in 1 M solution of methanol under the presence of NGOT (LGOT) and chloroplatinic acid (H_2PtCl_6). The reduction of NGO (LGO) and the platinum deposition in Pt/r-LGOT (Pt/r-LGOT), were achieved simultaneously under UV irradiation for 45 minutes from a mercury lamp of 200 W. Finally, among TiO_2, r-LGOT, and r-NGOTs, the amount of Pt was adjusted at 0.05 wt percent (Kim et al. 2011).

2.2 Preparation of nanographene sheets

Wang et al. (Wang et al. 2015) prepared nanographene sheets (NGs) according to the burn-quench method. Briefly, a 500 mL filter flask was used in order to transfer 200 mL of deionized H_2O. After that, enough CO_2 was bubbled into the 500 mL filter flask, meanwhile 1.8 g Mg ribbons were ignited under the CO_2. Then, the resultant solution was centrifuged, and precipitate was obtained. To remove Mg and MgO impurities, the precipitate was washed several times using 1 M HCl and distilled H_2O. Finally, after vacuum filtration and drying for 12 hours at 70°C, NGs were ready. It is worth noting that the described experimental conditions provide 1 g of NG from 25 g Mg (Wang et al. 2015).

2.2.1 Fabrication of hybrid a-MnO₂ nanoparticles (MPGs)

At first, 10 mmol of $(NH_4)_2 SO_4$ and $KMnO_4$ were diluted in 200 mL deionized H_2O under stirring for 1 hour. The obtained mixture was transferred into a 500 mL filter flask. Then, a satisfactory quantity of CO_2 was bubbled into the filter flask, while 1.8 g Mg ribbons were ignited in the CO_2 enriched atmosphere and kept for 12 hours at 40°C. Then, the solution was centrifuged and the precipitate was collected. Mg and MgO impurities were removed by washing the precipitate several times using 1 mol/L HCl and distilled H_2O. Finally, the yielded powder was dried under vacuum for 12 hours at 70°C (Wang et al. 2015). The synthesis method of MTGs is same as that of MPGs, except for the fact that the solution was introduced into a Teflon-coated container and autoclaved for 12 hours at 140°C (1°C/min) (Wang et al. 2015).

2.3 Preparation of nanographene mesoporous architectures

In a study by Zhang et al. (Zhang et al. 2013), the preparation of nanographene mesoporous architectures was achieved by using a burn-quench method. Briefly, a flask was filled with NH_4HCO_3 aqueous solution and CO_2 gas (0.1 M, 100 mL), and Mg ribbons were ignited with the use of an alcohol lamp. This simple burn-quench procedure results in a black mixture which consists of carbon and other impurities. MgO and Mg impurities were dissolved in a 2 M HCl solution by ultra-sonication dispersion. In order to obtain nanographene mesoporous architectures, vacuum filtration and drying procedure were followed for 12 hours at 608°C. Additionally, in order to achieve further elimination of the defects and improve the crystallinity of the as-prepared materials, thermal annealing was performed at 800°C for 2 hours under Ar atmosphere with a heating rate of 58°C min^{-1} (Zhang et al. 2013).

2.4 Nanographene platelet-type carbon nanofiber (CNF)

Mitani et al. (Mitani et al. 2012) preferred to use a platelet-type carbon nanofiber (CNF) over graphite powder as the starting material for nanographene preparation. The CNF finally consisted of stacked

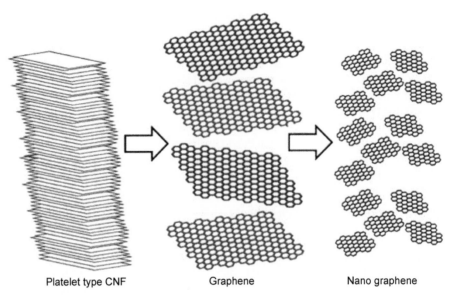

| Platelet type CNF | Graphene | Nano graphene |

Figure 1.1: Image of Scheme from CNF to nano-graphene. Reprinted with permission taken by Elsevier (Mitani et al. 2012).

graphite platelets in a perpendicular direction to the fiber axis. Therefore, CNF structure is more suitable for preparing nanographene by exfoliation, compared to the nanographene that is synthesized from at least micrometer-sized graphite powder. Figure 1.1 displays a schematic representation of graphene and nanographene fabrication from platelet-type CNF. The specific work proposes an inexpensive and convenient method for high-yield preparation of graphene nanosheets. The modified Hummers method was used in order to exfoliate the platelet-type CNF. This process is mainly used for energy storage devices, such as EDLC electrodes, catalytic fuel cell supports, and lithium-ion secondary batteries (Mitani et al. 2012).

2.4.1 Nanographene synthesis from platelet-type CNF

Platelet-type CNF was also proposed in another study by the research team of Mitsubishi Materials Electronic Chemicals Co. Ltd. Firstly, CNF was mixed with sulfuric acid H_2SO_4 (98%) and stirred at room temperature. After 5 minutes of stirring, potassium permanganate $KMnO_4$ was added to the mixture in ice water. Then, the solution was heated up to 35°C and stirred for additional 30 minutes. After H_2O addition, the mixture was stirred further for 1 hour at 100°C. In order to stop oxidation, H_2O and H_2O_2 were inserted before filtrating and the mixture was washed with HCl and H_2O. The final mixture was filtrated one more time, and sonicated for 2 hours. According to this procedure, the produced graphene oxide (GO) was reduced by H_2NNH_2 and NH_3 at 100°C. Finally, the resulted reduced GO (rGO) was dried in a vacuum oven at 160°C (Mitani et al. 2012).

2.5 Catalyst-free growth of nanographene films on various substrates

A catalyst-free graphene growth approach is proposed by Zhang et al. (Zhang et al. 2011). In the relevant study, nanocrystalline graphene is reported to grow directly on various substrates at a relatively low temperature (550°C), resulting in crystals of 10 nm. A remote chemical vapor deposition system (r-PECVD) was used for the graphene growth. Commonly, the temperature applied for CVD graphene growth is about 900–1000°C, but in the case of absence of catalyst, the growth and the growth temperature are much lower, allowing the use of substrates like glass. The precursor in this case was pure methane at a substrate temperature of 550°C (Zhang et al. 2011).

2.6 Preparation of precursor a-Ni(OH)₂-DS powders

In a study by Peng et al. (Peng et al. 2014), a-Ni(OH)$_2$-DSs were prepared by hexamethylenetetramine (HMT) hydrolysis of an aqueous solution of sodium dodecyl sulfate (SDS) and nickel chloride NiCl$_2$. In a typical preparation procedure, 0.20 M SDS, 0.30 M HMT, and 0.10 M NiCl$_2$ 6 H$_2$O were blended in a 100 mL three neck bottle. The procedure also involved mild stirring for 24 hours at 95°C. After the reaction was isolated, a suspension containing a green precipitate was washed with distilled H$_2$O in order to remove the impurities from the excess surfactant. The obtained product finally dried at room temperature (Peng et al. 2014).

For the preparation of hollow graphene nanoshells (HGNs), a-Ni(OH)$_2$-DS powders were first heated for 2 hours up to 200°C, with a rate of 5°C min^{-1} at Ar atmosphere, and in a second phase, for another 3 hours to 800°C. The sample was synthesized after treating 100 mg of the carbonized product with a 37 wt% HCl solution of 100 mL for 1 hour at room temperature. Next, the same solution was further left for a period of 12 hours to 70°C in order to obtain high pure HGNs. To remove the impurities, the yielded HGNs were filtered and rinsed with distilled water, and finally the sample was dried at 80°C (Peng et al. 2014).

For the following methodology, sulfur powder purchased from Alfa Aesar with 99% purity was used without extra purification. In order to accommodate the sulfur into the HGNs inner space, a coheating procedure was implemented. Typically, HGNs and sulfur powder with a mass ratio 3:7 were mixed in an agate mortar and strongly grounded for 10 minutes. The as-yielded Hollow graphene nanoshells-Sulfur (HGN-S) mixture was transferred to a sealed quartz bottle and heated for 4 hours at 155°C. The new composite cathodes HGN-S were obtained after the sulfur was successfully impregnated under vacuum condition into the hollow space of the HGNs (Peng et al. 2014).

2.7 Well-defined thiolated nanographene

A thiolated nanographene was used to increase the efficiency and the stability of Perovskite solar cells by Cao et al. (Cao et al. 2015). Specifically, a thiolated nanographene perthiolated tri-sulfur annulated hexa-peri-hexabenzocoronene (TSHBC) was used as the Hole Transporting Material (HTM) in the pristine form of Perovskite Solar Cells (PSCs). Pb-S coordinator bonds were formed at the interface of HTM and perovskite due to the presence of the thiol groups at the periphery. Furthermore, due to the tight binding of TSHBC, a rapid extract charge from perovskite is observed, preventing a high energy loss at the interface. In order to enhance the hole transporting property within HTM, the material was doped with graphene sheets into TSHBC. By adjusting the HOMO level of HTMs, the HOMO level of TSHBCs seems to influence the overall performance of HTMs in PSCs significantly. The biggest challenge of commercialization of PSCs is the long-term stability. To this end, a large number of techniques have been proposed and investigated. One promising approach to improve the stability of PSCs as an effective molecular sealing is the use of hydrophobic thiolated nanographenes.

As HTM for the fabrication of PSCs, TSHBC was used, which was prepared according to a new method. The new fabricated PSCs had a typical p-i-n configuration of FTO/TiO$_2$/CH$_3$NH$_3$PbI$_3$/TSHBC. In this case, a typical fabrication process was followed, where the mesoporous TiO$_2$ layer and the compact TiO$_2$ layer are applied on a fluorine-doped tin oxide. For electron extraction and export, a conducting glass substrate was preferred. In this study, Spiro-OMeTAD was replaced with TSHBC. This new HTM, of 20 nm thickness, was put on top of the perovskite layer. The process was entirely completed under ambient environment, and a relative humidity of 45% was achieved in the entire fabrication process (Cao et al. 2015).

2.8 Graphene molecules on TiO₂

Ji et al. (Ji et al. 2014) used a polyphenylene as a precursor (P1), which was modified with one or more carboxylic acid functional groups. FTIR spectroscopy revealed that the exposure of nanocrystaline

TiO_2 film for 15 hours at room temperature into a dichloromethane solution of P1 (conc. 0.1 mM) led to adsorption of P1 onto the film. The TiO_2/P_1 film was first washed with dichloromethane, and then it was exposed to an Ar saturated mixture of $FeCl_3$ (10 mg/mL) in CH_3NO_2/CH_2Cl_2 (1:3 v/v) at room temperature. The film became colored in a few seconds, confirming the coupling of the aromatic rings of the epiphany adsorbed P_1. After 1 hour under the above experimental reaction conditions, the film was washed with methanol and dichloromethane, following air drying. The dark red color of the film declares the deposition of graphene molecules (GMs) on the surface of TiO_2 (Ji et al. 2014).

2.9　NG films growth

According to the study of Jing Zhao et al., a remote plasma enhanced chemical vapor deposition (RPECVD) was used in order to grow nanographene (NG) films on fluor-phlogopite mica. This growth procedure was achieved on substrates at low temperatures, promoting large scale (up to 4 in) growth of graphene, without additional transfer process.

A typical growth approach was achieved using pure methane as a precursor, plasma power of 100 W, at 525°C, and with a gas pressure of 0.2 Torr. The size of nanographene islands that formed on the surface of the film was 10 nm. The different growth durations contributed massively to the thickness of the NG islands. Furthermore, nanographene films presented controllable conductivities by adjusting the thickness of their layers from one to a few layers (Zhao et al. 2012).

The fabrication process of NG film was achieved using mica as a starting material. The prepared samples were spin coated at 4500 rpm by a S1813 series photo resist before they were formed into strips by UV lithography (Karl Suss, MA6). The fabricated NG patterns were modified by a reactive ion etching system (Oxford Instruments Company, Plasma Lab 80 Plus), using O_2 pressure of 0.1 Torr, plasma power of 100 W, and plasma etching time of 11 seconds. The step following the fabrication of NG film is the preparation of contact electrodes. This process was achieved by implementing standard lifting off techniques during the metal deposition of Ti/Au (2 nm/30 nm) via the e-beam evaporation procedure. In order to achieve the optimization of contacts before any electrical measurements, the as-fabricated samples were annealed at 380°C for 1 hour in H_2 atmosphere. The NG strips presented zig zag morphology. The sheet resistance was 5.5 MΩ/sq, with the NG strips sized at 1.4 mm long with 2 μm width (Zhao et al. 2012).

2.10　Nanographene-conducted hollow carbon spheres—Experimental

In this study by Shubin Yang et al., silica/space/mesoporous-silica spheres were synthesized. The specific material was chosen as template because it consists of straight channels in mesoporous shells and tunable spacing between the shell and the core. Hexadodecyl was employed as the building block substituting HBC (HBC-C12). Hexadodecyl is distinct for its feasible formation of stable columnar liquid-crystalline phases and its good solubility. In addition, the silica/space/mesoporous-silica spheres were impregnated three times with a tetrahydrofuran (THF) solution of HBC-C12, under magnetic stirring at 40°C, with a weight ratio of HBC-C12 template fixed at 1:1. After that, the specimens were thermally treated for 5 hours at 400°C, and then further treated for 5 hours in argon at 700°C or 1000°C. Finally, the nanographene-constructed hollow carbon spheres (NGHCs) afforded by etching in a NaOH solution, named NGHCs-700 and NGHs-1000, respectively (Yang et al. 2010).

2.11　Synthesis of reduced graphene oxide

A direct synthesis of reduced graphene oxide (rGO) by chemical exfoliation of graphite was reported by Betancur et al. In order to prepare the reduced graphene oxide, mineral graphite without previous treatment was employed. Initially, 0.4 g of graphite was added to a 95% H_2SO_4 and 68–70% HNO_3 concentrated solution (3:1 vol. ratio). Before beginning the oxidation process, the solution was sonicated at constant temperature of 60°C for 90 minutes at 40 KHz. Thermal oil was used to heat the mixture up to 70°C, in order to prevent contraction between layers of graphene due to temperature loss. After that, the oxidation began with constant and intense stirring at 90°C for 24 hours. Then, the mixture was cooled down to room temperature and sonicated for 30 minutes. Finally, the resultant oxidized aliquot solution was slowly inserted in a separation funnel containing 5 mL of 28 v/v NH_4OH. The yielded mixture was recirculated several times to homogenise the reaction. Finally, exfoliation of the graphene sheets and reduction of functional groups were achieved by ammonium hydroxide (Betancur et al. 2018).

For the preparation of rGO, a possible mechanism process is proposed. The proposed process is based on the assumption that the mixture H_2SO_4/HNO_3 reacts similar to as it does in Hummers method, forming graphite intercalation compounds (GIC) complexes. It is claimed that this mechanism induces the exfoliation process, as it is presented in step 1 of Figure 1.2. The GIC that was formed during the process was mainly bisulfate graphite (HSO_4-Graphite), which is the key factor for the separation and oxidation of graphene sheets from the graphite. Furthermore, the electrochemical potential and the surrounding medium play an important role for the rate of formation, ranging from 3 to 5 minutes. According to Dimiev's study, GIC are formed following step 2, and is also described in Figure 1.2. A fundamental factor for the formation of an intercalation component of HSO_4-graphite as well as for the transformation of pristine graphite oxide (PGO) is the sulfuric acid reaction with an oxidizing agent, such as HNO_3. The specific oxidizer reacts in a similar way to $KMnO_4$ in the Hummers method, as it is shown in steps 1–3 (Figure 1.2). The above reaction has a positive enthalpy formation ($\Delta H+$) because of the graphite bisulfate that is needed for an oxidizing agent.

Another way to obtain graphene sheets is by adding other oxidizing agents or increasing the temperature up to 80–90°C. In this manner, Van Der Waals forces can be broken more efficiently. Steps 4 to 5 correspond to the reduction process that is used to minimize the hydroxyl and epoxy functional

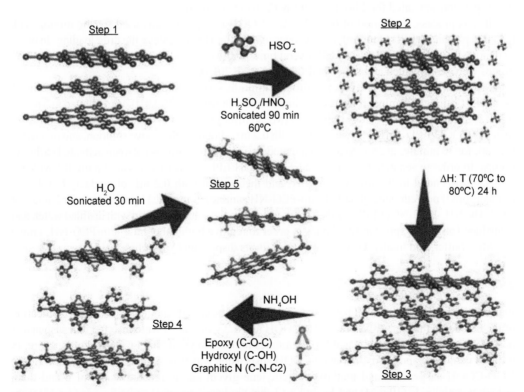

Figure 1.2: Reaction mechanism during the chemical exfoliation process. Reprinted with permission taken by Elsevier (Betancur et al. 2018).

groups after the oxidation process. During the process, the exothermic neutralization reaction was generated with ammonium hydroxide. Ammonium hydroxide reacts with GIC compounds, increasing the reaction temperature around 70°C. This reaction provides the necessary amount of energy in order to separate, reduce, and functionalize the graphene sheets from graphite. Following this, water culminates the separation of each layer for the removal of the remaining compounds. Therefore, GO has not so many application compared to rGO, which exhibits properties closer to those of a graphene sheet. The reduction of GO functional groups is achieved by using compounds, such as borane, sodium borohydride, hydrazine, ammonia, and others (Betancur et al. 2018).

2.12 Preparation of NGO for targeted chemotherapy and photothermal therapy

According to the study by Hung-Wei Yang et al., NGO was synthesized by a modified Hummer's method. At first, 0.125 g $NaNO_3$, 250 mg graphite platelet, and 12 mL 98% H_2SO_4 were mixed under magnetism in a flask with volume 100 mL, keeping temperature under 5°C in an ice bath, while 0.75 g $KMnO_4$ was slowly added. In addition, the mix was heated to 100°C, followed by slow insertion of 12 mL DI. For the incubation period (30 minutes), the temperature of the mix was increased 2°C. The other step includes 10% H_2O_2 in a 50 mL volume flask, which was inserted to the solution until the cessation of evolution of gas. In order to obtain large-scale GO (LGO), the solution was centrifuged at 11,000 rpm. Then, in order to remove the impurities, the precipitate (LGO) was washed several times with DI water. In addition, an ultrasonic liquid processor 2020 was used to ultrasonicate the precipitate at 800 W for 8 hours, and then centrifuged for 30 minutes at 14,000 rpm. After this process was used, syringe filters 0.2 μm to filtrate three times the supernatant, and the collected NGO was subjected to further modification. Then, an aqueous suspension of 10 mL NGO (2 mg/mL) was used for the addition of 10 mL NaOH (12 mg/mL). Therefore, in order to obtain NGO with COOH groups, the suspension sonicated for 2 hours at 800 W (Yang et al. 2013).

In this process, a solution of 0.5 mL NGO-COOH (2 mg/mL) was used for the mixing with 0.5 mL Epirubicin (EPI) (4 mg/mL) at 4°C overnight with pH 8 to achieve the drug loading. In order to remove the unbound EPI was rinsed several times with DI water. The obtained NGO-COOH/EPI groups were resuspended and stored at 4°C (Yang et al. 2013).

Moreover, 27 mg of N-hydroxysulfosuccinimide sodium salt (sulfo-NHS) and 24 mg of 1-ethyl-3-(3-dimethylaminepropyl) carbodiimide hydrochloride (EDC HCl), were diluted in 2 mL of 0.5 M 2-(N-morpholino)ethanesulfonic acid (MES buffer pH = 6.3) in the dark. In addition, 0.2 mL of NGO-COOH/EPI (3 mg/mL) was mixed with 0.2 mL aliquot of the above solution, and reacted in the dark for 30 minutes at 25°C to permit the modulation of amine bonds between activated carboxyl groups. In order to separate the activated NGO-COOH/EPI, it was rinsed with 0.8 mL 0.1 M MES buffer, resuspended in 0.2 mL MES buffer, and then mixed with 0.2 mL of pentaerythritol tetra (aminopropyl) polyethylene glycol (4-arm-PEG-NH_2-amine) (3 mg/mL) by vortexing for 2 hours at 25°C. The NH_2-PEG-NGO/EPI was then segregated from the solution, rinsed with distilled water, and centrifuged at 14,000 rpm for 30 minutes in order to remove both unbound 4-arm-PEG-NH_2-amine and MES buffer, and finally, DI water was used for its dispersion (Yang et al. 2013).

2.13 Synthesis of NGO-PEG

In this study by Chen et al., NGO-PEG was prepared by a modified Hummer's method by sonication of the GO flake. In brief, 1.0 g of Cl-CH_2-COOH and 1.2 g of NaOH were inserted to a suspension of 2 mg/mL of NGO, and sonicated for 30 minutes in order to yield NGO-COOH. Furthermore, in order to neutralize and purify, the nGO-COOH suspension was washed and filtrated several times. NGO-COOH suspension and methoxypolyethylene glycol amine reacted at pH of 6 and then, N-hydroxyccinimide (10 mM) and 1-ethyl-3-(3-dimethyllaminopropyl) carbodiimide (4 mM) were inserted to the above suspension. Then, in order to purify the obtained NGO-PEG, it was centrifuged

Table 1.1: Preparation methods for nanographenes.

Nanographene	Preparation method	Reference
NGO-100; NGO-200; NGO-300; NGO-TiO$_2$; NGO-TiO$_2$; r-NGOT; r-LGOT; Pt/r-NGOT; Pt/r-LGOT	Hummers method with two-step oxidation process	(Kim et al. 2011)
Nanographene sheets and hybrid a-MnO$_2$ nanoparticles	Burn-quench method	(Wang et al. 2015)
Nanographene mesoporous architectures	Burn-quench method	(Zhang et al. 2013)
Nanographene platelet-type carbon nanofiber (CNF)	Modified Hummers Method	(Mitani et al. 2012)
a-Ni(OH)$_2$-DS powders and Hollow graphene nanoshells (HGNs)	Catalytic self-limited assembly of nanographene on *in situ* formed nanoparticles (self-assembly method)	(Peng et al. 2014)
Hollow graphene nanoshells–Sulfur cathodes (HGN-S)	Catalytic self-limited assembly of nanographene on *in situ* formed nanoparticles (self-assembly method)	(Peng et al. 2014)
Nanocrystalline graphene growth	Remote chemical vapor deposition system	(Zhang et al. 2011)
Grow nanographene (NG) films	Remote plasma enhanced chemical vapor deposition	(Zhao et al. 2012)
Nanographene strain sensors	UV lithography	(Zhao et al. 2012)
Nanographene strain sensors	Reactive ion etching system	(Zhao et al. 2012)
NGO for targeted chemotherapy and photothermal therapy	Hummers method	(Yang et al. 2013)
NGO-PEG	Modified Hummers method	(Chen et al. 2014)
NGO-PEG solution	Modified Hummers method	(Chen et al. 2014)

with 30 kDa ultracentrifuge tube and then diluted against distilled water overnight with the purpose of removing any ions. After that, the process contained centrifuging of the mixture solution at 12.000 × g for 30 minutes in order to remove any unstable aggregates (Chen et al. 2014).

In order to synthesize NRGO-PEG solution, 20 mL of NGO-PEG (0.5 mg/mL) was used, and transferred to a sealed glass bottle. After that, it was bathed at 90°C for different times or bathed at different temperatures for 24 hours. The obtained NRGO-PEG solution was centrifuged for 30 minutes at 6.000 × g in order to eliminate unstable aggregates. Then, the final step is the storage of NRGO-PEG at 4°C for further use (Chen et al. 2014).

In order to label the yielded NR-GO-PEG, fluorescein isothiocyanate (FITC) was used. In brief, the solution of NRGO-PEG (0.5 mg/mL) was mixed with 0.1 mL FITC (13 mM) diluted in Dimethyl sulfoxide (DMSO), and kept under stirring overnight at room temperature. The FITC-labeled obtained mixtures were filtrated through 30 kDa filters in order to remove excess unbound FITC. Then, the yielded mixtures were centrifuged for 30 minutes at 12,000 × g for the elimination of solid aggregated FITC. The yielded NRGO-PEG/FITC was re-dispersed in distilled water. The whole process was achieved in the dark place. Then, 100 μg/L of NRGO-PEG/FITC and free FITC were used for 2 hours, respectively, in order to incubate A 549 cells (1 × 10^5 cells). The process was achieved in the dark. After this step, the cells were washed by buffered saline (PBS) five times (Chen et al. 2014).

For cytotoxicity assay, approximately 5,000 A549 cells/well in 96-well plate were plated with 100 μL medium, and then cultured for 24 hours. After this step, various concentrations of NRGO-PEG and NGO-PEG were inserted to the wells. In order to achieve the photo thermal therapy, adherent cells were used in order to incubate them at various concentrations of NRGO-PEG and NGO-PEG for 2 hours. In addition, a semiconductor laser (808 nm) was used in order to irradiate the samples with the power density of 0.6 W/cm^2 for 5 minutes. At the end of the process, the dead and living cells in the medium were gathered and rinsed with Phosphate-buffered saline (PBS) (Chen et al. 2014).

3. Characterizations

3.1 Graphene/TiO₂ composites

Atomic force microscopy (AFM) was one of the techniques for LGO on the characterization of prepared NGOs 100, 200, and 300. AFM results revealed that by increasing the concentration of the oxidant ($KMnO_4$), the thickness of the GO sheets is reduced to 0.9 nm, indicating a single-layered GO sheet. Also, the thickness of NGOs varies from 0.8–2.0 nm, which confirms that a number of NGO sheets are stacked as double layers.

Cross-sectional height profiles of LGO, NGO-100, NGO-200, and NGO-300 were created, and the results were analyzed (with AFM) regarding the size distribution. Size distribution is provided in the form of histograms, where the average sizes of LGO, NGO-100, NGO-200, and NGO-300 are 520 (± 480), 164 (± 128), 59 (± 47), and 36 (± 27) nm, respectively, while the initial GO (LGO) performs a wide size distribution of 520 (± 480) nm. Thence, all NGOs reveal narrower size distributions and smaller average sizes, compared to LGO (Kim et al. 2011).

In the same study, measurements of UV visible spectrometry confirm the absorption coefficient (α) spectra of LGO and NGOs, with a maximum absorption coefficient around 5.6 eV (~ 222 nm), indicating that the π→π transition of C=C is in the two-dimensional matrix of GO. From the same measurement, it seems that both the absorption edge region and the absorption maximum position vary slightly after the oxidation of GO. Also, in both cases, the degree of oxidation gave the blue shift. This blue shift may be attributed to the decreased size of the conjuration domain π on the GO sheet (Kim et al. 2011). The absorption peaks of LGO, NGO-100, NGO-200, and NGO-300 were confirmed to be 5.33 (~ 233 nm), 5.62 (~ 221 nm), 5.69 (~ 218 nm), and 5.75 (~ 216 nm) eV, respectively.

AFM images are in good agreement with the results from the UV visible absorption measurements. As substrate for fabricating a NGOT, NGO-300 was preferred over the other two because of its high degree of oxidation, as well as NG0-300 possessing the smallest size. Finally, NGOTs with different concentrations of NGO-300 were further reduced, implementing photocatalytic reduction procedure under UV irradiation (Kim et al. 2011). The reduced r-NGOTs, with concentration of NGO-300 ranging from 0 to 6 wt%, were characterized by Reflectance UV/Visible absorption Spectroscopy (DRUVS). According to the outcome, the doping of the surface of TiO_2 with r-NGO induced the optical property of TiO_2. All r-NGOTs present absorption to all visible ranges. It can be concluded that the increase of NGO content also increases the r-NGOTs visible absorption background, and this fact is attributed to the high number of r-NGOs created on TiO_2 surface. The photoreduction process seems to restore the π-electron conjuration within GO sheets, enhancing the visible range of absorption background. The surface charge of TiO_2 also varies in accordance to the r-NGO content. As acidic functional groups are present, the zeta (ζ) potential of the sample acquires negative values compared to pure TiO_2. Normally, the zero-point charge (ZPC) of r-NGOT is located at a lower pH compared to pure TiO_2 (Kim et al. 2011).

Two samples of r-NGOTs with different concentrations of NGO (panels a-d for r-NGOT and panels e-h for r-NGOT-4) were characterized by transmission electron microscopy (TEM). For carbon presence on r-NGOT, energy loss spectrometry mapping technique was implemented. It is observed that carbon concentration on r-NGOT-0.7 is much lower than that on r-NGOT-4, and carbon layers, with interlayer d-spacing of 0.5 nm, surround the surface of TiO_2. At this point, it should be noticed that the interlayer d-spacing of carbon layers is smaller compared to that of graphene oxide (~ 1 nm), but larger than pristine graphite (~ 0.34 nm). The above observation can be correlated to the partial removal of oxygen-containing groups during reduction process. Another conclusion that can be stated while comparing the TEM images is that the higher the NGO amount, the thicker the carbon layer on TiO_2 (Kim et al. 2011).

3.2 Nanographene sheet hybrid a-MnO₂ nanotube and nanoparticle

The crystal structure of the NGs, MPGs, and MTGs was evaluated by X-ray diffraction (XRD). For the case of NGs sample, two peaks are observed at 42.9 and 25.9°, corresponding to the (100) and (002) diffracted planes of NGs, respectively. The calculated interplanar d spacing accords with that of Bragg's law, which is about 3.46. The sharp diffraction peak (002) confirms that there is a graphite-like structure to the as-grown nanographene sheets, which allows the successful electrical conductivity of the samples. For the case of MPGs and MTGs, the 2θ values of diffraction peaks 60.31°, 50.01°, 41.98°, 37.55°, 28.80°, 17.90°, and 12.7° can be identified as the phases (411), (310), (440), (400), (310), (220), (200), and (110), respectively, of the a-MnO₂. The poor crystallinity of MPGs samples, occurring due to low reaction temperature (~ 40°C), was confirmed from the weak diffraction peaks. Commonly, when the hydrothermal temperature is increased up to 140°C, the growth mechanism of MnO_2 crystals is also increased, and this fact can be confirmed by the strong diffraction peaks (Wang et al. 2015).

Morphological images of the same samples were obtained by TEM and Scanning Electron Microscope (SEM). Concerning the TEM images, NG materials have partially scrolled and sheet-like structures, as it can be observed in Figure 1.3a. The wrinkles of NGs create the scrolled structure, which can offer an aggregation preventing characteristic and good mechanical properties to graphene-based materials. Figure 1.3b displays the micromorphology of the MPGs. The images reveal that the nanoparticles of MnO_2 might grow on the sheet surface of NGs or may be embedded in it, resulting in an optimal MnO_2 electrical conductivity. On the other hand, an uneven folding on the surface results in a high specific surface area. The aforementioned features seem to improve capacitance performance

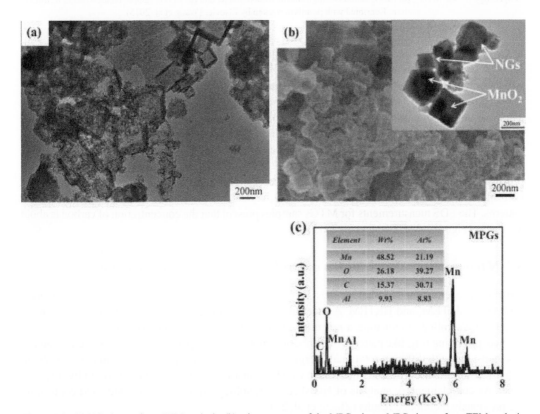

Figure 1.3: (a) NGs image from TEM analysis, (b) microstructure of the MPGs, inset: MPGs image from TEM analysis, and (c) MPGs sample measurements from EDS pattern. Reprinted with permission taken by Elsevier (Wang et al. 2015).

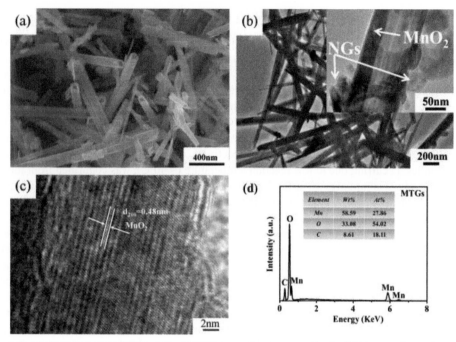

Figure 1.4: (a) MTGs image from SEM analysis, (b) MTGs image from low-magnification TEM process inset: microstructure morphology for the MTGs, (c) MnO$_2$ nanotubes high-magnification image, and (d) MTGs sample measurements from EDS pattern. Reprinted with permission taken by Elsevier (Wang et al. 2015).

of the material. Analysis of supplementary EDS showed that the average content of carbon in MPGs is about 17.06 wt% (Figure 1.3c). The same analysis revealed the presence of Al elements as well, which probably comes from the sample stage (Wang et al. 2015).

Figures 1.4a and b depict the microstructure morphology of MTGs. The samples of MTGs present a hollow tubular structure, with a wall thickness of about 10 nm and length about 1.0 μm. Due to the hollow tubular structure, the density of the material is reduced and the specific area is increased. The electrical conductivity between tubes improved because some NGs sheets are attached to the hollow tube's surface. Figure 1.4c illustrates images of MTGs with the use of high-magnification TEM process. An interplanar d spacing of 0.48 nm is presented from the lattice fringes. This measurement corresponds to a lattice plane (200) of an MnO$_2$ crystal, and is in agreement compared to XRD patterns. The EDS measurements for MTGs samples present that the concentration of carbon is about 8.61 wt% (Figure 1.4d) (Wang et al. 2015).

3.3 Nanographene sheets with a controlled mesoporous architecture

Structural properties and morphology of NG are provided by a number of different imaging techniques, such as AFM, TEM, and HRTEM. A sheet-like structure is presented with wrinkles and seems to be partially scrolled, eliminating a potential aggregation. Electro diffraction (SAED) method was applied, providing ring-like patterns of diffraction spots images. Specifically, it is depicted as regular hexagons, but with different rotational energies between them. This phenomenon may be attributed either to the formation of folds that cause overlays, or to a random overlay of the individual NG sheet. As complementary analysis of NG sheet morphology, AFM was also conducted confirming the previously mentioned results (Zhang et al. 2013).

The cyclic voltammetry (CV) slopes are present at a scanning rate of 100 mV s^{-1} in the potential windows of 3.5 and 4 V. Relatively pure NG containing low amounts of impurities, such as

N and O, affects the potent ideal double electrode layer behavior. Specific capacitance, at a scanning rate of 10 mV s⁻¹ with potential window at 4 V, reached maximum value of 95.2 F g⁻¹. It is important to note that in potential windows of 3.5 and 4 V, the capacitance retentions are maintained at 71 and 61%, respectively for the increase of scanning ratios from 10 to 200 mV s⁻¹, indicating the high capability rate of the NG electrodes. Low inner resistance (IR) was calculated to be ~ 9 Ω, due to the good conductivity of the NG. The specific capacitance was determined by a galvanostatic charge-discharge procedure. The symmetrical sketch of the charge-discharge slopes confirmed that the NG, when applied in IL electrolytes, has enhanced electrochemical capacitive characteristics. The reduction in IR, in the potential windows of 3.5 and 4 V, is at 1 A g⁻¹ 0.132 and 0.138 V, respectively. The specific capacitance is calculated from the discharge slopes at different charge-discharge current densities. The obtained Ragone plot provided information about maximum specific energy, which was 51.5 Wh kg⁻¹ for 1 Kw kg⁻¹ power density, while at values of 20 kW kg⁻¹, the specific energy decreased to 23.1 Wh kg⁻¹. However, those values are higher when compared to conventional ECs (Zhang et al. 2013).

3.4 Characterization of nanographene from carbon nanofibers

SEM imaging of CNF and size distribution are illustrated in Figures 1.5a and b, respectively. CNF size ranges from 40 to 380 nm, with a mean value of 80 nm. Figure 1.6 exhibits AFM analysis of a graphene oxide (GO) prepared from platelet CNF. After 2 hours of sonication, the thickness was reduced to 1 nm and the size to 60 nm. Many reports have suggested that if the yielded thickness is bigger than the ideal value (0.33 nm), it is an indication that functional groups and water/gas molecules are adsorbed on the surface (Mitani et al. 2012). High magnification TEM of graphene is displayed in Figure 1.7. The results of AFM and TEM are consistent for the case of graphene before reduction, while aggregation of RGO was clearly indicated in low magnification TEM image (Mitani et al. 2012).

The structures of CNF and RGO were confirmed by Raman spectroscopy. Raman spectra range from 1200 to 1800 cm⁻¹. Raman spectra of graphitic materials in this range reveal three peaks at 1350, 1580, and 1620 cm⁻¹, which are attributed to D, G, and D′ bands, respectively. The G peak is observed due to the lattice vibrations in the plane of the graphite-like rings. The D′ and D band peaks result from the edge of the graphene and from the defects. These D′ and D bands are not significant in the CNF compared to RGO, leading to larger D/G intensity after exfoliation (Mitani et al. 2012). Furthermore, a cyclic voltammogram measurement of RGO at a scan rate of 100 mV s⁻¹ in oval shape confirms that the charge/discharge behavior happens in the ion diffusion rate-limiting regime. The above measurements were conducted with a potential window from −1.75 to 0.75 V (ΔV: 2.5 V). On

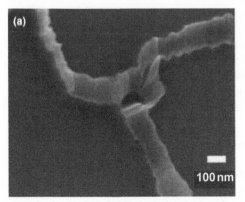
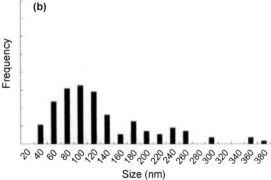

Figure 1.5: (a) SEM microstructure image of carbon nanofibers, and (b) the diameter distribution of carbon nanofiber from SEM image. Reprinted with permission taken by Elsevier (Mitani et al. 2012).

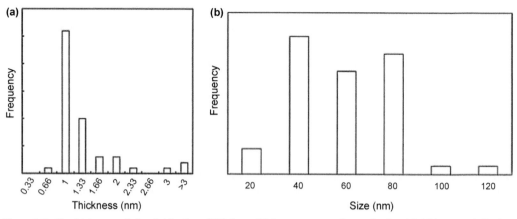

Figure 1.6: The thickness and size distribution of GO from AFM measurement after sonication: (a) thickness and (b) size. Reprinted with permission taken by Elsevier (Mitani et al. 2012).

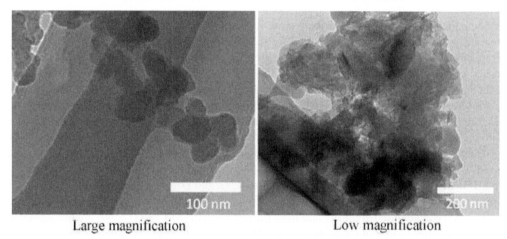

Figure 1.7: TEM image of RGO. Reprinted with permission taken by Elsevier (Mitani et al. 2012).

the other hand, the voltammograms at scan rates of 10 mV s⁻¹ and 1 mV s⁻¹ were quasi-rectangular. It is worth pointing out that at a higher voltage, the capacitance is lower than it is at a lower voltage. This observation supports that the exfoliation procedure may increase the number of defects or edge sites. Analyzing the chemical structure of EMI-TFSA, it is observed that cation (EMI⁺) has a plane structure, while the cation (EMI⁺) has a less steric structure compared to the anion (TFSA⁻). The capacitance values, calculated at a scan rate of 1 mV s⁻¹ from the anodic and cathodic scans were 50 and 60 F g⁻¹, respectively. The results indicate that even at a scan rate of 10 mV s⁻¹, the rectangular shape is distorted. This happens because the specific ionic liquid of this process is more viscous than the conventional organic electrolytes used in EDLCs (Mitani et al. 2012).

3.5 Catalyst free growth nanographene films

As-grown NG on SiO₂ substrate with various growth durations was characterized by AFM (Figure 1.8). The nucleation of NG all over the SiO₂ surface in the early stage of growth process is clear in Figure 1.8a. According to AFM analysis, the larger graphene islands are of 1.2–1.5 nm thickness, probably due to multi-layering of the material. NG continuous growth on the surface of SiO₂ yields more nucleation sites, leading to a continuous and uniform NG film. In addition, resistance

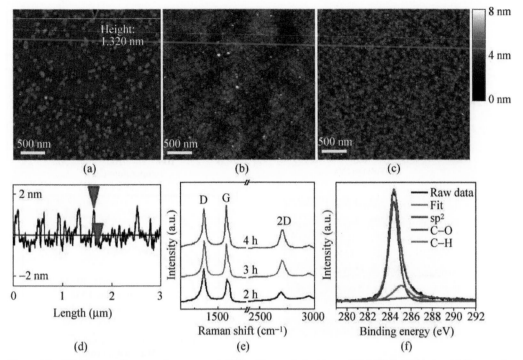

Figure 1.8: AFM images of nanographene grown on SiO$_2$ with growth durations of (a) 2, (b) 3, (c) 4 hours, (d) Topography height profile along the line present in image (a). (e) Raman spectra of the NG samples shown in (a), (b), and (c). The excitation laser had λ = 633 nm, and beam spot size ≈ 1 μm. (f) XPS characteristics of the as-grown NG, with the binding energies values ranging from 280–290 eV. Reprinted with permission taken by Springer (Zhang et al. 2011).

measurements suggest that sheet resistance increases as the thickness of nanographene is increased; for a typical sample, resistance is about 25 kΩ, while for a 3–4 layers sheet, the resistance is 2 kΩ. Figure 1.8e presents the Raman spectra of the specimen shown in Figures 1.8a and c. The peaks of G (graphitic mode), D (disorder mode), and 2D (D mode overtone) are common for graphitic structure. Figure 1.8f displays the spectrum of XPS process of the as-grown nanographene film on SiO$_2$ (Zhang et al. 2011).

In addition to SiO$_2$, this study suggests that a successful growing of films can also be achieved on other substrates as well, including, sapphire, mica, SiC, Si, atomic layer deposited (ALD) Al$_2$O$_3$, quartz, and even 4-inch glass wafers. Moreover, low surface roughness and less lattice mismatch are two main factors affecting both the size of formed crystals and the growth rate (Zhang et al. 2011). The same study examines, as well, the optical transparency and sheet resistance of NG films grown directly onto glass wafers or quartz. Figure 1.9c displays the growth of NG films on substrates of 4-inch glass wafers with excellent uniformity. Figure 1.9d displays the transmittance of various samples with different sheet resistances at wavelengths in the range of 200–1700 nm. Nanographene films of smaller crystal size have higher conductivity than films prepared from graphene oxide (Zhang et al. 2011).

3.6 Graphene nanoshells

SEM and TEM images show a flower-like morphology of the R-Ni(OH)$_2$-DS precursor assembled by R-Ni(OH)$_2$ layered nanosheets with high curvature and a thickness of 30–50 nm. XRD patterns show a regular arrangement of (001) planes and a layered structure of R-Ni(OH)$_2$. Basal peaks shift to lower angles when DS chains are introduced into the interlayer of R-Ni(OH)$_2$, recommending the

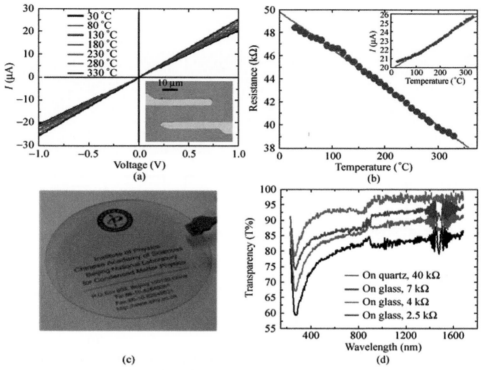

Figure 1.9: (a) In the inset, a typical device and I–V curves are presented, (b) Various temperatures and resistance changes for this device. The red line displays a linear fit. The inset shows the I–V curve. (c) 4-inch wafer scale nanographene film and the optical image. (d) Sheets with different resistance and the transmittance spectra of samples. Reprinted with permission taken by Springer (Zhang et al. 2011).

intercalate of DSs into the interlayer galleries. The asymmetric nature of (101) and (110) planes, as well as the d-spacing of the (003) and (006) planes, can also be observed, providing evidence of various topological status of intercalated DS anions (Peng et al. 2014).

3.7 Thiolated nanographene

The structure of a thiolated nanographene perthiolated tri-sulfur annulated hexa-peri-hexabenzocoronene (TSHBC) is presented in the study. The deposition of TSHBC onto fabricated perovskite devices is seen from cross-sectional SEM image of the material. I–V curve was used in order to compare the efficiency of measured devices according to the thickness of TSHBC. The average efficiency of 30 devices is $10.6 \pm 2.2\%$ (Cao et al. 2015).

3.8 Graphene molecules on TiO_2

The UV/Vis absorption spectra results of TiO_2/P1 film before and after the Scholl reaction conditions were obtained. The film appears to change from colorless to dark red for cases treated chemically in a School reaction. In contrast, for the case of pure TiO_2, there was no color change. Inset presents that the TiO_2/P1 film reflects the complexity of the chemical reaction when it is exposed to the $FeCl_3$ solution by changing absorption rates. In less than 30 minutes, the absorption changes were completed by 99% for all performed reactions (Ji et al. 2014). Quantum efficiency (QE or IPCE) measurement of a typical GM-coated device was conducted, providing the absorption spectrum of the corresponding TiO_2/GM film. The IPCE curve indicated that the photocurrent was generated via the photo-excitation of GMs. Conclusions of the specific study include that the power conversion efficiency of 0.87% is the better value for a sensitized solar cell using GM as a component of light harvesting. Also, the performance of the devices synthesized by the above method was found to have a high reproducibility, with the variations of Voc, FF, Jsc, and n being less than 10 percent. Finally, the devices show no degradation for a period of at least three weeks (Ji et al. 2014).

3.9 Piezoresistive nanographene films

In this present study, AFM images of packed NG islands are similar to those grown on SiO_2. Typical Raman spectra of these NG films are presented. The characteristic 2-D peak around 2708 cm^{-1}, G peak around 1594 cm^{-1}, and D peak around 1351 cm^{-1} suggest that NG consists mainly of sp^2-bonded carbon (Zhao et al. 2012). In this study, it was determined that current increases linearly with increased bias voltage. In addition, by bending the device up and down, production of strains from –0.29 to 0.37% was achieved. Additionally, during the above step, the device resistance (R) could present either an increasing or decreasing response. The reproducibility of the piezoresistive effect was also examined. Finally, experimental and theoretical data were fitted with a high level of agreement (Zhao et al. 2012). Strain sensor devices are completely based on NG. The sheet resistance varies from R_0 = 9.3 kΩ to 575 MΩ. In addition, a maximum Gauge factor (GF) of more than 300 is reached as a subsequence of sheet resistance increase. The suggestion that the sensitivity of the sensor is improved by the larger tunneling gap is confirmed by the obtained data (Zhao et al. 2012).

3.10 Nanographene-constructed hollow carbon spheres

Monodispersed spheres with a mean diameter of 380 nm were imaged by TEM. Broken spheres were also imaged, and an additional spacing between the core and the shell structure was revealed. TEM characterization images show the eyelike shell-core structure with shell thickness estimated to be around 70 nm and core diameter of 200 nm.

Mesoporous channels are aligned in the shells, with an inner diameter of ca. 2 nm. The Brunauer-Emmett-Teller (BET)-based nitrogen adsorption/desorption analysis reveals a surface area of more than 970 m^{2}g^{-1}. The mesoporous domain is confirmed by the Type-IV isotherm obtained. During the fabrication of NGHCs, the unique structure of the synthesized spheres may provoke the arrangement and impregnation of NGs within the nanochannels (Yang et al. 2010).

NGHCs and their dual wall structure is revealed under TEM characterizations. The nanochannels are aligned in an orderly manner in the exterior wall, with an inner diameter of around 2 nm, and orientated perpendicular to the surface of curved spheres. Also, this study demonstrates the selected area electron diffraction (SAED), along with the cross-sectional high-resolution (HRTEM) image of the material (Yang et al. 2010). Capacity measurements of NGHC revealed a higher value (713 m Ah g^{-1}) at a rate of C/5 (one lithium per 6 units (LiC6) in 5 hours) than the theoretical value of graphite, which is 372 m Ah g^{-1}. This configuration suggests that apart from the classical mechanism of graphite intercalation compound (GIC), there is also a storage mechanism of the added lithium. Differential capacity versus cell-voltage plots of NGHCs show peaks at 0.1 to 1.4 V, and slopes at 2.5 to 3, which are notably different from those of graphite. The peak at 0.1 V can be attributed to the Li extraction from graphene layers, while the peak at 1.4 V can be attributed to the Li extraction from the nanocavities. The slope from 2.5 to 3 could be a result of the residual H atoms at the edge of nanographenes (Yang et al. 2010).

3.11 Reduced graphene oxide

In order to examine the composition and morphology of graphene layers using NH_4OH, SEM, and complementary EDX, measurements were conducted. Figures 1.10a and c display morphologies

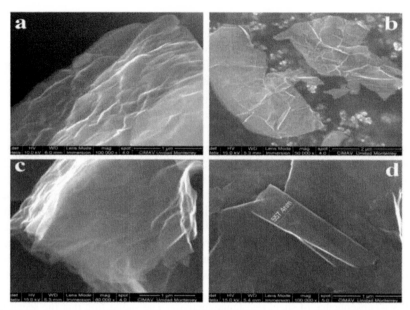

Figure 1.10: FE-SEM image of exfoliated graphite using NH_4OH. Reprinted with permission taken by Elsevier (Betancur et al. 2018).

with a rough surface and low transparency, and such results are related to sheets compounded by a few graphene layers. Figure 1.10b confirms graphene multi-layering by exhibiting white zones and little roughness. The structure with noticeable transparency contains the lowest carbon and oxygen concentrations, indicating an efficient reduction process of graphite to graphene (Betancur et al. 2018). The exfoliation without exothermic neutralization reaction induces structural and elemental changes in concentration of the material (Figure 1.11). The resulting average ratio value of C/O without using NH_4OH is 1.05. In this case, due to skipping last reaction, there is a higher oxygen concentration. All images in Figure 1.11 confirm that the graphene sheets are very large and thick due to the multi-layer formation (Betancur et al. 2018).

Additionally, C1s spectra of GO provide information about binding energy of the material. Results obtained by this analysis suggest that with the presence of rGO, the binding energy is at 284.3 eV, which corresponds to sp^2 hybridization of C=C/C-C in the aromatic rings. Peaks obtained at 286.1 and 285.3 eV can be attributed to C=O and C-O, respectively, while peaks at 286.9 and 288.6 eV can be attributed to O-C=O and to π-π transitions, indicating the restoration of sp^2 hybridization. Binding energy with peak at 285.96 eV is related to two types of bond—C-N from graphitic and pyrrolic nitrogen, and C-O from hydroxyl and epoxy groups (Betancur et al. 2018).

A big sheet (area around 9 μm^2), with some impurities over a silicon substrate is presented in Figure 1.12. A region of the sheet containing some holes was used in order to measure the thickness, which was found to be around 3 nm. According to the study of Botas et al., the thickness value for a GO monolayer is between 0.40 nm and 2.04 nm. The thickness for one graphite oxide layer is found to be 0.93 nm, according to the study of Dubin et al. A complete exfoliation procedure, suggested by Stankovich et al., is revealed when the thickness of the sheet is around 1 nm, which is attributed to the presence of oxygen covalent bonds, as well as a displacement of sp^3 hybridized atoms slightly below and above the original graphene plane. In addition, if the chemical exfoliation procedure is not complete, it could be considered that the graphene thickness derives from graphite to 0.34 nm, in order to confirm the number of layers inside some graphene sheets. According to the studies by Sorokina (presented the importance of graphite bisulfate and the possibility of obtaining GO in the presence of any oxidant agent and H_2SO_4), Botas, and Dubin, the yielded 3 nm thick sheet in this experiment corresponds to three, four, or ten graphene layering (Betancur et al. 2018).

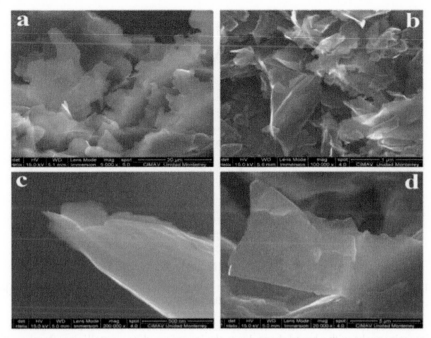

Figure 1.11: FE-SEM image from exfoliated graphite without NH₄OH. Reprinted with permission taken by Elsevier (Betancur et al. 2018).

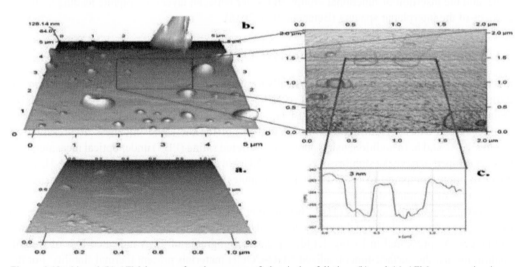

Figure 1.12: (a) and (b) AFM images after the process of chemical exfoliation, (b) and (c) AFM cross-section image and thickness measurements showing several sheets with thickness of 3 nm. Reprinted with permission taken by Elsevier (Betancur et al. 2018).

Color version at the end of the book

In Figure 1.13a, XRD characteristic peaks of the material are displayed at 26.62° and 43.27°, which are yielded from (002) and (100) reflection planes, respectively. In addition, these characteristic peaks are in agreement with the study by Kwon et al. for rGO formed by thermal exfoliation and hydrazine. Furthermore, the peaks revealed at about 7.27° and 9.12° correspond to (001) plane form GO, and are in agreement with the results that are reported by other authors. Figure 1.13b displays

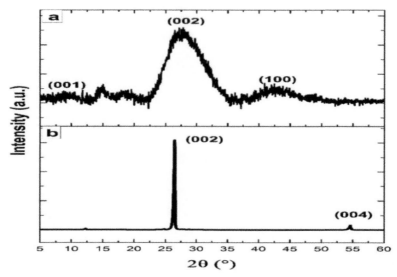

Figure 1.13: XRD spectra for chemical exfoliation with (a) NH_4OH and (b) Graphite source. Reprinted with permission taken by Elsevier (Betancur et al. 2018).

the low intensity peaks with a broad shape due to the loss of graphite crystallinity after the exfoliation procedure. These results confirm the amorphous morphology, the particle reduction to nanometric scale, and the insertion of functional groups and water between layers of graphite leading finally to changes of the interplanar spacing (Betancur et al. 2018).

3.12 Characterizations of PEGylated NGO (PEG-NGO)

Stable PEG-NGO and PEG-LGO were physically formed by carbodiimide-catalyzed amide formation during sonication of an aqueous solution of NGO-COOH or LGO-COOH with 4-arm-PEG-NH_2 amine. The thickness of PEG-LGO was 4 nm, with the lateral width about 800 nm, compared to PEG-NGO, whose width appears less than 100 nm for the same thickness as presented in Figures 1.14a and b. In addition, using phosphate buffered saline (PBS) under optical measurements was determined to have the stability of NH_2-PEG-NGO in serum. It is worthy to note that NH_2-PEG-NGO solution, compared to NGO-COOH (30 minutes), remains a homogeneous suspension for 240 minutes of standing storage.

Thermogravimetric analysis (TGA) was used at a temperature range of 200–500°C in order to examine the thermal stability of the graphene-based materials and confirm the mass of grafted organic substances. As presented in Figure 1.14d, typical thermograms of TGA indicated loss of weight, confirming that the surface-functionalized NGO-COOH materials present thermal stability. For the characterization of PEG-NGO and NGO-COOH, samples used X-ray photoelectron spectroscopy (XPS) in order to elucidate the surface composition, structure, and functional groups of GO. In addition, as presented in Figure 1.14e, C1s XPS spectrum of NGO-COOH displayed a significant degree of oxidation at different peaks, which can be attributed to carbon atoms in C=C/C-C, C-C, C=O, and O-C=O bonds. Figure 1.14f presents the successfully grafted 4-arm-PEG-NH_2 from the peaks onto NGO carboxyls with conjugated ratio around 24.3 percent. The experimental data displayed the preparation of NH_2–PEG-NGO homogeneous suspension in physiological solutions, thus blocking major aggregation in blood vessels (Yang et al. 2013).

In order to investigate the distribution of 99mTc-labeled PEG-NGO and PEG-LGO in mice from 30 minutes to 24 hours after injection, Single-Photon Emission Computer Tomography/Computer

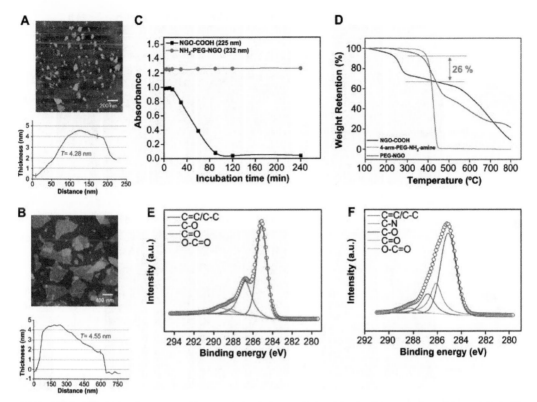

Figure 1.14: (A) AFM image of PEG-NGO deposited on substrate of mica. (B) AFM image of mica substrate grafted with PEG-LGO. (C) Optical stability of NH$_2$-PEG-NGO and NGO-COOH in PBS at a concentration of 0.2 mg/mL. NGO-COOH compared to graphite presents a strong UV absorption peak at 225 nm, corresponding to the π-π transitions of aromatic C-C bonds. Formation of NH$_2$-PEG-NGO was determined by a shift of the peak to 232 nm. (D) TGA curves of NGO-COOH, 4-arm-PEG-NH$_2$-amine PEG, and NH$_2$-PEG-NGO with a heating rate of 10°C/min in N$_2$. (E) The C1s peak in the XPS spectrum of NGO-COOH. (F) The C1s peak in the XPS spectrum of NH$_2$-PEG-NGO. Reprinted with permission taken by Elsevier (Yang et al. 2013).

Tomography (SPECT/CT) was used. Moreover, as presented in Figure 1.15a, the ^{99m}Tc-labeled PEG-NGO accumulated after injection at a low level in the lungs, relatively high in spleen, and at a high level in the liver, and was being eliminated at 6 hours after injection. It is worthy to note that erythrocytes are ~ 7 μm and PEG-NGO particles are less than 100 nm, with the result of their quick sequestration in the Reticuloendothelial system (RES). Figure 1.15b presents high accumulation of B ^{99m}Tc-labeled PEG-LGO in lungs, resulting in the risk of respiratory failure. In contrast with the elimination of PEG-NGO after 6 hours from the injection, the PEG-LGO injection killed the mice after 6 hours, with the result to choose PEG-NGO as a possible drug carrier (Yang et al. 2013).

In order to choose a targeting cell model, we examined Epidermal Growth Factor Receptor expression (EGFR) in glioma cell lines U87, C6, T 98, and GL261, as presented in Figure 1.16a. The analysis (Quantitative Western blot analysis with normalization to Actin) indicated that U87 cells expressed strongly with EGFR (1.2-fold higher than GL261 cells), in contrast with the weak expression of T98 cells. In Figure 1.16b, the expression of EGFR on U87 cells was confirmed by immunofluorescence. Therefore, U87 cells were chosen in order to investigate the targeting, accumulation, and interdiction of the cell growth signal (Yang et al. 2013).

In order to maximize targeting and therapeutic efficacy, the drug loading process and C225 antibody-conjugation were optimized. Figure 1.16c depicts that the insertion of EPI (1500 μg) had a resultant saturating concentration of EPI/mg NGO-COOH up to 87% conjugation, in

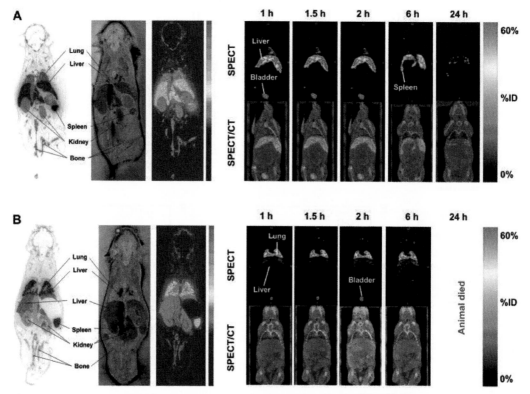

Figure 1.15: Whole body SPECT/CT tomography and volume-rendering image of ^{99m}Tc-labeled (A) PEG-NGO and (B) PEG-LGO. Reprinted with permission taken by Elsevier (Yang et al. 2013).

Color version at the end of the book

contrast with only 50%, as reported previously. The quantity of immobilized C225, as displayed in Figure 1.16d, reached saturation at 75 μg of inserted C225 (conjugation up to 71%). The simple π-stacking under controlled pH is the main mechanism for the EPI loading onto NGO-COOH (Yang et al. 2013).

As presented in Figure 1.17a, the *in vitro* cytotoxicity study observed a decrease in expression of EGFR after the treatment of U87 cells with PEG-NGO-C225, in contrast with a slight decrease with C225 alone, confirming that C225 conjugation to PEG-NGO greatly enhanced its capability to downregulate EGFR. Furthermore, activation of ERK 1/2 and EGRF with phosphorylation in response to treatment by EGF was greatly blocked to induce apoptosis. This result confirmed that the PEG-NGO-C225 can effectively target the cells of the tumor with decreasing of growth signal of EGFR-positive cancer cells. It is worthy to note that C225 presents lower efficacy to downregulate EGFR in contrast with PEG-NGO-C225. It is possible that after PEG-NGO-C225 treatment, the recycling of EGFR is hampered, with the result of reducing the quantity of C225 that is required for therapy. In addition, Figures 1.17b and c present the NIR absorbance of graphene oxide. In order to investigate the cytotoxicity of drugs, the temperature of PEG-NGO solution was increased by an 808 nm laser (2 W/cm^2, 120 seconds) from 36 to 94°C, with ΔT = 58°C, compared to a ΔT of only 25°C for PEG-LGO solution. In order to investigate the cytotoxicity of drugs, as presented in Figure 1.17d, the temperature of PEG-NGO solution was increased by an 808 nm NIR laser (2 W/cm^2, 120 seconds) from 36 to 94°C (ΔT = 58°C), compared to a ΔT of 25°C for PEG-LGO solution (Yang et al. 2013).

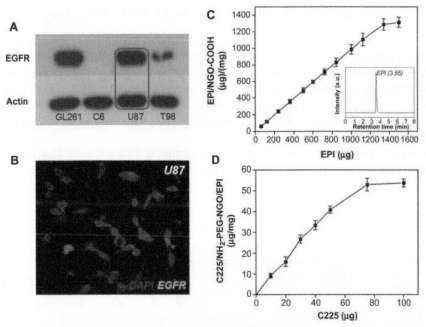

Figure 1.16: (A) Western blot of EGFR in glioma cell lines. (B) Expression of nuclei and EGRF (grey) of U87 cells. (C) Quantity if EPI bound per 1 mg of NGO-COOH vs. quantity of EPI inserted. Inset: Analysis of EPI by HPLC. Values are means ± SD (n = 6). (D) Quantity of C225 bound per 1 mg of NH_2-PEG-NGO/EPI vs. amount of C225 inserted. Reprinted with permission taken by Elsevier (Yang et al. 2013).

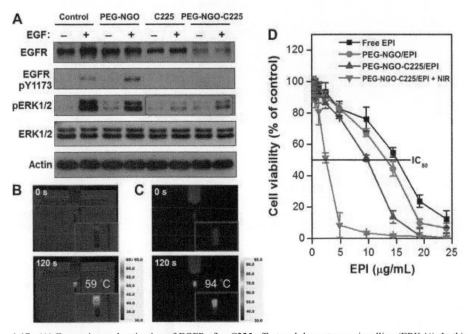

Figure 1.17: (A) Expression and activation of EGFR after C225 effect and downstream signalling (ERK ½). In this step, 100 µg/mL PEG-NGO-C225 or PEG-NGO or to 19.5 µg/mL C225 were used for 24 hours in order to expose the cells U87, and then treated with 10 µg/mL EGF for 30 minutes. In addition, 15 µg total cell lysates were separated by SDS-PAGE, followed by Western blotting. Thermal images after irradiation at 2 W/cm² for 120 seconds of eppendorf vials containing solutions of (B) PEG-LGO and (C) PEG-NGO. (D) U87 cells and their relative cell viability after drug treatment for 24 hours. The process time was 120 seconds, with NIR at 2 W/cm². Values are expressed as means ± SD (N = 8). Reprinted with permission taken by Elsevier (Yang et al. 2013).

3.13 AFM images and other results for GO, NGO, NGO-PEG, and NRGO-PEG

Figure 1.18a presents images of GO, NGO, NGO-PEG, and NRGO-PEG. The sheet diameters, as presented in Figures 1.18b, are 300, 68, 71, and 52 nm for GO, NGO, NGO-PEG, and NRGO-PEG, respectively, assuming that their shape was round. Also, Figure 1.18b presents that the lateral width of NRGO-PEG was much smaller than NGO-PEG, indicating that the breakage of chemical bond between graphene sheets during reduction process was achieved. As presented in Figure 1.18c, the sheet thickness of NGO and GO was approximately 1.0 nm, with the NRGO-PEG approximately 1.5 nm (Chen et al. 2014).

Figure 1.19a presents the temperature of vials containing 3 µg/mL at 55°C, 6 µg/mL at 60°C, 10 µg/mL at 65°C, 20 µg/mL at 68°C, and 30 µg/mL at 70°C, within 5 minutes of laser irradiation of 808 nm at 0.6 W/cm². Figure 1.19b presents vials containing water and their thermal images after

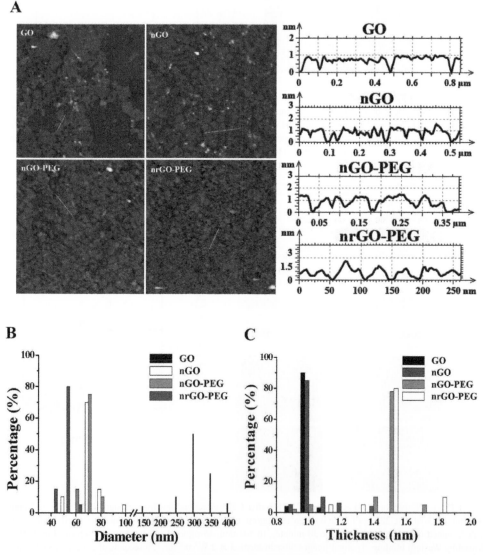

Figure 1.18: (A) AFM images of GO, NGO, NGO-PEG, and NRGO-PEG. AFM histograms of (B) diameter and (C) thickness for GO, NGO, NGO-PEG, and NRGO-PEG. The diameter was assumed to be in round shape. Reprinted with permission taken by Springer (Chen et al. 2014).

irradiation for 8 minutes, for 3 µg/mL of NGO-PEG solution, and 3 µg/mL of NRGO-PEG solution, respectively. It is worthy to note that NRGO-PEG with concentration 20 µg/mL had transferring efficiency of 50% in contrast with the solution of NGO-PEG at 3 µg/mL, with transferring efficiency of about 11.7 percent. In addition, the employed green reduction approach and the obtained NRGO-PEG may be a great photosensitizer for PTT (Chen et al. 2014).

Figure 1.20a presents the UV-vis spectrometer measurements in order to verify the formation of stable NRGO-PEG suspension. A good linear relationship with $R^2 = 0.999$ between the concentration of NRGO-PEG and the absorbance at 261 nm was found, as presented in the inset of Figure 1.20a, confirming the great dispersibility of NRGO-PEG in water. In addition, the dispersions NGO-PEG and NRGO-PEG containing 10% of fetal bovine serum and two kinds of DMSO and ethanol were evaluated, and it was confirmed that NRGO-PEG stayed stable in these solvents for three months at storage temperature 4°C, as presented in Figure 1.20b, indicating that this green reduction route

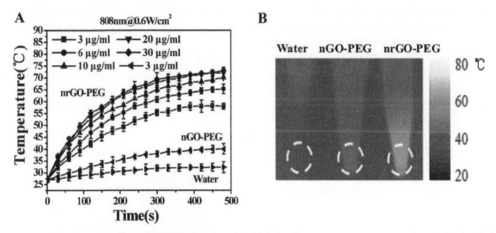

Figure 1.19: Photothermal effect of NGO-PEG and NRGO-PEG (A) Temperature curves versus time during irradiation with 808 nm laser (0.6 W/cm²) for vials containing water, NGO-PEG (3 µg/mL), and various concentrations of NRGO-PEG, respectively. (B) Thermal images of vials with water, NGO-PEG, and NRGO-PEG (3 µg/mL) after irradiation by 808 nm laser (0.6 W/cm²) for 8 minutes. Reprinted with permission taken by Springer (Chen et al. 2014).

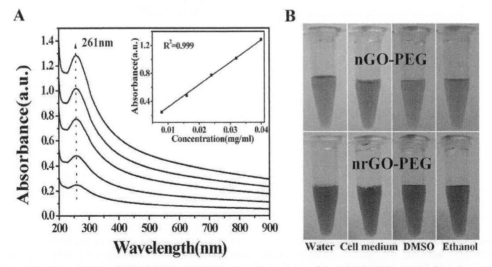

Figure 1.20: Dispersibility of NRGO-PEG (A) UV-vis absorption spectra of the NRGO-PEG aqueous dispersion. Inset: Correlation of absorbance at 261 nm against concentration. (B) Photographs of NGO-PEG and NRGO-PEG solution in several solvents after storage at 4°C for 3 months. Reprinted with permission taken by Springer (Chen et al. 2014).

of NRGO-PEG did not disrupt his dispersion. Finally, the yielded NRGO-PEG by this green reduction process may be used as promising biomaterial due to the long-term stability and the great biocompatibility under high concentrations in water (Chen et al. 2014).

In order to confirm the NRGO-PEG uptake by tumor cells, Fluorescein isothiocyanate (FITC) was employed to label NRGO-PEG by physical adsorption. UV-vis spectra in Figure 1.21a reveals the binding of FITC to NRGO-PEG, and presents an approximate adsorption peak at 450 nm. After this process, complex NRGO-PEG/FITC was incubated by A549 cells for 2 hours. After removing the extracellular NRGO-PEG/FITC, each cell was observed fluorescence microscopy. Figure 1.21b indicates the cellular uptake of NRGO-PEG with enough strong fluorescence inside the cells. In addition, in order to examine the cellular uptake ratio of NRGO-PEG, FCM analysis was used. Furthermore, Figure 1.21c presents that the cells that are cultured with NRGO-PEG and 100 µg/mL of free FITC had 71.5 and 2.1% of uptake ratio, respectively, confirming that NRGO-PEG nanoparticles could easily insert A549 cells, likely via endocytosis, in contrast with liposoluble FITC, which could not enter in A549 cells. In addition, laser irradiation could be used at 808 nm, in order to enable locally rapid temperature and selective photothermal heating at high accumulation of NRGO-PEG in tumor cells (Chen et al. 2014).

A main concern for all biomaterials is toxicity. Figure 1.22a indicates that after the exposure of A 549 cells to 5, 10, 30, 50, and 100 µg/mL with NGO-PEG and NRGO-PEG for 24 hours did not present cytotoxicity to A 549 cells. Moreover, Figure 1.22c confirms that after the irradiation with laser, NRGO-PEG displays much more dead cells compared to NGO-PEG (Chen et al. 2014).

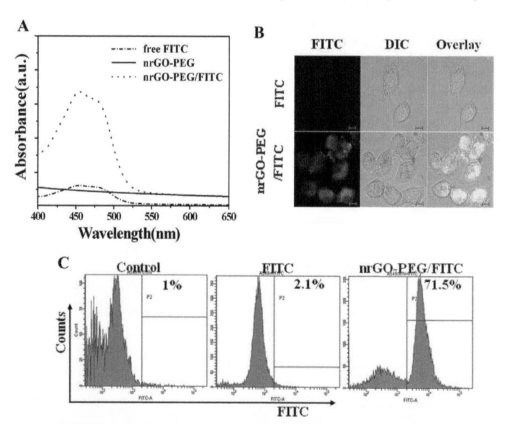

Figure 1.21: Cellular uptake of NRGO-PEG in A549 cells. (A) UV-vis absorbance spectra of free NRGO-PEG/FITC, NRGO-PEG, and FITC. (B) Fluorescence images of cells cultured with free FITC and NRGO-PEG/FITC, respectively, were imaged for 2 hours by confocal microscope. (C) Cellular uptake ratio of NRGO-PEG. The cells were cultured with 100 µg/mL of NRGO-PEG/FITC and free FITC, respectively, for 2 hours before FCM analysis. Reprinted with permission taken by Springer (Chen et al. 2014).

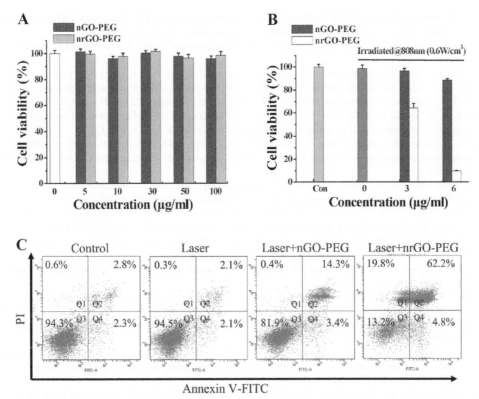

Figure 1.22: Cytotoxicity of NGO-PEG and NRGO-PEG in A549 cells (A) Relative cell viability after treatment with different concentrations of NGO-PEG and NRGO-PEG for 24 hours. (B) Relative cell viability after 808 nm laser irradiation (0.6 W/cm²) for 5 minutes for control cells and the cells cultured with different concentrations of NGO-PEG and NRGO-PEG. (C) FCM analysis of cell death induced by laser irradiation for control cells and the cells cultured with 6 μg/mL of NGO-PEG or NRGO-PEG for 2 hours. Reprinted with permission taken by Springer (Chen et al. 2014).

4. Conclusions

The most interesting conclusion of this chapter is that nanographene materials have superior properties compared to pure graphene. Properties such as efficiency, transparency, and conductivity of graphene films grown on 4-inch quartz and glass wafer are superior to those of graphene oxide or chemically exfoliated graphene. In addition, with thiolated nanographene molecules as whole transporting materials, the Perovskite solar cell had 12.8% best efficiency according to the reported results. Also, this chapter confirms that a nanocrystalline TiO_2 film injected with polyphenylene compounds, via the Scholl reaction, can be oxidized to graphene molecules. All these results opens a new way for the use of nanographene molecules into devices and photoactive structures against cancer, batteries, and other potential applications.

References

Avouris, P. and C. Dimitrakopoulos. 2012. Graphene: Synthesis and applications. Mater. Today 15(3): 86–97.

Betancur, A.F., N. Ornelas-Soto, A.M. Garay-Tapia, F.R. Pérez, Ángel Salazar and A.G. García. 2018. A general strategy for direct synthesis of reduced graphene oxide by chemical exfoliation of graphite. Mat. Chem. Phys.

Boehm, H.P., R. Setton and E. Stumpp. 1986. Nomenclature and terminology of graphite intercalation compounds. Carbon 24(2): 241–245.

Cao, J., Y.M. Liu, X. Jing, J. Yin, J. Li, B. Xu et al. 2015. Well-defined thiolated nanographene as hole-transporting material for efficient and stable perovskite solar cells. J. Am. Chem. Soc. 137(34): 10914–10917.

Chen, J., X. Wang and T. Chen. 2014. Facile and green reduction of covalently PEGylated nanographene oxide via a 'water-only' route for high-efficiency photothermal therapy. Nanoscale Res. Lett. 9(1): 86.

Iupac. 1995. Recommended Terminology for the Description of Carbon as A Solid (IUPAC Recommendations 1995) 67: 491.

Ivanovskii, A.L. 2012. Graphene-based and graphene-like materials. Russ. Chem. Rev. 81(7): 571–605.

Ji, Z., R. Wu, L. Adamska, K.A. Velizhanin, S.K. Doorn and M. Sykora. 2014. *In situ* synthesis of graphene molecules on TiO$_2$: Application in sensitized solar cells. ACS App. Mater. Interfaces 6(22): 20473–20478.

Kemp, K.C., H. Seema, M. Saleh, N.H. Le, K. Mahesh, V. Chandra et al. 2013. Environmental applications using graphene composites: Water remediation and as adsorption. Nanoscale 5(8): 3149–3171.

Kim, H., G. Moon, D. Monllor-Satoca, Y. Park and W. Choi. 2011. Solar photoconversion using graphene/TiO$_2$ composites: Nanographene Shell on TiO$_2$ core versus TiO$_2$ nanoparticles on graphene sheet. J. Chem. Phys. 116: 1535–1543.

Mitani, S., M. Sathish, D. Rangappa, A. Unemoto, T. Tomai and I. Honma. 2012. Nanographene derived from carbon nanofiber and its application to electric double-layer capacitors. Electrochim. Acta 68: 146–152.

Nemes-Incze, P., Z. Osváth, K. Kamaras and L.P. Biro. 2008. Anomalies in thickness measurements of graphene and few layer graphite crystals by tapping mode atomic force microscopy. Carbon 46(11): 1435–1442.

Novoselov, K.S., A.K. Geim, S.V. Morozov, D. Jiang, Y. Zhang, S.V. Dubonos et al. 2004. Electric field in atomically thin carbon films. Science 306(5696): 666–669.

Novoselov, K.S., V.I. Fal'Ko, L. Colombo, P.R. Gellert, M.G. Schwab and K. Kim. 2012. A roadmap for graphene. Nature 490(7419): 192–200.

Peng, H.J., J. Liang, L. Zhu, J. Huang, X. Cheng, X. Guo et al. 2014. Catalytic self-limited assembly at hard templates: A mesoscale approach to graphene nanoshells for lithium–Sulfur batteries. ACS Nano 8(11): 11280–11289.

Shenderova, O.A., V.V. Zhirnov and W.D. Brenner. 2002. Carbon nanostructures. Crit. Rev. Solid State 27(3-4): 227–356.

Wang, C., F. Li, Y. Wang, H. Qu, X. Yi and Y. Lu. 2015. Facile synthesis of nanographene sheet hybrid α-MnO$_2$ nanotube and nanoparticle as high performance electrode materials for supercapacitor. J. Alloys Compd. 634: 12–18.

Yang, H.-W., Y.-J. Lu, K.-J. Lin and S.-H. Hsu. 2013. EGRF conjugated PEGylated nanographene oxide for targeted chemotherapy and photothermal therapy. Biomaterials 34(29): 7204–7214.

Yang, S., X. Feng, L. Zhi, Q. Cao, J. Maier and K. Müllen. 2010. Nanographene-constructed hollow carbon spheres and their favorable electroactivity with respect to lithium storage. Adv. Mater. 22(7): 838–842.

Zhang, H., X. Zhang, X. Sun, D. Zhang, H. Lin, C. Wang et al. 2013. Large-scale production of nanographene sheets with a controlled mesoporous architecture as high-performance electrochemical electrode materials. Chem. Sus. Chem. 6(6): 1084–1090.

Zhang, Li, Z. Shi, Y. Wang, R. Yang, D. Shi and G. Zhang. 2011. Catalyst-free growth of nanographene films on various substrates. Nano Res. 4(3): 315–321.

Zhao, J., H. Congli, Y. Rong, S. Zhiwen, C. Meng, Y. Wei et al. 2012. Ultra-sensitive strain sensors based on piezoresistive nanographene films. Appl. Phys. Lett. 101(6).

Zhu, Y., M. Shanthi, C. Weiwei, L.-J. Xuesong, S. Won, R. Jeffrey et al. 2010. Graphene and graphene oxide: Synthesis, properties, and applications. Adv. Mater. (Weinheim, Germany) 22(35): 3906–3924.

CHAPTER 2

Bio Realization of the Synthesis of Diamond-like Carbon Thin Film by Low Cost Electrodeposition Technique

S. Gupta

1. Introduction and Overview

Compared to any other gemstone, diamonds probably feature more predominantly in the history and cultural heritage of the human race. They were prized for their scarcity for centuries, and remain a symbol of wealth and prestige. The word diamond came from the Greek Adams, meaning indestructible. Diamonds were first discovered from mines in India as early as the 4th century BC, but the modern diamond era only started in 1866, when huge diamond deposits were discovered experimentally in Kimberley, South Africa. Apart from their appeal as gemstones, diamonds possess an outstanding range of physical and chemical properties. Among all the elements in the periodic table, very few elements exhibit the remarkable importance and versatility of carbon. From biomacromolecules to waxy hydrocarbons to diamond-like carbons or diamonds, carbon compounds display a huge range of chemical and physical properties unrivalled by any member of the periodic kingdom. Equally impressive is the ability of carbon compounds to amalgam their physical and chemical properties to an astonishing degree. Carbon materials always promote the development of humanity's civilization. Carbon nanomaterials, such as fullerenes, carbon nanotubes, nanodiamond, and graphene, are among the most important building blocks in modern nanoscience and nanotechnology. Nanodiamond particles or nanodiamond (ND), which was recently rediscovered, attracted special research interest owing to the combination of many remarkable properties, such as hardness, crystallinity, chemical stability, non-porosity, low toxicity, dopability, wide bandgap, narrow particle size distribution (typically of 4–6 nm), and the possibility of varying the particle properties by means of chemical functionalization of its surface (Basiuka Elena et al. 2013). The best example of this case is one of the most industrially important carbon compounds, Diamond-like carbon (DLC). It is a common name, which represents all kinds of carbon and related material in between graphite and diamond. In the past fifty years, DLC has reached the level of semi-industrialization. From the

The Oxford Educational Institutions; The Oxford College of Engineering; Bangalore-560068, India.
Email: guptasnigdha646@gmail.com

unintentional discovery of DLC to the booming study all over the world, and then to the wide use in various fields, DLC has become a symbol of the era. In order to study the relationship between the structure and properties of DLC, researchers have been working on its classification and putting forward its theoretical model. Preparation technologies of DLC have been improving continuously. Indeed, a glance at any compendium of material data properties will prove that diamond is usually 'the biggest and best'. A selection of some of these properties is given in Table 2.1. Among other remarkable properties, diamond is the hardest material in the universe, has the highest thermal conductivity at room temperature, is transparent over a wide range of wavelengths, the stiffest material, the least compressible, and is inert to most of the chemical reagents. Due to this wide range of outstanding properties, it is not surprising at all that diamond has sometimes been called 'the ultimate engineering material'. With band gap energy of 5.4 eV, diamond is a very good electrical insulator (May 2000). Generally, thin films of diamond are produced through a variety of various physical vapor deposition (PVD) and chemical vapor deposition (CVD) methods, including ion beam deposition, pulsed laser deposition, plasma enhanced CVD, etc., onto a substrate held at elevated temperatures of around

Table 2.1: Some of the outstanding properties of diamond (Field 1992, Spear and Dismukes 1994).

- Cubic in structure and space group is $Fd\overline{3}m$
- Extreme mechanical hardness (90 GPa) and wear resistance
- Young's modulus (1050 GPa)
- Density (3515 kg/m^3)
- Highest bulk modulus ($1.2 \times 10^{12} Nm^{-2}$)
- Lowest compressibility ($8.3 \times 10^{-12} m^2 N^{-1}$)
- Highest room temperature thermal conductivity ($2 \times 10^3 Wm^{-1}K^{-1}$)
- Thermal expansion coefficient at room temperature is very low (1×10^{-6} K)
- Broad optical transparency from the deep ultraviolet to the far infrared
- Highest sound propagation velocity (17.5 km s^{-1})
- Very good electrical insulator (room temperature resistivity is $10^{13} \Omega$ cm)
- Diamond can be doped, becoming a semiconductor with a wide band gap of 5.4 eV
- Very resistant to chemical corrosion
- Zero residual stress
- Biologically compatible
- Some surfaces exhibit very low or 'negative' electron affinity
- Diamonds are excellent electrical insulators except some specialized applications include the use of blue diamond as semiconductors
- The toughness of natural diamond is 3.4 MN m$^{-3/2}$
- Constant specific gravity of single crystal diamond (3.52)
- Diamond has a high refractive index (2.417) and moderate light dispersion (0.044), properties
- Oxidation stability is 600°C
- Reactivity with ferrous materials is very high
- Dielectric constant at 300 K is 5.7
- Sound propagation velocity is 18.2 km/s
- Carrier lifetime is 10^{-10}s
- Breakdown voltage is 10^7 v/cm
- E-mobility is 1600 cm^2/vs and H-mobility is 1900 cm^2/vs
- Density is 3.51 gm/cm^3
- Melting point is 4100 K at 12.5 GPa
- Graphitization (transformation from sp^3 to sp^2) -1400°C
- Atomic density -1.77×10^{23} cm^{-3}
- Most often colorless to yellow or brown, rarely pink, orange, green, or blue
- Refractive index (as measured via sodium light of 589.3 nm) is 2.417

900°C at very slow growth rates of the order of microns per hour. However, the deficiencies of these techniques are equipment that is more complex, higher temperature, higher cost, and more difficult in deposition uniform films with large area and irregular surface (Guo et al. 2007).

The cost of diamond is such ($1.00/carat) that quality films are now being produced commercially for use as die cutters for nonferrous materials. However, if the cost could be lowered further while maintaining quality, applications such as heat conductors to help cool electronics and optics might turn to diamond films (May 2000). Preparation of the thin films by electrodeposition at ambient or sub-ambient temperatures and pressure rather than by higher temperature vapor deposition is thought to be a promising avenue for lowering the costs of diamond and diamond-like thin films. As electrodeposition is experimentally conducted in a liquid, as opposed to a low-density gas, faster deposition rates are anticipated. The equipment required for electrodeposition is simpler and less expensive than that needed for vapor deposition methods. In addition, electrodeposition is not a line of sight process; hence, it is not impossible to make uniform coatings on objects of irregular size and shape. Maissel and Glang (Maissel and Glang 1970) in their research pointed out that the film materials prepared in the gas phase can also be obtained from electrochemical deposition in the liquid phase, and vice versa. Being enlightened by this theoretical concept, Namba first pioneered the experiment of electrodeposition from solution as opposed to gas phase for the production of diamond thin films in 1992 (Namba 1992). After that, there has been a tremendous improvement and activity in this area. From the pioneering work by Namba in 1992 to date, a survey of literature on selection of experimental conditions and characterization techniques investigated by various researchers clearly reveal that the synthesis and characterization of diamond-like carbon film by electrodeposition has quite a good impact and demand in the scientific community. As would be expected, type of the film, the substrate, counter electrode, electrolyte/carbon source, temperature, voltage, and current; all these parameters affect the properties of the resulting carbon film. For example, pulsed electrodeposition of dimethylformamide (DMF) at 1600 V produces a higher quality DLC film than that from methanol or acetonitrile at 600–1600 V or DMF at 600 V (Cai et al. 2000). Now it is almost a well-established fact, and literature supports, that liquids with a higher dipole moment and dielectric constant produce a more diamond-like film, as do methyl groups containing organic liquids over those with ethyl groups. As per the literature survey this can be demonstrated in a comparison that ranked acetonitrile > DMF > nitromethane > methanol > nitroethane > ethanol in terms of their ability to generate a more diamond-like film by electrodeposition (Fu et al. 2000). However, by changing processing conditions, others have shown that it is possible to obtain high quality films from ethanol (Sun et al. 2000). The majority of the works were carried out at high voltages, but some works have been carried out at low voltages, including electrodeposition at room temperature from a solution of lithium acetylide prepared *in situ* in dimethylsulfoxide DMSO (Kulak et al. 2003, Shevchenko et al. 2001), and also from acetic acid and formic acid in aqueous solutions (Gupta et al. 2003, 2004). Deposition from acetylene-saturated ammonia has also been carried out at low temperatures of −33°C and below (Chen et al. 2003, Novikov and Dymont 1997a, Novikov and Dymont 1997b). It is believed that in vapor-phase deposition, the active species is a CH_3 radical, and the mechanism proceeds by deposition of both diamond and graphitic carbon. Concurrently, excess of H radicals are produced preferentially with the graphitic phase of carbon over the diamond phase (May 2000). To date, no methods of deposition from solutions have been able to achieve the high quality of diamond film that is possible through vapor phase deposition. However, methods to increase the population of CH_3 radicals at the surface of the electrode are very appealing because they may lead to a higher percentage of sp^3 vs. sp^2 carbon. The unique combination of structure, properties, and applications of DLC films are discussed within the framework of some review articles (Robertson 1992, Bewilogua and Hofmann 2014, Mochalin et al. 2011, Jastrzębski et al. 2017). Material characterizations, especially Raman spectroscopy, which is considered the key property of diamond and related materials, were discussed elsewhere (Ferrari and Robertson 2004). There are quite a good number of books and book chapters (Spitsyn et al. 2005, Alexander et al. 2003) also where the structure, property, process developement, and applications of diamond and its related materials are discussed in detail. In this book chapter, I have concentrated only on the synthesis of DLC film by electrodeposition technique. I have mainly focused on the mechanism in detail of both the cathodic and anodic electrodeposition of DLC. An emphasis is given on the discussion of the application, especially on biomedical application of this particular material, and the possible impact of electrodeposition technique is discussed here.

2. A Brief History of DLC by Electrodeposition

Liquid deposition technique has many advantages, such as availability for large area deposition on intricate surfaces, low bath temperature, low consumption of energy, and the simple experimental setup, over the vapor deposition technique, with a view to the practical applications. Besides that, electrodeposition has the potential for the room temperature or near room temperature synthesis of many kinds of solid-state materials in thin film form, but films deposited by this method are mainly metallic or ionic compounds through the reaction of ions (Zhitomirsky et al. 1997, Fleury 1997). This technique was not very famous for the synthesis of covalent compound materials. Only during the last two or three decades, several attempts have been made by various researchers to deposit covalent carbon or carbon nitride films by the electrodeposition technique from organic solvents under high voltages, medium voltages, as well as low voltages.

There is theoretical evidence that most materials which can be deposited from the vapor phase can also be deposited in the liquid phase (Maissel and Glang 1970). Enlightened by this idea, Namba (Namba 1992) first attempted to grow diamond phase carbon films using ethanol solution as electrolyte at a temperature less than 70°C on silicon substrates. In their pioneering work, substrates were negatively biased with sufficient dc potential to simulate ionized deposition conditions used in physical vapor deposition, and they have confirmed that the film is composed of small grains of diamond phase or diamond-like structures. In his article, however, only x-ray photoelectron spectroscopy results had been shown to indicate that there was carbon phase in the films, and no obvious evidence could be seen to confirm that diamond or DLC films had been obtained. Thereafter, Suzuki et al. (Suzuki et al. 1995) deposited carbon films by electrolysis of water-ethylene glycol solution. Graphite carbon had been obtained according to their Raman spectra. An anodic deposition of diamond-like carbon film based on electrochemical process was developed by Novikov et al. (Novikov and Dymont 1997). They used acetylene in liquid ammonia as an electrolyte, and two types of films were produced at the anode. Type I was transparent and fragile, whereas type II was black and plastic. The aggregation of diamond, diamond-like carbon, and crystalline graphite was reported by the electrolysis of heated methanol by Tosin et al. (Tosin et al. 1999). They observed an inhomogeneity in the film thickness in the vertical direction. They tried to explain this effect by relating it to an inhomogeneous turbulent 'boiling layer' present near the cathode. They have also mentioned in their study that the growth rate of the film on the wetted capillary area is higher than that on the immersed cathode. Almost at the same time, Fu et al. (Fu et al. 1999) have mentioned in their work that Diamond-like carbon and carbon nitride (CN_x) films could be deposited simultaneously on the cathode and the anode from acetonitrile liquid at ambient pressure. In their study, the bath temperature was 80°C, and they used Si-substrate at both the anode and cathode. After that, different attempts have been made by different workers to deposit DLC or different types of carbon materials by electrodeposition technique using alcohol or water-alcohol system and different types of organic solutions (Fu et al. 2000, Cao et al. 2000, Guo et al. 2000, Guo et al. 2001, Kiyota et al. 1999). Guo et al. (Guo et al. 2000) in their work have reported electron field emission properties from DLC films deposited by electrolysis of methanol for the first time.

Yan et al. (Yan et al. 2004) in their work attempted to prepare the DLC nanocomposite films containing polycrystalline SiO_2 nanoparticles, making use of the electrolysis of methanol-dimethylethoxydisilane (DDS) solution. This novel method could provide a new strategy for electrochemical synthesis of DLC nanocomposites. Furthermore, the SiO_2 nanocrystals incorporated in a suitable carbon matrix may have potential applications in microelectronic devices and optical apparatus. Again in the same year, Yan et al. (Yan et al. 2004) reported the synthesis of Diamond-like carbon (DLC) and nitrogen doped DLC films on Si (100) substrate by the electrolysis of another new electrolyte methanol-urea solution under high voltage, atmospheric pressure, and low temperature. In their work, they have proposed a growth mechanism also based on their analysis. They have mentioned the reactions of the CH_3 groups with each other to form carbon network, and the reactions of the CH_3 and NH_2 groups to form carbon nitride network, respectively to form DLC and nitrogen doped DLC films on Si substrates. So far, as reported in the literature, all the researchers have synthesized diamond-like carbon film applying higher potential between the electrodes. Almost at the same time, many workers reported the electrodeposition of DLC using medium voltage range as well as low voltage range (Jiang et al. 2004, Jackson et al. 2012, Gupta et al. 2003, 2004). Jackson et al. (Jackson et al. 2012), in their paper, reported electrodeposition of carbon structures at medium voltage range and at room temperature using aqueous solutions of ethanol at different concentration levels. They maintained a constant electric potential between the electrodes in the range of 80 to 300 V at a current density of approximately 2.0 mA/cm². One interesting feature in their study was mirror-polished pieces of n-type silicon (100) with resistivity of about 1 Ω-cm, and dimensions of $30 \times 10 \times 0.3$ mm³ were used as electrodes (anode and cathode). Gupta et al. (Gupta et al. 2003), in their paper, reported that Diamond-like carbon (DLC) films could be deposited by electrodeposition technique onto SnO_2 coated glass substrates by using a mixture of acetic acid and water as electrolyte. They have also pointed out that the applied potential between the electrodes was remarkably low-merely 2.1 V. Afterwards, many researchers have reported the synthesis of diamond-like carbon at lower applied potential. So, it is well established, and literature also supports that diamond-like carbon films can be deposited by electrodeposition using high voltage, medium voltage, as well as low voltage using different type of electrolytes and different deposition conditions. Now the recent trends are going on to explore the functional properties of the electrodeposited diamond-like carbon film. Paul et al. explained in their paper the hydrophobicity in DLC films by the electrodeposition technique (Paul et al. 2008). They have judged the hydrophobicity in these films by measuring the contact angles of water droplets with the films, and shown that the films were extremely hydrophobic in nature. Afterwards, Falcadea et al. (Falcadea et al. 2012) have reported in their paper the corrosion protection and wear resistance of diamond-like carbon coating on titanium alloy using organic liquids such as acetonitrile (ACN) and N,N-dimethylformamide (DMF). Besides that, they have mentioned a significant reduction in coefficient of friction. Sirk and Sadoway (Sirk and Sadoway 2008), in their paper, demonstrated the extremely low temperature electrodeposition of diamond-like carbon films through oxidation of acetylides prepared both *in situ* and *ex situ*, dissolved in dimethylsulfoxide DMSO. The synthesis was carried out onto gold substrates at a temperature of −33°C. They have studied the potentiodynamic, potentiostatic, and galvanostatic properties of the films. In the year 2009, Manhabosco and Muller (Manhabosco and Muller 2009) also demonstrated the synthesis of Diamond-like carbon films (DLC) on titanium substrates by the liquid-phase electrodeposition technique at ambient pressure and temperature using acetonitrile and N, N-dimethyl formamide (DMF) as electrolyte. They applied quite a high voltage (1200 V) between the electrodes. In their investigation, Manhabosco et al. (Manhabosco et al. 2013) described the cell response and corrosion behavior of electrodeposited diamond-like carbon films. They measured the surface roughness and contact angle of bare nanostructured titanium samples and the electrodeposited diamond-like carbon films (DLC) on the nanostructured samples. In addition, they also evaluated the mechanical properties and the corrosion resistance of the samples in phosphate buffered saline (PBS) in their investigation. Recently, Ismail et al. (Ismail et al. 2016) explored the synthesis of diamond-like carbon films by electrodeposition technique for solar cell applications. In their paper, they have studied that a high transparent diamond-like carbon (DLC) film could be deposited on ITO and silicon substrates by electrodeposition technique, using a combination of methanol and ethanol as electrolyte. They investigated optical and microstructural properties of DLC films by varying the volume percentage of ethanol and methanol.

So, since the discovery of the synthesis of carbon material by electrodeposition to date, there are many researches going on, and in the near future, many new investigations will come as their natural way in this area.

3. Amorphous Carbon Films Prepared by Cathodic Deposition Mode and Growth Mechanism

3.1 Application of high voltage (1000–3000 V)

A number of organic liquids were utilized as electrolytes by different workers for the deposition of DLC films by electrodeposition technique. The dielectric constant (ε) and dipole moment (σ) of the electrolytes are key parameters in liquid phase electrodeposition technique. Viscosity of the electrolyte also plays an important role in liquid electrodeposition technique. Thus, a favorable compromise between all these parameters is very essential to find out an efficient and effective electrolyte in electrodeposition of DLC. Some experiments have proved that hydrogenated amorphous carbon films with obvious characteristics of diamond structure can be prepared using liquid phase deposition under atmospheric pressure and low temperature. As these organic solvents possess a lower electrical conductivity, hence a higher voltage in the range of 1000–4000 V should be applied between two electrodes. In all the cases, the deposition rate is very slow. Here, the two electrodes system can be regarded as an ideal parallel plate capacitor filled with dielectric medium. A solution with a dielectric constant ε is polarized under the action of an applied external electric field. Induced charges created on the electrodes and the density (also known as the polarization ability) of charges is determined by the formula mentioned in equation (1).

When the capacitor is full of a medium with a dielectric constant of ε, the medium can be polarized by applying an electric field between the plates of the capacitor. There is induction charge on the surface of the plate, and the density of the induction charge (also called polarization ability) P, can be represented by the equation

$$P = \sum \sigma / \Delta V = (\varepsilon_r - 1) \, \varepsilon_0 \, E = (\varepsilon - \varepsilon_0) \, E \tag{1}$$

where P, σ, E, and ΔV represent polarization, dipole moment, electric strength, and volume element of the dielectric, respectively, and ε, ε_0, and ε_r represent dielectric constant of the medium, absolute permittivity, and relative permittivity of the medium respectively. E is the electric field, which is proportional to the potentials applied between the two electrodes as E = U/d, where U is the applied potentials and d is the separation between the two electrodes. Under the same deposition conditions, P is in direct proportion to the dielectric constant ε only. Obviously, under the same deposition conditions, the polarization P depends only on the dielectric constants. It is therefore necessary to select organic solutions with high dielectric constants (Jiu et al. 1999). Carbon sources used frequently contain methanol (Mayama et al. 2010), alcohol (Sreejith et al. 2005, Izake et al. 2005), methyl cyanide, DMF, DMSO (Kulak et al. 2003, Jiang et al. 2004), and so on, for high voltage deposition method.

Reaction mechanism

The electrochemistry process for liquid phase electrodeposition of DLC films is not very simple, but rather complicated. After going through the literature survey, it has been observed that most of the work that supports the growth of diamond-like carbon film is based on the polarization reaction mechanism. All molecules from the carbon sources used in deposition of the DLC films are polar molecules. The electron distribution in polar molecules is not symmetric, that is, the centers of the positive charge do not coincide with the center of negative charge. Here, the polarization is mainly due to the electronic as well as orientation or dipolar. The polarization produced due to the displacement of electrons is known as electronic polarization. The dipolar or orientation polarization is produced only when an electric field is applied to a polar molecule. The dipole experiences a torque and tries to align parallel to the applied field, which results in a rotation of the dipoles. The total polarization can be expressed here

$$P = P_e + P_o = N\alpha_e E_i + N\alpha_o E_i \qquad (2)$$

where α_e is the electronic polarizability, α_o is the orientation polarizability, E_i is the internal or local field which is equal to the sum of applied electric field plus the field produced by the surrounding dipoles, and N is the total number of particles (dipoles) per unit volume. The orientation polarizability can be written as

$$\alpha_o = \frac{\mu_p^2}{3KT} \qquad (3)$$

where μ_p is the dipole moment of the material and this is the micro parameter in polarization, and T is the absolute temperature. Again, for elemental solid dielectrics such as diamond, the electronic polarizability can be expressed as

$$\frac{N\alpha_e}{3\varepsilon_0} = \frac{\varepsilon_r - 1}{\varepsilon_r + 2} \qquad (4)$$

The above equation is known as Clausius–Mosotti equation. In this equation N is the number of dipoles per unit volume, ε_r is relative permittivity or dielectric constant of the material, and this is the macro parameter in polarization. ε_o is the absolute permittivity in free space, which is a constant parameter. Again, one more thing we have to consider is that when the potential V is applied, the energy of the capacitor is

$$W = \frac{1}{2}(CV^2) = \frac{\varepsilon_0\varepsilon_r S}{2dV^2} \qquad (5)$$

where S is the area of plate, and d is the distance between the plates. When the potential is increased, the energy of the capacitor will also be increased. Then the degree of polarizability of the medium will be enhanced. If we compare equations (1), (2), (3), (4), and (5), we can easily observe that even though the dielectric constant and dipole moment of the electrolyte are the main parameters, other factors, such as the number of dipoles per unit volume, temperature, distance between the cathode and the anode, and their surface area and effective electric field all are important factors in the polarization mechanism. So, the rate and fate of the electrodeposition technique depends on all these parameters and that is the reason the reaction mechanism in electrodeposition is not so easy, but rather complicated. In the experiment by liquid electrodeposition for the deposition of DLC films, the ability depends mainly on the polarization of the medium in these organic liquids. Under higher dielectric constant and higher applied potentials, the polarization of medium is enhanced and as a result, the reaction current density increases. The chemical bond energy and bond length of correlated molecules in electrolyte are also important things to consider, and have been listed in Table 2.2. From the table, we can see that the chemical bond between the CH_3 group and polar group is the weakest.

For DLC films in a chemical reaction, these weak bonds are always broken first. Therefore, to choose a proper electrolyte, the first priority is given to the electrolyte containing CH_3 radical. According to the analysis above, we can infer that the high voltage deposition of DLC films by

Table 2.2: The bond energy and bond length of relative chemical bonds.

Chemical bonds	CH_3-OH	CH_3O-H	$(CH_3)_2$-N-COH	$(CH_3)_2$N-COH	CH_3-CN	C-N
Bond energy (Kcal/mol)	91	104	------	-------	119	204
Bond length (nm)	0.143	0.097	0.147	0.132	0.149	0.116

electrolysis follows a polarization reaction mechanism. Moreover, the detailed reaction process is as follows (He et al. 2005, Zhu et al. 2003)

1. In case of polar molecules the centers of the negative and the positive charge are not coincident and the molecules are partially polarized.

2. Under high potentials, polar molecules again try to reorient themselves along the field directions, and the distance between the positive and negative charge centers gradually increases. As a result, the polar molecules turn into energized molecules:

$$CH_3 \rightarrow \frac{high\ potential}{(energized\ molecules)} CH_3. \ldots \ldots \ldots \ldots \ldots .X \tag{6}$$

3. When a high potential is applied to the surface of the electrode, the surface is also activated and becomes the reaction-activated site.

4. When energized molecules CH_3X^* move to the surface of the electrode, they are absorbed on the activated sites and turn into activated molecules

$$CH_3X^* \rightarrow \frac{adsorbed\ in\ the\ active\ sites}{(activated\ molecules)} CH_3X^{\neq} \tag{7}$$

5. Activated molecules which are brought about by the oxidation-reduction reaction on the electrode turn to carbon and other products

$$CH_3X^{\neq} \rightarrow \frac{electrochemical}{reaction} C + other\ products \tag{8}$$

6. Continuous formation of sp^2 and sp^3 bonded carbon leads to the growth of diamond-like carbon films.

The above reaction mechanism can be thought of as simply being based on the experimental conditions and measurement results obtained from DLC films. However, many problems are still there and need to be solved. For instance, the growth and nucleation of the methyl group adsorbed on the cathode surface, and the decomposition and formation of sp^2 or sp^3 carbons in the growth process, etc. These problems must be addressed further in order to deposit perfect and high quality DLC or diamond films in the liquid phase by high voltage cathodic deposition mode.

3.2 *Application of mid voltage (200–800 V)*

Literature supports that atomic hydrogen effectively etches graphite phase in vapor phase deposition of diamond film. Inspired by this concept, many researchers attempted to add deionized water into an organic reagent in the electrodeposition process. The results showed that the voltage required for the synthesis of diamond-like carbon film reduced, or growth rate of the reaction increased, and the sp^3 content in the films also increased (Zhang et al. 2007), and even nanodiamond particles have been found in the deposited films. A thermodynamic coupling model has been proposed to explain the above phenomenon (Li et al. 2012). The entire growth process of carbon film can be divided into three steps: (i) organic molecules are decomposed into the groups containing carbon, (ii) they are captured by the electrode, and (iii) carbon films grow through a series of reactions by making and breaking of carbon and hydrogen bonds. Using methanol as an example, carbon films are grown by generating methyl ions and degenation-dehydrogenation conversion. The procedure is simply expressed as follows

$$CH_3OH \rightarrow CH_3 + OH \qquad \qquad \nabla G_1 > 0 \tag{9}$$
$$(T = 300K, P = 101\ KPa)$$

$$CH_3 \rightarrow C(sp^3, sp^2) + \frac{3}{2}H_2 \qquad \nabla G_2 < 0 \tag{10}$$

$$(T = 300K, P = 101\ KPa)$$

where ∇G is the change of Gibbs free energy. The increment of Gibbs free energy is a criterion for judging whether a reaction occurs spontaneously at constant temperature and constant pressure. $\nabla G_1 > 0$ means the reaction is happening spontaneously in the right hand direction with lower input energy, which means that by applying lower input voltage, the methyl radicals in thermodynamics can spontaneously transform into carbon and hydrogen atoms. Note that this reaction usually needs to overcome an energy barrier; otherwise, the resultant may be in metastable phase, such as ethylene chain, etc. So, the fact is that the electrolyte plays a very important role in the film formation. When it consists of deionized water with alcohol, water interacts in two ways: (i) it can participate in the formation of dynamic hydrogen-bonded chains, thereby raising the polarizability, and (ii) it can form relatively stable structures, such as $H_2O\ (ROH)_4$, which has zero net dipole moment and consequently diminishes volume polarizability, and can perhaps help to form sp^3 carbon hybridization in liquid phase electrodeposition technique (Aprano et al. 1979). As H_2O is a polar molecule, it can be easily decomposed into H ion and hydroxyl ion by inputting a lower electrical energy, that is, lower voltage. H ion obtains electron to change into atomic hydrogen. Hydrogen atoms near the cathode surface should have two functions: interrupting the C-H bonds in the methyl to form C-C bond, which is better for increasing the growth rate of the film, and breaking chemical bonds in organic molecules owing to large amounts of energy from the association of hydrogen atoms.

$$H \rightarrow 0.5H_2 \qquad \nabla G_3 < 0 (T = 300K,\ P = 101\ KPa) \tag{11}$$

It is believed that three reactions (9), (10), and (11) could be thermodynamic coupled when the two processes above occur at the same time. The overall coupling reaction is as follows

$$CH_3OH + xH = C(sp^3) + OH + (0.5x + 1.5)H_2 \tag{12}$$

$$(T = 300K,\ P = 101\ KPa)$$

$$\Delta G_4 = \Delta G_1 + \Delta G_2 + x\Delta G_3 \tag{13}$$

Here x represents the coupling coefficient, increased with concentration of hydrogen atoms near the substrate. The larger the coupling coefficient x, the smaller the ∇G_4, the reaction (13) moves more easily to the right. Therefore, we can conclude that the growth of carbon films can occur under a low applied potential between the electrodes because of strong polarization and spontaneous decomposition of water molecules.

Sreejith et al. (Sreejith et al. 2005) and Jackson et al. (Jackson et al. 2012) proposed a mechanism based on ethanol as an organic electrolyte in the medium voltage range (80–300 V) on the Si substrate, which is quite similar to the proposed mechanism in the high voltage range (He et al. 2005, Zhu et al. 2003). Using ethanol as the electrolyte, there is a strong possibility that ethanol molecules are polarized under an applied potential, as can be demonstrated by the following equations

$$CH_3CH_2OH \rightarrow CH_3CH_2^{\delta+} - OH^{\delta-} \rightarrow CH_3CH_2^{\delta+} \ldots\ldots\ldots OH^{\delta-} \tag{14}$$

$$CH_3CH_2^{\delta+} \ldots\ldots\ldots OH^{\delta-} \xrightarrow{\ \equiv Si\text{-}H\ } CH_3CH_{2(solution)}^{+} + H_2O + \equiv Si^{-}$$

$$CH_3CH_{2(solution)}^{+} + \equiv Si^{-} \longrightarrow\ \equiv Si - CH_2CH_{3(substrate)} \tag{15}$$

In the cathode (substrate), the major step is the dehydrogenation of ethyl that occurs via electrooxidation reaction

$$CH_3CH_2^+ + ne^- \longrightarrow 2C + {}^5/_2 H_2 \tag{16}$$

The electrochemical anodic dehydrogenation of the alkyl groups attached to the surface leads to the formation of diamond and diamond-like phases. Subsequent dehydrogenation leads to the formation of DLC. According to their proposed mechanism, the bonds could also be formed between one carbon and another adjacent carbon atom sitting on the substrate, thus propagating and extending the chain both along the vertical and horizontal directions of the substrate.

3.3 *Application of low voltage (2–20 V)*

Further attempts had been made by other researchers to prepare amorphous carbon films using aqueous solution of acetic acid or formic acid or acetylene saturated ammonia as electrolyte (Cao et al. 2000, Gupta et al. 2004, Chen et al. 2003, Novikov and Dymont 1997) under the applied voltage of less than 20 V. This result is easy to explain based on the above discussion because of the spontaneous formation of hydrogen ions generated from the ionization of aqueous solution of acetic acid or formic acid as electrolyte. Therefore, the reaction (12) can proceed in the right direction only by applying a low input power, which means low voltage. That means a large amount of free ions of hydrogen in liquid phase guarantees to grow amorphous carbon films under a low external electric field. The effect of voltage on the carbon film growth can be explained and summarized in Table 2.3.

Even though the mechanism of electrodeposition of amorphous carbon film is more complex, Gupta et al. (Gupta et al. 2003) put forward a rather simple and concise mechanism of depositing DLC films at low applied potential between the electrodes from ionic electrolytes, such as acetic acid in water. The acetic acid initially was dissociated into CH_3^+ and $(COOH)^-$ ions. Then CH_3^+ ions migrate to the cathode along with H^+ ions generated from the hydrolysis of water present in the electrolyte. Then, through the process of making and breaking bonds, DLC films are formed on the cathode.

A generalized mechanism for electrolytes containing COOH functional groups, therefore, is represented as below

$$R - COOH \rightarrow R^+ + COOH^- \tag{17}$$

where R represents alkyl or aryl group

At the cathode

$$R^+ + e^- \rightarrow R \tag{18}$$

$$R + R \rightarrow R - R \, (film) \tag{19}$$

At the anode

$$COOH^- - 2e^- \rightarrow CO_2 + H^+ \tag{20}$$

Thus, the cathode surface would experience the presence of a very high concentration of methyl radicals and hydrogen ions, which are the basic requirements for the formation of DLC films by liquid phase electrodeposition. Now, the substrate surface (SnO_2) having oxygen deficient sites would provide anchorage to the CH_3^+ radicals along with H^+, which would take part in breaking and making of sp^2 and sp^3 bonded carbon, respectively, to ensure the growth of DLC film on the SnO_2 coated glass. Formation of critical clusters subsequently leads to the formation of continuous films. Thus, one may conceive a model of DLC by electrochemical process, as described above. The simplified model is illustrated schematically in Figure 2.1 (Gupta et al. 2003).

Table 2.3: Effect of voltage on the growth of carbon film by electrodeposition.

Electrolytes	Voltage (V)	Mechanism
Pure organic reagents	1000–4000	Polarization and decomposition of organic compounds
Organic reagents with deionized water	200–800	Polarization and decomposition of deionized water
Electrolyte with a large amount of H ions	2–20	Polarization and decomposition of deionized water and the trapping of hydrogen ions

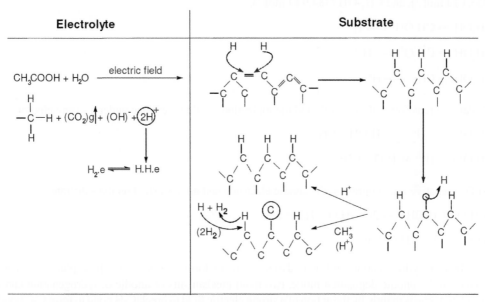

Figure 2.1: Simplified model for the growth of DLC films by electrodeposition technique.

Alternatively, electrolytes having OH functional groups will also undergo dissociation under high voltage, producing R^+ according to the equation given below

$$R - OH + H_2O \rightarrow R^+ + H^+ + 2OH^- \tag{21}$$

4. Amorphous Carbon Films Prepared by Anodic Deposition Mode

So far, since the pioneering work of DLC by electrodeposition, only a few researchers attempted to prepare diamond-like carbon films by anodic deposition mode. The first attempt was made by Novikov and Dymont (Novikov and Dymont 1997, Novikov and Dymont 1997) in the year 1997. A brittle carbon film was successfully synthesized by them using anodic oxidation of aqueous solution of ammonia saturated acetylene under 2.0–3.0 V. Meanwhile, in 2001, they gained nanodiamond particles using 50 mol% ammonium acetate in acetic acid solution as electrolyte at the anodic basement (Aublanc et al. 2001). In other reports, such as Shevchenko et al. (Shevchenko et al. 2001) and Li et al. (Li et al. 2012), they used C_2HLi/DMSO and methanol as solutions for the anodic deposition of diamond-like carbon film by liquid electrodeposition.

As per the mechanism discussed in the cathodic deposition technique, methanol can be decomposed in three ways to form positive charged particles such as CH_3^+ and H^+, and negative groups such as OH^-, CH_3O^-, and CH_2OH^-, etc. If deionized water is added into the electrolyte, the water molecules also decompose into H^+ and OH^-. The positive ions (CH_3^+ and H^+) move to the cathode under the electric field and react with the surface to form carbon films. If the above opinion is true, the negative groups (OH^-, CH_3O^-, and CH_2OH^-) might move to the anode and form carbonaceous films too.

According to the literature reported by Li et al. (Li et al. 2013), the methanol molecules are decomposed under high electric, and the negative charge species CH_3O^- and CH_2OH^- are captured on the anode. The polarization-reaction process for methanol can be described as follows

(1) Methanol can decompose in three possible ways under high applied potential depending on the available energy and similarity among bond energies of H-CH_2OH (401.7 kJ mol^{-1}), CH_3O-H (435.1 kJ mol^{-1}), and CH_3-OH (384.9 kJ mol^{-1}).

$$CH_3OH \rightarrow CH_2OH^{\delta-} ----- H^{\delta+}$$

$$CH_3OH \rightarrow CH_3O^{\delta-} ----- H^{\delta+}$$

$$CH_3OH \rightarrow CH_3^{\delta+} ----- H^{\delta-}$$

(2) Partial breakdown of the bonds and the ionization would take place under a strong electric field

$$CH_2OH^{\delta-} ----- H^{\delta+} \rightarrow CH_2OH^{\delta-} + H^{\delta+}$$

$$CH_3O^{\delta-} ----- H^{\delta+} \rightarrow CH_3O^{\delta-} + H^{\delta+}$$

(3) The negative ions migrate to the anodic electrode and are absorbed on the substrate

$$2CH_3O^{\delta-}/CH_2OH^{\delta-} \rightarrow 2C + H_2O + H_2\uparrow + 2e$$

$$4OH^{\delta-} \rightarrow O_2\uparrow + H_2O + 4e$$

Then, dehydrogenation and deoxidation occur to form carbon films. To explain the growth mechanism by anodic deposition mode, two main mechanisms of anodic dehydrogenation can be considered: one is mainly by the action of a strong electric field to break C-H bond to form C-C bonds, another mainly by the movement of the negatively charged hydroxyl to the anode and reaction with the hydrogen atoms in C-H bonds. The anode deoxygenation, however can only rely on the action of electric field to interrupt C-H to form C-C bonds, so a certain amount of oxygen elements will remain in the carbon film. Many studies have demonstrated that the existence of oxygen elements in the carbon will cause the downward shift of G peak (Mckindra et al. 2011, Zhou et al. 2011). The etching effect of O$^-$ ions on sp^2 carbon is 50 times higher than that of H$^+$ ions (Ramesham et al. 1997, Urruchi et al. 2000), which might cause some change in surface morphology of the film by the anode deposition.

5. Synthesis of Carbon Nanotubes (CNTs) by Electrodeposition Technique

Carbon nanotubes (CNTs) are cylindrical carbon molecules with outstanding mechanical, electrical, thermal, optical, and chemical properties. The special nature of carbon combines with the molecular perfection of buckytubes (single wall carbon nanotubes) to endow them with exceptionally high material properties, such as electrical and thermal conductivity, strength, stiffness, and toughness. No other element in the periodic kingdom bonds to itself in an extended network with the strength of the carbon-carbon bond. Graphite has each carbon atom linked to three other carbon atoms by covalent bonds to form hexagonal rings, which are connected to each other to form plane sheets known as graphenes. The structure of carbon nanotube (CNT) is a cylinder formed by rolling a graphene sheet and the ends closed by fullerene hemispheres. If we imagine the rolling up of graphene into a seamless tube, it can be thought of in various ways. For example, carbon-carbon bonds can be parallel or perpendicular to the tube axis, resulting in a tube where the hexagons circle the tube like a belt, but are oriented differently. Alternatively, the carbon-carbon bonds need not be either parallel or perpendicular, in which case the hexagons will spiral around the tube with a pitch depending on how the tube is wrapped. In most respects, the properties of tubes of different types are essentially the same. The exception to this is in their electrical conductivity, where these subtle structural differences can have profound effects. There is literature where the structure and the properties of carbon nanotubes are explained nicely (Qiu and Yang 2017, Odom et al. 1998, Salvetat et al. 1999). They have the potential as low loss connectors in nanoscale integrated circuits and field-effect transistors, photovoltaic cells, in materials, and more. Carbon nanotubes (CNTs) have the potential to revolutionize materials engineering, making it possible to create materials with novel properties. Since the tubes with different orientation of carbon-carbon bonds are molecularly distinct, there exists the possibility of chemically separating different type of structures, and even of growing only selected types of tube structure, although currently, all production processes produce a random mixture of different tube structures. At present, the growth mechanisms of carbon nanotube are not fully understood, and therefore it has been difficult to manipulate the synthesis of these nanostructures. In fact, synthesis processes thus far have only been optimized by trial and error. Another roadblock is that carbon nanotubes, although they have great potential for many applications, have not been produced in large enough quantities and at low enough cost to replace any currently available materials in the market. A large part of this cost (approximately 80%) comes from the synthesis step alone. Clearly, improvements in the efficiency of the synthesis process must be implemented before CNTs can become a viable technology. As a low cost alternative, researchers also started working on the synthesis of diamond-like carbon film in the form of carbon nanotubes (CNTs) by simple electrodeposition technique in opposition to the high cost vapor deposition technique. In the year 2005, Pal et al. (Pal et al. 2005) reported for first time the synthesis of carbon nanotubes by electrodeposition on Si substrate using a mixture of acetonitrile and water as electrolyte at an applied d.c. potential ~ 20 V. They have pointed out in their work that the presence of magnetic field played a crucial role in the growth of carbon nanotubes and hence the electronic properties. Thereafter, in 2006, Zhou et al. (Zhou et al. 2006) reported the synthesis of CNTs by electrochemical deposition at room temperature using the applied potential between the electrodes of the order of 1000 V. In their work, a standard three-component electrochemical cell was used for the depositions of carbonaceous films using a mixture of 40 vol% methanol (CH_3OH) and 60 vol% benzyl alcohol (C_6H_5CHOH) as the electrolyte. There are some reports on polymer/carbon nanotubes composite films (Jiang et al. 2004) also by liquid electrodeposition. In this investigation, the nanocomposite films of polypyrrole (PPy) and multiwalled carbon nanotubes (MWCNT) were synthesized by direct oxidation of pyrrole in 0.1 M aqueous solution of dodecylbenzene sulfonic acid containing a certain amount of MWCNT.

Here we are trying to realize the synthesis of CNT by magnetic field assisted electrodeposition technique using acetonitrile as the electrolyte (Pal et al. 2005).

Motion of Charge Particle in a Combined Parallel Electric and Magnetic Fields

When an electric charge with charge q, velocity V, and mass m, is moving in an electromagnetic field, it will be subjected to a force called Lorentz force given as

$$F = q\,E + q\,(V \times B) \tag{22}$$

where E is electric field along Y-axis, and B is a magnetic field. The electric field repels the charge upward parabolically in the y direction, but the magnetic field pulls it down so its direction will be along X-axis. When Fe = Fm on the charge q, its direction will be perpendicular at E along X-axis.

The charge will move along the path of electric field (which is parallel to the magnetic field). Magnetic field has no effect when there is no perpendicular component of motion. However, with the slightest perturbation, the Lorentz force will cause the trajectory to bend, as shown in Figure 2.2. The motion of a charged particle in a combined parallel electric and magnetic field could be explained by the following

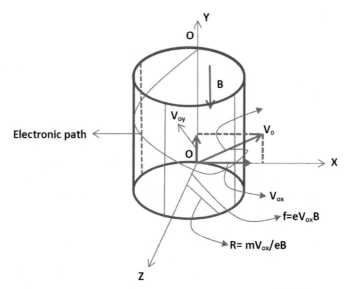

Figure 2.2: The helical path of an electron at an angle (not 90°).

(1) When both electric and magnetic fields act simultaneously on an electron, no force is exerted due to the magnetic field and the motion of the electron is only due to the electric field intensity. Since the electron moves parellel to the magnetic field; the value of

$$\varphi = 0 \; ; \; \therefore f_m = q. \; B \times V = q.BV \sin 0^\circ \; ; \; f_m = 0 \tag{23}$$

(2) The electron moves in a direction to the fields with a constant acceleration.

(3) If the electric field is along the Y-axis and magnetic field is also along the Y-axis, the motion of an electron is specified by

$$V_y = V_{0y} ; \; y = V_{0y} \, t - \tfrac{1}{2} \, at^2 \tag{24}$$

where, $a = q. \; E/m$ = the acceleration due to the magnetic field.

(4) If a component of velocity V_{0x} perpendicular to the magnetic field exists, initially this component along with the magnetic field will set the electron in a circular motion.

(5) The radius of the circular path is independent of the electric field, but the velocity along the field changes with time. As a result of this, the electron travels in a helical path with the pitch changing with time.

Figure 2.3 shows the electrolytic bath with parallel electric and magnetic fields, and Figure 2.4 shows the SEM micrographs of a representative film. The film was deposited onto Si substrate, attached to a copper cathode when the deposition was carried out at a magnetic field of ~ 0.1 T applied to the same direction as the electric field. CNTs were synthesized by electrolysis using acetonitrile (1% v/v) and deionized water as electrolyte. Electrolysis was carried out at atmospheric pressure and the bath temperature was kept at ~ 300K. Graphite was used as the counter electrode (anode). The electrodes were separated by a distance of ~ 8 mm. The detailed experimental section is described elsewhere (Pal et al. 2005).

The applied potential between the electrodes was ~ 20 V. As per the analysis of Pal et al. (Pal et al. 2005), the carbon nanotubes obtained without using magnetic field are randomly arranged, while the application of a magnetic field culminated in aligning the nanotubes in the plane of the substrate surface. The SEM micrographs show the circular trajectories of the charge particle in a combined parallel electric and magnetic field, which supports the established science behind the

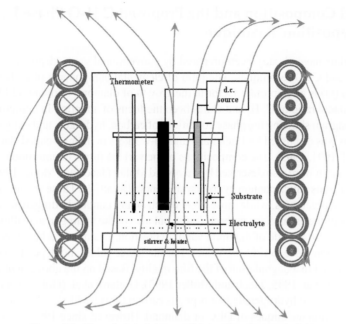

Figure 2.3: Electrolytic bath in a parallel electric and magnetic field.

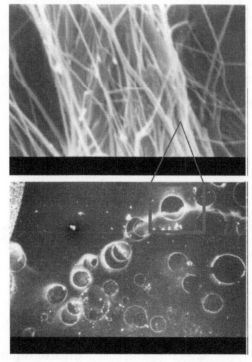

Figure 2.4: Micrograph showing circular pattern for growth in combined electric and magnetic field.

motion of a charged particle in a combined electric and magnetic field. Once the circular trajectories are magnified, that shows the tube-like structure. Magnetic field, undoubtedly, plays an important role in the alignment of carbon nanotubes.

6. Chemical Composition and the Proposed C-H-O Phase Diagram for Electrodeposition Technique

Carbon phase films have already been prepared by conventional vapor deposition for many years, including PVD and CVD techniques. The ratio of elemental components of carbon (C), hydrogen (H), and oxygen (O) is an important factor in the growth of diamond or diamond-like carbon films. Thus, the component ratio of C, H, and O has been the focus of attention in numerous experiments. The "C-H-O" diagram showing the phase conditions of carbon, hydrogen, and oxygen for the growth of diamond films was discussed by Bachmann and co-workers in their so-called Bachmann diagram (Bachmann et al. 1991, 1995). The compositions of the various forms of amorphous C-H alloys on a ternary phase diagram were also described by Jacob and Moller (Jacob and Moller 1993). According to their analysis, the films which contain hydrogen are denoted as a:C-H. These are necessarily deposited using hydrocarbon precursors, while those films which do not contain hydrogen are called a:C. DLC consists not only of the amorphous carbons (a-C), but also of the hydrogenated alloys, a-C:H. These diagrams have become a guide of diamond vapor deposition. Diamond carbon deposits can only be obtained in the region close to C/O = 1 values. For the lower values, no deposition occurs, whereas the higher values lead to a degradation of the film quality. Based on the abovementioned theoretical models (Bachmann et al. 1995, Jacob and Moller 1993), Gottardi et al. (Gottardi et al. 2008) tried to explain the deposition of hydrogenated amorphous carbon films by PECVD. However, all these are about vapor deposition techniques of DLC or diamond. However, since 1992, no theoretical model has ever been discussed so far in liquid electrochemical deposition of DLC films. Here, the C-H-O phase diagram is introduced into liquid phase method, and an attempt to discover the relationship between electrolyte composition and DLC films formation based on the concept of Bachmann diagram (Bachmann et al. 1991, 1995) and the concept of Jacob & Moller phase diagram (Jacob and Moller 1993) is made. The analytically pure organic chemicals and mixed water and electrolyte solutions have been used as electrolyte to prepare DLC films. Here, the positions of some common electrolytes are shown in the C-H-O diagram in Figure 2.5. Ethanol, acetone, and 2-propanol are all in the non-diamond carbon region, but DLC films were successfully prepared by electrolysis

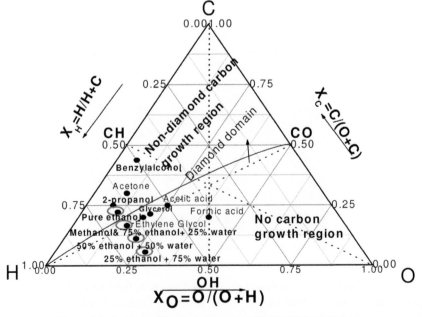

Figure 2.5: Positions of the electrolytes as per the C-H-O phase diagram.

of these organic chemicals, respectively. This is the same with the vapor deposition. Methanol is at the CO line, the boundary of the diamond domain. While ethylene glycol and glycerol are also on the same boundary as methanol, no films were deposited from glycerol and very few were reported from ethylene glycol (Kulak et al. 2003). The main reason is their high viscosity, which is not appropriate for DLC films formation in liquid electrochemical deposition. Afterwards, mixed organic solution was further introduced into liquid phase deposition as an electrolyte. Deionized water is added into analytically pure ethanol to form mixed organic solution. Composition of the electrolyte varies with the ratio of deionized water, as does the elemental component ratio of C, H, and O. For example, using a mixture of ethanol and water has been shown in the diagram (Figure 2.5). With the increase of deionized water ratio, the positions of the mixed solutions start from the non-diamond carbon region and go through the diamond domain, and then go into the no growth region in the Bachmann diagram.

Thus, it is possible to prepare DLC films using electrolytes located in the no growth region for vapor deposition of diamond. The best example is ethanol. There are many reports for the synthesis of diamond-like carbon films using a mixture of water and ethanol as an electrolyte (Sun et al. 2000, Jackson et al. 2012, Pang et al. 2010). In some of the reports, the researchers have used pure ethanol as an electrolyte (Namba 1992, Wan et al. 2009), and this electrolyte lies in the non-diamond carbon growth region. Some other examples such as acetylene, lie in the non-diamond carbon growth region, but there are researchers who have used acetylene saturated ammonia (Chen et al. 2003, Novikov and Dymont 1997a, b) as an electrolyte for the deposition of diamond-like carbon films by electrolysis. There are reports for the synthesis of diamond-like carbon films using electrolytes, such as acetonitrile (CH_3CN), nitro-methane (CH_3-NO_2), DMSO, etc. In these electrolytes, in addition to C, H, and O, nitrogen (N), sulfur, etc. are also the constituent elements, which have to be considered as one of the parameters in the phase diagram. Therefore, it is clearly understandable that the model and explanation in liquid electrodeposition will be quite different from that of vapor phase diamond synthesis. Accordingly, the morphology and microstructure of the DLC films will be affected by the electrolyte composition even when the electrolytes are in the same area in the Bachmann diagram. Therefore, the mechanism of liquid electrochemical deposition of DLC films is not simple, but a rather complicated one that will be certainly different from the vapor phase deposition process. The growth of DLC films can be affected by many factors. For example, the conductivity of the electrolyte, dielectric constant, viscosity, temperature, applied potential between the electrodes, the distance between the electrodes, and temperature—all are important parameters in deciding the rate and fate of the electrochemical reaction. Above that, the addition of additive can change the composition of the solution, and at the same time, the conductivity of the electrolyte will be changed. Moreover, this might affect the formation of DLC films to some extent at the same time.

Since the Bachmann diagram and Jacob and Moller's phase diagram came from the statistics of vapor phase deposition techniques, the results of liquid deposition do not exactly follow the same component classification. Thus, it is necessary to develop a proper model and explanation for the depositions of DLC films by liquid electrochemical technique, both experimentally and theoretically. A better understanding of the change of electrolyte composition and the correlation with the film's properties needs to be developed to guide the experimental researcher in the near future.

7. Applications and Biomedical Applications of DLC & Impact of Electrodeposition

The unique properties of DLC films, such as high hardness, low friction coefficient, chemical inertness, infrared transparency, and high electrical resistivity, combined with smoothness and low deposition temperatures, make them suitable for a large number of applications. Some common applications of DLC based on the corresponding property are shown in Table 2.4. Depending on which form is applied, DLC is as hard, or even harder, than natural diamond. In ta-C form, DLC typically

Table 2.4: Some common applications and corresponding requisite properties of DLC.

Applications	Property requirements
Tribology	
Wear resistant coating for cutting tools	Hard, corrosion resistant, wear resistant, and chemically inert
Impact resistant coating for high-density computer disks	Thin films with high hardness, corrosion resistance, and chemical inertness
Decorative coatings	Hard, transparent in the visible range
Optics	
Protective coatings for IR optics (windows and lenses of Ge, ZnS, or ZnSe) and different type of optical mirrors	Hard, corrosion resistant, and chemically inert, transparent in the IR region of the EMS spectrum
Anti-reflection coatings for Ge	Same as above and refractive index ~ 2
Protective layer for solar cells used in space	Low radiation damage, transparent in visible range of the spectrum
X-ray windows	Self-standing films with high transmission and low damage threshold for X-rays, Smooth surface tropography
Optoelectronic	
Protective layer for electronic devices	Hard, chemically inert, and corrosion resistant, insulating, and high breakdown voltage
Diffusion barriers for optical fibers and optoelectronic devices	Moisture resistant, resistant to oxygen, hard
Protective coatings for photolithography masks and optical disks	Low sputtering yield
Passivating layers in semiconductor	Thermal stability and electrical resistance
Field emission for flat panel displays	Low dielectric constants < 4
Biomedical Applications	
Rotary blood pumps	High hardness, low coefficient of friction, excellent surface finish of amorphous DLC
Artificial heart	Do
Mechanical heart valves	Do
Coronary artery stents	Do
Coatings for metallic orthopedic pins Artificial heart valves	Chemically inert

measures between 5000–9000 HV. Other forms range from 1000–4000 HV. The high hardness of DLC coatings reduces the likelihood of hard particle penetration into tools or parts. Low friction co-efficients and high hardness make them suitable in tribological applications as solid lubricants, in micro-electromechanical systems (MEMS) storage devices, and in automotive applications as hard coating for years (Murakawa et al. 1999), whereas chemical inertness and bio- and hemo-compatibility make them suitable as coating in biomedical implants. They can be employed for new applications— ranging from wear resistance to electronic devices or protective antireflection coating for solar cells (Bursikova et al. 2002, Franta et al. 2004).

The modern world could not exist without continuously developing and modifying the existing health care system. One of the crucial aspects of an interaction of a biomaterial with a tissue is through its surface. For this purpose, the use of biocompatible coatings seems to be a reasonable solution. The rapid developments in this field signify mostly the use of diamond-like carbon (DLC) coatings. These coatings act as a diffusive barrier, preventing tissue near implantation site from penetrating other body parts. Graphene and carbon nanotubes are widely spread carbon allotropes for advanced micro and nano technological applications. However, they are not very popular for biomedical applications due to their cytotoxic effects. In this respect, diamond-like carbon (DLC), being mechanically stable and non-cytotoxic, is spreading rapidly, especially for protective coatings in orthopedic and stent research. Very recently, DLC through proper functionalization is emerging as a potential material for advanced applications, such as, biomolecular monitoring, cancer therapy, and neural cell culture, etc. (Jastrzębski et al. 2017). Two properties of diamond mainly make it a highly desirable candidate for biomedical applications: first, it is bio inert, meaning that there is minimal immune response when diamond is implanted into the human body, and second, its electrical conductivity can be altered in a controlled way, from insulating to near-metallic material. Recent studies even advanced one-step further. Through doping of DLC with various elements, new biologically significant features are being explored day by day. Among dopants that have a high potential for improving the properties of coatings for implants, there are Ag, Si, Cu, Ti, Ca, and P (Olejnik et al. 2017). Nistor and May have concluded in their paper that diamond emerges as one of the major new biomaterials of the 21st century that could shape the way medical treatment will be performed, especially when invasive procedures are required (Nistor and May 2017). DLC in the form of ND suggest the potential application in biological systems for adsorptive separation, purification, and analysis of proteins, design of delivery vehicles for drugs, genes, and antibodies, the fluorescence labeling with luminescent nanodiamond particles. There are some review articles on nanodiamond or DLC for its biomedical applications (Roy and Lee 2007, Auciello 2017). However, the general functionality of DLC films depends on the sp^2/sp^3 ratio, which can be varied within a certain limit depending on the synthesis technique and conditions. Depending on a particular application, the properties of DLC could be tailored or altered greatly by varying the hydrogen content and fraction of SP^3 (shown in Table 2.5). Different metal nanoparticles (Au, Ag, Si, Cu, Cr, Ti, Ni, Zn, Fe, W, V, etc.) and non-metal elements (H, N, F, B, O, etc.) are doped into DLC films to tailor the physical and chemical properties suitable for different biomedical applications (Paul 2017). The bottom line is that DLC has been proved to be safe and effective for implanted medical devices, such as stents, hip, and knee joints. DLC coatings allow implants to maintain integrity, avoid formation of debris, prevent uncontrolled cell growth, and to not cause infections.

For all practical applications of diamond related materials, the manufacturers mainly use physical vapor deposition, chemical vapor deposition such as sputtering, arc discharge, etc. for producing high quality diamond or diamond-like carbon films. So far, there is no industrial application reported based on electrodeposited DLC films. Since after the discovery of DLC by electrodeposition, so far researchers mainly concentrated on the experimental parameters, deposition conditions, etc. In recent years, there are some reports on electrodeposited DLC to explore its bio functional properties

Table 2.5: Characteristics of DLC films and other carbon-based materials (Robertson 2002).

Carbon Material/Property ⟶	sp^3 (%)	H (%)	Hardness (GPa)
Diamond	100	0	100
Graphite	0	0	
Glassy C	0	0	3
Evaporated C	0	0	3
Sputtered C	5	0	
ta-C	80–88	0	80
a-C:H hard	40	30–40	10–20
a-C:H soft	60	40–50	< 10
ta-C:H	70	30	50
Polyethylene	100	67	0.01

(Manhabosco et al. 2013, Zeng et al. 2014). Biological substrates, such as protein, DNA, etc. are very soft in nature. High temperature and high-pressure deposition techniques, such as chemical vapor deposition and physical vapor deposition techniques may not be very well-suited for biomaterials. The common drawbacks of these high temperature and high pressure DLC coatings are: high internal stress (Sikora et al. 2010, Modabberasl et al. 2015) and degradation of mechanical properties due to heat treatment (Tallant et al. 1995). It should also be remembered that because of high internal stresses, an additional problem concerning the coating's delamination may appear. There exists a relation between the level of internal compressive stress and the sp^2/sp^3 ratio. It shows that the presence of high amount of sp^3 bonds in the material (above 80%) leads to higher stress values that can even reach 10 GPa (Dearnley et al. 2010). Wet chemical and liquid electrochemical deposition technique could be a better alternative for diamond-like carbon or nanodiamond coating on biological substrate, and to develop new bio-conjugate material. Now it is well-established that DLC could be deposited by liquid electrodeposition technique. Henceforth, researchers should concentrate more on the functional and bio-functional properties of DLC films by electrodeposition, and based on that, the number of patents and papers should be published for the sake of future applications of electrodeposited DLC films.

8. Conclusions and Future Prospects

High voltages, mid voltage, and low voltage deposition of diamond-like carbon by liquid electrodeposition are well supported by the literature using different types of electrolytes, but so far, no methods of deposition from solutions have been able to achieve the high quality of diamond film that is possible through CVD as far as the hardness and proper adhesion to the substrate are concerned. However, methods to increase the population of sp^3 carbon radicals at the surface of the electrode and proper anchorage to the substrate are very appealing because they may lead to a higher percentage of sp^3 vs. sp^2 carbon. With further research on growth of amorphous carbon films by electrodeposition, the theory and technology in this field will gradually improve. The technique will possess a huge development potential and be widely used to prepare new materials because of unique properties and low cost. Even though there is a significant development and progress for the electrochemical synthesis of diamond-like carbon film, true to my understanding, research work in near future should be more intensive in the following aspects: (1) selection of carbon sources, appropriate additives and dopants, (2) mechanism of formation and stability of amorphous or diamond-like carbon with diamond phase structure, (3) proper theoretical model and its explanation, which will help the experimentalist to find out proper electrolytes, and finally (4) application of the films using the electrodeposition technique.

References

Alexander, V., V. Dolmatov and O. Shenderova. 2003. Detonation nanodiamonds and related materials. Bibliography Index, Issue 1; The Bibliography Index is published in the framework of the First International Symposium Detonation Nanodiamonds: Technology, Properties and Applications. St. Petersburg, Russia.

Aprano, A.D., D.I. Donato and E. Caponetti. 1979. The static dielectric constant of solutions of water in n-alcohols at 15, 25, 35, and 45°C. J. Sol. Chem. 8: 135–146.

Aublanc, P., V.P. Novikov, L.V. Kuznetsova and M. Mermoux. 2001. Diamond synthesis by electrolysis of acetates. Diam. Relat. Mater. 10: 942–946.

Auciello, O. 2017. Novel biocompatible ultrananocrystalline diamond coating technology for a new generation of medical implants, devices, and scaffolds for developmental biology. Biomater. Med. Appl. 1: 1–11.

Bachmann, P.K., D. Leers and H. Lydtin. 1991. Towards a general concept of diamond chemical vapour deposition. Diam. Relat. Mater. 1(1): 1–12.

Bachmann, P.K., H.J. Hagemann, H. Lade, D. Leers, D.U. Wiechert and H. Wilson et al. 1995. Thermal properties of C/H-, C/H/O-, C/H/N- and C/H/X-grown polycrystalline CVD diamond. Diam. Relat. Mater. 4(5): 820–826.

Basiuka Elena, V., A. Santamaría-Bonfila, V. Meza-Lagunaa, T.Y. Gromovoyb, E. Alvares-Zaucoc, F.F. Contreras-Torresa et al. 2013. Solvent-free covalent functionalization of nanodiamond with amines. Appl. Surf. Sci. 275: 324–334.

Bewilogua, K. and D. Hofmann. 2014. History of diamond-like carbon films—from first experiments to worldwide applications. Surf. Coat. Technol. 242: 214–225.

Bursikova, V., P. Sladek, P. Stahel and L. Zajickova. 2002. Improvement of the efficiency of the silicon solar cells by silicon incorporated diamond-like carbon antireflective coatings. J. Non-cryst. Solids 299: 147–1151.

Cai, K., D. Guo, Y. Huang and H.S. Zhu. 2000. Evaluation of diamond-like carbon films deposited on conductive glass from organic liquids using pulsed current. Surf. Coat. Technol. 130: 266–273.

Cao, C.B., H.S. Zhu and H. Wang. 2000. Electrodeposition diamond-like carbon films from organic liquids. Thin Sol. Film. 368(2): 203–207.

Chen, A.M., C. Pingsuthiwong and T.D. Golden. 2003. Electrodeposition of diamondlike carbon films on nickel substrates. J. Mater. Res. 18: 1561–1565.

Dearnley, P.A., A. Neville, S. Turner, H. Scheibe, R. Tietema and R. Tap et al. 2010. Coatings tribology drivers for high density plasma technologies. Surf. Engg. 26: 80–96.

Falcadea, T., T.E. Shmitzhaus, O.G. Reis, A.L.M. Vargas, R. Hüblerc and I.L. Müller et al. 2012. Electrodeposition of diamond-like carbon films on titanium alloy using organic liquids: Corrosion and wear resistance. Appl. Surf. Sci. 263: 18–24.

Ferrari, A.C. and J. Robertson. 2004. Raman spectroscopy of amorphous, nanostructured, diamond-like carbon, and nanodiamond. Phil. Trans. R. Soc. London Ser. A. 362: 2477–512.

Field, J. 1992. Properties of Natural and Synthetic Diamond (1st Edition). Academic Press, Elsevier.

Fleury, V. 1997. Branched fractal patterns in nonequilibrium electrochemical deposition from oscillatory nucleation and growth. Nature 390: 145–148.

Franta, D., I. Ohlidal, V. Bursikova and L. Zajickova. 2004. Optical properties of diamond-like carbon films containing SiOx studied by the combined method of spectroscopic ellipsometry and spectroscopic reflectometry. Thin Sol. Film 455-456: 393–398.

Fu, Q.A., J. Jiu, C. Cao and H.S. Zhu. 2000. Electrodeposition of carbon films from various organic liquids. Surf. Coat. Technol. 124(2): 196–200.

Fu, Q., J.T. Jiu, H. Wang, C.B. Cao and H.S. Zhu. 1999. Simultaneous formation of diamond-like carbon and carbon nitride films in the electrodeposition of an organic liquid. Chem. Phys. Lett. 301: 87–90.

Gottardi, G., N. Laidani, R. Bartali, V. Micheli and M. Anderle. 2008. Plasma enhanced chemical vapor deposition of a C:H films in CH_4–CO_2 plasma: Gas composition and substrate biasing effects on the film structure and growth process. Thin Sol. Film. 516: 3910–3918.

Guo, J., H. Wang and H. Yan. 2007. Recent developments in the preparation of diamond-like carbon films by the liquid. Chemistry Online 7: 521–526.

Guo, D., K. Cai, L.T. Li and H.S. Zhu. 2000. Preparation of hydrogenated diamond-like carbon films on conductive glass from an organic liquid using pulsed power. Chem. Phys. Lett. 325(5-6): 499–502.

Guo, D., K. Cai, L.T. Li, Y. Huang, Z.L. Gui and H.S. Zhu. 2001. Evaluation of diamond-like carbon films electrodeposited on an Al substrate from the liquid phase with pulse-modulated power. Carbon 39: 1395–1398.

Gupta, S., R.K. Roy, B. Deb, S. Kundu and A.K. Pal. 2003. Low voltage electrodeposition of diamond-like carbon films. Mater. Lett. 57(22): 3479–3485.

Gupta, S., M.P. Chowdhury and A.K. Pal. 2004. Synthesis of DLC films by electrodeposition technique using formic acid as electrolyte. Diam. & Relat. Mater. 13(9): 1680–1689.

He, W.L., R. Yu, H. Wang and H. Yan. 2005. Electrodeposition mechanism of hydrogen-free diamond-like carbon films from organic electrolytes. Carbon 43: 2000–2006.

Ismail, R.A., A.M. Mousa, A.H. Mustafa and K.H. Walid. 2016. Synthesis of diamond-like carbon films by electrodeposition technique for solar cell applications. Opt. Quant. Elect. 48: 16–24.

Izake, E.L., T. Paulmier, J.M. Bell and P.M. Fredericks. 2005. Characterization of reaction products and mechanisms in atmospheric pressure plasma deposition of carbon films from ethanol. J. Mater. Chem. 15: 300–306.

Jackson, T., Z. Hudson, C.A.B. Luiz, A.S. Gilmare, J.C. Helder and C.P. Alfredo et al. 2012. Electro-deposition of carbon structures at mid voltage and room temperature using ethanol/aqueous solutions. J. Electrochem. Soc. 159: D159–D161.

Jacob, W. and W. Moller. 1993. On the structure of thin hydrocarbon films. Appl. Phys. Lett. 63: 1771–1773.

Jastrzębski, k., A. Jastrzębska and D. Bociąga. 2017. A review of mechanical properties of diamond-like carbon coatings with various dopants as candidates for biomedical applications. Acta Innovations 40(22): 40–57.

Jiang, H., L. Huang, S. Wang, Z. Zhang, T. Xu and W. Liu. 2004. Synthesis of DLC films by electrolysis of dimethyl sulfoxide. Electrochem. Solid-State Lett. 7: D19–D21.

Jiu, J.T., K. Cai, Q. Fu, C.B. Cao and H.S. Zhu. 1999. Liquid deposition of hydrogenated carbon films in N, N-Dimethyl formamide solution. Mater. Lett. 41(2): 63–66.

Kiyota, H., H. Araki, H. Kobyashi, K. Kitaguchi, M. Lida and H. Wang et al. 1999. Electron field emission from diamond-like carbon films deposited by electrolysis of methanol liquid. Appl. Phys. Lett. 75(15): 2331–2333.

Kulak, A.I., A.I. Kokorin, D. Meissner, V.G. Ralchenko, I.I. Vlasov and A.V. Kondratyuk et al. 2003. Electrodeposition of nanostructured diamond-like films by oxidation of lithium acetylide. Electrochem. Commun. 5: 301–305.

Li, Y., G.F. Zhang, Y.Y. He and X.D. Hou. 2012. Electrical double layer model and thermodynamic coupling for electrochemically deposited hydrogenated amorphous carbon films. J. Electrochem. Soc. 159: 918–920.

Li, Y., G.F. Zhang, X.D. Hou and D.W. Deng. 2012. Synthesis and tribological properties of diamond-like carbon films by electrochemical anode deposition. Appl. Surf. Sci. 258: 6527–6530.

Li, Y., G.F. Zhang, X.D. Hou and D.W. Deng. 2013. Growth mechanism of carbon films from organic electrolytes. J. Mater. Sci. 48: 3505–3510.

Maissel, L.I. and R. Glang (eds.). 1970. Handbook of Thin Film Technology. McGraw-Hill: New York.

Manhabosco, T.M. and I.L. Muller. 2009. Electrodeposition of diamond-like carbon (DLC) films on Ti. Appl. Surf. Sci. 255: 4082–4086.

Manhabosco, T.M., L.A.M. Martins, S.M. Tamborim, M. Ilha, M.Q. Vieira, F.C.R. Gumac et al. 2013. Cell response and corrosion behavior of electrodeposited diamond-like carbon films on nanostructured titanium. Corros. Sci. 66: 169–176.

May, P.W. 2000. Diamond thin films: a 21st-century material. Philos. Trans. R. Soc. Lon. A. 358: 473–495.

Mayama, N., H. Yoshida, T. Iwata, K. Sasakawa, A. Suzuki, Y. Hanaoka et al. 2010. Characterization of carbonaceous films deposited on metal substrates by liquid-phase electrodeposition in methanol. Diam. Relat. Mater. 19(7–9): 946–949.

Mckindra, T., M.J. O'keefe and R. Cortez. 2011. Reactive sputter-deposition of oxygenated amorphous carbon thin films in Ar/O$_2$. Diam. Relat. Mater. 20: 509–515.

Mochalin, V.N., O. Shenderova, D. Ho and Y. Gogotsi. 2011. The properties and applications of nanodiamonds. Nature Nanotechnol. 209: 1–13.

Modabberasl, A., P. Kameli, M. Ranjbar, H. Salamati and R. Ashiri. 2015. Fabrication of DLC thin films with improved diamond-like carbon character by the application of external magnetic field. Carbon 94: 485–493.

Murakawa, M., T. Komori, S. Takeuchi and K. Miyoshi. 1999. Performance of a rotating gear pair coated with an amorphous carbon film under a loss-of-lubrication condition. Surf. Coat. Technol. 121: 646–652.

Namba, Y. 1992. Attempt to grow diamond phase carbon films from an organic solution. J. Vac. Sci. Technol. A. 10: 3368–3370.

Nistor, P.A. and P.W. May. 2017. Diamond thin films: giving biomedical applications a new shine. J. R. Soc. Interf. 14: 20170382(1–15).

Novikov, V.P. and V.P. Dymont. 1997a. Mechanism for electrochemical synthesis of diamond-like carbon. Tech. Phys. Lett. 23: 350–351.

Novikov, V.P. and V.P. Dymont. 1997b. Synthesis of diamondlike films by an electrochemical method at atmospheric pressure and low temperature. Appl. Phys. Lett. 70: 200–202.

Odom, T.W., J.-L. Huang, P. Kim and C.M. Lieber. 1998. Atomic structure and electronic properties of single-walled carbon nanotubes. Nature 391: 62–64.

Olejnik, A., L. Świątek and D. Bociąga. 2017. Biological evaluation of modified DLC coatings—a review. World Scientific News 73(1): 61–71.

Pal, A.K., R.K. Roy, S.K. Mandal, S. Gupta and B. Deb. 2005. Electrodeposited carbon nanotube thin films. Thin Sol. Film. 476: 288–294.

Pang, H., X. Wang, G. Zhang, H. Chen, G. Lv and S. Yang. 2010. Characterization of diamond-like carbon films by SEM, XRD and Raman spectroscopy. Appl. Surf. Sci. 256(21): 6403–6407.

Paul, R., S. Dalui, S.N. Das, R. Bhar and A.K. Pal. 2008. Hydrophobicity in DLC films prepared by electrodeposition technique. Appl. Surf. Sci. 255: 1705–1711.

Paul, R. 2017. Uniformly dispersed nanocrystalline silver reduces the residual stress within diamond-like carbon hard coatings. Nano-Structures & Nano-Objects 10: 69–79.

Qiu, H. and J. Yang. 2017. Industrial Applications of Carbon Nanotubes Micro and Nano Technologies; 1st Edition, pp. 47–69. Chapter 2—Structure and Properties of Carbon Nanotubes; Elsevier.

Ramesham, R., W. Welch, W.C. Neely, M.F. Rose and R.F. Askew. 1997. Plasma etching and patterning of CVD diamond at 100°C for microelectronics applications. Thin Sol. Film. 304: 245–251.

Robertson, J. 1992. Mechanical properties and structure of diamond-like carbon. Diam. Rel. Mater. 1: 397–406.

Robertson, J. 2002. Diamond-like amorphous carbon. Mater. Sci. Eng. R 37: 129–281.

Roy, R.K. and K.R. Lee. 2007. Biomedical applications of diamond-like carbon coatings: a review. J. Bio. Mater. Res. Part B Appl. Biomater. 83(1): 72–84.

Salvetat, J.-P., J.-M. Bonard, N.H. Thomson, A.J. Kulik, L. Forró, W. Benoit et al. 1999. Mechanical properties of carbon nanotubes. Appl. Phys. A 69: 255–260.

Shevchenko, E., E. Matiushenkov, D. Kochubey, D. Sviridov, A. Kokorin and A. Kulak. 2001. Synthesis of carbon films with diamond-like structure by electrochemical oxidation of lithium acetylide. Chem. Commun. 317–318.

Sikora, A., F. Garrelie, C. Donnet, A. Loir, J. Fontaine and J.C. Sanchez-Lopez. 2010. Structure of diamond-like carbon films deposited by femtosecond and nanosecond pulsed laser ablation. J. Appl. Phys. 108(11): 113516–9.

Sirk, A.H.C. and D.R. Sadoway. 2008. Electrochemical synthesis of diamondlike carbon films. J. Electrochem. Soc. 155(5): E49–E55.

Spear and Dismukes. 1994. Synthetic Diamond-Emerging CVD Science and Technology. Wiley, New York.

Spitsyn, B.V., M.N. Gradoboev, T.V. Galushko, T.A. Karpuhina, N.V. Serebryakova, I.I. Kulakova et al. 2005. Purification and functionalization of nanodiamond. pp. 241–252. *In*: Shenderova, O.A., D. Gruen and Ya A. Vul' (eds.). Synthesis, Properties and Applications of Ultrananocrystalline Diamond. Springer, Dordrecht.

Sreejith, K., J. Nuwad and C.G.S. Pillai. 2005. Low voltage electrodeposition of diamond like carbon (DLC). Appl. Surf. Sci. 252: 296–302.

Sun, Z., Y. Sun and X. Wang. 2000. Investigation of phases in the carbon films deposited by electrolysis of ethanol liquid phase using Raman scattering. Chem. Phys. Lett. 318: 471–475.

Suzuki, T., Y. Manita, T. Yamazaki, S. Wada and T. Noma. 1995. Deposition of carbon films by electrolysis of a water-ethylene glycol solution. J. Mat. sci. 30: 2067–2069.

Tallant, D.R., J.E. Parmeter, M.P. Siegal and R.L. Simpson. 1995. The thermal stability of diamond like carbon. Diam. Relat. Mater. 4: 191–199.

Tosin, M.C., A.C. Peterlevitz, G.I. Surdutovich and V. Baranauskas. 1999. Deposition of diamond and diamond-like carbon nuclei by electrolysis of alcohol solutions. Appl. Surf. Sci. 44-145: 260–264.

Urruchi, W.I., M. Massi, H.S. Maciel, C. Otani and L.N. Nishioka. 2000. Etching of DLC films using a low intensity oxygen plasma jet. Diam. Relat. Mater. 9: 685–688.

Wan, S., L. Wang and Q. Xue. 2009. An electrochemical strategy to incorporate iron into diamond like carbon films with magnetic properties. Electrochem. Commun. 11: 99–102.

Yan, X., T. Xu, G. Chen, Q. Xue and S. Yang. 2004. Synthesis of diamond-like carbon/nanosilica composite films by an electrochemical method. Electrochem. Commun. 6: 1159–1162.

Yan, X., T. Xu, G. Chen, S. Yang and H. Liu. 2004. Study of structure, tribological properties and growth mechanism of DLC and nitrogen-doped DLC films deposited by electrochemical technique. Appl. Surf. Sci. 236: 328–335.

Zeng, A., V.F. Neto, J.J. Gracio and Q.H. Fan. 2014. Diamond-like carbon (DLC) films as electrochemical electrodes. Diam. Relat. Mater. 43: 12–22.

Zhang, G.F., J.Y. Du, Y.Y. He, G.Q. Li and X.D. Hou. 2007. Surface morphology of diamond-like carbon films prepared by liquid deposition. J. Chin. Elect. Micros. Soc. 26: 19–23.

Zhitomirsky, I., L. Galor, A. Khan and M.D. Spang. 1997. Electrolytic PZT films. J. Mater. Sci. 32: 803–07.

Zhou, H., C. Qiu, F. Yu, H. Yang, M. Chen and L. Hu. 2011. Raman scattering of monolayer graphene: the temperature and oxygen doping effects. J. Phy. D; Appl. Phys. 44: 185404–185406.

Zhou, D., E.V. Anoshkina, L. Chow and G. Chai. 2006. Synthesis of carbon nanotubes by electrochemical deposition at room temperature. Carbon 44: 1013–1024.

Zhu, H.S., J.T. Jiu, Q. Fu, H. Wang and C.B. Cao. 2003. Aroused problems in the deposition of diamond-like carbon films by using the liquid phase electrodeposition technique. J. Mater. Sci. 38: 41–145.

CHAPTER 3

Synthesis of Coal-based Nanocarbon
Advances and Applications

Manoj B.

1. Introduction

Carbon, well-known from the primordial time, is known as the king of elements, owing to its usefulness and multiplicity in fields such as advanced materials and energy harvesting applications, which is indisputable (Manoj 2014). It is extensively dispersed in nature, from molecules of life to materials in outer space. Profusely existing allotropic forms of carbon are graphite, diamond, fullerenes, and nanotubes. Amorphous carbon basically exists in the form of hydrocarbon fuel, soot, and other adulterated forms of carbonaceous constituents, which have structures of either graphite or diamond. Graphite has sp^2 hybridized bonds, while diamond has a meta-stable form with a significant amount of sp^3 bonds (Ferrari and Robertson 2001, 2004, Sadezky et al. 2005). The notable properties of these materials are high mechanical strength, flexibility, etc. by virtue of which, they have utility in thin film technology and in nano-scale electronic devices in abundance. Carbonaceous materials have remarkable features, making it a potential material which can revolutionize the future of numerous areas of utilizations. The discovery of carbon nanotubes (CNTs) by Iijima (1991) and graphene extraction by Novoselov et al. (2004) unveil amazing tensile strength and ballistic conductivity, and outstanding physicochemical properties. These make them vital constituents in applications, such as catalysts, sensors, electrochemical cells, and electronic devices (Manoj et al. 2017a). Of late, carbon dots (CDs) and graphene quantum dots (GQDs) find applications such as bio imaging, photovoltaic, opto-electronic sensing, targeted drug delivery, etc. (Hoang et al. 2018). Their unique properties, such as inertness to chemical change, size-dependent luminescence, and less toxicity make their biocompatibility excellent (Thomas et al. 2018, Manoj et al. 2017a). Regardless of the advancement of numerous fabrication approaches, gram scale production of environment-benign carbon-derived nanomaterials at an affordable cost faces a bottleneck. Precursors for producing nanocarbon play a crucial role for a feasible fabrication technology. Most conservative techniques employ expensive carbon feed-stock, for example, graphite and hydrocarbons (camphor, sugar, acetylene, xylene, methane, etc.) for graphene and CNT synthesis (Zhu et al. 2018, Cao and Rogers 2009, Moothi et al.

Department of Physics & Electronics, Christ (Deemed to be University), Bengaluru, India 560029.
Email: manoj.b@christuniversity.in

2015, Avouris and Dimitra Kopoulas 2012), while carbon nanostructures, such as graphite, fullerenes, carbon fibers, carbon black, graphene, and graphene oxide have been used as carbonaceous precursors to synthesize GQDs and CDs (Deng et al. 2016, Xu et al. 2014, Hoang et al. 2018, Bronikowski 2006). The exorbitant cost of fabrication has restricted the commercialization of these materials. Hence, facile green manufacturing methods to prepare carbonbased nanomaterials via economical means have unlimited possibility. In the present work, we are collating the research carried out in recent years on large scale synthesis of nanostructures from widely available precursors which are otherwise considered hazardous to the environment.

Coal is a sedimentary rock carbonaceous fuel usually found in the rock strata as layers or veins. Its elemental composition is made up of hydrogen, sulfur, oxygen, and nitrogen, with a significant presence of trace elements (Van Krevelen 1957, Speight 2005, Given 1984). For centuries, the utility of coal has been appreciated only as an energy resource. It is predominantly used for the generation of power and is used for industrial utility, such as refining metals, textile and cement manufacturing. Formation of coal is a biological and geological process that takes place over a long period where dead vegetation is transformed into peat, which in turn changes to lignite, and finally to anthracite. As coal is a composite heterogeneous solid, its chemical and physical properties vary considerably. Understanding the physical and chemical structure of this solid fuel, otherwise considered as a low quality fuel, the overuse of which degrades the environment, is essential in exploring the novel application of this material.

The discovery of fullerene from coke by Pang et al. in 1991, witnessed a tremendous progress in research on the transformation of solid hydrocarbon into various class of novel carbon materials, such as graphite (Pang et al. 1991, Cameán et al. 2010, González et al. 2003, 2004, Huang et al. 2013), activated carbon (Zeng et al. 2013, Zhang et al. 2018a, b), graphene (Anu et al. 2016) carbon nanoparticle, and nanospheres (Anu Mohan and Manoj 2012), graphene-tin oxide composite (Anu et al. 2019a, b), and lately GQDs and CDs (Manoj et al. 2017a, 2018a, b). However, the presence of inherent minerals embedded in the matrix of coal plays a major role in the redox ability of carbon-based nanomaterials such as CNTs and graphene (Wang et al. 2016). Chua et al. (2016) described that minerals, during the ball-milling process, contaminate the graphene nanomaterials, which in turn leads to lowering of catalytic peak (at -0.65 V), and poorer onset oxidation potential (Hoang et al. 2018). The presence of impurities, such as Mn, significantly alters the onset potential of reduced graphene oxide, as reported by Wang and co-investigators (2016). The same group reported that traces of heavy metal ions in the precursor graphite considerably lower the oxygen reduction reaction (ORR) potentials of the obtained nanostructures.

The present review also brings out the advancement in the utilization of various types of coal, which is considered an energy resource which is an environmental hazard. The development of this material as a precursor for various types of nanostructure is discussed. Here we outline various demineralization strategies adopted on coal to enhance its quality. Various methods of value addition to the carbon resources-coal, which is otherwise considered a fuel source whose use generates excessive ash and particulate matter, are discussed.

2. Physical Characteristics

Coal is a breakable, ignitable, solid hydrocarbon fuel molded by the decay and modification of flora under high temperature, pressure, and compaction. The specific gravity of coal varies with ash content and rank; it is highest for anthracite (1.5) and lowest for lignite (1.2). The hardness value increases from 2.5 to 3 units for hard bituminous coal and anthracite, respectively (Speight 2005). The main component of coal is vegetal matter, which is chemically modified to different degrees and intermixed with inorganic matter in varying proportions. Over a period of time, as variation in geological conditions puts pressure on dead biotic material, it is transmuted into different forms of coal. The widely used ASTM (American Society for Testing and Materials International Classification

System) categorizes coal into high, medium, low-volatile bituminous coal, sub-bituminous, lignite, anthracite, and graphite according to their heating value, percentage of volatile matter, and fixed carbon content (Speight 2012). Anthracites have more than 91% of elemental carbon and resemble the structure of graphite; they possess better alignment of aromatic rings with a little contribution from aliphatic carbon (Speight 2005, Van Krevelen 1957). Most coals are amorphous in nature and do not show any periodicity, whereas bituminous coal shows crystalline structure with amorphous background, due to the development of aromatic ring systems similar to that of graphite. This makes them a valuable resource of different types of carbon allotropes. One can easily extract these structures by simple oxidative treatment and greener synthesis.

Graphite is a compact structure of carbon which is anisotropic and has a large value of thermal conductivity. It consists of hexagonal rings having spacing of 0.142 nm and the distance between crystallographic planes is about 0.335 nm (Delhaes 2001), forming thin parallel plates called graphene. Every single carbon atom undertakes sp^2 hybridization with the un-hybridized p-orbital overlapping sideways to give a massive π system, with both sides of the plane of atoms. The fourth valence electron which is not part of the covalent bonds may be easily dislocated from the shell by applying an electric field. It provides electrical conductivity to graphite and becomes de-localized over the single layer of atoms. The valence electrons which are free from any bonding freely wander through the graphene layer. However, there is no direct interaction between the de-localized electrons in one layer of sheet and the neighboring one. These atoms are bound together by strong covalent bonds, whereas the graphene layers are held to each other by weak van der Waals forces (Lavrakas 1957, Geim 2009). Scientific research on the properties associated with the planar, hexagonal arrangement of carbon atoms in graphene is the onset for the development of carbon nanotubes and fullerenes, which in turn led to the revolutionary progress in nanotechnology.

3. Chemical Characteristics

Due to its origin, coal is a composite and assorted mineral having both organic (carbon) and inorganic (mineral) components. The organic form comprises aromatic structures whose aromaticity increases from low to high rank. The inorganic component of coal contains various minerals in minor and trace forms. It is known that, coals (barring anthracite) are made up of units consisting of 1–3 condensed benzene rings, which are linked by carbon chains having different functionalities. Coal is a complex carbonaceous material with cross-linked linkage which is made up of aromatic and hydro aromatic units connected by short aliphatic and ether linkages (Speight 2005, Manoj et al. 2009, Manoj 2016, Hoang et al. 2018). They are nano-sized sp^2 carbon allotropes whose sizes vary with the carbon content (in the range 70 wt% in lignite to 75 and 85 wt% in sub-bituminous and bituminous coal, respectively, and reaching 94 wt% in anthracites) (Haenel 1992, Vasireddy et al. 2011, Mathews and Chaffee 2012). Another important factor is that coal is low cost and abundantly available in nature, thereby making it an ideal carbon source for the preparation of nanomaterials (Manoj et al. 2017, Binoy et al. 2007). Coal is reported to have a turbostratic structure consisting of aromatic ring stacking, aliphatic side-chain entanglement, hydrogen bonds, cation bridges, and charge transfer interactions through oxygen containing functional groups (Binoy et al. 2009, Speight 2005, Balachandran 2014). With the aid of modern spectroscopic techniques, one can gain the knowledge of functional groups, which helps to elucidate the structure and functional groups of this carbonaceous source. The mineral matter (inherent) impurities are attached to the functional groups, which adversely affect most of its utilization and processing, and need to be eliminated through a systematic procedure of demineralization (Pratima et al. 2012).

The mineral matters may be present as a distinct entity or dispersed within the organic matrix of coal, which can be removed by physical or chemical methods. The former is economic, but not very effective in separating the finely distributed minerals, while in the latter, inherent as well as absorbed mineral species can be leached out. The existing reports revealed that the enrichment of coal efficiency through different practices such as chemical and biological means is a topic of unending research (Mukherjee and Borthakur 2003, Wijaya et al. 2011, Alvarez et al. 1997).

4. Chemical Leaching of Coal

Chemical leaching of coal with the aid of mineral acid, organic acid, or alkaline solutions helps to lower ash making minerals to a great level. These practices, either alone or in combination, tailed by a physical technique, are largely followed for the manufacture of clean coal. The general tactic followed to upgrade low rank coal involves the process of leaching under a range of environments. Chemical cleaning of coal with acidic/basic reagents is reported to be effective in reducing significant amounts of ash-forming minerals and sulfur. Most often used chemical beneficiation procedures include leaching with alkalis NaOH and KOH, $Ca(OH)_2$, HNO_3, and sequential washing with various organic acids (Mukherjee and Borthakur 2003, Alvarez et al. 1997, Shakirullah et al. 2006, Manoj et al. 2009, Manoj 2013). Since coal has become one of the universal sources of energy and will continue to be so for many more decades, intensive research work has been carried out to explore its properties by several authors (Wu and Steel 2007, Rubiera et al. 2003). The mineral solubilization of a UK-based bituminous coal using two stage chemical leaching by hydrofluoric acid (HF) and ferric ions was discussed by Wu and Steel (2007). The study reported the production of ultra-clean coal with the inorganic acid, which leads to the decrease of ash content to a very low value (0.1 wt%) from the initial. Rubiera et al. (2002) have probed the effect of demineralization technique on the combustion behavior of coal, and reported lowering of ash content by 0.3 wt percent.

Conventional techniques to remove minerals have not yielded the desired result, due to which chemical processes are considered. Direct leaching is a prevailing method to eliminate minerals from coal. Coal minerals generally contain quartz and kaolinite as minor elements, which form ash and traces of calcium. HF can effectively sequester this ash, forming minerals to a very low level. Even though mineral acids can lower them to a large extent, there are a few disadvantages, such as deteriorating the surface morphology, reducing the carbon content, and increasing the environmental degradation owing to their robust oxidizing power. Chemical cleaning of coal with strong acid and basic solutions was found to be effective in lowering substantial quantity of ash-creating minerals and sulfur.

Coal comprises of heterogeneous aromatic configuration, whose aromaticity increases with alteration of rank from lignite to anthracite. The investigation of the organic and inorganic constituents of coal before and after leaching is important in understanding the functional group modification. Vibrational spectroscopy is a novel analytical tool which helps in elucidating the functional groups. This technique will be able to identify carbo-hydrogenated structure and hetero atomic functions, as well as reveal the existence of minerals. This is presently one of the proven techniques for coal characterization. Most prominent bands in the FT-IR spectra coal and its chemical leached products are identified and presented in Figure 3.1. One could observe changes in the fingerprint region and functional group region. Sharp bands were noticed at 540 cm^{-1}, which arise due to silicates, while the bands at 3620 cm^{-1} and 3700 cm^{-1} were attributed to minerals present in the coal matrix (Andreas et al. 2003, Van Krevelen 1993). The intensity of the mineral bands lowered with leaching concentration of HF, as reported by Manoj et al. (2009). The distinct peaks at 2850 and 2920 cm^{-1} are assigned to symmetric and asymmetric $-CH_2$ stretching correspondingly (Kaur 2008). With variation in concentration of HF, the absorption due to this band became clear and distinct. The C=C or graphitic band could not be extracted after the leaching process.

The spikes at 1014, 1036, and 1108 cm^{-1} originated from the silicate minerals, whose intensity decreased and became least upon leaching with 30% HF, as reported by the authors (Manoj et al. 2009). The absorption at 1465 and 1380 cm^{-1} corresponding to methyl group are associated with amorphous carbon in the sample, which showed a reduction in intensity with rise in concentration of HF. Aliphatic bands and oxygenated functional groups showed lower transmittance in the non-

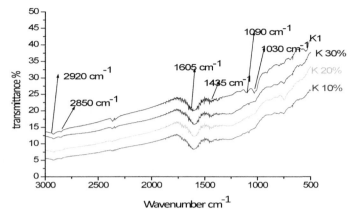

Figure 3.1: FT-IR analysis of the functional groups in coal.

extractable material of coal, revealing the effectiveness of extraction with HF. The intensity of the mineral band showed a remarkable increase, confirming its potency in demineralizing coal.

With regard to mineral dissolution, HF is found to be a more effective reagent than any other chemical, resulting in products with higher organic content. It effectively leaches out all the minerals associated with coal, except sulfide minerals (Wu and Steel 2007, Manoj and Kunjomana 2010a). Manoj (2012) reported the use of successive leaching of coal with dilute acid followed by strong acid to understand the modification in structure. Dilute nitric acid (dil. HNO_3) was used for treating coal, before HF leaching. To study the effect of successive organic acid leaching, coal was also simultaneously treated with EDTA prior to acetic acid solubilization. The use of mild acid would lower the binding of minerals to the matrix without oxidizing coal. The evaluation of the structure after demineralization reveals the oxidation of coal matrix and loss of calorific value with leaching. The making of ultra-clean coal by chemical leaching in acidic and basic media has been widely reported (Nabeel et al. 2009, Wu and Steel 2007).

Solubilization of minerals in coal with mineral acids not only modifies the surface morphology and deteriorates the carbon structure, but also reduces the calorific value. These acids have strong oxidizing power, and the safe disposal of the spent liquid is a major environmental concern. In order to overcome such drawbacks, some mild organic leachants were used by many research groups for de-ashing coal (Manoj and Elcey 2010b). When conditions become conducive, the microorganism secretes organic acids which are capable of degrading the minerals associated with the coal minerals in an eco-friendly manner (Lerato et al. 2012). Manoj (2013), Manoj and Elcey (2010b) reported the utility of various carboxylic acids for demineralization. The efficiency and mechanism of mineral dissolution have been discussed in detail.

Upon leaching the coal with acetic acid, oxalic acid, citric acid, and EDTA, the traces of sulfur were reduced to zero. Calcium was eliminated completely, while a remarkable reduction in silicates and aluminates were observed. The carbon content has been raised from an initial value of 80.91 to 82 wt% (EDTA leaching) and 82.22 wt% (gluconic acid leaching), respectively. However, on leaching with acetic acid and oxalic acid, the calorific value of coal was not increased, as a result of the oxidation of organic structure. The dissolution with citric acid and EDTA caused a lowering in ash content to 1.94 wt% and 1.81 wt%, respectively. Citric acid leaching curtailed the presence of aluminates without affecting the calorific value of coal. EDTA and gluconic acid annihilated the mineral content to an appreciable amount without much oxidation to the coal matrix. The results thus indicated that mild leachants are efficient as inorganic reagents (e.g., HCl, HNO_3, HF, etc.) in demineralization. These acids contain chelating agents to mobilize nitric acid-insoluble oxides/hydroxides, which in turn considerably lessen the ash forming constituents in coal (Shakirullah et al. 2006, Manoj 2012). However, on carboxylic acid washing, a meager change in the volatile matter was noticed. Enhancing the organic matter content in coal is gaining worldwide attention in recent days. Enhancing the microbial activity in coal is gaining the interest of researchers, and is a potential area to be explored.

The two step chemical leaching process enabled substantial removal of sulfur, and thus modified the chemical composition of samples. With successive leaching, structural parameters, *viz.* aromaticity (*fa*) and stacking height (*L$_c$*) were increased, while the interlayer spacing (d$_{002}$) was decreased. The average number of carbon atoms and carbon layers in the lamellae were estimated to be 21 and 8, respectively, when leaching coal with hydrofluoric acid (HF). On leaching with HNO$_3$ and HF, a significant change in the functional groups was seen, as the content of silicates, calcites, and aluminates were removed. The mineral content was enhanced considerably on acetic acid and Ethylenediamine tetra acetic acid (EDTA) leaching. Minerals of calcites and alumino-silicates were totally washed out by citric acid leaching. Reduction of oxygen to a greater extent is noted upon leaching with acetic acid, citric acid, and EDTA leaching. The functional groups were modified after leaching along with the enhancement of graphitic structure, as confirmed by spectroscopic investigation and XRD. The demineralization procedures could remove the unwanted minerals from the coal matrix to a great extent, and ordering of carbon lattice (graphitization) is observed.

5. Structural Parameters and Coal Rank

Generally, C-NMR spectroscopy is applied for the identification of carbon atoms in different chemical environments. Coal aromaticity can be estimated by the aid of quantitative X-ray analysis. Aromaticity (*fa*) is used as a measure to describe the ratio of carbon atoms in aliphatic side chains to aromatic rings (Binoy et al. 2009, Manoj and Kunjomana 2012a). The values of aromaticity (*f$_a$*), lattice spacing (*d$_{002}$*), stacking height (*L$_c$*), and rank of coal (-ratio of aliphatic carbon to aromatic carbon-I$_{20}$/I$_{26}$) were plotted against carbon content, as shown in Figure 3.2 (Reproduced with permission). An appreciable degree of agreement between elemental carbon content and structural parameters (R^2 = 0.97, 0.92, 0.95, and 0.95) were found in the coal samples. With increase in *fa*, *d$_{002}$* decreases, while I$_{20}$/I$_{26}$ enhances. Coal becomes more condensed after the effect of chemical induced orientation and the inter-planar spacing *d$_{002}$* approaches that of graphite (3.35Å). The graphitization degree of coal was found to be changed with variation of rank and concentration of leachant (Manoj et al. 2009, Manoj 2012, Manoj 2014). A high correlation is reported with structural parameters and volatile matter (R^2 = 0.80, 0.89, 0.988, and 0.86), which is in agreement with the reported result for the elemental carbon.

Another study on Nigerian coals by Sonibare et al. (2010) reported the utility of various spectroscopic techniques in structural elucidation of coals of different ranks. The structural parameter,

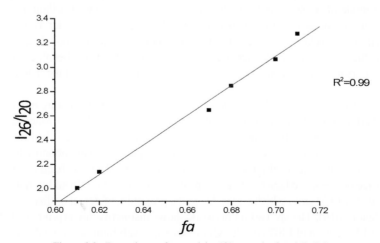

Figure 3.2: Dependence of aromaticity (*fa*) on rank of coal (I$_{20}$/I$_{26}$).

such as average lateral sizes (L_a), stacking heights (L_c), and interlayer spacing (d_{002}) of the crystallite structures were reported to be ranging from 1.65 to 2.57, 0.81 to 1.33, and 0.348 to 0.358 nm, respectively. The FT-IR analysis of coals indicated the stretching vibrations of -OH bonds, aliphatic -CH, -CH$_2$, and -CH$_3$ absorptions, C=C, -CH aromatic structure, and C=O stretching vibrations of carbonyl groups (Ponni Narayanan and Manoj 2013). Coal is a nano-structured material, whose characteristics can be elucidated by the multiplicity of local structure configurations. The presence of branched porous structure makes this solid hydrocarbon a favorable, economic material for storing hydrogen in fuel cells. Comprehensive information on the intricacy and heterogeneity of coal can be obtained by relating the information attained from diverse techniques (Manoj and Kunjomana 2012a). The turbostratic structure of crystallites in coal is well supported by the appearance of peaks owing to (002), (100), and (110) reflections of graphite (Sonibare et al. 2010, Lu et al. 2001, Darmstadt et al. 2000, Manoj and Kunjomana 2012). Pioneer work in quantifiable X-ray diffraction study of carbonaceous material was carried out by Warren in 1934. Many researchers have reported structural analysis of highly defected carbonaceous substance and established Warren's equations (Franklin 1950, Alexander and Sommer 1956, Warren 1934). Elaborative efforts have been made to characterize inherent mineral matter and organic matrix of coal by XRD (Takagi et al. 2004, Manoj and Kunjomana 2010a). Lu et al. (2001) elucidated the aromaticity and size of nanocrystallite Australian black coals from the x-ray profile analysis. Manoj (2012, 2014) reported the existence of turbostratic carbon in the demineralized sub-bituminous coal samples when leached with hydrofluoric acid (HF). The x-ray profile of the carbon was de-convoluted with two Gaussian peaks for the band around 26° (Figure 3.3).

The graphitic-like domain in coal gives rise to diffused peaks. In addition to the crystalline structure, coal has a significant amount of highly disordered amorphous carbon, which contributes to the high background intensity. The asymmetric band (002) around 26° indicated the existence of γ-band at 20°, thereby confirming the existence of saturated hydrocarbon like aliphatic side chains attached to the edge of crystallites (Takagi et al. 2004, Sonibare et al. 2010, Manoj and Kunjomana 2012). Tuinstra and Koenig (1970) noted that the ratio of intensity of defect to and graphitic modes (I_D/I_G or I_{20}/I_{26}), varies inversely with the in-plane correlation length or grain size. Manoj (2014) reported the dependence of I_{20}/I_{26} versus lateral dimension (La) of coals of various ranks (Figure 3.4). It is reported that with the evolution of rank of coal from low to high, the reliance of rank on in-plane lattice parameter is saturated. This is attributed to the rise in particle size of sp^2 clusters (Figure 3.4).

For disordered carbon structure like that in coal, two leading aspects are to be taken into deliberation. Primarily, the T-K relation is valid only for confined sp^2 structure, whose defectiveness is due to the 3-D confinement of graphitic plane and uniform distribution of nanocrystallites. The other factor is relevant for highly disordered graphitic plane with prominent defect peak (γ-peak). The breathing mode of A$_{1g}$ symmetry involves phase near the K-point of brillouine zone, which is forbidden in graphite and active only when adjacent to the presence of disorder. Hence, the degree of disorder has an inverse relation to L_a and is attributed to the presence of six-fold aromatic structure in graphite. As the disorder in lattice increases, a large number of clusters become smaller and more disoriented and open up. As a result, I_{20} decreases with respect to I_{26} and the T-K relation will be violated.

Raman spectroscopy is an analytical tool used for understanding the nature of carbon in coal in hydrocarbon (Ferrari and Robertson 2000, Manoj 2014, 2015, Ramya et al. 2016, Manoj and Kunjomana 2010a). Raman analysis of graphitic carbon exhibits the signature of G and 2-D band at 1580 cm^{-1} and 2700 cm^{-1}, respectively. The G band originates from the in-plane optical phonon vibration. G mode has E$_{2g}$ symmetry, which occurs at all sp^2 sites (Anu et al. 2016). The 2-D band originates from the double resonance process, and hence appears to be dispersing in nature. In addition to these bands, a D band is noticed at 1350 cm^{-1}, which arises due to the proximity of defect or diamond like structure (Elcey and Manoj 2016). The sp^3 carbon network shows distinctive fingerprint at 1148 cm^{-1}, with minor peaks at 1274 cm^{-1} and 1307 cm^{-1} (ta-C) in the case of sub-bituminous coal. Other researchers designated this band to carbon structures which are rich in sp^3 type carbon, such as hexagonal diamond

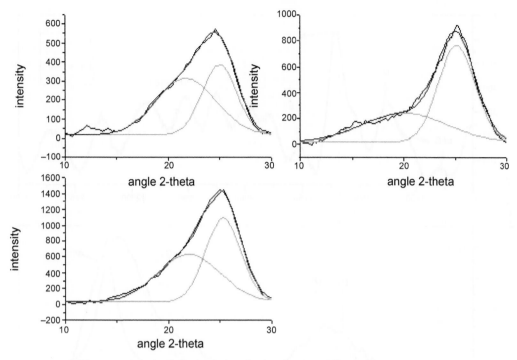

Figure 3.3: Gaussian fit for the demineralized coal at (002) diffraction peak (Manoj 2014).

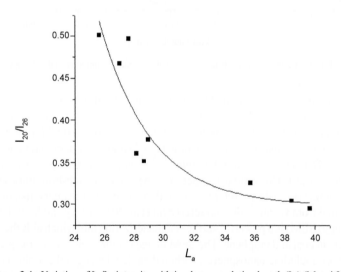

Figure 3.4: Variation of I_{20}/I_{26} intensity with in-plane correlation length (L_a) (Manoj 2014).

or nanocrystalline diamond. The origin of this band is also attributed to the co-existence of sp^2-sp^3 structure (Manoj 2015, Manoj et al. 2018b). The 2-D region is broad and bumpy with numerous high intensity absorption peaks for the coal samples having wrinkled multilayer graphene structures with 4–5 layers (Figure 3.5).

The G band in the Raman spectrum shows the presence of sp^2 carbon network. As the layer thickness increases, the position of this band shifts to lower energy, resulting in softening of the bonds with addition of graphene layers. Other than the G and 2-D peaks, D and D' peaks are also observed in the spectrum of graphene and are attributed to defect-induced Raman features, which

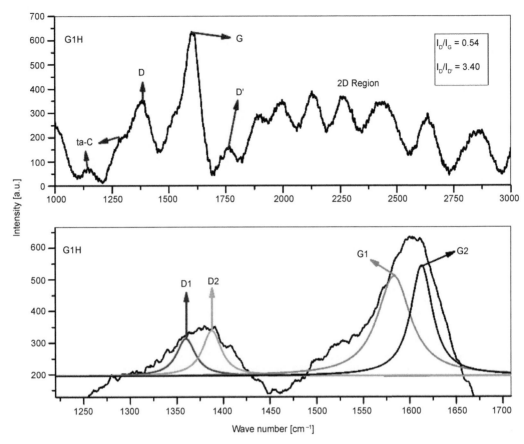

Figure 3.5: Raman spectrum of nanostructure obtained from Sub-bituminous coal (Manoj et al. 2018).

are inactive in highly crystalline carbon. The ratio of intensity of defective peak to graphitic peak, (I_D/I_G) is a commonly adopted standard for estimating and elucidating the nature of defect in graphitic materials. This ratio is reported to be ~ 0.54 for sub-bituminous coal, ~ 0.58 for bituminous coal, and ~ 0.63 in the case of tertiary coal. The intensity ratio (I_D/I_D') is another parameter one can obtain from the Raman analysis. D' peak is generally less intense in comparison to D peak (normally appears as a small shoulder to G peak). This ratio was reported to be ~ 3.40 for sub-bituminous coal (indicating the boundary-like defects), ~ 3.16 for bituminous coal (attributed to boundary-like defects) and ~ 6.8 for lignite (originated from vacancy-like defects) (Anu et al. 2016, Elcey and Manoj 2016). Scanning Electron Microscopy (SEM) is a powerful surface characterization tool which aids the surface analysis of the carbon layers in coal and graphene. The SEM image (Figure 3.6) reveals the presence of mixed phase nanostructures, including nanospheres, carbon dots, few layer graphene sheets, and multilayer graphene in HF leached bituminous coal, whose diameter ranges from 116–182 nm (Manoj 2016). The EDS analysis (Energy dispersive spectrograph) revealed the presence of carbon (78.29%) and oxygen (20.73%) with traces of sulfur.

Scanning Electron Microscopy (SEM) analysis is a tool which confirms the formation of layered structure in the coal. The bituminous coal and sub-bituminous coal have few layer graphene and dots embedded in the structure, while the lignite is analogous to stack like bulk graphite. They are the ideal precursor for the synthesis of graphene quantum dots and graphene layers. Yet another important surface characterization tool used for investigating the surface morphology of the coal derived nanocarbon is Transmission electron microscope (TEM). The surface morphology of the nanostructure derived from coal is presented in Figure 3.7.

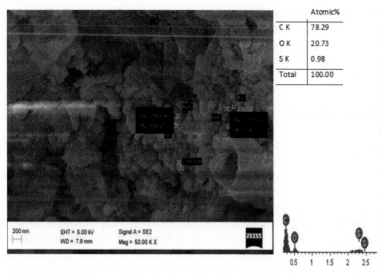

	Atomic%
C K	78.29
O K	20.73
S K	0.98
Total	100.00

Figure 3.6: SEM-EDS analysis of nanostructure obtained from bituminous coal (Manoj et al. 2018a).

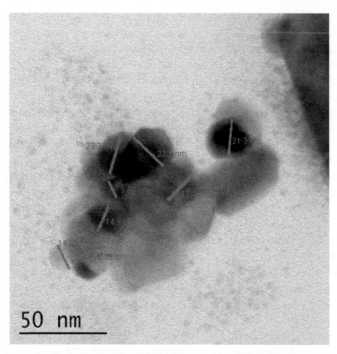

Figure 3.7: TEM micrograph of quantum nanodots obtained from coal-exhibiting hexagonal and pentagonal geometry (Manoj 2017).

Color version at the end of the book

The TEM image of the nanocarbon obtained from the coal reveals the formation of graphene structure with hexagonal, spherical, and corn shaped carbon nanotubes. In some parts, graphene layers, dots embedded in layers, and multilayers are also formed due to the agglomeration. The TEM analysis confirms the formation of mixed phase structure of carbon nanomaterial in the leached product of sub-bituminous coal (Manoj et al. 2018b, Akshatha et al. 2017).

It concluded that solubilization of minerals in coal is very effective with chemical leaching. One could extract the nanocarbon structure from coal in a very simple method. Coals of different qualities are found to be a better precursor for the novel carbon material. The leaching not only eliminates the mineral content, but also enhances the aromaticity, carbon content, and degree of ordering of nanolayers in coal.

6. Bioleaching of Coal

The method of solubilization with microbes in the presence of dilute acid has been reported in recent days, and is highly economic when the presence of minerals is very low. Minerals are extracted from their respective ores with the aid of living microorganisms, where numerous species of fungi are used. This technique is much more eco-friendly than the conventional heap leaching employing various chemicals. Microorganisms are widely exploited since they have the potential to degrade petroleum/oil, and can be used in various bioleaching processes as single organisms or in consortium. There are two reported levels of interaction between microorganisms and the mineral surface (Boon 2001, Castro et al. 2000). The first one is physical sorption mainly attributed to electrostatic forces. The reduced pH normally occurs during leaching environment, and the microbial cell envelopes develop a positive charge, enhancing the electrostatic exchanges with the mineral phase. This electrostatic force decreases the cohesive/adhesive interactions of mineral matter with coal matrix, and in turn lowers the binding of minerals. Another level is considered as chemical sorption, where chemical binding of cells and minerals gets established. In the presence of minerals, microorganisms such as silicate oxidizing fungi in the treated sample form a filamentous matrix, which solubilises the mineral. Bioleaching of metals from fly ash using *Aspergillus niger* (*A. niger*) has been reported by Deenan and Ting (2005) and Aung and Ting (2005). *A. niger* metabolises sucrose to citric and oxalic acids, and the increase in concentration of these acids along with other unidentified products constitutes a wide variety of transformation in the substrate. Bayraktar (2005) reported that, on treating with *A. niger*, there was a significant decrease in dehydrogenation reactions, i.e., the hydrogen to methane molar ratio and the coke yield. Acharya et al. (2003) studied the reaction mechanism of bioleaching of manganese ore with manganese reducing fungi *Penicillium citrium* and various acids, such as oxalic acid, citric acid, and sulfuric acid. The microbial solubilization of selected Indian bituminous coals carried out by Sharma and Wadhwa (1997) also revealed the removal of minerals by mixed culture of bacterium. The changes in mineral matter composition and functional groups of these coals were reported with the aid of FT-IR spectroscopy and X-ray analysis. The efficacy of some white-rot fungi on the bio-liquefaction of selected low-rank Turkish coals was explored by Başuran et al. in 2003. The chemical configuration of the extract and the microbial pathway of coal conversion were deliberated with the aid of FT-IR. It is reported that as the aromatic content rises, the resistance of the coal structure towards the action of microbial organisms also increases. Acharya and co-workers (2005) made an attempt to desulfurise the coal from North Eastern coalfields of Assam in India, which has a high total concentration of sulfur (2.13%) by using the indigenous fungal culture (*Aspergillus* sp.) isolated from coal. With the aid of natural fungal strain, desulfurisation was carried out on coal sample of size of ~ 74 μm. An elimination, as high as 70–80% of total sulfur, established the convenience of using the indigenous culture of *Aspergillus* sp. for desulfurisation. The results advocate an attack of the fungal strain on the organic sulfur. The study also reported the lowering of ash content due to acidic conditions imposed by the experimental process. Ghorbani et al. in 2008 carried out biological leaching of aluminum by isolated fungi from low grade bauxite. The secretion of citric and oxalic acids by the fungi *A. niger* was stated to be the cause for the removal of aluminates from bauxite minerals.

Biodegradation of coal and the inherent mineral phase is a complex process which is driven by extracellular enzymes in the existence of numerous chelators released by different fungi. Despite the slow conversion rates during the biological breakdown of coal, optimization of the process on a large scale could precipitate the development of technology for remediation of low rank coals. Manoj and Elcey (2010) reported the complete removal of the mineral phase in coal samples with *A. niger* and its mixed cultures with *Pencillium* spp. leaching by forming the respective mineral salts. High Performance Liquid Chromatogram (HPLC) analysis confirmed the secretion of oxalic acid, gluconic acid, and citric acid during the bio-solubilization (Manoj 2012). The study reports reduction of ash by 98.5% by leaching with *A. niger*, while a lowering of 90% of ash content was noticed with mixed culture leaching (with *Aspergillus niger* and *Penicillium* spp.). The semiquantitative analysis (SEM-EDS) of the solubilized coal established the removal of silicates, aluminates, and calcites. From the x-ray analysis, intensity ratio (I_{20}/I_{26}) (which is a measure of disorder in the amorphous carbon) was found to be 0.19 when coal is leached with *Penicillium* spp., establishing high graphitization and purity. The stacking height along the *c*-axis (L_c) of bioleached coal was found to vary from 22.32 to 2.25 nm, whereas lateral dimension (L_a) was ranging from 34.36 to 0.15 nm. Upon treating with *Penicillium* spp., the interlayer spacing of the carbon layers in coal is changed to 0.337 nm, and that of ordered graphite to 0.335 nm. The FT-IR spectra of samples showed more lignin content, which is usually present in immature coals. The presence of methyl symmetric bending vibrations in tertiary butyl groups was found in low rank coals. The fungal leaching was most beneficial in the region of aromatic, out of plane C-H bending, with different degrees of substitution. The intensity of the bands due to carbonyl groups was increased in intensity when treated with fungi, whereas the oxygen functional groups showed a reverse trend. The mineral band due to silicates also decreased in intensity when treated with fungal culture.

It is evident that bioleaching could alter the structure of carbon layers of coal samples and enhance the graphitization. Bioleaching is an upcoming field in nanocarbon synthesis from coal of different ranks which have inherent graphenic structure embedded in it. The elimination of mineral and ash in it is a complex and tedious procedure. Ultra-pure coal is a novel material for synthesis of graphene, reduced graphene, fullerene, quantum dots, carbon dots, and carbon nanotubes.

7. Synthesis of Carbon Nanostructures from Coal

Coal by nature is an abundantly available solid fuel, and delivers most of the power requirements for the living and fiscal advancement of a country. Besides the contribution to energy, it is used as a vital and imperative reserve for the manufacture of chemicals, such as creosote oil, naphthalene, phenol, and benzene, to name a few. Synthesis of coal-derived materials, such as activated carbon, carbon blacks, and electrodes, has been reported widely in recent times (Saha et al. 2005, Zhou et al. 2012). With the development of nanocarbon materials, scientific examination of the novel application of coal-derived products has gained tremendous progress (Li et al. 2006, Zhou et al. 2012, Binoy et al. 2009, Vijapur et al. 2013, Manoj 2014). Carbon precursors, such as graphite, hydrocarbons, natural biomaterials, and even polymer waste, have been widely reported as precursors for the preparation of graphene (Rao et al. 2007). Zhou et al. (2012) reported the synthesis of chemically modified graphene and graphene-noble metal composites with anthracite as the precursor material by means of catalytic graphitization, chemical oxidation, and dielectric barrier discharge plasma assisted deoxygenation. As coal possesses large number of poly-aromatic structures similar to sp^2 structure of graphene, it can be used as a precursor for the manufacture of carbon nanomaterials. Though research on mineral solubilization and structural elucidation of coal has been widely reported, substantial work on coal-based carbon nanostructure is still scarce.

This section provides the general approaches followed for the synthesis of nanocarbon from cheap and abundantly available precursor coal. Nanocarbon materials & graphene-based nanostructures have attracted tremendous research interest in recent days (Manoj et al. 2017a, 2018b). Graphene and carbon nanotubes exhibit unique physical and chemical properties which find applications in the area of electronic devices, catalyst, sensors, and supercapacitors. Of late, carbon dots (CDs) and graphene quantum dots (GQDs) are in the limelight. They have engrossing characteristics, such as steady fluorescence (FL), chemiluminescence (CL), and electrochemical luminescence originated from size and defect effect (Xiao et al. 2015, Zhu et al. 2011, Zhang et al. 2012). Low toxicity, high stability in colloidal forms, and little photo quenching make these carbon-based organic nanoparticles superior to their inorganic counterparts. Fluorescent carbon dots (CDs) have enormous usefulness in the fields such as bio-imaging, fluorescence probing, labeling, and energy conversion in photovoltaic, to name a few (Li et al. 2013, Yongqiang et al. 2015, Pan et al. 2010). Nanocarbon structures have been made from carbon-based precursors, such as fullerene, glucose, graphite or graphene oxides, carbon nanotubes and carbon fiber by chemical methods, solvo-thermal, hydrothermal, microwave assisted techniques, or electrochemical paths. They are facile approaches but the precursors are costly. Tailoring of dimension, configuration, and optical properties of the carbon dots by tuning the sp^2 structure in the carbon background in the synthesis is a major task. Large scale fabrication of the carbon dots in a facile way is possible with the selection of materials with a large amount of sp^2 domains in it.

Remarkable work has been employed on developing carbon dots and nanodots with unique and fascinating properties. Top-down synthesis is found to be facile, which has the benefit of being low cost and making a highly economic use of abundant raw materials (Manoj 2015, Phitsini et al. 2016, Sun et al. 2013). It also facilitates scalable production and has simple operation procedures. Top-down approach is the synthesis technique commonly adopted by many research groups for large scale synthesis of GQDs, CDs, and other carbon nanomaterials (CNMs). During top-down synthesis, GQDs and carbon nanostructure are produced from the cleavage of carbonaceous precursors (Zhou et al. 2012, Ye et al. 2013, Manoj 2012, 2014, Manoj et al. 2017, Binoy et al. 2009, Yongqiang et al. 2014, Jun et al. 2015). Acidic oxidation is one of the commonly adopted techniques for large scale synthesis of carbon nanostructure from coal. In the course of acid oxidization, epoxy groups appear to be aligned linearly on the carbon lattice, and such cooperative configuration induces a breach of the underlying C-C bonds (Figure 3.8). The formation of an epoxy chain is an energetically favorable condition for it to be further oxidized into epoxy pairs, which then convert to more stable carbonyl pairs at room temperature. The existence of these linear defects makes the graphitic domains delicate and easily fragmented. Some ultrafine pieces, which are sub-nanometer sized aromatic sp^2 domains bounded by the mixed epoxy lines and/or edges, may further break up to finally form the Graphene Quantum Dots (GQDs).

Production of CDs by pyrolysis or carbonization of low molecular weight carbonyl acids, fruits, or plants has also been popular in current years (Hu et al. 2014, Gaddam et al. 2017). They are bottom-up methods involving intricate synthesis techniques, making use of a range of toxic and volatile organic solvents. They also require expensive catalysts to initiate the condensation reactions along with precursors which are costly, making the synthesis complex (Gaddam et al. 2017, Manoj 2016, 2017b). In spite of good performance, the cost of the nanosensors is very high because they

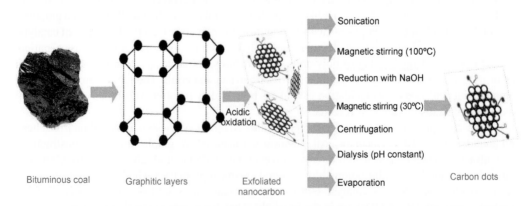

Figure 3.8: Schematic representation of synthesis of nanocarbon (Thomas et al. 2018).

are prepared through complex synthesis roots, and the production yield is very low (Manoj et al. 2017a). In addition, many of them use the precursors as complex dyes, which causes environmental pollution. An environmentally benign and facile synthesis method is very essential and has attracted the attention of coal chemists worldwide.

There have been several studies on the facile synthesis of carbon nanostructures with unique properties from coal, agricultural waste products, and humic substances. Das et al. (2016) reported the extraction of onion-like fullerene and graphene-like nanosheets from low value coal by oxidation process. These carbon nanomaterials exhibit photo-degradation of 2-nitro phenols in water solvent at various pH conditions. Kovalchuk and co-investigators (2015) reported the blending of coal-derived GQDs with PVA. The synthesized product dispersed excellently in water without any surface passivation agent. The PVA/GQD composite was exhibiting broad photoluminescence emission in the visible range. Coal-based GQDs have the quoted advantage of low cost and are viable for extensive industrial applications. These materials can be used as an economical and sustainable substitute to widely used inorganic dots (Thomas et al. 2018). Yongqiang et al. (2014) extracted carbon nanostructure of the dimension 10 nm with simple oxidation of coal of different ranks. As the rank of coal increases, the production yield was also found to decrease (Manoj and Kunjomana 2012a). Low rank coal is an ideal candidate for the synthesis of carbon dots and nanocrystals. Shengliang Hu (2016) reported a facile green, mild, and cost-effective method to synthesize carbon dots from coal with selective oxidation of H_2O_2 on organic carbon in coal. These CDs exhibit excellent photo catalytic activity in the visible light, and have promising utility in the photo degradation of organic pollutants without creating any secondary pollution. Hu et al. (2014) adopted a top-down technique to chemically alter the abundant coal structure to fluorescent CDs with a combination of carbonization and oxidation. The synthesized CDs exhibited two distinct emission modes, where the high energy emission was majorly affected by partial reduction. The CDs display substantial fluorescent quenching with heavy metal ions such as Cu^{2+} ions exhibiting the limit of detection of 2 nM.

There are reported studies of synthesis of natural nanodots from natural resources such as coal of different rank and humic substances derived from plants (Yongqiang et al. 2015). These abundant carbon matrices are formed during the 'humification' or coalification process (Manoj et al. 2017, Yongqiang et al. 2015). These nanostructures can be tailored into carbon dots and GQDs of defined shape, size, and properties (Ye et al. 2013, Manoj and Kunjomana 2014). Low grade coal, such as sub-bituminous and lignite, are energy sources with low heat value, but have large number of nanocarbon crystallites entrenched in their inherent 3-D matrix. These novel carbon materials can be extracted in a systematic fashion for many applications, such as fluorescent sensing and electrochemical sensing. The prominent techniques available at present to make nanocarbon structure from the precursor coal are primarily focused on thermal plasma, chemical vapor deposition (CVD), and arc discharge methods, to name a few. Synthesis of multi-walled CNTs (MWCNTs) by thermal plasma method was established by Tian et al. in 2004a, b. Arc sputtering of the copper anode at high temperature leads to the creation of copper particles, which work as catalysts for the formation of CNTs. The work testified the creation of multi-wall carbon nanotube (MWCNTs) with interlayer spacing of 0.343 nm and dimension of ~ 7 nm. The Raman analysis unveiled peaks at 1600 and 1291 cm^{-1} of G (graphitic order) and D (defect signal) peaks, respectively. However, the low value of I_G/I_D ratio (or high value of I_D/I_G), indicate the low-graphitic content in MWCNTs (Anu Mohan et al. 2016a, b, Manoj 2015). The role of Cu, Fe, and Co catalysts on the materialization of CNTs was explored by the same group (Tian et al. 2004). The diameter of the synthesized MWCNTs has the dimension of 50–70 nm, and enhanced catalytic activity of Cu compared to Fe and Co, leading to a 5% yield of CNTs with Cu compared to 1% yield for Fe or Co. Coal exhibited great potential as a replacement for graphite electrode in arc discharge technique (Williams et al. 1999). Kumar et al. (2013) adopted this route to synthesize high quality single-walled CNTs (SWCNTs) from bituminous coal with the aid of Zr/Ni catalyst. The diameter of the CNTs ranged from 1 to 2 nm with bundle length and diameter of 50–140 nm and 4–10 nm, respectively. The I_D/I_G ratio is 0.33, indicating that CNTs have few defects without amorphous carbon. The study also reported that the concentration of Zr/Ni significantly affected the growth of SWCNTs, and the highest yield of the product was obtained at the Zr/Ni ratio of 3:1. In a similar attempt, Awasthi et al. (2015) reported synthesis of carbon nanotubes having size of 1.7 nm to 1.2 nm using annealed bituminous coal electrode in the presence of Fe and Ni Y catalysts. The average lengths of these SWCNTs were 0.2 nm. In the absence of catalyst, MWCNTs with diameter and length distributions of 8–20 nm and 5–10 nm, respectively were synthesized by the same group. The Raman results exhibited I_D/I_G ratio of ~ 0.29 for SWCNTs and ~ 0.32 for MWCNTs, confirming the high graphitization and good quality of the products (Hoang et al. 2018).

Of late, Li et al. (2017) made an elaborate attempt to make CNTs (B-CNTs) having bamboo structure with bitumite via an arc discharge route with $Ni-Sm_2O_3$ catalyst. The high value of G to D band intensity ratio (I_G/I_D = 1.72) showed high degree of graphitization for the structure of B-CNTs. Zhang et al. (2017) utilized coal-derived GQDs having a size distribution of 2–3.2 nm to accumulate graded porous carbon nanosheets (HPCNs) using a $Mg(OH)_2$ template combined with KOH activation. The product contained thin layer nano sheets which were stacked lightly, and has poor crystallinity with the intensity ratio I_D/I_G value ranging between 0.83–0.91. They possess micro/mesoporous configuration with surface area of 1450–1882 m^2g^{-1} and have a pore volume of 3.397–5.222 cm^3g^{-1}. Subsequently, the researchers testified to the enrichment of strength and flexibility of carbon nanofiber fabrics with the addition of coal-derived GQDs to a polyacrylonitrile (PAN) solution, followed by electro spinning and carbonization at 1000°C under N_2 gas (Hoang et al. 2018). A tensile strength of ~ 2.2 MPa and Young's modulus of ~ 70 MPa were attained at the ratio of GQDs to PAN = 1:1 (wt%) with I_D/I_G of 1.01. This is ascribed to enhance the viscosity of the electro spinning solution activated by strong interface GQDs to PAN composite (Hoang et al. 2018).

Manoj et al. (2017, 2018a) extracted oxygenated carbon dots from the lignite by oxidation-reduction techniques. TEM image of the products shows particles of the order 3–4 nm having circular, rectangular, and hexagonal shapes (Figure 3.9). The fringe spacing is enumerated to be 0.51 nm, attributed to the integration of oxygen to the in-plane of graphene. The SAED pattern reveals the nanocrystalline nature of the synthesized CDs. The coal nanostructure is refluxed with chloroform, and is presented in Figure 3.9. One could observe the stacking of heterogeneous structures attributed to the formation of poly nanocrystalline carbon. The same group recently reported the extraction of graphene-like nanostructure from the lignite by simple reflux with nitric acid, followed by heating at 200°C. The surface layers are separated and one could easily obtain the graphene-like nanostructure. The TEM image reveals the formation of a layered structure where the darker region indicates the formation of multilayer, and the lighter shade is attributed to the formation of few-layer graphene oxide. The SAED pattern of the graphene oxide layers indicates the hexagonal atomic structure and crystalline nature of the product.

The synthesized GQDs and CDs from low grade coal forms stable suspension with high colloidal stability. These dots exhibit high fluorescence and show green fluorescence when exposed to UV-irradiation. The acidic oxidation followed by neutralization and dialysis process yields GQDs of varying sizes from the sample. One can easily obtain different particle sizes from the same precursor. Depending on the particle size, defect, and extent of oxygen functional group, their fluorescence excitation maxima will change from UV to green region. The extraction of onion-like fullerene and graphenelike nanosheets from low quality coal by oxidation process was reported by Das et al. in 2016. These carbon nanomaterials exhibit photo-degradation of 2-nitro phenols in water solvent at various pH conditions. Ye et al. (2013) synthesized GQDs of the size range ~ 29 nm and ~ 5.8, respectively from anthracite and coke, respectively, with moderate defect to graphitic ratio. The chemical and physical structures of CDs are diverse and totally dependent on the synthesis route and precursors adopted. CDs and GQDS possess few or multilayer graphene and are connected by chemical groups on the edges or boundaries. They have a very large lateral dimension compared to their stacking height. They exhibit core and shell structure of carbon core with certain crystallinity (nanocrystalline), with lattice spacing ~ 0.24 nm, which arises from the (100) plane of graphene. These GQDs can be of different shapes, such as pentagonal, hexagonal, triangular, and spherical, and so on, while CDs will always have a spherical shape. CDs are broadly classified as nanoparticles without

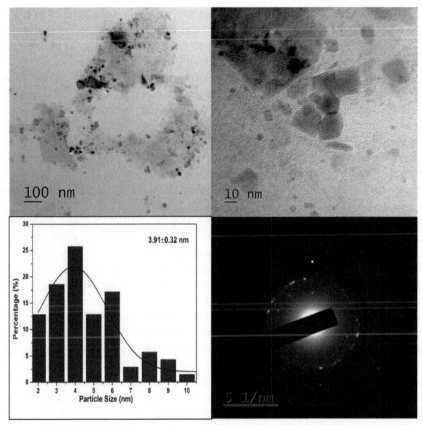

Figure 3.9: TEM image of the nanocarbon dots and graphene layers extracted from lignite by acid reflux (Manoj et al. 2018).

proper periodicity and particles with clear crystal structure. CDs normally have interlayer distance of 0.34 nm, attributed to the (002) plane of graphite (Manoj et al. 2017). They have many functional groups attached to their core, such as oxygen functional groups, amino-based, or carboxyl groups. Yin et al. (2014) reported the facile synthesis of GQDs with tunable emission, which was ascribed to the presence of oxygen functional groups. During chemical oxidation, oxygenated groups form on the edges and are inserted between the planes of the graphite layers, which render the layered structure of graphite easily ruined. The epoxy groups tend to formulate a line on the carbon lattice, which induces the underlying C-C bonds to break. The existence of these linear defects makes the graphite sheets fragile, making them fragment easily during oxidation. Once the epoxy chain appears, it is energetically preferable for it to be further oxidized into epoxy pairs that are even transformed into steadier carboxyl pairs at room temperature. The particles size is of the order of 10–17 nm, with majority of them having about 17 nm. The hexagonal nanoparticles are piled to form nanosheets with cubic geometry. The lattice constant is measured to be ~ 0.21 nm–0.24 nm, and the SAED pattern exhibit ring pattern. This can be accredited to the polycrystalline nature of nanocarbon. The study reported scalable production of spherical nanoparticles with a size of the order of 5 nm with chemical oxidation and dialysis. These sphere-shaped dots are homogeneous and have uniform size, and have an inclination to agglomerate. Das et al. (2016) extracted various advanced carbon materials from sub-bituminous coal. The diameter of the carbon nanotubes (CNTs) formed is in the range of 18–24 nm, while that of the branch carbon nanotubes (BCNTs) is in the range of 35–92 nm. The study reported that alkaline leaching of coal with subsequent acid treatment preferred the creation of the carbon nanoballs, carbon nanotubes (CNTs), and branch carbon nanotubes (BCNTs). Yongqiang and group (2014) used coal precursors of different qualities to obtain single-layer graphene quantum dots (S-GQDs). Chemical oxidation followed by particle segregation with a series of centrifugation separated coal into $Coal_A$ and $Coal_B$ fractions. Up on characterizing with the analytical tools, such as TEM, AFM, XRD, Raman, and FTIR, $Coal_A$ was found to mainly consist of S-GQDs with an average height of about 0.5 nm and plane dimension of ~ 10 nm. The synthesized S-GQDs revealed excitation-dependent fluorescence and superior electrochemiluminescence. $Coal_B$ was found to be a mix of carbon nanomaterials (CNMs), which include agglomerated GQDs, graphene oxide, carbon quantum dots, and nanocrystals of carbon. They reported that low rank coals have more short domains or nanosheets which are more ideal for the preparation of S-GQDs. The study reported decrease of production yield of S-GQDs from 56.30% to 14.66% when the coal rank increased gradually. In contrast, high-ranked coals had a high production yield of $Coal_B$, and might be more suitable for synthesizing other carbon nanostructures that were inherent in $Coal_B$.

Hu et al. (2014) established a top-down approach to chemically modify the low-cost coal to luminous CDs by a collective process of carbonization, followed by chemical oxidation and etching. The physical and photosensitive properties of the as-prepared CDs are altered by altering the configurations of carbon crystallites embedded in the precursor. These CDs exhibited characteristic emission modes, whose intensity at the short-wavelength was pointedly enriched by incomplete reduction. The advancement of the electronic structure and the study of surface states shared two attributes with the fluorescence, nano-sized sp^2 carbon domains, and surface defects are accountable for the observed emission features. The obtained CDs were proven as an effective fluorescent-sensing material for label-free and selective detection of Cu(II) ions with a detection limit as low as 2.0 nM, revealing a great potential for real time sensor applications. In a related work, Chao Hu et al. (2014) described the synthesis of carbon dots with controlled size via carbonization of anthracite at high temperature, and reported that the size of CDs is not affected by the carbonization temperature. It is also worthwhile to mention that optical properties were not affected significantly with the particle size variation. With the variation of temperature (from 0 to 1500°C), the average size and height of anthracite-derived CDs also altered in the range of 1.96 ± 0.73 to 3.10 ± 0.80 nm, and 1.04 to 1.42 nm, respectively. Besides, the synthesized dots exhibited a crystalline structure with a lattice spacing of ~ 0.21 nm, which is due to the (100) graphite facet (Hoang et al. 2018).

Tonkeswar Das and Binoy (2017) reported the materialization of nanodiamonds from low-grade coals with the aid of low-power ultrasonic stimulus in hydrogen peroxide (H_2O_2) trailed by dialysis. The size of the carbon nanocrystals are found to be of the order of 4–15 nm, with planar spacing of the crystal lattice in the range of 0.20–0.23 nm. Like graphene quantum dots, carbon dots (CDs) are also produced with coal as precursor with top-down strategy (Manoj et al. 2017, 2018). Yongqiang et al. (2014) refluxed coal in the presence of nitric acid at 130°C for 12 hours. The product was vacuum-dried to eliminate un-reacted acid and re-dispersed in distilled water, which was further neutralized with liquid ammonia and the supernatant was separated by centrifugation. The obtained mono dispersed spherical CDs have the size of 3–5 nm with a graphite lattice spacing of 0.212 nm.

Yet another popular technique for nanocarbon synthesis is Hydrothermal and Solvo-thermal method. It is a cost-effective technique for the preparation of GQDs in the presence of robust alkali (such as NaOH and ammonia) as scissoring agent to fragment the inherent nanocarbon structure present in the carbon-based precursors into colloid GQDs. It involves oxidation of the hydrocarbon or coal precursor in concentrated H_2SO_4 and HNO_3, followed by mild sonication for 15–20 hours and hydrothermal de-oxidation of the oxidized sample (200°C for 10 hours) under strong alkaline hydrothermal conditions (pH > 12). Well-powdered sample is refluxed in acid mixture with microwave sonication, followed by hydrothermal reaction. The obtained product is further purified by dialysis. One can synthesize GQDs in the diameter range of ~ 10 nm with an average of 2–3 layers. Ye et al. 2015b in their recent work reported the modification of bandgap of the photoluminescent GQDs from anthracite. The synthesized nanodots have the bandgap of 2.9 eV to 2.05 eV with emission in the blue-green to orange–red region.

Yong in 2014 carried out the facile hydrothermal synthesis of highly crystalline GQDs, which display almost excitation-independent UV and blue intrinsic emissions. These dots were synthesized from glucose, sulfuric acid, and water as reagents. The UV emission is initiated due to the recombination of electron-hole pairs localized in the C-C bonds, whereas the blue fluorescence is attributed to the transition of electron in sp^2 domains. Phitsinsi et al. (2016) reported the single-step hydrothermal carbonization process to synthesize photoluminescent carbon nanodots from limeade. The dots exhibit multicolor emission depending on the excitation wavelength. The fluorescence quenching of the nanomaterials under UV light is made use for the effective sensing of Ferric ion sensing in aqueous media. Senthil and group (2016) synthesized CDs of diverse sizes and functionalities from lignite. They followed the techniques such as refluxing, microwave radiation, and laser ablation in the ethylenediamine solution. The fluorescence and production yield exhibit changes with the synthesis route adopted. One-step solvo-thermal route for the preparation of strongly green-photoluminescent GQDs from coal involves the dissolution of coal powder in DMF with the concentration of 270 mg mL⁻¹ and sonication for 30 minutes and then heating at 200°C for 5 hours to get few layer GQDs with an average height of 2 nm. Geng et al. (2017) reported the low cost solvo-thermal synthesis of orange-colored fluorescent GQDs from coal tar. The synthesized dots showed bright and stable emission at 605 nm. Upon encapsulation with liposome, these GQDs become highly soluble in polar solvents and exhibit a redshift to 640 nm.

Doping with nitrogen considerably enhances the optical behavior of CDs, which changes UV–Vis absorption with increase of excitation wavelength, further enhancing the Quantum Yield (QY). There are different forms of nitrogen, such as pyridinic, pyrrolic, or graphitic, and its presence changes the fluorescence and emission behavior of the CDs (Sudolská and Otyepka 2017, Holá et al. 2017). Li et al. (2013) testified the high production yield of N-doped CDs (25.6%) with particle size of 4.7 nm from precursor anthracite with solvo-thermal technique in the solvent DMF at 180°C for about 12 hours. The N-doped CDs exhibit an absorption peak at 270 nm originated from the n-π* transition of C-O structure and an excitation-dependent Photo luminescent (PL) emission with cyan color when excited under 365 nm UV light with 47% quantum yield. The excitation and emission maxima of N-doped CDs were at 310 and 445 nm, respectively. It is reported that the solvo-thermal method could produce carbon dots having lower oxygen functionalities and enhanced the sp^2 conjugated domains. This could efficiently constrain the non-radiative electron-hole recombination, leading to high QY. In another report, coal tar, upon refluxing with nitric acid at 80°C for 24 hours, followed by solvo-thermal treatment in toluene at a temperature of 200°C for 12 hours produces CDs with an average size of 1.5 to 4.5 nm and lattice spacing of 0.21 nm (Geng et al. 2017). The CDs have the graphitic to defect intensity ratio (G band at 1582 and D band at 1372 cm⁻¹ and I_D/I_G) of 1.03, and exhibited orange fluorescence when excited with 365 nm light. With variation in excitation wavelength (495 to 575 nm), the PL emission peak is red shifted between 598 and 612 nm. Further, to enhance the solubility of CDs, a composite of liposome with carbon dots having a size range of 50–100 nm was prepared. They also show visible fluorescence (predominantly white) in aqueous solution, with PL maxima shifting to 640 nm with a quantum yield of 10.7 percent. The PL emission was stable against variation in pH.

Microwave-assisted synthesis route has been widely used to synthesize materials, as it has both the advantages of hydrothermal and microwave techniques. It involves fast and homogeneous heating, leading to much more uniformly sized particles. This method involves the dispersion of the sample in water or mild acid at desired concentration in a closed glass bottle and heating it in a microwave oven at a certain power for a period of time. During the process of microwave treatment, the solution changes its color with the formation of GQDs. The production yield by this technique from hydrocarbon precursor is very high. The ultrasound sonication produces alternate low and high pressure waves in liquid. This forms cavities in the liquid and leads to high speed impinging liquid jets, resulting in separation of the graphene layers and formation of GQDs. Liu and group (2017) reported the synthesis of highly crystalline CDs by microwave-assisted pyrolysis from the precursor tomato. The product exhibited remarkable quantum confinement with fluorescence in the UV region.

8. Optical Properties of the Nano Carbon Dots

Despite the multiplicity of the structures, the carbon dots (CDs) have some similar optical properties on the absorption, and to a great extent, fluorescence. CDs usually show a distinctive π-π^* transition of the C=C bond and an n-π^* transition of the C=O bond at about 230 nm and 270–300 nm, respectively. Characteristic of graphene is a single π-π^* emission of the C=C bond at 270 nm (Manoj et al. 2017). Band gap of the nanocarbon can be tailored by adopting different synthesis techniques. It is reported that the coal-derived oxygenated carbon dots have energy of the order of 2–3.5 eV (Manoj et al. 2017, Ruquan et al. 2015). These values of band gap indicate that the obtained nanocarbons are direct band semiconductors in nanoscale size (Tauc 1966). Thomas et al. (2018) very recently synthesized coal-based nanocarbons from bituminous coal of different origin. The group reported that upon chemical treatment, particle dimension is lowered to nano size, which could significantly enhance the energy gap. During chemical oxidation, the oxygen groups are added to the basal plane of the carbon matrix. The presence of functional groups of oxygen coupled with particle size reduction alters the band gap. Upon making composite with red phosphorous, the energy gap is found to be decreased to 1.37 eV and 1.43 eV for 1:2 and 2:1 of rGO red phosphorus complex, correspondingly. The incorporation of red phosphorous in different proportions to the graphene oxide alters the energy gap of the composite. With a change in concentration of the red phosphorous in the composite, the energy gap of nanocarbon is dropped. This indicates the tailoring of the energy band by controlling the concentration of phosphorous during the composite formation. More systematic work in this aspect is warranted.

The fluorescence emission of the nanocarbon structure derived from lignite under various excitation wavelengths is presented Figure 3.10. The carbon dots derived from lignite are reported to exhibit excitation-dependent fluorescence behavior. As the excitation energy is more than the energy gap (or lower wavelength), emission from the energy gap transition plays a primary role. At a higher wavelength, one could see that surface and defect state emission plays a vital role, leading to excitation-dependent fluorescent emission.

The photo luminescent (PL) emission in carbon nanoparticle upon excitation can be due to two reasons. Li et al. (2014) attributed this to two leading states-energy gap and surface state transitions. If the luminescence mechanism is affected by the size distribution, the Stokes shift approaches linearly to zero with increasing excitation wavelength (Yin et al. 2014, Ye et al. 2015, Wang et al. 2016, Manoj et al. 2017, Zhang et al. 2018a,b). The Stokes shift originates due to the difference between the positions of the band maxima of the excitation and the emission spectra, and has been widely used to confirm the quantum confinement effect in nanocrystals, indicating the existence of band gap of the bulk material. Li et al. (2014) reported that one can passivate the surface state and control the fluorescence emission from excitation dependent to excitation independent. This excitation-independent PL behavior in carbon dots is originated from the radiative transition of sp^2 carbon core. If the surface state is not passivated, one can obtain the excitation-dependent FL behavior mainly due to the C-O, C=O, and O=C-OH functional groups. Short chain nitrogen containing organic compounds are reported to be a better passivation agent. Amino group is an ideal candidate which can control the surface passivation of the luminescent carbon dots and can determine the character of luminescence. Ray et al. (2009) reported the synthesis and characterization of fluorescent carbon nanoparticles, and their utility in bio-imaging application.

The luminescent performance of the nanocarbon is altered by varying the composition of phosphorus in the composition (Thomas et al. 2018). Luminescence properties originated from the surface state and functional groups are considerably abridged with the variation of phosphorous to nanocarbon background. The intensity of fluorescence emission is slightly varied with 2:1 composite. The authors reported significant variation in the bandgap of graphene oxide system with the addition of phosphorus to the carbon matrix.

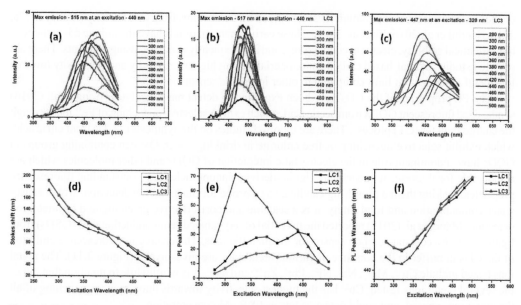

Figure 3.10: Optical properties of nanocarbon with excitation wavelengths from 280 to 500 nm in intervals of 20 nm (a) nanostructure with particle size 21 nm-LC1 (b) nanostructure with particle size 17 nm-LC2 (c) nanostructure with particle size 21 nm-LC3 (d) The stokes shift with excitation wavelength (e) PL peak intensity versus excitation wavelength (f) PL peak wavelength versus excitation wavelength (Manoj 2017).

Color version at the end of the book

9. Dependence of Fluorescence Effect on Solvent pH

The role of solvent pH on fluorescence property of the nanoparticle is very essential in advanced utilizations, such as sensing biochemical and pharmaceutical applications. Manoj and coworkers (2017a, b, 2018b) investigated the dependence of fluorescence property on variation in pH. It is observed that at neutral and basic pH, the position of the spectra remains unaltered. However, at acidic pH (pH-4), there was a shift in the position of the fluorescence spectra. When the fluorescence property was studied with varying pH of solvent, it was observed that the emission intensity was comparatively more at alkaline pH. A similar pH dependent fluorescence behavior of coal-derived GQDs was reported by Ye et al. (2013) in higher-ranked coals. The surface groups, especially the carbonyl and carboxyl groups, introduce trapping states with different energy levels in CDs, making them emit light that varies with excitation energy.

10. Fluorescent Sensing Applications: Detection of Mineral Ions

Fluorescence sensing with nanocarbon is gaining interest in recent days due to the remarkable advantages, such as superior sensitivity, economical, and short response. Stable fluorescence, chemiluminescence, and electrochemical luminescence of CDs are a few prominent properties of interest (Gaddam et al. 2017, Zhu et al. 2012, Manoj et al. 2017). Carbon-based materials are extensively used in electroanalysis and electrocatalysis for biosensing. Graphene and carbon dots have prime importance in the electrochemical sensing, mainly owing to their large surface area, biocompatibility, and excellent stability. Exceptional conductivity coupled with ultra-small size make them an ideal candidate for nano-electrodes, which are highly efficient in electron-transfer reactions. Graphene-based biosensors have an advantage over transition metal sensors, as they exhibit fast electron transport, good biocompatibility, and good thermal conductivity. There are reported studies on the synthesis of tunable photoluminescent carbon nanostructures from coal recently (Manoj 2017a, 2018, Senthil et al. 2016, Ye et al. 2014). These carbon nanomaterials have the unique advantages of high chemical stability, electrical conductivity, and are frequently used as sensing elements. There are reported studies of coal-based carbon nanomaterial having higher selectivity and sensitivity towards low metal ion detection limit compared to other inorganic sensors.

The PL properties of the CDs have been exploited to detect many analytes. Wang et al. (2016) synthesized fluorescent GQDs from acetic acid, which exhibit selective quenching of Fe^{3+} ions through the charge transfer (CT) process. There are reported studies on the synthesis of CDs from citric acid, which exhibit selective quenching of free chlorine in drinking water. Oxygen containing groups in GQDs have a dominant role in the electrostatic interaction of GQDs and other molecules which are responsible for fluorescent quenching. Lignite-derived nanoparticles have outstanding fluorescence properties, making them a promising candidate as a probe for the metal ion detection, which causes water contamination and food safety. It is a scalable and cost-effective green method for metal ion detection. Manoj et al. (2017) reported the potential of oxygenated semiconducting dots (OSDs) from lignite as a low cost nanocarbon sensor for the metallic ions. The change in fluorescence emission of the CDs in buffer in the presence of various metal ions are investigated (Figure 3.11). The metal ions tested include Cu^{2+}, Mn^{2+}, Na^+, Ni^{2+}, Pb^{2+}, Zn^{2+}, and Cr^{2+}. It is worthwhile to mention that out of the several metal ions tested, Cu^{2+} has the highest selectivity towards fluorescence quenching of all the metal ions sensed, and could detect as low as 08.9 pM concentration.

Hu et al. (2014) in their report attributed this effect to high thermodynamic affinity of Cu^{2+} ions for the NO^- chelate groups on the surface of carbon dots and the metal-ligand binding kinetics. The Cu^{2+} ions suppress the fluorescence of the nanodots via electron or energy transfer. The type of fluorescence is non-dynamic in nature and reported by Gaddam et al. (2017) and Yue Wang & group in their recent studies (2015). Yet another development in the sensing application of lignite-based carbon dots is in dextrose sensing using the fluorescence property. Manoj et al. (2018) reported that the coal-based carbon dots could detect the dextrose level as low as 0.125 mM with good linearity. Graphene synthesized by chemical routes have plenty of functionalities in the form of carboxyl,

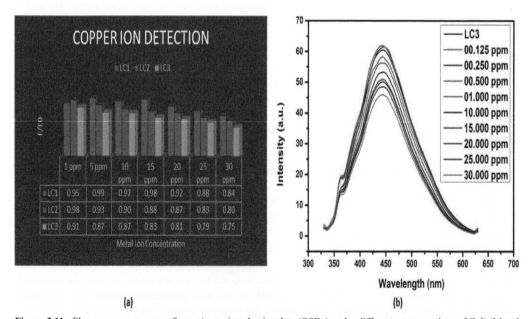

(a)   (b)

Figure 3.11: Fluorescence response of organic semiconducting dots (OSDs) under different concentrations of Cu^{2+} (Manoj et al. 2017).

Color version at the end of the book

epoxy, and hydroxyl groups on the surface. This renders them hydrophilic and soluble in water or organic solvents. These oxygenated groups play a major role and behave as an energetic site for the dextrose molecules, which have a major influence on the PL behavior of the CDs.

Carbon materials are considered one of the most promising electrode materials due to high porosity, large specific area, and low cost. Nanocarbons derived from high rank coal and coke are investigated for their novel electrochemical properties—both in pure and composite form. Anthracite has the highest energy and more carbon content among all coals. More carbon content in the source implies more carbon clusters or ions available for NCD formation. However, apart from these factors, microscopic organic components (called as macros) also play a major role. The energy storage applications in portable or remote devices, such as batteries and conventional capacitors are acquiring a major concentration of researchers. Electrochemical performance of carbon-based electrode materials are affected by the variation of specific surface area (Wang et al. 2008), structure and functional group (Xing et al. 2006), and pore size distribution (Bandaru and Prabhakar 2007). These features suggest that nanocarbon derivatives are suitable materials for polarizable electrodes. PANI-nanocarbon composite for the polyelectrolyte application is an upcoming field of research. Proper interface of polymer with nanocarbon matrix enhances the charge-discharge capacity and cyclic stability. The conductivity of such a system is heavily dependant on the percentage of polymer, reduced graphene, and the metal ions used (Pandey et al. 2019, Virginia et al. 2019). Recent studies show that anthracite-based nanocarbon products, on intercalating with Li-ion, enhance their electrochemical and optical properties.

11. Conclusion

The urge to synthesize carbon nanomaterials via sustainable and cost effective strategies has been a driving spirit worldwide among researchers. Nanocarbon materials have wide applications in pollution control, especially in heavy metal sensing and food quality control. They are the potential substitute for the inorganic nanoparticles usually used for such a cause. Employed precursors meant for the synthesis of carbon-based novel material have a vital role in the success of the emerging technology. Scientific research on utility of coal has received tremendous importance due to its effective use as an energy source. Since the discovery of carbon nanotubes, researchers are looking for an economically viable and efficient method for obtaining carbon nanomaterials and nanodots. There is a need to find cheap and simple techniques to prepare carbon nanomaterials. Coal is a heterogeneous carbon resource, whose features are determined by the diversity of possible local structure conformation. The existence of the branched porous structure makes coal a promising, cost-effective candidate for various advanced carbon materials. Coal crystallites have turbostratic structure, whose crystallite parameters increase with rank. The nanomaterials in coal contain crystalline amorphous carbon with nanocrystalline structure and were found to obey Tuinstra-Koening relation. The Raman analysis by curve fitting provided the evidence of different types of carbon allotropes in the nanocarbon extracted from coal. The aromaticity, degree of graphitization, and lateral dimensions could be altered by chemical leaching of coal with the aid of mineral acid and successive leaching. It is also noticed that the sample, when leached with hydrofluoric acid (HF), shows more ordering or stacking of carbon layers and is attributed to the substantial decrease of mineral content while enhancing the content of carbon.

With organic acid and fungal leaching, the impurity level in coal is substantially eliminated with significant enhancement in the carbon and oxygen presence. Upon treating with *Penicillium* spp., more graphitization takes place in the sample. Upon comparison with pure graphite diffraction (PG), it is evident that leaching with the fungus penicillium has introduced more of a graphitic ordered structure in the coal. Bio-solubilization is potential field to be explored in enhancing the quality of coal as a novel material in energy harvesting application.

Coal is also a potential resource for graphene, such as nanocarbon, GQDs, CDs, which have potential applications. These carbon dots are highly stable, water soluble, have a tailor-made band gap, and have tunable luminous properties from bluish to greenish region. These properties are due to the amalgamation of quantum size effect and defect state emission instigated from the particle size and oxygen functional groups. They exhibited solvent-dependent luminescence in aqueous solutions without the aid of any external passivation agent. These oxygenated CDs could be exploited as a label for the detection of heavy metals in water having high sensitivity and lower detection limit, exhibiting great prospective in nanocarbon sensing application. Coal and its derivatives are materials which are least explored as a potential novel material. Their application ranges from the conventional resources to sources of novel carbon material for energy applications, sources of biocompatible nanomaterials, and biomedical applications. Nanocarbons derived from coal of various rank and its composites have novel electrochemical applications, and this is an upcoming field in carbon research.

References

Acharya, C., R.N. Kar and L.B. Sukla. 2003. Studies on reaction mechanism of bioleaching of manganese ore. Miner. Eng. 169(10): 1027–1030.

Akshatha, A.R., M.R. Ashlin and B. Manoj. 2017. Extraction and characterization of preformed mixed phase graphene sheets from graphitized sub-bituminous coal. Asian J. Chem. 29(11): 2425–2428.

Alexander, L.E. and E.C. Sommer. 1956. Systematic analysis of carbon black structure. J. Phys. Chem-US 60: 1646–1649.

Alvarez, R., C. Clemente and D. Gomez-Limon. 1997. The influence of nitric acid oxidation of low rank coal and its impact on coal structure. Fuel 82: 2007–2015.

Andreas, G., I. Andreas and Victoria, Kapina. 2003. Study of low rank Greek coals using FT-IR spectroscopy. Energ. Source 25: 995–1005.

Anton, K, Kewer Huang, Changsheng Xiang, Angel A. Martin and James M. Tour. 2015. Luminescent polymer composite films containing coal-derived graphene quantum dots. ACS Appl. Mater. Inter. 7: 26063–26068.

Anu, N. Mohan and B. Manoj. 2012. Synthesis and characterization of carbon nanospheres from hydrocarbon soot. Int. J. Electrochem. 7(10): 9537–9549.

Anu, N. Mohan, A.V. Ramaya and B. Manoj. 2016a. Extraction and characterization of wrinkled graphene nano crystals from commercial graphite. Asian J. Chem. 28(5): 1031–1034.

Anu, N. Mohan, A.V. Ramaya and B. Manoj. 2016b. Synthesis and characterization of sp2-sp3 bonded disordered graphene like nanocarbon from coconut shell. Adv. Sci. Eng. Med. 8(2): 112–116.

Anu, N. Mohan, B. Manoj and S. Panicker. 2019a. Facile synthesis of graphene–tinoxide nanocomposite derived from agricultural waste for enhanced antibacterial activity against *Pseudomonas aeruginosa*. Sci. Rep.-UK 9(1): 4170.

Anu, N. Mohan and B. Manoj. 2019b. Surface modified graphene/SnO_2 nanocomposite from carbon black as an efficient disinfectant against *Pseudomonas aeruginosa*. Mater. Chem. Phys. 232: 139–144.

Avouris, P. and C. Dimitrakopoulos. 2012. Graphene: synthesis and applications. Mater. Today 15: 86–97.

Awasthi, S., K. Awasthi, A.K. Ghosh, S.K. Srivastava and O.N. Srivastava. 2015. Formation of single and multi-walled carbon nanotubes and graphene from Indian bituminous coal. Fuel 147: 35–42.

Balachandran, M. 2014. Role of infrared spectroscopy in coal analysis—an investigation. Am. J. Anal. Chem. 5(6): 367–372.

Bandaru, R. and Prabhakar. 2007. Electrical properties and application of carbon nanotube structures. J. Nanosci. Nanotechno. 7(4): 1239–1267.

Başaran, Yeşim, Adil, Denizli, Billur Sakintuna, Alpay Taralp and Yuda Yürüm. 2003. Bio-liquefaction/solubilization of low-rank Turkish lignites. Energ. Fuel 17: 1068–1074.

Bijiang Geng, Dewen Yang, Fengfeng Zheng, Chen Zhang, Jing Zhan and Zhen Li et al. 2017. Facile conversion of coal tar to orange fluorescent carbon quantum dots and their composite encapsulated by Liposomes for bioimaging. New. J. Chem., doi: 10.1039/C7NJ030005C.

Binoy, K. Saikia, R.K. Boruah and P.K. Gogoi. 2007. FT-IR and XRD analysis of coal from Makum. J. Earth Syst. Sci. 116: 575–579.

Binoy, K.S., R.K. Boruah and P.K. Gogoi. 2009. A X-ray diffraction analysis on graphene layers of Assam coal. J. Chem. Sci. 121(1): 103–106.

Boon, M. 2001. The mechanism of direct and indirect bacterial oxidation of sulphide mineral. Hydrometallurgy 62(1): 67–70.

Bronikowski, M.J. 2006. CVD growth of carbon nanotube bundle arrays. Carbon 44: 2822–2832.

Cameán, I., P. Lavela, J.L. Tirado and A.B. García. 2010. On the electrochemical performance of anthracite-based graphite materials as anodes in lithium-ion batteries. Fuel 89: 986–991.

Cao, Q. and J.A. Rogers. 2009. Ultrathin films of single-walled carbon nanotubes for electronics and sensors: a review of fundamental and applied aspects. Adv. Mater. 21: 29–53.

Castro, I.M., J.L.R. Fietto, R.X. Vieira, M.J.M. Tropia, L.M.M. Campos and E.B. Paniago. 2000. Hydrometallurgy 57(1): 39–90.

Celin Acharya, L.B. Sukla and V.N. Misra. 2005. Biological elimination of sulphur from high sulphur coal by Aspergillus-like fungi. Fuel 84: 1597–1600.

Chao Hu, Chang Yu, Mingyu Li, Xiuna Wang, Junyu Yang and Zongbin Zhao et al. 2014. Chemically tailoring coal to fluorescent carbon dots with tuned size and their capacity for Cu (II) detection. Small 10(23): 4926–4933.

Chua, C.K., Z. Sofer, B. Khezri, R.D. Webster and M. Pumera. 2016. Ball-milled sulfur-doped graphene materials contain metallic impurities originating from ball-milling apparatus: their influence on the catalytic properties. Phys. Chem. Chem. Phys. 18: 17875–17880.

Darmstadt, H., C. Roy and K. Serge. 2000. Solid state C^{13}-NMR spectroscopy and XRD studies of commercial and pyrolytic carbon blacks. Carbon 38: 1279–1287.

Das, T. Binoy, K. Saikia and B.P. Baruah. 2016. Formation of carbon nano-balls and carbon nano-tubes from northeast Indian Tertiary coal: value added products from low grade coal. Gondwana Res. 31: 295–304.

Deenan, S. and Yen-Peng Ting. 2005. Bioleaching of spent refinery processing catalyst using Aspergillus niger with high yield oxalic acid. J. Biotechnol. 116(2): 171–184.

Delhaes, P. 2001. Graphite and Precursors. CRC Press. 256.

Deng, J., Y. You, V. Sahajwalla and R.K. Joshi. 2016. Transforming waste into carbon-based nanomaterials. Carbon 96: 105–115.

Donald L. Pavia, Gary M. Lampan and George S. Kriz. 2008. Introduction to Spectroscopy, Brooks Cole, 5th edn.

Dong, Yongqiang Jianpeng Lin, Yingmei Chen, Fengfu Fu, Yuwu Chi and Guonan Chen. 2014. Graphene quantum dots, graphene oxide, carbon quantum dots and graphite Nano crystals in coals. Nano Scale 6: 7410–7415.

Elecy, C.D. and B. Manoj. 2010. Demineralization of coal by stepwise bioleaching: as study of sub-bituminous Indian coal by FTIR and SEM. J. Uni. Chem. Technol. Metall. 45(4): 385–390.

Elcey, C.D. and B. Manoj. 2016. Graphitization of coal by Bio-solubilization: Structure probe by Raman spectroscopy. Asian J. Chem. 28(7): 1557–1560.

Ferrari, A.C. and J. Robertson. 2000. Interpretation of Raman Spectra of disordered and amorphous carbon. Phys. Rev. B. 61: 14095.

Ferrari, A.C. and J. Robertson. 2001. Resonant Raman spectroscopy of disordered, amorphous and diamond like carbon. Phys. Rev. B 64: 075414-1–075414-13.

Ferrari, A.C. and J. Robertson. 2004. Raman spectroscopy of amorphous nanostructured m diamond-like carbon and nano diamond. Phil. Trans. 362: 2477–2512.

Franklin, R. 1950. The interpretation of diffuse X-ray diagrams of carbon. Acta Cryst. 3: 107–121.

Gaddam, R.R. Sudip Mukherjee, N. Punugupati, D. Vasudevan, C.R. Patra and R. Narayan. 2017. Facile synthesis of carbon dots and residual carbon nanobeads: Implications for ion sensing, medicinal and biological applications. Mat. Sci. & Eng. 73(1): 643–652.

Geim, A.K. 2009. Graphene: Status and Prospects. Science 324: 1530–1534.

Ghorbani, Y., M. Oliazadeh and A. Shavedi. 2008. Aluminum solubilization from red mud by some indigenous fungi in Iran. J. Appl. Biosci. 7: 207–213.

Given, P.H. 1984. In Coal Science. 3, M.L. Gorbaty, J.W. Larsen and I. Wender (eds.). Academic Press, San Diego, CA.

González, D., M.A. Montes-Morán and A.B. Garcia. 2003. Graphite materials prepared from an anthracite: a structural characterization. Energ. Fuel 17: 1324–1329.

González, D., M.A. Montes-Morán, A.B. Garcia and Suárez-Ruiz. 2004. Structural characterization of graphite materials prepared from anthracites of different characteristics: a comparative analysis. Energ. Fuel 18: 365–370.

Haenel, M.W. 1992. Recent progress in coal structure research. Fuel 71: 1211–1222.

Holá, K., M. Sudolská, S. Kalytchuk, D. Nachtigallová, A.L. Rogach, M. Otyepka et al. 2017. Graphitic nitrogen triggers red fluorescence in carbon dots. ACS Nano 11: 12402–12410.

Huang, S., H. Guo, X. Li, Z. Wang, L. Gan, J. Wang et al. 2013. Carbonization and graphitization of pitch applied for anode materials of high power lithium ion batteries. J. Solid State Electrochem. 17: 1401–1408.

Hu Yin, He Da-Wei, Wang Yang-Sheng and Duan Jia-Hua. 2014. An approach to controlling the fluorescence of graphene quantum dots: From surface oxidation to fluorescent mechanism. Chin. Phys. B 23(12): 128103.

Iijima, S. 1991. Helical microtubules of graphitic carbon. Nature 354: 56.

Jun Xiao, Pu Liu, Lihua Li and Guowei Yang. 2015. Fluorescence origin of nano diamonds. J. Phys. Chem. C. 119: 2239–2248.

Kaur, H. 2008. Analytical Chemistry 4th edn. Pragati Prakashan Publishers, Meerut, 116–205.

Khin Moh Aung and Yeng-Peng Ting. 2005. J. Biotechnol. 166(2): 159–170.

Kumar, K., R.K. Singh, A.K. Ghosh, R. Sen, S.K. Srivastava, R.S. Tiwari et al. 2103. Synthesis of coal-derived single-walled carbon nanotube from coal by varying the ratio of Zr/Ni as bimetallic catalyst. J. Nanopart. Res. 15.

Lerato, M.S., E.I. Eric and A.C. Keith. 2012. Biological degradation and solubilization of coal. Biodegradation 24(3): 305–318.

Li, J., Y. Cao, L. Wang and D. Jia. 2017. Cost-effective synthesis of bamboo-structure carbon nanotubes from coal for reversible lithium storage. RSC Adv. 7: 34770–34775.039.

Li, X., J. Hayashi and C.Z. Li. 2006. FT-Raman spectroscopic study of the evolution of char structure during the pyrolysis of a Victorian brown coal. Fuel 85: 1700–1707.

Lingling Li, Gehui Wu, Guohai Yang, Juan Peng, Jianwei Zhao and Jun-Jie Zhu. 2013. Focusing on luminescent graphene quantum dots: current status and future perspectives. Nanoscale 5: 4015–4039.

Lu, L., V. Sahajwalla, C. Kong and D. Harris. 2001. Quantitative X-ray diffraction analysis and its application to various coals. Carbon 39(12): 1821–1833.

Manoj, B., A.G. Kunjomana and K.A. Chandrasekharan. 2009. Chemical leaching of low rank coal and its characterization using SEM/EDAX and FTIR. J. Miner. Met. Mater. Eng. 8(10): 821–828.

Manoj, B. and A.G. Kunjomana. 2010b. Chemical solubilization of coal using HF and characterization of products by FTIR, FT Raman, SEM and elemental analysis. J. Miner. Met. Mater. Eng. 9(10): 919.

Manoj, B. and A.G. Kunjomana. 2010a. FT-Raman spectroscopic study of Indian-bituminous and sub-bituminous coal. Asian Journal of Material Science 2(4): 204–210.

Manoj, B. and C.D. Elcey. 2010. Demineralization of coal by stepwise bioleaching: a study of sub-bituminous Indian coal by FTIR and SEM. J. Uni. Chem. Technol. Metall. 45(4): 385–390.

Manoj, B. and A.G. Kunjomana. 2012a. Study of stacking structure of amorphous carbon by X-ray diffraction technique. Int. J. Electrochem. Sci. 7(4): 3127–3134.

Manoj, B. and A.G. Kunjomana. 2012b. Structural characterization of selected Indian coals by X-ray diffraction and spectroscopic techniques. Trends Appl. Sci. Res. 7(6): 434–444.

Manoj, B. and A.G. Kunjomana. 2012c. Chemical leaching of an Indian bituminous coal and characterization of the products by vibrational spectroscopic techniques. Int. J. Min. Met. Mater. 19(4): 279–283.

Manoj, B. 2012. Chemical leaching of high volatile Indian bituminous coal by carboxylic acid and characterization of the products by SEM/EDS. Journal of Environmental Research and Development 6(3A).

Manoj, B. 2013. Bio-demineralization of Indian bituminous coal by Aspergillus niger and characterization of the products. Res. J. Biotechnol. 8(3): 49–54.

Manoj, B. and C.D. Elcey. 2013. Demineralization of sub-bituminous coal by fungal leaching: A structural characterization by X-ray & FTIR analysis. Res. J. Che. Environ. 17(8): 11–15.

Manoj, B. and A.G. Kunjomana. 2014. Systematic investigations of graphene layers in sub-bituminous coal. Russ. J. Appl. Chem. 87(11): 1726–1733.

Manoj, B. 2014. Investigation of nanocrystalline structure in selected carbonaceous materials. Int. J. Miner. Metall. Mater. 21(9): 940–946.

Manoj, B. 2015. Synthesis and characterization of porous, mixed phase, wrinkled, few layer graphene like nanocarbon from charcoal. Russ. J. Phys. Chem. A+ 89(13): 2438–2442.

Manoj, B. 2016. A comprehensive analysis of various structural parameters of Indian coals with the aid of advanced analytical tools. Int. J. Coal Sci. Technol. 3(2): 123–132.

Manoj, B. M.R. Ashlin and G.T. Chirayil. 2017a. Tunable direct band gap photoluminescent organic semiconducting nanoparticles from lignite. Sci. Rep.-UK 7(1): 18012.

Manoj, B. 2017b. Extraction of preformed mixed phase graphene nano sheets from graphitized coal by fungal leaching. Handbook on Research on Inventive Bioremediation Technique 1: 281–299.

Manoj, B., M.R. Ashlin and G.T. Chirayil. 2018a. Facile synthesis of preformed mixed nanocarbon structures from low rank coal. Mater. Sci. Poland 36(1): 14–20.

Manoj, B., M.R. Ashlin and G.T. Chirayil. 2018b. Tailoring of low grade coal to fluorescent nanocarbon structures and their potential as a glucose sensor. Sci. Rep.-UK 13891.

Mathews, J.P. and A.L. Chaffee. 2012. The molecular representations of coal—a review. Fuel 96: 1–14.

Mayank Pandey, M. Balcandran, G.M. Joshi, N.N. Ghosh and A.S. Vendan. 2019. Superior charge discharge ability of reduced graphene oxide/Li-ion embedded polymer composite films. J. Mater. Sci.: Mater. Electron. 30(3): 2136–2145.

Mukherjee, S. and P.C. Borthakur. 2003. Effect of leaching high sulphur sub-bituminous coal by potassium hydroxide and acid on removal of mineral matter and sulphur. Fuel 82: 783–788.

Nabeel, A., T.A. Khan and D.K. Sharma. 2009. Studies on the production of ultra-clean coal by alkali-acid leaching of low grade coals. Energ. Source Part A 31: 594–601.

Novoselov, K.S., A.K. Geim, S.V. Morozov, D. Jiang, Y. Zhang, S.V. Dubonos et al. 2004. Electric field effect in atomically thin carbon films. Science 306: 666–669.

Oguz Bayraktar. 2005. World J. Microb. Biot. 21(5): 661–665.

Pan, D., Jingchun Zhang, Zhen Li and Minghong Wu. 2010. Hydrothermal route for cutting graphene sheets into blue-luminescent graphene quantum dots. Adv. Mater. 222: 734–738.

Pang, L.S.K., A.M. Vassallo and M.A. Wilson. 1991. Fullerenes from coal. Nature 352: 480.

Phitsinsi, S., C.S. Tiwary, J. Wetcharungsri, S. Porntheeraphat, R. Hoonsawat, P.M. Ajayan et al. 2016. Blue photo luminescent carbon nanodots from limeade. Mater. Sci. Eng. C 69: 914–921.

Ping Yong, Ligang Zhou, Shenil Zhang, Neng Wan and Wei Pan. 2014. Facile synthesis and photoluminescence of graphene quantum dots. J. Appl. Phys. 116: 244306.

Ponni Narayanan and B. Manoj. 2013. Study of changes to the organic functional group of a high volatile bituminous coal during organic acid treatment process by FTIR spectroscopy. J. Miner. Met. Mater. Eng. 1(02): 39.

Pratima, M., M.K. Sinha and S.K. Sahu. 2012. Chemical beneficiation of low grade coal—A review. 16th International Conference on Non-ferrous Metals; Eds: Abhilash and Prabhakar Landge.

Ramya, A.V., A.N. Mohan and B. Manoj. 2016. Wrinkled graphene: synthesis and characterization of few-layer graphene-like nanocarbon from kerosene. Mater. Sci. Poland 34(2): 330–336.

Rao, C.N.R., A.K. Sood and K.S. Subrahmanyam. 2007. graphene: The new two dimensional nanomaterials. Anew. Che. Int. Ed. 88: 7752–7777.

Ray, S.C., Arindam Saha, Nikhil R. Jana and Rupa Sarkar. 2009. Fluorescent carbon nano particels: synthesis, characterization and bioimaging application. J. Phys. Chem. C 113(43): 18546–18551.

Riya Thomas, E. Jayaseeli, N.M.S. Sharma and B. Manoj. 2018. Opto-electric property relationship in phosphorous embedded nanocarbon. Results. Phys. 10: 633–639.

Rubiera, F., A. Arenillas, C. Pevida, G.F. Gacia, K.M. Steel and J.W. Patrick et al. 2002. Coal structure and reactivity changes induced by chemical demineralization. Fuel 79: 273–279.

Rubiera, F., A. Arenillas and J.J. Arias. 2003. Combustion behavior of ultra clean coal obtained by chemical demineralization. Fuel 82: 2145–2151.

Sadezky, A.H., H. Muckenhuber, Grothe, R. Niessner and P. Ulrich. 2005. Raman micro spectroscopy of soot and related carbonaceous materials: spectral analysis and structural information. Carbon 43(8): 433–442.

Saha, B., D. Chingombe and R.J. Wakeman. 2005. Surface modification and characterization of coal based activated carbon. Carbon 43(15): 3132–3143.

Senthil, K.T., T.S. Raghupathy, D. Palanivel, K. Raji and P. Ramamurthy. 2016. Fluorescent carbon nanodots from ignite: unveiling the impeccable evidence of quantum confinement. Phys. Chem. Chem. Phys. 18: 12065–12073.

Shakirullah, M., I. Ahmad, H. Rehman, M. Ishaq, U. Khan and H. Ullah. 2006. Effective chemical leaching and ash depletion of low rank coal with EDTA and Citric acid. Jour. Chem. Soc. Pak 28(1): 56–61.

Sharma, D.K. and G. Wadhwa. 1997. Demineralization of coal by stepwise bioleaching: a comparative study of three Indian coals by Fourier transform infrared and X-ray diffraction techniques. World J. Microb. Biot. 13(1): 29–36.

Shengliang, Hu., W. Zhijia, Ch. Quing, T. Adrian and J. Yang. 2016. A facile and green method towards coal-based fluorescent carbon dots with photocatalytic activity. App. Surf. Sci. 378: 402–407.

Shujun Wang, Ivan S. Cole, D. Zhao and Qin Li. 2016. The dual roles of functional groups in the photoluminescence of graphene of graphene quantum dots. Nanoscale 8: 7449–7458.

Sonibare, O.O., H. Tobias and F.F. Stephen. 2010. Structural characterization of Nigerian coals by X-ray diffraction, Raman and FTIR spectroscopy. Energy 35: 5347–5353.

Speight G. James. 2005. Handbook of Coal Analysis. John Wiley & Sons. 166.

Speight G. James. 2012. The Chemistry and Technology of Coal. 3rd edn, John Wiley & Sons.

Sudolská, M. and M. Otyepka. 2017. Exact roles of individual chemical forms of nitrogen in the photoluminescent properties of nitrogen-doped carbon dots. Appl. Mater. Today 7: 190–200.

Sun, Y. Shiqi Wang, Chun Li, Peihui Luo, Lei Tao and Gaoquan Shi et al. 2013. Large scale preparation of graphene quantum dots from graphite with tunable fluorescence properties. Phys. Chem. Phys. 15: 9907–9913.

Takagi, H., K. Maruyama, N. Yoshizawa, Y. Yamada and Y. Sato. 2004. XRD analysis of carbon stacking structure in coal during heat treatment. Fuel 83: 2427–2433.

Tauc, J., R. Grigorovici and A. Vancu. 1966. Optical properties and electronic structure of amorphous germanium. Phys. Stat. Sol. 15: 627–637.

Tian, Y. Zhang, B. Wang, W. Ji, Y. Zhang and K. Xie. 2004a. Coal-derived carbon nanotubes by thermal plasma jet. Carbon 42: 2597–2601.

Tian, Y., Y.L. Zhang, Q. Yü, X.Z. Wang, Z. Hu, Y.F. Zhang et al. 2004b. Effect of catalysis on coal to nanotube in thermal plasma. Catal. Today 89: 233–236.

Tonkeswar Das, P.K. Buruah, M.R. Das and B.K. Saikia. 2016. Formation of onion-like fullerene and chemically converted graphene-like nanosheets from low quality coals: application in photocatalytic degradation of 2-nitrophenol. RSC Advances 42: doi: 10.1039/C6RA04392E.

Tonkeswar Das and B.K. Saikia. 2017. Nanodiamonds produced from low-grade Indian coals. ACS Sustainable Chem. Eng. 5(11): 9619–9624.

Tuinstra, F. and J.L. Koenig. 1970. Raman spectrum of graphite. T Chem Phys. 53: 1126–30.

Van Chinh Hoang, Mahbub Hasaan and Vincent G. Gomes. 2018. Coal derived carbon nnaomaterials—Recent advances and applications. Applied Materials Today 12: 342–358.

Van Krevelen, D.W. 1957. Coal. 3rd edition, Elsevier, 109–117.

Vasilis Lavrakas. 1957. The lubricating properties of graphite. J. Chem. Edu. 34(5): 240.

Vasireddy, S., B. Morreale, A. Cugini, C. Song and J.J. Spivey. 2011. Clean liquid fuels from direct coal liquefaction: chemistry, catalysis, technological status and challenges. Energy Environ. Sci. 4: 311–345.

Viginia, M., A.K. Sunitha, Riya, Thomas, M. Pandey and B. Manoj. 2019. An investigation on structural, electrical and optical properties of GO/ZnO nano composite. Int. J. Electrochem. 14: 3752–3763.

Vijapur, S.H., Dang Wang and G.G. Botte. 2013. Raw coal derived large area and transparent graphene films. ECS Solid State Letters 7: M45–M47.

Wang, L., C.K. Chua, B. Khezri, R.D. Webster and M. Pumera. 2016. Remarkable electrochemical properties of electrochemically reduced graphene oxide towards oxygen reduction reaction are caused by residual metal-based impurities. Electrochem. Commun. 62: 17–20.

Wang, L.H. Toyoda and M. Inagaki. 2008. Dependence of electric double layer capacitance of activated carbons on the types of pores and their surface areas. Carbon 23(2): 111–115.

Warren, B.E. 1934. X-ray diffraction study of carbonblack. J. Chem. Phys. 2(9): 551–555.

Warren, B.E. 1941. X-ray diffraction in random layer lattices. Phys. Rev. 59(9): 693–698.

Weijian Liu, Chun Li, Xianho Sun and Jinping Wang. 2017. Highly crystalline carbon dots from fresh tomato: UV emission and quantum confinement. Nanotechnology 28(48): 485705.

Wijaya, N., T.K. Choo and L. Zhang. 2011. Generation of ultra-clean coal from Victorian brown coal. Sequential and single leaching at room temperature to elucidate the elution of individual inorganic elements. Fuel Process. Technol. 92: 2127–2137.

Williams, K.A., M. Tachibana, J.L. Allen, L. Grigorian, S.C. Cheng, S.L. Fang et al. 1999. Single-wall carbon nanotubes from coal. Chem. Phys. Lett. 310: 31–37.

Wu, Z. and K.M. Steel. 2007. Demineralization of a UK bituminous coal using HF and ferric ions. Fuel 86: 2194–2200.

Xiaoming Li, Shengli Zhang, Sergei A. Kulinich, Yanli Liu and Haibo Zeng. 2014. Engineering surface states of carbon dots to achieve controllable luminescence for solid-luminescent composites and sensitive Be^{2+} detection. Sci. Rep.-UK 4: 4976.

Xing, W.S.Z., S.Z. Qiao and R.G. Ding. 2006. Superior electric double layer capacitors using ordered mesoporous carbons. Carbon 44(2): 216–221.

Xu, H., Q. Lin, T. Zhou, T. Chen, S. Lin and S. Dong. 2014. Facile preparation of graphene nanosheets by pyrolysis of coal-tar pitch with the presence of aluminum. J. Anal. Appl. Pyrol. 110: 481–485.

Ye Ruqvan, Changsheng Xiang, Jian Lin, Zhiwei Peng, Kewei Huang, James M. Tour et al. 2013. Coal as an abundant source of graphene quantum dots. Nat. Commun. 4: 2943–2947.

Ye Ruquan, Zhiwei Peng, Andrew Metzger, Jian Lin and James M. Tour. 2015. Bandgap engineering of coal-derived graphene quantum dots. Applied Mat. & Interfaces 7: 7041–7048.

Yin Hu He Da-Wei, Wang Yong-Sheng, Duan Jia-Hua, Wang Su-Feng and Fu Ming. 2014. An approach to controlling the fluorescence of graphene quantum dots: From surface oxidation to fluorescent mechanism. Chin. Phys. B 23(12): 128103.

Yongqiang Dong, Jianpeng Lin, Yingmei Chen, Fengfu Fu, Yuwu Chi and Guonan Chen. 2014. Graphene quantum dots, graphene oxide, carbon quantum dots and graphite nanocrystals in coals. Nanoscale 6: 7410–7415.

Yue Wang, Wen-ting Wu, Ming-boWu, Hong-di Sun, Hui Xie, Chao Hu et al. 2015. Yellow visual fluorescent carbon quantum dots from petroleum coke for the efficient detection of Cu^{2+} ions. New Carbon Mater. 30: 6.

Zeng, C., Q. Lin, C. Fang, D. Xu and Z. Ma. 2013. Preparation and characterization of high surface area activated carbons from co-pyrolysis product of coal-tar pitch and rosin. J. Anal. Appl. Pyrol. 104: 372–377.

Zhang, M., Linling Bai, Weihu Shang, Wenjing Xie, Hong Ma, Yingyi Fu et al. 2012. Facile synthesis of water-soluble, highly fluorescent graphene quantum dots as a robust biological label for stem cells. J. Mater. Chem. 22: 7461–7467.

Zhang, S., J. Zhu, Y. Qing, C. Fan, L. Wang, Y. Huang et al. 2017. Construction of hierarchical porous carbon nanosheets from template-assisted assembly of coal-based graphene quantum dots for high performance supercapacitor electrodes. Mater. Today Energy 6: 36–45.

Zhang, C., S. Jiang and W. Zhang. 2018a. Adsorptive performance of coal-based magnetic activated carbon for cyclic volatile methylsiloxanes from landfill leachate. Environ. Sci. Pollut. Res. Int. 25: 4803–4810.

Zhang, C., J. Li, Z. Chen and F. Cheng. 2018b. Factors controlling adsorption of recalcitrant organic contaminant from bio-treated coking waste water using lignite activated coke and coal tar-derived activated carbon. J. Chem. Technol. Biotechnol. 93: 112–120.

Zhou Quan, Zongbin Zhao, Yating Zhang, Bo Meng, Anning Zhou and Jieshan Qiu. 2012. Graphene sheets from graphitized anthracite coal: Preparation, decoration, application. Energy Fuels 26: 5186–5192.

Zhu, J., S. Zhang, L. Wang, D. Jia, M. Xu, Z. Zhao et al. 2018. Engineering cross-linking by coal-based graphene quantum dots toward tough, flexible, and hydrophobic electro spun carbon nano fiber fabrics. Carbon 129: 54–62.

Zhu, S. Junhu Zhang, Chunyan Qiao, Shijia Tang, Yunfeng Li, Wenjing Yuan et al. 2011. Strongly green-photo luminescent graphene quantum dots for bio imaging applications. Chem. Commun. 47: 6858–6860.

Zhu, Z., L. Garcia-Gancedo, A.J. Flewitt, H. Xie, F. Moussy and W.I. Milne. 2012. A critical review of glucose biosensors based on carbon nanomaterials: Carbon nanotubes and graphene. Sensors 12: 5996–6022.

CHAPTER 4

Biomass-derived Carbon Compounds and their Potential Application for Electronic and Magnetic Materials

Retno Asih, Malik Anjelh Baqiya, Yoyok Cahyono and *Darminto**

1. Introduction

Carbon is one of the most abundant elements in the earth's crust, as well as the second most abundant element in the human body after oxygen. Carbon is located at the 6th group of the periodic table, giving it four free valence electrons to bond covalently with other elements. It yields the electronic-ground-state configuration of $1s^2 2s^2 2p_x^{\,1} 2p_y^{\,1}$. Carbon occurs naturally in several forms (allotropes) having different physical properties. The ability of carbon to form many allotropes and compounds relies on the orbital's hybridization, which results in different forms of bonding. The allotropes of carbon are characterized by the type of hybridized bonding forming its structure, ranging from pure sp^2 as in graphene, carbon nanotubes, and fullerenes, to pure sp^3 as in diamond. The variation in hybridization structures of carbon materials enables a wide spectrum of properties, such as high bulk mechanical hardness, tribological properties, and chemical inertness (Mansour 2011).

One of carbon's allotropes that has recently attracted growing interest is graphene. Researches on graphene have been conducted massively in the last decade since its first report in 2004 (Novoselov et al. 2004). Due to its exceptional mechanical, electrical, thermal, and optical properties, graphene has been suggested to be a very promising candidate for spin electronic devices (Tombros et al. 2007, Rao et al. 2009, Singh et al. 2011, Qin et al. 2014). With such remarkable properties, however, a large production of graphene is challenging. Accordingly, graphene derivatives have been developed into alternative materials for industrial applications. One of them is reduced graphene oxide (rGO), a graphene oxide with reduced oxygen-containing functional groups. rGO is not the same as pristine graphene because some oxygen functional groups and defects remain on the graphene sheet (Sarkar et al. 2014). The defects and oxygen functional groups that are partially decorated on the sheet with p-p bonds act as active sites to the analyte molecules, bringing in a better performance for

Department of Physics, Institut Teknologi Sepuluh Nopember, Kampus ITS Sukolilo, Surabaya 60111, Jawa Timur, Indonesia.
* Corresponding author: darminto@physics.its.ac.id

various applications, such as chemical sensors (Yavari and Koratkar 2012, Kim et al. 2015, Minitha et al. 2018), supercapacitors (Ma et al. 2014, Rasul et al. 2017, Jayaraman et al. 2017), microwave absorption (Cao et al. 2018, Ma et al. 2018), energy storage (Wang et al. 2009, Zhao et al. 2012, Kim et al. 2014), and photovoltaic cells (Czerniak-Reczulska et al. 2015, Barpuzary and Qureshi 2015). Moreover, defect states in rGO lead to various magnetic features, which make rGO a candidate for molecular-based soft magnets (Qin et al. 2014).

Another allotrope of carbon that has been widely studied for its potential application for semiconductor technologies is carbon thin films. Thin films of amorphous carbon (a-C), diamond-like carbon (DLC), and carbon nitride (a-C:N) are particularly fascinating. They offer excellent tribological, optical, and electrical properties, with the additional advantage that carbon is a biocompatible material (Robertson 2001, Coleman et al. 2006, Mangolini et al. 2013). The a-C films consist of a mixture of sp^2 and sp^3 hybridizations, thus it has a broad range of optical and electrical properties. This is due to the ability of carbon to form different types of interatomic bonds, to take up different sites, and to adopt different structures. However, for each application, there are different requirements on the film properties, for example, the achievable adhesion level, and coating cost. Therefore, several methods have been developed to control the structure, composition, and thus the properties of a-C films, such as variation of ion energy, plasma treatment, pressure, and the use of different gases in the film's preparation (Marton 2012). Chemical vapor deposition (CVD), spray pyrolysis, anodization, electrodeposition, spin coating, thermal and electron beam evaporations are some techniques performed to prepare carbon films (Mohagheghpour et al. 2016, Milenov et al. 2019, Marcinauskas et al. 2019, Aghamir et al. 2019).

Carbon compound of rGO is commonly produced from graphene oxide (GO) by a reduction method. Due to the relative ease in creating large quantities, rGO has become the most obvious solution in the production of graphene to desired quality levels. However, its production requires graphite, which is usually obtained by exploiting mining materials. It provides more guarantees in giving result of a high-quality graphene, but it is not environmentally friendly and relatively difficult to renew a mined graphite resource. Consequently, researchers currently attempt to find a new green-method in producing graphene and its derivatives from natural carbon-rich resources, such as biomass (Shams et al. 2015, Jain et al. 2015b, Deng et al. 2016, Lu and Zhao 2017, Jung et al. 2018, Liu et al. 2018). Several strategies, including direct pyrolysis, hydrothermal carbonization, and ionothermal carbonization have been widely performed for synthesizing biomass-derived carbons (Tsai et al. 2006, Sevilla and Fuertes 2009, Hoekman et al. 2011, Liu et al. 2012). All these strategies aim to satisfy specific applications and to elevate the value of biomass.

The biomass-derived carbons have been used in developing several applications, such as fuel cells (Winter and Brodd 2004), water splitting (Kalyani et al. 2013), supercapacitors (Frackowiak and Beguin 2002, Wang et al. 2014), and lithium-ion and/or sodium-ion batteries (Goodenough and Park 2013, Roberts et al. 2014, Su and Schlögl 2010, Jain et al. 2013, Sennu et al. 2016). Nevertheless, a big challenge comes from the utilization of crude biomass, in which derived carbons usually suffer from ill-defined morphologies and a poor porous structure (Deng et al. 2016). Among biomass used to produce carbon compounds, coconut shell is popular as it possesses high volatility, low ash content (Yalcin and Arol 2002, Daud and Ali 2004, Parikh et al. 2005, Tsai et al. 2006), and high hardness (Yalcin and Arol 2002). Carbon compounds derived from coconut shells can yield high specific surface areas and exhibit favorable electrochemical properties, such as high specific capacity, good cyclability, and availability (Liu and Zhang 2011, Jain et al. 2015a).

Coconut (*Cocos Nucifera*) is somewhat naturally abundant in Indonesia. Its production reached 18 million tons in 2017, which makes Indonesia the largest producer of coconut in the world (Hendaryati and Arianto 2017). Wastes of coconut farming, including coconut shell and husk, are usually just burned, and only some of them are used to produce charcoal for cooking purposes and handicraft products. By biorefinery process, the utilization of coconut shell, which contains a great amount of carbon, can be elevated to chemicals, energy, and materials, replacing the need for petroleum, coal, natural gas, and other non-renewable energy and chemical sources. Consequently, we synthesized biomass-derived carbon compounds from coconut shell and coconut sap by a relatively simple and energy-efficient thermal method. Structures of the obtained carbon compounds, including functional groups, defect concentration, and morphology were examined. The dominant phase of rGO was obtained in carbon compounds synthesized from coconut shell, and a-C was found in the compounds prepared from coconut sap. Carbon compounds thermally reduced at higher temperature show lesser functional groups but higher concentration of defects. The effect of temperatures applied during the reduction process on the magnetic, electrical, and photoluminescence (PL) properties was also investigated. Enhanced magnetism induced by defects was confirmed in the rGO samples. Moreover, a preliminary study on the potential applications of the synthesized carbon compounds as radar absorbing materials and as electrodes for supercapacitors is briefly discussed.

2. Methodology

Carbon compounds were prepared from biomass of coconut shell and sap by a thermal method. The obtained carbon compounds were then characterized to examine their phases, structures, magnetic, and electrical properties. The characterizations were performed using x-ray diffractometer (XRD), Fourier transform infrared (FTIR) and Raman spectroscopies, electron microscopies (SEM and TEM) to investigate phases and structures, and four-point probe method, magnetometer, photoluminescence (PL) spectroscopy, cyclic voltammetry, and vector network analyzer (VNA) to investigate electronic, magnetic, PL properties of the samples and their properties related to the applications as a supercapacitor and radar absorbing material (RAM), respectively.

2.1 Samples preparation

Coconut shells and coconut sap are the raw materials used in the synthesis of carbon-based materials. Old coconut shells are chosen as the raw material, since they contain a higher concentration of carbon (C) compared to young ones. Old coconut shells have a darker color and harder texture than young coconut shells. A coconut consists of approximately 35% mesocarp and epicarp, 12% endocarp, i.e., shells, 28% endosperm, and 25% water. Chemical compositions of coconut shells include cellulose (34%), hemicellulose (21%), and lignin (27%). Moreover, coconut shells have carbon (C) as the main element, which is about 74%, oxygen (O; 22%), and other primary metals reached 4%, including silicon (Si), potassium (K), sulfur (S), and phosphorus (P). Old coconut shells contain smaller amount of non-carbon elements compared to the young ones. The burning process of coconut shell can increase the concentration of carbon up to 50 percent. Mozammel et al. (2002) reported that raw coconut shell contains 18.3% carbon, 10.5% moisture, 67.7% volatile components, and 3.5% ash, while the charcoals of coconut shell contain 76.3% carbon, 10.6% volatile components, and 13.1% ash. Accordingly, the burning process was initially performed on raw coconut shells to make charcoal powders before performing thermal reduction.

Husks or fibers adhered on the surface of coconut shells were trimmed and removed to get clean shells with refined surface. The shells were then burned at 80–100°C in the air for 1 hour until they turned to charcoal. Furthermore, the charcoal was ground and sieved to get homogeneous charcoal powder. Thermal reduction process was done on the charcoal powder by heating the powder at designated temperatures, i.e., 400°C, 600°C, 800°C and 1000°C, for 5 hours in the air. The designated temperatures were determined based on thermal analyses of coconut shell powder examined using Differential Thermal Analysis (DTA) and Thermogravimetric Analysis (TGA). Evaporations of moisture contents come off at temperatures of 100–120°C. A significant weight loss occurs in the temperature range of 200–600°C, when the decomposition of cellulose, hemicellulose, and lignin (aliphatic and aromatic hydrocarbon) takes place. At this temperature range, some oxygen and hydrogen atoms are away from their initial interatomic bonds to form CO_2, CO, and CH_4. A reduction on impurity compounds carried by the drying of gases transpires at 600–1000°C, so that it undergoes a state of purification and an increase in carbon content. It should be noted that ideal carbonization process can be achieved not only at high temperature, but also at the controlled atmosphere and pressure.

The resultant powder from thermal reduction process was then ultrasonically exfoliated to form a thinner graphitic layer. The exfoliation was carried out for 10 hours in the media of distillated water, HCl 1 M, and H_2SO_4 1 M, respectively. Finally, the synthesized compounds were characterized to examine their phase, structure, and properties.

Synthesis of carbon-based materials from coconut sap was also performed by thermal method. Coconut sap is the fluid contained in the flower buds of a coconut plant, which can be collected by a tapping process. It is widely processed as coconut sugar. Sucrose is the main sugar constituent of coconut sap, at approximately 13–17 percent. It also contains small amounts of glucose and fructose. These sugar constituents along with other saccharides determine the sweetness of the sap. However, because of high sucrose content, coconut sap can be a suitable medium for microbial growth, and thus it is easily rotten. A rotten sap is characterized by a decrease in pH due to the change of sugar into organic acids by microbes. Besides sugar content, coconut sap also has substances of protein, fat, minerals (calcium and phosphor), and a small amount of vitamin C.

As the first step of the synthesis process, coconut sap was heated at 100°C for 8 hours to reduce the water content, and hence it formed caramel. The caramel was further heated at 250°C for 1.5 hours to get charcoal. The charcoal was ground to get charcoal powder. Further heat treatment was performed at 400 and 600°C for 1 hour. The obtained powder was then ultrasonically exfoliated in distilled water and dimethyl sulfoxide (DMSO) solution, respectively, for 2 hours. Furthermore, centrifugation was conducted at 4000 rpm for 30 minutes to get carbon solutions. The solutions were then deposited on an Indium-Tin-Oxide (ITO) glass substrate to prepare thick films of amorphous carbon (a-C) via spin coating technique. Referring to the heating temperature applied to prepare charcoal powders of coconut sap, we further labelled the obtained a-C thick films as a-C-400 and a-C-600.

2.2 Characterization

Elements presented in the obtained samples were examined using an X-ray fluorescence (XRF) spectrometer. Carbon was found to be the main element in charcoal powder of coconut shell with a relative percentage of about 82%, as displayed in Table 4.1. The relative contents of oxygen (O), hydrogen (H), and nitrogen (N) were found to be approximately 9%, 8%, and 0.5%, respectively. Other primary metals, including S, P, K, Fe, Ni, Zn, Rb, Ba, and Cu, remain in the charcoal powder with a total concentration of approximately 1 percent.

Phases and structures of the samples were characterized by the XRD (PANalytical X'Pert Pro) diffractometer using CuKα radiation ($\lambda = 0.154056$ nm). To identify functional groups presented in the samples, Fourier Transform Infrared (FTIR) measurement was performed in the wavenumber range of 4000–500 nm^{-1}. Structures of the samples were also examined by Raman spectroscopy (Bruker Senterra R200-785) with the incident wavelength of 785 nm. This measurement is particularly

Table 4.1: Elemental contents of charcoal powder of coconut shell obtained by x-ray fluorescence (XRF) analysis.

Element	Percentage (%)	Element	Percentage (%)
Carbon (C)	81.84	Iron (Fe)	0.04
Hydrogen (H)	7.61	Nickel (Ni)	0.03
Oxygen (O)	9.02	Zinc (Zn)	0.01
Nitrogen (N)	0.53	Rubidium (Rb)	0.01
Sulfur (S)	0.01	Barium (Ba)	0.03
Phosphor (P)	0.02	Copper (Cu)	0.05
Potassium (K)	0.87		

substantial for carbonaceous materials with ordered and disordered states (Ferrari and Robertson 2000, Rao et al. 2009).

The absence of a bandgap in graphene makes all wavelengths of incident radiation resonant, and accordingly, the Raman spectrum contains information on both atomic structure and electronic properties (Ferrari and Basko 2013). Ferrari and Robertson (2000) found that the visible Raman spectra depend on the configuration of the sp^2 sites in sp^2-bonded clusters. Therefore, in the case of the sp^2 clustering controlled by the sp^3 fraction, as in tetragonal a-C (ta-C) or hydrogenated a-C (a-C:H) films, the observable Raman parameters can be used to derive the sp^3 fraction. The Raman spectra of graphite, graphene, and its derivatives show two sharp modes, i.e., the G-peak at around 1580–1600 cm^{-1} and the D-peak at around 1350 cm^{-1}. The former is assigned to zone center phonons of E_{2g} symmetry, and the latter is attributed to K-point phonons of A_{1g} symmetry (Ferrari and Robertson 2000). Concentration of disorder/defects presented can be qualitatively estimated from Raman spectra by determining the intensity ratio of D and G peak (I_D/I_G). The in-plane graphitic crystal size (La) of the samples can be calculated based on Raman data by the formula presented in equation (1) (Cançado et al. 2006), where λ is the wavelength used for the measurement.

$$\text{La (nm)} = (2.4 \times 10^{-10})\lambda^4 \left(I_D/I_G\right)^{-1} \tag{1}$$

X-ray photoelectron spectroscopy (XPS) measurements were carried out to clarify the hybridized bonding structure, i.e., sp^2 and sp^3 forming in the obtained samples and to qualitatively estimate their relative ratio. The measurements were performed in the kinetic energy (KE) range of 600 to 40 eV at beam line BL3.2Ua, Synchrotron Light Research Institute (SLRI), Thailand. To get information on the bonding characteristics and the composition of sp^2 and sp^3 hybridizations in the obtained carbon compounds, the C1s spectra were fine-scanned and then deconvoluted into several subpeaks. Scanning electron microscopy (SEM, EVO® MA 10) and transmission electron microscopy (TEM, JEOL-1400) were used to observe the morphology of the carbon compounds.

In addition to those structure characterizations, magnetic properties of the samples prepared from coconut shell were also examined using Superconducting Quantum Interface Device (SQUID) magnetometer (Quantum Design MPMS-XL) by measuring isothermal magnetization at 300 K under an applied field up to 10 kOe. Temperature dependency of magnetic susceptibility was also evaluated under the applied field of 100 Oe in the zero-field cooled (ZFC) and field-cooled (FC) conditions. Electrical conductivities of the samples were measured using four-point probe method at room temperature. A thick film of a-C on ITO substrate with an area of 1×1 cm^2 was prepared for the measurements. Gap energy of the a-C film was roughly evaluated from the electrical conductivity data and optical property examined by the UV-Vis spectrometer.

To evaluate microwave absorption, the reflection loss (RL) of the sample is calculated according to transmission line theory, as described in equation (2). Frequency dependence of RL of the samples is examined using vector network analyzer (VNA) in the range of X-band frequencies (8–12 GHz) at room temperature.

$$R_L = 20 \log \frac{|Z_{in} - Z_0|}{|Z_{in} + Z_0|} \tag{2}$$

Here, Z_0 describes the impedance of free space ($\sim 377\ \Omega$), and Z_{in} describes the input impedance of microwave absorption layer, which can be expressed as equation (3)

$$Z_{in} = \sqrt{\frac{\mu_r}{\varepsilon_r}} \tanh\left\{ j\left(\frac{2\pi fd}{c}\right)\sqrt{\mu_r \varepsilon_r} \right\} \tag{3}$$

where f is the microwave frequency, d is the thickness of the absorber, and c is the light velocity in vacuum. μ_r and ε_r are the relative permeability and permittivity of the sample, respectively. Z_{in} is obtained from the experimental data.

The electrochemical properties of the samples were investigated under three-electrode cells at room temperature to evaluate the performance of the obtained carbon electrodes for supercapacitors. The working electrodes were prepared by mixing the carbon compounds prepared from coconut shell and glucose as a spacer in acidic medium (HCl 1 M) at various weight ratios to form a homogeneous slurry. The obtained slurry was pressed to get a pellet form with a diameter of 1 cm. A silver (Ag) paste was used to coat surfaces of the pellet. Electrochemical measurements were performed with 6 M KOH aqueous solution as the electrolyte. The specific capacitance (*C*) of the electrode was estimated from cyclic voltammetry (CV) curves and calculated according to equation (4).

$$C = \frac{Z}{mv} \int_{V_1}^{V_2} I(V) dV \tag{4}$$

Z is the impedance factor, *m* is the mass of the working electrode, and *v* is the scan rate (50 mV/s). The mass of the working electrode was between 120 and 170 mg. *I(V)* is the response current as a function of applied potential in the range of V_1 (–1 V) to V_2 (1 V). The energy density (*W*) of the electrode was further calculated from the specific capacitance as equation (5).

$$W = \frac{1}{2} C V^2 \tag{5}$$

Photoluminescence (PL) properties of the sample were investigated by PL spectroscopy at excited wavelengths between 296–352 nm. The investigations were done in the suspension form of the sample, which were obtained by dissolving the sample in distillated water at various concentrations 0.001 mg/ml, 0.002 mg/ml, and 0.003 mg/ml. Small concentrations of suspension are used to prevent non-uniform absorption radiation. The fixed excited wavelength of 343 nm was applied to examine PL spectra as functions of suspension concentrations and the heating temperature applied in the reduction process.

3. Structure of Biomass-derived Carbon Compounds

3.1 Carbon compounds synthesized from coconut shell

Figure 4.1 displays XRD patterns of the samples calcined at various temperatures in the air. Three main peaks at 2θ of approximately 15°, 23°, and 35° were observed in the XRD pattern of charcoal powder, which are indicated as characteristic peaks of cellulose (PDF 00-056-1718) (Park et al. 2010). The peaks were no longer observed in the calcined samples. Instead, two broad peaks with less intensities were detected at 2θ ≈ 24° and 43°, which belong to peaks of rGO (Mishra et al. 2014, Fu et al. 2013, Drewniak et al. 2016). This means that a phase transformation to rGO prevails by the heating process. The XRD patterns of the obtained samples are matching with the pattern observed in the commercial rGO (®Graphenea) (Asih et al. 2019). The broad peaks indicate amorphous features, which originate from disordered multilayer states. This suggests that the resultant rGO remains disordered. In the case of graphite, a 3-D system of hexagonal carbon layer, a sharp peak is observed at 2θ of 26.5°, which indicates higher crystallinity compared to graphene oxide (GO) and rGO (Mishra et al. 2014). Due to defects and other functional groups in its sheets, rGO consists of smaller domain of sp^2, and hence has broad and less intense XRD peaks compared to those of graphene oxide (Sarkar et al. 2014, Mishra et al. 2014). The main peak of (002) at 2θ ≈ 24° gives the separation between rGO layers as 0.3705 nm, which is significantly larger than the interlayer separation of 0.3340 nm in graphite (Sarkar et al. 2014), indicating that the rGO layers have not combined to form graphite. With an increase in the heating temperature, the (002) peak shifts to higher 2θ region, and the (10) peak at ~ 43° becomes more noticeable. These results suggest that the interlayer separation becomes smaller, and the degrees of staking disorder becomes higher. Based on the heating temperature and for convenience in future discussions, we labelled the samples prepared from coconut shell rGO-400, rGO-600, rGO-800, and rGO-1000.

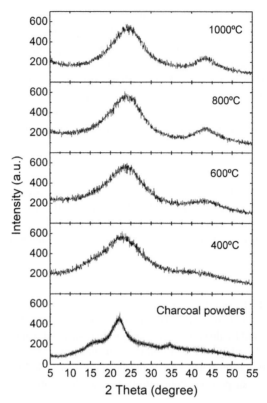

Figure 4.1: XRD patterns of the synthesized samples from coconut shell calcined at 400°C (rGO-400), 600°C (rGO-600), 800°C (rGO-800), and 1000°C (rGO-1000) for 5 hours in the air.

To further confirm functional groups presented in the samples, FTIR measurements were performed. Figure 4.2 shows FTIR spectra of charcoal powder, rGO-400, rGO-600, rGO-800, and rGO-1000. In addition to those spectra, FTIR spectra of graphite (®Merck) and a thin film of graphene oxide (®XFNANO) are also presented in Figure 4.2 as a comparison. Functional groups, including C=C, =C-H, C-O, C-C, C=O, C-H and O-H, were confirmed to exist in charcoal powders prepared from coconut shell, as well as in the commercial products of graphite (®Merck) and the thin film of graphene oxide (®XFNANO).

Some functional groups disappear by increasing heating temperature, in which rGO-1000 has less functional groups compared to other samples. Stretching vibrations from C=C at ~ 1500 cm^{-1} and C-O at 1148 cm^{-1} remain, while O-H stretching vibrations at ~ 3400 cm^{-1} disappear with an increase in the heating temperature, indicating the occurrence of de-oxygenation process.

Analyses on C1s XPS spectra of rGO samples confirm the presence of various types of oxygen functional groups, including C=O, C-OH, C-O-C, and O-C=O. The C=C component with sp^2 hybridization is the main assignment to be observed on the spectra. Figure 4.3 shows C1s XPS spectrum of rGO-1000. The deconvolution of the C1s peak of rGO-1000 displays four peaks located at binding energies of 284.2, 286.5, 288.3, and 292.5 eV, which can be assigned to the C=C, C-O-C, O-C=O and π-π* functional group, respectively. The C=C assignment of rGO-1000 was estimated to be ~ 63%.

The measurement was also done on the commercial rGO (rGO-c, ®Graphenea) for comparison. Four peaks assigned as C=C (284.2 eV), C-OH (285.5 eV), C=O (287.8 eV), and π-π* (289.9 and 291.6 eV) are observed in rGO-c. Table 4.2 presents the summary of C1s XPS data of the samples obtained from fitting calculations. The functional group of C=C with sp^2 hybridization generally increases by increasing the temperature applied during the heating process, in which the maximum

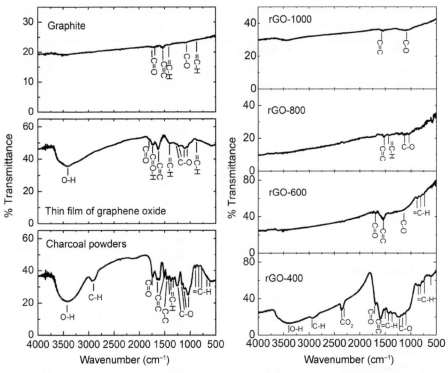

Figure 4.2: Fourier Transform Infrared (FTIR) transmittance spectra of charcoal powder, rGO-400, rGO-600, rGO-800, and rGO-1000 prepared from coconut shell, graphite (®Merck), and a thin film of graphene oxide (®XFNANO).

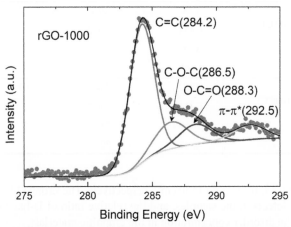

Figure 4.3: The C1s x-ray photoelectron spectroscopy (XPS) spectrum of rGO-1000.

Color version at the end of the book

percentage of ~ 77% was found in rGO-800. The results also confirmed that the synthesized rGO from coconut shell has comparable or even higher C=C component as well as rGO-c.

Figure 4.4a presents Raman spectra of rGO samples obtained by the calcination process of charcoal powder at various temperatures for 5 hours, followed by an exfoliation process for 10 hours. Two main bands of D- and G-band were observed in all samples. D band is attributed to either defects or to the breakdown of translational symmetry, and G band is related to the first-order scattering of the E_{2g} mode of graphitic sp^2 domain (Tuinstra and Koenig 1970, Fu et al. 2013). The D- and G-band

Table 4.2: Summary of XPS C1s data obtained from fitting calculations regarding the binding energy (BE) values, full width at half maximum (FWHM, in eV), percentage of each component, and their respective assignments.

Samples	Assignments	BE (eV)	FWHM	at%
Commercial rGO (rGO-c, ®Graphenea)	C=C	284.2	2	50.9
	C-OH	285.5	2.4	21.7
	C=O	287.8	3	16.6
	π-π*	289.9	2.4	8.3
		291.6	2	2.4
Charcoal powder (CP)	C=C	284.2	2.2	45.3
	C-OH	285.8	2	12.4
	C=O	287.4	2	7.2
	π-π*	289.6	3	25.8
		292.6	2.2	9.2
rGO-400 (CP heated at 400°C)	C=C	284.4	2.3	46.0
	C-O-C	286	3	12.4
	O-C=O	288.2	3	11.6
	π-π*	289.6	3	14.9
		292.1	3	15.1
rGO-600 (CP heated at 600°C)	C=C	284.3	2.1	75.2
	C-O-C	286.8	3	15.9
	O-C=O	289	2	5.8
	π-π*	292.7	2	3
rGO-800 (CP heated at 800°C)	C=C	284	2.2	77.1
	C-O-C	286.2	2	11.2
	C=O	287.9	2	7.3
	π-π*	292.7	2	4.3
rGO-1000 (CP heated at 1000°C)	C=C	284.2	2.2	62.6
	C-O-C	286.5	3	16.3
	O-C=O	288.3	3	12
	π-π*	292.5	3	9.1

appear at approximately 1350 cm^{-1} and 1550 cm^{-1}, respectively. The strong D-band suggests that there are great amounts of defects in the samples, and the relative ratio of D-band compared to G-band depends strongly on the disorder concentration in the graphitic materials.

The D/G intensity ratio (I_D/I_G) of the samples is displayed in Figure 4.4b. The I_D/I_G values were found to be greater than 1 for all samples. The ratio, in general, increases with increasing temperature at varied holding temperatures. This implies that the heating process enhances the concentration of defects and disorder presented in the samples. The in-plane graphitic crystal size (La) was then calculated using equation (1). Depending on I_D/I_G values, the La varies, ranging from 80–50 nm. The increase in the I_D/I_G value reduces the in-plane graphitic crystal size. This indicates that the graphitic structures with sp^2 hybridization increase, but their domain size is reduced by the heating process.

Defects presented in the graphene-based materials, e.g., rGO, GO, can be in the forms of vacancies, adatoms, mixed sp^2/sp^3 hybridization, zigzag-type edges, local topology perturbations, Stone-Wales defects, pentagonal-octagon pairs, and substitution with non-carbon atoms in the lattice (Tuček et al. 2018). Thus, it is important to clarify types of defects presented in the samples, in particular, to study its fundamental physics.

Microstructure analyses on the obtained powder using SEM and TEM indicate the presence of flake-like particles. Figure 4.5 displays the SEM and TEM images of rGO-400, showing stacked layers. The flake particles likely consist of many layers of graphitic structure. They are entangled with each other. It has been reported that rGO can easily suffer agglomeration and restacking due to the Van der Waals interactions between neighboring sheets (Wu et al. 2010). Mechanical exfoliation in acidic medium has been suggested to effectively reduce the thickness of rGO layers prepared from coconut shell (Nugraheni et al. 2017). However, types of defects presented in the samples could not

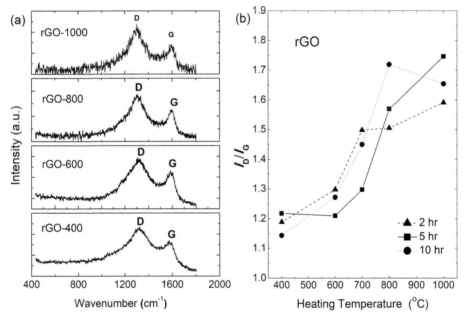

Figure 4.4: (a) Raman spectra of rGO-400, rGO-600, rGO-800, and rGO-1000 prepared from coconut shell by thermal reduction process for 5 hours, followed by sonic exfoliation for 10 hours. The D- and G-band are indicated as D and G, respectively. (b) The D/G intensity ratio (I_D/I_G) of the samples as a function of heating temperature with holding times of 2, 5, and 10 hours.

Figure 4.5: (a) SEM image and (b) TEM image of rGO-400 prepared from coconut shell by thermal reduction process for 5 hours, followed by sonic exfoliation in distilled water for 10 hours.

be clarified so far from only obtained microstructure images. The presence of single- or few-layer rGO sheets was not clearly confirmed in the current TEM image. Therefore, further analyses using high-resolution TEM (HR-TEM), scanning TEM (STEM), field-emission SEM (FESEM), atomic force microscopy (AFM), and Scanning Tunneling Microscope (STM) are required to examine types of defects in the samples and to precisely clarify structures and dimension of the samples.

3.2 Carbon compounds synthesized from coconut sap

In the case of carbon-based materials prepared from coconut sap, XRD analysis signifies the presence of amorphous carbon (a-C), as displayed in Figure 4.6a. A broad peak with less intensity at ~ 20° is a characteristic of a-C. Impurity phase of *sylvite* (KCl) salt was also detected in the XRD pattern at $2\theta \approx 28°$ and 40°. This salt easily dissolves in water; thus, it can be removed by washing the sample in distilled water. The peaks of KCl disappear in XRD pattern of the sample through washing process. No clear hump at ~ 43° was observed in the XRD pattern of a-C, unlike disordered rGO. As shown in Figure 4.6b, FTIR spectrum of the obtained sample confirmed the existence of oxygen functionalities of C-O at 1045, 1080, and 1217 cm^{-1} and O-H with a sharp transmittance peak at 3408 cm^{-1}. Stretching vibrations from C=C were observed at 1587 cm^{-1}. Two types of C-H stretching vibrations were seen at ~ 1400 cm^{-1} and ~ 2900 cm^{-1}. The former is a symmetric stretching, while the latter is an asymmetric one. At smaller wavenumbers, 675–995 cm^{-1}, bending vibrations out of plane from alkene functional group, =C-H, appear. The resultant a-C was then used to make a-C thick film on ITO substrate.

a-C is one of the carbon allotropes that has mixed hybridizations, which are commonly sp^2 and sp^3. This leads a-C to have a broad electrical property in between conducting (as graphite with pure sp^2 hybridization) and insulating (as diamond with pure sp^3 hybridization). a-C can be classified into many types depending on the concentration of those two hybridization states, which hence would determine their properties. The higher the concentration of sp^2 hybridization, the more a-C behaves like graphite. On the other hand, a-C will have characteristics more like diamond when it contains higher concentration of sp^3 hybridization. The presence of mixed hybridization of sp^2 and sp^3 was confirmed in the synthesized a-C by XPS at C1s spectrum, as reported by Diaz et al. (1996).

Figure 4.7 displays C1s XPS spectrum of the a-C-600 thick film. The deconvolutions on the C1s XPS spectra of thick films prepared from coconut sap indicate the presence of three peaks at 284.37, 285.7, and 286.5 eV for a-C-400, and at 284.25, 285.2, and 286.55 eV for a-C-600, which are assigned as sp^2 C=C, sp^3 C-C, and C-O, respectively. Relative ratios of sp^2 and (sp^2+sp^3) hybridizations were estimated to be 75% for a-C-400 and 80% for a-C-600.

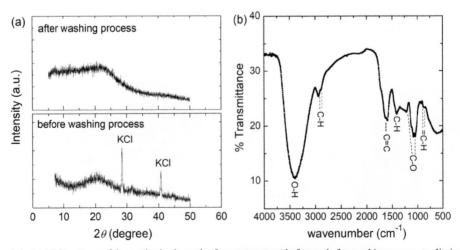

Figure 4.6: (a) XRD patterns of the synthesized samples from coconut sap before and after washing process to eliminate the impurity phase of KCl. (b) FTIR transmittance spectrum of the obtained sample.

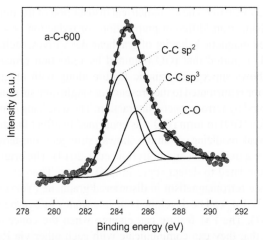

Figure 4.7: The C1s x-ray photoelectron spectroscopy (XPS) spectrum of the a-C-600 thick film.

Figure 4.8: SEM image of a-C powder prepared from coconut sap.

Figure 4.8 shows a SEM image of a-C prepared from coconut sap. Agglomerated particles with a poorly porous structure were observed. Energy-dispersive x-ray (EDX) analysis on the SEM images of a-C powders shows that the elemental compositions of a-C consist of carbon (51.9%), oxygen (42.7%), chlorine/Cl (1.3%), and other primary metals of Na (0.4%), K (3.3%), and S (0.4%).

Carbon-based compounds in the form of disordered multilayered rGO and amorphous carbon (a-C) have been successfully synthesized from coconut shell and coconut sap, respectively. XRD, XPS, FTIR, and Raman spectroscopies confirmed a single phase of rGO and a-C with different types of oxygen-containing functional groups and large amounts of defects. However, the obtained samples still have a thick layer due to agglomeration and restacking. An improvement in the exfoliation techniques is necessary to get a thinner single- or few-layered rGO sheets with a large surface area. Further characterizations are also required to clarify their structures and types of defects presented in the samples, and hence can enhance sample qualities to aim for a goal of large-scale production of graphene-based compounds with desired quality levels from biomass wastes and products.

4. Magnetic and Electrical Properties of Biomass-derived Carbon Compounds

While pristine graphene is diamagnetic in nature, rGO has been reported to exhibit varied magnetic properties, including superparamagnetic (Sarkar et al. 2014), paramagnetic (Sepioni et al. 2010, Sun et al. 2014), and even ferromagnetic at room temperature (Khurana et al. 2013, Qin et al. 2014,

Bagani et al. 2014, Sun et al. 2014). The asymmetrical distribution of the functional groups presented on graphene surface arising from different preparations or reduction processes is suggested to be responsible for the varied magnetic behavior of graphene derivatives, including rGO.

Sarkar et al. (2014) reported that rGO obtained by reduction process of graphite oxide by the chemical method shows superparamagnetic feature along with magnetic hysteresis at room temperature. This behavior is attributed to the presence of single domain composed of defect induced magnetic moments coupled by ferromagnetic interaction. The oxidizing conditions during annealing process of graphene oxide (GO) to form rGO was confirmed to affect the magnetization strength, in which rGO prepared in weak oxidizing conditions shows a stronger room-temperature ferromagnetism than that prepared in strong oxidation conditions (Qin et al. 2014). The origin of ferromagnetic (FM) feature is believed to be related to defect states in rGO sheets. Theoretical studies suggested that a single defect can induce ferromagnetism in disordered graphene (Yazyev 2008), and a magnetic moment of approximately 1 μ_B can be developed due to vacancy or hydrogen chemisorption defects (Yazyev and Helm 2007). The FM feature can emerge when the concentration of defects in the sample is large enough that they can communicate with each other via Ruderman-Kittel-Kasuya-Yosida (RKKY) or exchange interaction. Sepioni et al. (2010) notified that a small concentration of defects leads to paramagnetic state as the induced magnetic moments are well-separated. Moreover, transitions among those magnetic features, i.e., paramagnetic, diamagnetic, and ferromagnetic states, can be achieved and depend on various defects, such as edge states, vacancies, and functional groups (Sun et al. 2014).

The rGO prepared from coconut shells shows an increase in the magnetization with the increase of defect concentration tuned by heating temperature during thermal-reduction process. A weak but appreciable magnetic hysteresis was confirmed in all samples—rGO-400, rGO-600, rGO-800, and rGO-1000 (Darminto et al. 2018). The saturation magnetization (M_S) increases with increasing heating temperature, as does the remanence magnetization (M_r) and coercivity (H_C). A significant enhancement of M_S was achieved when rGO was thermally reduced at 1000°C, i.e., rGO-1000 (Darminto et al. 2018).

Table 4.3 presents the values of M_S, M_r, and H_C of the samples. The M_S of rGO-1000 is approximately 35 times as large as those prepared at 400°C (rGO-400) and 600°C (rGO-600). Furthermore, elemental analysis using x-ray fluorescence (XRF) confirms that magnetic impurities are likely to be ruled out to explain a drastic increase of M_S in rGO-1000. No significant increase in magnetic impurities (i.e., Fe and Ni) was observed by the variation of heating temperature. The maximum concentration of 0.07% of magnetic impurities found in the samples will contribute to $\sim 0.3 \times 10^{-2}$ emu/g of magnetization. This indicates that the main contribution of the observed magnetization is due to intrinsic properties of the samples, and hence suggests that defects presented in the samples play a key role in the observed magnetization. This suggestion is consistent with the increment of defect concentration as indicated from Raman spectra, in which rGO-1000 has the highest ratio of I_D/I_G. The results further indicate defect induced magnetism in rGO, as it has been previously predicted and reported by Yazyev and Helm (2007), Yazyev (2008, 2010), and Khurana et al. (2013).

Table 4.3: The values of the saturation magnetization (M_s), the remanence magnetization (M_r), and coercivity (H_C) of the rGO samples determined from hysteresis curves.

Sample	M_S (10^{-2} emu/g)	M_r (10^{-2} emu/g)	H_C (Oe)
rGO-400	6	0.4	50
rGO-600	7	0.5	55
rGO-800	3	0.6	100
rGO-1000	220	29	140

A mechanism by which magnetic moments at the defect sites communicate is still under investigation. The types of defects presented in the rGO layer should be clarified first. In the case of vacancies, dangling bonds on them which have spin unit will contribute to the magnetism (Yazyev and Helm 2007). It has theoretically been predicted that magnetic moments in the presence of vacancies are due to the formation of localized states at Fermi level (E_F) because of the lattice distortions generated by the defect and consequent electron-electron interactions (Tuček et al. 2018). Depending on their positions in the graphene layer, vacancies can induce ferromagnetism and/or antiferromagnetism. If the presence of vacancies is the main reason, then the increase in magnetization with defects can be understood from the fact that positions on one sublattice are preferentially vacant than positions on other sublattices. However, in the existence of various defects, such as adatoms, mixed sp^2/sp^3 hybridization, and zig-zag type edges, the mechanism of magnetism in rGO could be more complicated. A comparative study on the magnetism of rGO prepared from coconut shells and that of commercial product (rGO-c, ®Graphenea) suggests that the former has larger M_S than the latter because of a higher amount of its functional groups and defect concentration (Asih et al. 2019).

Temperature dependence of magnetization, $\chi(T) = M/H\ (T)$, of rGO-1000 under applied field of 10 Oe shows the divergence between ZFC and FC curves at 300 K, which is speculated to be due to either the disordered spins/magnetic anisotropy or ferromagnetic clusters. A magnetization of about 40×10^{-5} emu/g was observed in rGO-1000 at 300 K (Darminto et al. 2018). In the case of commercial rGO (rGO-c), a small magnetization of approximately 10^{-5} emu/g can still be found at room temperature with a prominent divergence between ZFC and FC curves at 100 K (Asih et al. 2019). A possible transition of diamagnetic to paramagnetic states is indicated in both rGO-c and rGO prepared from coconut shell when the temperature is lowered (Retno et al. 2019, Darminto et al. 2018). Defect state and functional group presented in the samples strongly affect the magnetic properties of rGO.

Electrical conductivity of a-C thick film prepared from coconut sap was evaluated. The conductivity was found to be 3.7 and 4.3 $S.m^{-1}$ for a-C-400 and a-C-600, respectively. By analyzing the UV-Vis spectra using Tauc plot method (Tauc 1968), energy gap (E_g) of the a-C films was determined to be approximately 1.1 eV. The values of the estimated conductivity and energy gap are in the range of those of semiconducting materials. The E_g of a-C can be varied and tuned depending on the concentration of sp^2 and sp^3 hybridization. With such advantage, a-C can be an alternative of semiconducting materials, which can be largely produced from biomass with relatively low cost.

5. Potential Applications of Biomass-derived Carbon Compounds

The applications of the obtained carbon compounds for microwave absorber, supercapacitor, and their photoluminescence (PL) properties are briefly evaluated. We show the frequency dependence of the reflection loss (RL) on the obtained rGO. A broad bandwidth covering almost all X-band frequency was observed. The specific capacitance of rGO-based electrodes is discussed based on voltammetry curves. PL properties of rGO solutions with various concentrations were also examined.

5.1 Application as microwave absorbing materials

Reduced graphene oxide (rGO) is a candidate for microwave absorber and high efficiency electromagnetic interference (EMI) shielding because of its high specific surface areas and carrier mobilities, coupled with abundant defects and hydroxyl, epoxy, and carboxyl groups. Moreover, the carrier hopping of carbon materials including rGO can be thermally activated; thus rGO has stable or even enhanced permittivity which can be used as microwave absorber and EMI shielding in harsh thermal environments (Wen et al. 2014).

For practical application as microwave absorbing materials, generally, rGO must possess specific desired properties, including wide bandwidth absorption and good thermal stability. Moreover, rGO has excellent electrical conductivity and high dielectric constant, and hence, results in a remarkable conductivity loss during the microwave absorbing process. These properties depend strongly on dielectric and magnetic loss characteristics. rGO has advantages in term of large interfacial polarization for the better absorption of the microwave. This is related to two relaxations of electronic dipole polarization resulting from functional groups and defect polarization generating from oxygen-containing functional groups combined with the presence of some defects incorporated by chemical oxidation and reduction methods, respectively.

In order to overcome the relatively low reflection loss in the microwave absorbing process, it is important to design the graphene-based compounds through the incorporation of other materials with specific absorption properties or by preparing the graphene materials with other structures to enhance the absorbing performance. One of them is the so-called graphene-based magnetic metal/metal oxide nanocomposites. These combinations could make novel electromagnetic properties and relatively high microwave absorbing behavior.

Residual defects and functional groups in rGO can create the synergistic effect of polarization loss and conduction loss. Therefore, the rGO can be used as a component of composites, which can be decorated with magnetic components to possess excellent microwave absorption (Wen et al. 2014, Ding et al. 2016, Zhang et al. 2018). Assembling the porous Co_3O_4 on rGO has been reported to increase the RL, reaching −61 dB with the broadened bandwidth, almost covering the whole investigated frequency (2–18 GHz) at elevated temperatures of 353–473 K (Ma et al. 2018). Zhang et al. (2018) reported that the selective-frequency microwave absorption can be achieved by implanting small $NiFe_2O_4$ clusters on rGO. Absorption peak shifts from 4.6 to 16 GHz by the gradual decrease in the concentration of $NiFe_2O_4$ clusters, covering almost 72% of the measured frequencies. This indicates that $NiFe_2O_4$-rGO can be an alternative candidate for multipurpose microwave absorption, ranging from commercial use to military and aerospace technologies.

Both strong attenuation and good impedance matching are required conditions to get an ideal microwave absorber. When the ratio of absorbing material in the protective layer is not enough, the attenuation capacity will weaken, causing penetrating waves which cannot be completely dissipated, and will be reflected by metal surface, and thus detected by radar. On the contrary, when the ratio of absorbing material in the protective layer is too high, it will lead to the emergence of strong signals of reflected waves that can be detected by radar as well. One of the solutions to get a light microwave absorber with high efficiency and broad bandwidth is by fabricating multilayered absorbing materials. This multilayered design is considered an effective way to achieve microwave absorbing materials with high efficiency, broad bandwidth, and light weight. Moreover, graphene hybrids also offer a great potential as microwave absorbing because of the synergy in their dielectric loss and magnetic loss induced by hybridization (Cao et al. 2018).

Here, we evaluated the RL of the pure rGO without any decorated magnetic components. Figure 4.9 shows the microwave absorption properties of the obtained rGO. A broad absorption peak was observed at a frequency of ~ 10.5 GHz, which indicates that the bandwidth is broadened covering all X-band frequencies (8–12 GHz). The maximum RL was found in rGO prepared without further thermal reduction and exfoliation processes, which is approximately −11 dB at 10.5 GHz. The RL decreases with increasing heating temperature applied during the reduction process. rGO-400 and rGO-700 have the RL of −8.4 dB and −5.7 dB at ~ 10.5 GHz, respectively.

Functional groups on rGO layers play an important role in creating polarization. Therefore, the decrease in functional groups via the reduction process could effectively reduce the RL. Reduction and exfoliation processes also result in smaller sp^2 domain, and increase the concentration of defects, which could facilitate the dissipation of microwave, hence weakening the attenuation capacity of the absorber. The current results are still far from an ideal microwave absorbing material. Considerable efforts are required to fabricate microwave absorbers with strong attenuation and good impedance matching from biomass-derived carbon compounds.

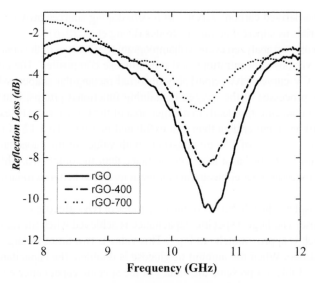

Figure 4.9: Reflection loss (RL) curves of the obtained rGO prepared by thermal reduction process at 400°C (rGO-400) and 600°C (rGO-600), followed by sonic exfoliation in HCl solution.

5.2 *Application as electrodes of supercapacitor*

Carbon-based supercapacitor is a promising alternative energy storage due to its high specific capacitance, high power density, and long cycle lifetime compared to conventional capacitors and batteries. Recently, one development of supercapacitor focuses mainly on the combination between graphene- and metal oxide-based nanocomposites as electrodes for achieving both high power and high density. This will lead to a significant development in eco-friendly energy storage materials (Jain et al. 2015a, Jayaraman et al. 2017). It is important to note that rGO must have some residual oxygen functional groups to introduce pseudocapacitance effect and enhance the energy storage capacity. Therefore, the presence of the residual oxygen groups on the graphene sheets is essential.

Reduce graphene oxide (rGO) is mostly used as an alternative to pristine graphene in graphene-based supercapacitors, because rGO can be produced at a large scale with low cost. However, a lower capacitance than expected is usually achieved when using rGO as an electrode material because rGO suffers from serious agglomeration and restacking due to the Van der Waals interactions between neighboring sheets (Wu et al. 2010). To address this disadvantage, a spacer material like glucose is usually introduced to inhibit the restacking of graphene sheets during calcination. Ma et al. (2014) reported that sugar-derived carbon/graphene composite as electrode supercapacitor exhibits a high specific capacitance with excellent electrochemical stability. Functionalized rGO with glucose and nickel oxide was found to have excellent cyclic stability and high specific capacitance, in which the capacitance is about five times higher than that of rGO, which has no glucose or nickel oxide (Tran et al. 2015).

High performance carbon-based materials for supercapacitors are usually achieved by preparing carbon materials with a high specific surface area, forming a special surface electronic structure and specific morphologies (Deng et al. 2016). High specific surface area can be obtained by post-activation treatment. Activated carbon (AC) prepared from coconut shell, which is added to $FeCl_3$ as a graphitic precursor and $ZnCl_2$ as an activating agent, posseses an extremely high specific surface area ($1847 \ m^2 \ g^{-1}$) and high performance in supercapacitor. The maximum specific capacitance achieved is $268 \ F \ g^{-1}$, which was measured under a scan rate of $1 \ A \ g^{-1}$ in 6 M KOH as an electrolyte (Sun et al. 2013). Composites based on carbon materials have also become an alternative to supercapacitors with high power density and high energy density. Besides pore structure, the existence of various elements in biomass-derived carbons serving as *in-situ* doping sources provides a bright future for carbon-based materials as supercapacitor electrodes (Deng et al. 2016).

In the current work, we analyzed cyclic voltammogram (CV) curves of the obtained rGO exfoliated in HCl solution at various concentrations and rGO/glucose composites. The curves are displayed in Figure 4.10. All CV curves have remarkably distorted-rectangular shapes, which can be related to the strong redox processes of the oxygen-containing functional groups. Ultrasonication in HCl solution with higher concentration leads to a larger area of the CV curve. rGO exfoliated in 10 wt% HCl 12 M (rGO-10) has a larger area than that exfoliated in 5 wt% HCl 12 M (rGO-5) and 1 wt% HCl 12 M (rGO-1). The areas of CV are consistent with values of the capacitance, as indicated in equation (4), and thus rGO-10 has a larger capacitance than the other two. This indicates that the rGO-10 electrode possesses a better electrical conductivity due to better exfoliation of rGO in a more acidic condition.

Figure 4.10b displays the CV curves of the electrodes of rGO/glucose composites with different ratios of rGO:glucose. The largest specific-capacitance is achieved when the ratio between rGO and glucose is 1:1, which is approximately 40 F g⁻¹. This value is two times as large as of that obtained at the rGO-10 electrode. When the amount of glucose is doubled, the capacitance decreases about 50%, to be 23 F g⁻¹. Table 4.4 presents the values of the specific capacitance and energy density of the electrodes, which are estimated from equations (4) and (5), respectively. The introduction of FeCl$_3$ as precursor and activating agent enhances the specific capacitance as well as the energy density.

The reason for lowering capacitance by further increase of glucose is still unclear. An inhomogeneity distribution of glucose could be a possible reason; thus, it might not optimally

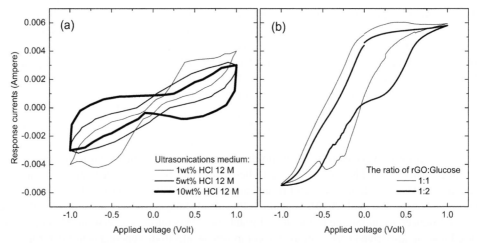

Figure 4.10: (a) CV curves of rGO exfoliated in HCl solution at various concentrations, which are recorded at 50 mV.s⁻¹ in 6 M KOH. (b) CV curves of rGO/glucose composites at different rGO:glucose ratio.

Table 4.4: The values of the specific capacitance and energy density of the rGO and rGO/glucose electrodes.

Electrode	Specific capacitance (F.g⁻¹)	Energy density (Wh. g⁻¹)
rGO-1	21	0.28
rGO-5	18	0.24
rGO-10	21	0.29
rGO/Glucose 1:1	40	0.54
rGO/Glucose 1:2	23	0.31
rGO/Glucose/FeCl$_3$ 1:1:1	84	1.14
rGO/Glucose/FeCl$_3$ 2:2:1	100	1.35

prevent the restacking of rGO and may not lead to the formation of layered and porous structures. Accordingly, a drastic enhancement of specific surface area is not achieved, and hence there is no increase in the specific capacitance. The obtained specific capacitance is much lower than that of the theoretical value of 550 F g^{-1} calculated for single-layer graphene (Xia et al. 2009). Further improvements on the preparation of rGO and fabrication of electrodes are required to significantly improve the specific capacitance.

5.3 *Photoluminescence properties*

Graphene oxide (GO) and reduced graphene oxide (rGO) exhibit interesting photoluminescence (PL) properties, varying from low-energy fluorescence in red to near-infrared wavelengths (Luo et al. 2009), to blue fluorescence with a narrow bandwidth (Pan et al. 2010). The variation in PL properties was related to defect states, which strongly depend on sample preparation and reduction methods. Studies on the evolution of the PL properties during gradual reduction of GO by Chien et al. (2012) suggested the tunable PL spectra of GO during gradual reduction are attributed to the variation of PL emission from two different types of electronically excited states due to the heterogeneous electronic states of GO and rGO with variable sp^2 and sp^3 hybridizations, in which PL emission shifts from the original yellow-red in GO to the blue PL in rGO.

Figure 4.11 shows the PL spectra of rGO-800 suspensions with various concentrations at the excited wavelength (λ_{ex}) of 343 nm. Emission peak of PL spectra was observed at 650–710 nm in all suspensions, which is in the range of red wavelength of visible light, without any shift in the spectra. PL emission intensity tends to increase by increasing concentration of the suspension. This increment signifies the increase in the number of electrons that are excited and/or emitted. The excited state is an unstable state, and electrons will return to the ground state, occupying the void that was originally abandoned (recombination or de-excitation process). In the recombination process, an amount of energy is used by electrons to experience vibrations before returning to the ground state, and consequently, electrons will be at energy levels within the energy gap. A radiative transition occurs in this process by emitting electromagnetic waves (emission spectrum). In this case, the increased emission intensity indicates that the electrons are in the highest state at the energy levels within the energy gap.

The blue shift in PL emission, however, was not observed in our obtained rGO, as illustrated in Figure 4.11. At the excited wavelength of 343 nm, all samples show red fluorescence with a peak around 680 nm. The highest intensity was found in rGO-1000, which has the largest concentration

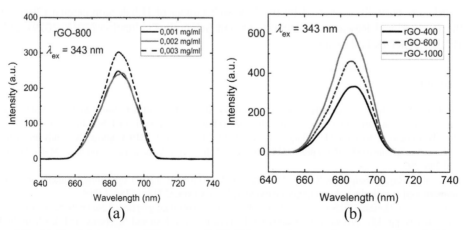

Figure 4.11: (a) Photoluminescence (PL) spectra of rGO-800 suspension at various concentrations (0.001 mg/mL, 0.002 mg/mL, and 0.003 mg/mL). (b) PL spectra of rGO suspension (0.003 mg/mL). The spectra were obtained at an excited wavelength of 343 nm.

of defects. Numerous disorder-induced defect states within the π-π* gap lead to the formation of PL spectrum centered at long wavelengths, i.e., in red wavelengths. The result indeed confirms that the PL characteristic of rGO varies depending on the preparation history, which influences heterogeneity of sp^2 and sp^3 hybridizations in the samples.

6. Conclusions

Biomass-derived carbon compounds have been successfully synthesized by a green synthesis method using biomass from coconut products, i.e., coconut shell and coconut sap. The synthesis procedure involves a thermal process. Homogeny charcoal powders from coconut shells were thermally reduced at various temperatures of 400°C, 600°C, 800°C, and 1000°C in the air for 5 hours, followed by mechanical exfoliation, which results disordered multilayers of reduced graphene oxide (rGO). Meanwhile, a lower temperature in the thermal process was used in preparing carbon-based materials from coconut sap, resulting in amorphous carbon (a-C). The a-C was further used to prepare a-C thick film on ITO substrate. The conductivity and energy gap of the a-C film were found to be 8.3 S.m^{-1} and 0.16 eV, respectively, which are in the range of those of semiconducting materials.

FTIR spectra confirmed that some functional groups disappear by increasing heating temperature, in which rGO-1000 contains less functional groups compared to other samples. On the other hand, the concentration of defects estimated from Raman spectra increases by thermal reduction. Microstructure analyses on the rGO powder indicate the formation of flake-like particles. Further investigations on magnetic properties suggest that various magnetic states, including diamagnetic, paramagnetic, and weak ferromagnetic, coexist in the obtained rGO. The magnetization increases by increasing defect concentration led by thermal reduction at higher temperature, which confirms defect induced magnetism in rGO. A preliminary study on the application of the obtained rGO as a radar absorbing material indicates that the reflection loss (RL) of the compound is relatively stable with a broad bandwidth, covering almost all *X*-band frequency. Furthermore, examinations on the capacitance and energy density of the rGO/glucose electrodes imply that rGO can be an alternative for graphene-based supercapacitors. Photoluminescence (PL) analysis also suggests the obtained rGO has a low-energy fluorescence in red wavelength (660–700 nm). Although it requires further improvements in the application, our works confirm that biomass-based carbon materials can be potential candidates for radar absorbing materials and supercapacitors.

Acknowledgment

This work was partially supported by the Research Grants of Hibah Kompetensi (2016–2018) and the International Research Collaborations and Scientific Publication (2017–2019), DRPM, KEMENRISTEKDIKTI, Indonesia.

References

Aghamir, F.M., A.R. Momen-Baghdadabad and M. Etminan. 2019. Effects of deposition angle on synthesis of the amorphous carbon nitride thin films prepared by plasma focus device. Appl. Surf. Sci. 463: 141–149.

Asih, R., E.B. Yutomo, D. Ristiani, M.A. Baqiya, T. Kawamata, M. Kato et al. 2019. Comparative study on magnetism of reduced graphene oxide (rGO) prepared from coconut shells and the commercial product. Mater. Sci. Forum 966: 290–295.

Bagani, K., M.K. Ray, B. Satpati, N.R. Ray, M. Sardar and S. Banerjee. 2014. Contrasting magnetic properties of thermally and chemically reduced graphene oxide. J. Phys. Chem. C 118: 13254–14259.

Barpuzary, D. and M. Qureshi. 2015. Graphene filled polymers in photovoltaic, graphene-based polymer nanocomposites in electronics. pp. 157–191. *In*: K.K. Sadasivuni, D. Ponnamma, J. Kim and S. Thomas (eds.). Cham: Springer International Publishing.

Cançado, L.G., K. Takai, T. Enoki, M. Endo, Y.A. Kim, H. Mizusaki et al. 2006. General equation for the determination of the crystallite size La of nanographite by Raman spectroscopy. Appl. Phys. Lett. 88: 163106.

Cao, M., C. Han, X. Wang, M. Zhang, Y. Zhang, J. Shu et al. 2018. Graphene nanohybrids: excellent electromagnetic properties for the absorbing and shielding of electromagnetic waves. J. Mater. Chem. C 6: 4586–4602.

Chien, C.T., S.S. Li, W.J. Lai, Y.C. Yeh, H.A. Chen, I.S. Chen et al. 2012. Tunable photoluminescence from graphene oxide. Angew. Chem. Int. Ed. 51: 6662–6666.

Coleman, J.N., U. Khan, W.J. Blau and Y.K. Gun'ko. 2006. Small but strong: A review of the mechanical properties of carbon nanotube-polymer composites. Carbon 44: 1624–1652.

Czerniak-Reczulska, M., A. Niedzielska and A. Jędrzejczak. 2015. Graphene as a material for solar cells applications. Advances in Materials Sciences 15: 67.

Darminto, R. Asih, Kurniasari, M.A. Baqiya, S. Mustofa, Suasmoro, T. Kawamata et al. 2018. Enhanced magnetism by temperature induced defects in reduced graphene oxide prepared from coconut shells. IEEE Trans. Mag. 54: 1–5.

Daud, W.M.A.W. and W.S.W. Ali. 2004. Comparison on pore development of activated carbon produced from palm shell and coconut shell. Bioresour. Technol. 93: 63–69.

Deng, J., M. Li and Y. Wang. 2016. Biomass-derived carbon: synthesis and applications in energy storage and conversion. Green Chem. 18: 4824.

Diaz, J., G. Paolicelli, S. Ferrer and F. Comin. 1996. Separation of the sp^2 and sp^3 components in the C1s photoemission spectra of amourphous carbon films. Phys. Rev. B 54: 8064–8069.

Ding, Y., Q.L. Liao, S. Liu, H.J. Guo, Y.H. Sun, G.J. Zhang et al. 2016. Reduced graphene oxide functionalized with cobalt ferrite nanocomposites for enhanced efficient and lightweight electromagnetic wave absorption. Sci. Rep. 6: 32381.

Drewniak, S., R. Muzyka, A. Stolarczyk, T. Pustelny, M. Kotyczka-Morańska and M. Setkiewicz. 2016. Studies of reduced graphene oxide and graphite oxide in the aspect of their possible application in gas sensors. Sensors 16(1): E103.

Ferrari, A.C. and J. Robertson. 2000. Interpretation of Raman spectra of disordered and amorphous carbon. Phys. Rev. B 61: 14095.

Ferrari, A.C. and D.M. Basko. 2013. Raman spectroscopy as a versatile tool for studying the properties of graphene. Nature Nanotechnology 8: 235–246.

Frackowiak, F. and F. Beguin. 2002. Electrochemical storage of energy in carbon nanotubes and nanostructured carbons. Carbon 40: 1775–1787.

Fu, C., G. Zhao, H. Zhang and S. Li. 2013. Evaluation and characterization of reduced graphene oxide nanosheets as anode materials for lithium-ion batteries. Int. J. Electrochem. Sci. 8: 6269–6280.

Goodenough, J.B. and K.-S. Park. 2013. The Li-ion rechargeable battery: a perspective. J. Am. Chem. Soc. 135: 1167–1176.

Hendaryati, D.D. and Y. Arianto. 2017. Tree crop estate statistics of Indonesia 2015–2017. Directorate General of Estate Crops: 1–98.

Hoekman, S.K., A. Broch and C. Robbins. 2011. Hydrothermal carbonization (HTC) of lignocellulosic biomass. Energy Fuels 25: 1802–1810.

Jain, A., V. Aravindan, S. Jayaraman, P.S. Kumar, R. Balasubramanian, S. Ramakrishna et al. 2013. Activated carbons derived from coconut shells as high energy density cathode material for Li-ion capacitors. Sci. Reports 3: 3002.

Jain, A., C. Xu, S. Jayaraman, R. Balasubramanian, J.Y. Lee and M.P. Srinivasan. 2015a. Mesoporous activated carbons with enhanced porosity by optimal hydrothermal pre-treatment of biomass for supercapacitor applications. Microporous and Mesoporous Materials 218: 55–61.

Jain, A., R. Balasubramanian and M.P. Srinivasan. 2015b. Production of high surface area mesoporous activated carbons from waste biomass using hydrogen peroxide-mediated hydrothermal treatment for adsorption applications. Chemical Engineering Journal 273: 622–629.

Jayaraman, S., A. Jain, M. Ulaganathan, E. Edison, M.P. Srinivasan, R. Balasubramanian et al. 2017. Li-ion vs. Na-ion capacitors: A performance evaluation with coconut shell derived mesoporous carbon and natural plant based hard carbon. Chemical Engineering Journal 316: 506–513.

Jung, S., Y. Myung, B.N. Kim, I.G. Kim, I.-K. You and T. Kim. 2018. Activated biomass-derived graphene-based carbons for supercapacitors with high energy and power density. Sci. Reports 8: 1915.

Kalyani, P., A. Anitha and A. Darchen. 2013. Activated carbon from grass—a green alternative catalyst support for water electrolysis. Int. J. Hydrogen Energy 38: 10364–10372.

Khurana, G., N. Kumar, R.K. Kotnala, T. Nautiyal and R.S. Katiyar. 2013. Temperature tuned defect induced magnetism in reduced graphene oxide. Nanoscale 5: 3346–3351.

Kim, I.-D., S.-J. Choi and H.-J. Cho. 2015. Graphene-based composite materials for chemical sensor application, electrospinning for high performance sensors. pp. 65–101. In: A. Macagnano, E. Zampetti and E. Kny (eds.). Cham: Springer International Publishing.

Kim, J.M., W.G. Hong, S.M. Lee, S.J. Chang, Y. Jun, B.H. Kim et al. 2014. Energy storage of thermally reduced graphene oxide. International Journal of Hydrogen Energy 39: 3799–3804.

Liu, S., L.P. Abrahamson and G.M. Scott. 2012. Biorefinery: Ensuring biomass as a sustainable renewable source of chemicals, materials, and energy. Biomass and Bioenergy 39: 1–4.

Liu, Y., J. Chen, B. Cui, P. Yin and C. Zhang. 2018. Design and preparation of biomass-derived carbon materials for supercapacitors: A review. J. of Carbon Research 4: 53.

Liu, Z. and F.-S. Zhang. 2011. Removal of copper(II) and phenol from aqueous solution using porous carbons derived from hydrothermal chars. Desalination 267: 101–106.

Lu, H. and X.S. Zhao. 2017. Biomass-derived carbon electrode materials for supercapacitors. Sustainable Energy Fuels 1: 1265–1281.

Luo, Z.T., P.M. Vora, E.J. Mele, A.T.C. Johnson and J.M. Kikkawa. 2009. Photoluminescence and band gap modulation in graphene oxide. Appl. Phys. Lett. 94: 111909.

Ma, J., T. Xue and Xue Qin. 2014. Sugar-derived carbon/graphene composite materials as electrodes for supercapacitors. Electrochimica Acta 115: 566–572.

Ma, J., X. Wang, W. Cao, C. Han, H. Yang, J. Yuan et al. 2018. A facile fabrication and highly tunable microwave absorption of 3D flower-like Co_3O_4-rGO hybrid-architectures. Chemical Engineering Journal 339: 487–498.

Mangolini, F., F. Rose, J. Hilbert and R.W. Carpick. 2013. Thermally induced evolution of hydrogenated amorphous carbon. Appl. Phys. Lett. 103: 161605.

Marcinauskas, L., V. Dovydaitis, A. Iljinas and M. Andrulevičius. 2019. Structural and optical properties of doped amorphous carbon films deposited by magnetron sputtering. Thin Solid Films 681: 15–22.

Mansour, A. 2011. Structural analysis of planar sp^3 and sp^2 films: Diamond-like carbon and graphene overlayers. Thesis. King Abdullah University of Science and Technology. http://hdl.handle.net/10754/205813.

Marton, M., M. Vojs, E. Zdravecká, M. Himmerlich, T. Haensel, S. Krischok et al. 2012. Raman spectrscopy of amorphous carbon prepared by pulsed arc discharge in various gas mixtures. J. of Spectroscopy 2013: 467079.

Milenov, T., A. Dikovska, G. Avdeev, I. Avramova, K. Kirilov, D. Karashanova et al. 2019. Pulsed laser deposition of thin carbon films on SiO_2/Si substrates. Appl. Surf. Sci. 480: 323–329.

Minitha, C.R., V.S. Anithaa, V. Subramaniam and R.T.R. Kumar. 2018. Impact of oxygen functional groups on reduced graphene oxide-based sensors for ammonia and toluene detection at room temperature. ACS Omega 3: 4105–4112.

Mishra, S.K., S.N. Tripathi, V. Choudhary and B.D. Gupta. 2014. SPR based fiber optic ammonia gas sensor utilizing nanocomposite film of PMMA/reduced graphene oxide prepared by *in situ* polymerization. Sensor and Actuators B 199: 190–200.

Mohagheghpour, E., M. Rajabi, R. Gholamipour, M. Lalarijani and S. Sheibani. 2016. Correlation study of structural, optical and electrical properties of amorphous carbon thin films prepared by ion beam sputtering deposition technique. Appl. Surf. Sci. 360: 52–58.

Mozammel, H.M., O. Masahiro and S.C. Bhattacharya. 2002. Activated charcoal from coconut shell using $ZnCl_2$ activation. Biomass and Bioenergy 22: 397–400.

Novoselov, K.S., A.K. Geim, S.V. Morozov, D. Jiang, Y. Zhang, S.V. Dubonos et al. 2004. Electric field effect in atomically thin carbon films. Science 306: 666–9.

Nugraheni, A.Y., D.N. Jayanti, Kurniasari, S. Soontaranon, E.G.R. Putra and Darminto. 2017. Structural analysis on reduced graphene oxide prepared from coconut shell by synchrotron x-ray scattering. IOP Conf. Ser.: Mater. Sci. Eng. 196: 012007.

Pan, D.Y., J.C. Zhang, Z. Li and M.H. Wu. 2010. Hydrothermal route for cutting graphene sheets into blue-luminescent graphene quantum dots. Adv. Mater. 22: 734.

Parikh, J., S.A. Channiwala and G.K. Ghosal. 2005. A correlation for calculating HHV from proximate analysis of solid fuels. Fuel 84: 487–494.

Park, S., J.O. Baker, M.E. Himmel, P.A. Parilla and D.K. Johnson. 2010. Cellulose crystallinity index: measurement techniques and their impact on interpreting cellulase performance. Biotechnology for Biofuels 3: 10.

Qin, S., X. Guo, Y. Cao, Z. Ni and Q. Xu. 2014. Strong ferromagnetism of reduced graphene oxide. Carbon 78: 559–565.

Rao, C.N.R., K. Biswas, K.S. Subrahmanyam and A. Govindaraj. 2009. Graphene, the new nanocarbon. J. Mater. Chem. 19: 2457–2469.

Rasul, S., A. Alazmi, K. Jaouen, M.N. Hedhili and P.M.F.J. Costa. 2017. Rational design of reduced graphene oxide for superior performance of supercapacitor electrodes. Carbon 111: 774–781.

Roberts, A.D., X. Li and H. Zhang. 2014. Porous carbon spheres and monoliths: morphology control, pore size tuning and their applications as Li-ion battery anode materials. Chem. Soc. Rev. 43: 4341–4356.

Robertson, J. 2001. Ultrathin carbon coatings for magnetic storage technology. Thin Solid Films 383: 81–88.

Sarkar, S.K., K.K. Raul, S.S. Pradhan, S. Basu and A. Nayak. 2014. Magnetic properties of graphite oxide and reduced graphene oxide. Phys. E Low-Dimesional Syst. Nanostruct. 64: 78–82.

Sennu, P., H.-J. Choi, S.-G. Baek, V. Aravindan and Y.-S. Lee. 2016. Tube-like carbon for Li-ion capacitors derived from the environmentally undesirable plant: *Propis juliflora*. Carbon 98: 58–66.

Sepioni, M., R.R. Nair, S. Rablen, J. Narayanan, F. Tuna, R. Winpenny et al. 2010. Limits on intrinsic magnetism in graphene. Phys. Rev. Lett. 105: 207205.

Sevilla, M. and A. Fuertes. 2009. The production of carbon materials by hydrothermal carbonization of cellulose. Carbon 47: 2281–2289.

Shams, S.S., L.S. Zhang, R. Hu, R. Zhang and J. Zhu. 2015. Synthesis of graphene from biomass: A green chemistry approach. Mat. Lett. 161: 476–479.

Singh, V., D. Joung, L. Zhai, S. Das, S.I. Khondaker and S. Seal. 2011. Graphene based materials: Past, present and future. Progress in Materials Science 56: 1178–1271.

Su, D.S. and R. Schlögl. 2010. Nanostructured carbon and carbon nanocomposites for electrochemical energy storage applications. ChemSusChem. 3: 136–168.

Sun, L., C.G. Tian, M.T. Li, X.Y. Meng, L. Wang, R.H. Wang et al. 2013. From coconut shell to porous graphene-like nanosheets for high-power supercapacitors. J. Mater. Chem. A 1: 6462–6570.

Sun, P.Z., K. Wang, J. Wei, M. Zhong, D. Wu and H. Zhu. 2014. Magnetic transitions in graphene derivatives. Nano Res. 7: 1507–1518.

Tauc, J. 1968. Optical properties and electronic structure of amorphous Ge and Si. Materials Research Bulletin 3: 37–46.

Tombros, N., C. Jozsa, M. Popinciuc, H.T. Jonkman and B.J. van Wees. 2007. Electronic spin transport and spin precession in single graphene layers at room temperature. Nature 448: 571–574.

Tran, M.H. and H.K. Jeong. 2015. One-pot synthesis of graphene/glucose/nickel oxide composite for the supercapacitor application. Electrochimica Acta 180: 679–689.

Tsai, W.T., M.K. Lee and Y.M. Chang. 2006. Fast pyrolysis of rice straw, sugarcane bagasse and coconut shell in an induction-heating reactor. J. Anal. Appl. Pyrol. 76: 230–237.

Tuček, J., P. Błoński, J. Ugolotti, A.K. Swain, T. Enoki and R. Zbořil. 2018. Emerging chemical strategies for imprinting magnetism in graphene and related 2D materials for spintronic and biomedical applications. Chem. Soc. Rev. 47: 3899–3990.

Tuinstra, F. and J.L. Koenig. 1970. Raman spectrum of graphite. J. Chem. Phys. 53: 1126.

Wang, H., Z. Li and D. Mitlin. 2014. Tailoring biomass-derived carbon nanoarchitectures for high-performance supercapacitors. ChemElectroChem. 1: 332–337.

Wang, L., K. Lee, Y.-Y. Sun, M. Lucking, Z. Chen, J.J. Chao et al. 2009. Graphene oxide as an ideal substrate for hydrogen storage. ACS Nano 3: 2995–3000.

Wen, B., M. Cao, M. Lu, W. Cao, H. Shi, J. Liu et al. 2014. Reduced graphene oxides: light-weight and high-efficiency electromagnetic interference shielding at elevated temperature. Adv. Mater. 26: 3484–3489.

Winter, M. and R.J. Brodd. 2004. What are batteries, fuel cells, and supercapacitors? Chem. Rev. 104: 4245–4270.

Wu, Z.S., D.E. Wang, W.C. Ren, J.P. Zhao and H.M. Cheng. 2010. Anchoring hydrous RuO_2 on graphene sheets for high-performance electrochemical capacitors. Adv. Funct. Mater. 20: 3595.

Xia, J., F. Chen, J. Li and N. Tao. 2009. Measurement of the quantum capacitance of graphene. Nat. Nanotechnol. 4: 505.

Yalcin, M. and A.I. Arol. 2002. Gold cyanide adsorption characteristics of activated carbon of non-coconut shell origin. Hydrometallurgy 63: 201–206.

Yavari, F. and N. Koratkar. 2012. Graphene-based chemical sensors. The Journal of Physical Chemistry Letters 3: 1746–1753.

Yazyev, O.V. and L. Helm. 2007. Defect-induced magnetism in graphene. Phys. Rev. B: Condens. Matter Mater. Phys. 75: 125408.

Yazyev, O.V. 2008. Magnetism in disorded graphene and irradiated graphite. Phys. Rev. Lett. 101: 037203.

Yazyev, O.V. 2010. Emergence of magnetism in graphene materials and nanostructures. Rep. Prog. Phys. 73: 056501.

Zhang, Y., X. Wang and M. Cao. 2018. Confinedly implanted $NiFe_2O_4$-rGO: Cluster tailoring and highly tunable electromagnetic properties for selective-frequency microwave absorption. Nano Research 11: 1426–1436.

Zhao, B., P. Liu, Y. Jiang, D. Pan, H. Tao, J. Song et al. 2012. Supercapacitor performance of thermally reduced graphene oxide. Journal of Power Source 198: 423–427.

CHAPTER 5

Bio-based Carbon Materials for Anaerobic Digestion

Kaijun Wang,[1,2] *Yasir Abbas*[1] and *Sining Yun*[1,*]

1. Introduction

The anaerobic digestion process, which is disposing of organic wastes biologically with the help of microbial community in the absence of oxygen, is an effective and widely used method to recover renewable biogas (methane), which may substitute fossil fuels for power generation (Huang et al. 2016, Li et al. 2018, Nowak et al. 2011, Yun et al. 2019, Zhang et al. 2017a). As depicted in Figure 5.1 (Lee et al. 2012, Paritosh et al. 2017), the anaerobic digestion process contains three phases, i.e., enzymatic hydrolysis, acid formation, and gas production, which take place through the mutual cooperation of four types of microorganisms: hydrolytic, fermentative, acidogenic, and methanogenic bacteria. The hydrolytic bacteria in the first phase play a role of degrading such macromolecules as carbohydrates, proteins, and fats to monomeric molecules, such as sugars, amino acids, and fatty acids. In the second phase, the fermentative bacteria are responsible for converting these monomers to organic acids, such as acetate, propionate, butyrate, and isobutyrate. Then, acidogenic bacteria transform fatty acids, other than acetate, into hydrogen, carbon dioxide, and acetate. In the last phase, methanogenesis happens, which is implemented by two types of methanogens, which produce methane via the fermentation of acetic acid and the reduction of carbon dioxide.

A schematic diagram of a representative experimental setup for the anaerobic digestion process is shown in Figure 5.2 (Yun et al. 2018, Zhang et al. 2017b). The digestion is performed in a glass reactor. The substrates and inoculum are added to the anaerobic reactor. After fully stirring by using a glass rod manually, the feeding inlet is sealed with a rubber bung, where a gas outlet is set up. The glass reactor is put into a digital thermostatic temperature-controlled chamber, and biogas volume is measured by the water displacement method.

Generally, crude biogas is mainly made up of methane (CH_4, 40–75%), carbon dioxide (CO_2, 15–60%), and trace amounts of other impurities, such as water (H_2O, 5–10%), hydrogen sulfide

[1] Functional Materials Laboratory (FML), School of Materials Science and Engineering, Xi'an University of Architecture and Technology, No. 13 Yanta Road, Xi'an, Shaanxi 710055, China.
[2] School of Science, Xi'an University of Architecture and Technology, No. 13 Yanta Road, Xi'an, Shaanxi 710055, China.
* Corresponding author: yunsining@xauat.edu.cn

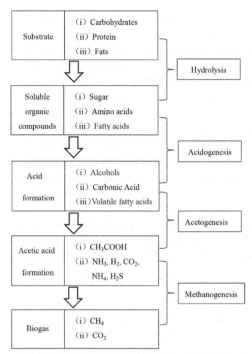

Figure 5.1: Schematic diagram of anaerobic digestion process. In this process, the anaerobic digestion phases take place through the synergistic action of four types of microorganisms: hydrolytic, fermentative, acidogenic, and methanogenic bacteria.

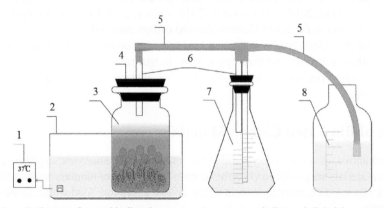

Figure 5.2: Schematic diagram of anaerobic digestion process: 1. temperature indicator, 2.digital thermostatic temperature-controlled chamber, 3. anaerobic reactor, 4. rubber bung, 5. rubber tube (gas tube), 6. glass tube, 7. gas collecting bottle, 8. water collecting bottle. Reproduced with permission from Yun et al. 2018, copyright 2018 the Elsevier.

(H_2S, 0.005–2%), ammonia (NH_3, < 1%), oxygen (O_2, 0–1%), carbon monoxide (CO, < 0.6%), and nitrogen (N_2, 0–2%). These impurities may prevent biogas from being utilized directly as vehicle fuel or injected to natural gas grid (Ryckebosch et al. 2011, Shen et al. 2017). Due to the instability of anaerobic micro-ecology, unstable process and low methane yield often happened in anaerobic digestion (Chen et al. 2008). The decrease in pH followed by the accumulation of volatile fatty acids can cause the breakdown of syntrophic association between microbial bacteria, which often results in negative effects in reactor performances (Shanmugam et al. 2018). In addition, ammonia (NH_3) and ammonium (NH_4^+) involved in the anaerobic digestion of organic substrate have a joint effect on the inhibition of acids (Lu et al. 2014).

From the perspective of practicality and environmental friendliness, bio-based carbon material, such as biochar and activated carbon, can be used as an effective additive for enhancing or stabilizing the anaerobic digestion process (Zhang et al. 2018). Bio-based carbon materials, prepared by carbonization or pyrolysis of biomass waste, are featured with many interesting characteristics (Schulz and Glaser 2012). For example, due to their porous structure, bio-based carbon materials could reduce ammonia inhibition and also immobilize methanogens (Luo et al. 2015). It has been demonstrated that bio-based carbon materials could be used for *in-situ* CO_2 sequestration with producing biogas of high CH_4 percentage in anaerobic digestion of sludge (Shen et al. 2016). Bio-based carbon materials have been proved to promote methane production and shorten lag phase in animal manure due to the reduction of inhibitive sulfide and organic loading impact during digester start-up (Aktas and Cecen 2007). Also, due to their conductivity, activated carbon and biochar could serve as "electron transport" to promote direct interspecies electron transfer among microbial communities (Zhao et al. 2016). In addition, when the biochar is added to the digester, the as-generated digestate, enriched with macro- and micro-nutrients, is beneficial for crop growth, which increases the fertilizer utilization of digestate (Shen et al. 2017). In general, these carbon materials derived from biomass waste as accelerants have been confirmed to be applied into anaerobic digestion systems to facilitate the improvement of fermentation environment, which will make the environment more harmonious and utilize energy more efficiently.

Recently, bio-based carbon materials have been widely concerned. In the United States, the market size of bio-based carbon (i.e., biochar) production was 3×10^5 tons in 2017, which is expected to reach 1.4×10^6 tons by 2023, with a growth rate of 24% (United States Biochar Market Report 2017). In addition, more than 80% of biochar is used as soil conditioner. At the present stage, the bio-based carbon materials will be primarily used as soil amendment and carbon-based fertilizer in the future for alleviating soil compaction (Deal et al. 2012), promoting the fertilizer utilization (Schulz and Glaser 2012), enhancing water retention, increasing ground temperature, and improving microbial environment (Jindo et al. 2012, Lehmann et al. 2011, Zhou et al. 2019). This is recent market trends prevailing with respect to the waste biomass-derived carbon materials.

This chapter provides a brief introduction to the source, preparation methods, characterization of bio-based carbon materials, and their applications in anaerobic digestion systems. Also, we give a summary and future perspective for bio-based carbon materials used as additives for anaerobic digestion systems.

2. Source of Bio-based Carbon Materials

As mentioned above, as a promising additive for improving the anaerobic digestion process, bio-based carbon materials are produced via carbonization or pyrolysis of biomass waste, which generally includes agriculture waste, municipal solid waste, animal excretion, food waste, forest residue, and industrial waste.

2.1 Renewable source

2.1.1 Agricultural waste

Agriculture countries, such as China, Pakistan, India, etc., produce agricultural waste in abundant quantity. The most common agricultural residues in these countries are the rice husk, wheat straw, cotton waste, maize waste, sugar cane, groundnut shell, fruit skin, and straw (Demirbas 2001). Traditionally, agricultural wastes are mainly used as feedstock in combustion processes. However, recently, the search of transformation of the agricultural waste by thermochemical conversion processes, producing renewable energy and high valued products, has been promoted significantly (Sun and Cheng 2002).

As an outstanding application, because of their availability and low price, agricultural wastes have proven to be promising raw materials for the production of carbon-based materials, such as biochar and activated carbons by pyrolysis and activation (Li et al. 2014, Rosas et al. 2010, Shen et al. 2015, Song et al. 2013, Yuan et al. 2011). Compared to feedstock, as its derived product, bio-based carbon materials are more environmentally friendly with a lower emission of greenhouse gases (Yuan et al. 2011). Furthermore, bio-based carbon materials, derived from agricultural residues which generally contain plenty of inorganic matter, mainly composed of potassium and calcium which remains on the surface of carbon materials (Rosas et al. 2010), are beneficial to soils. Interestingly, as one of the most effective elements in the chemical activation of carbonaceous resources, potassium (mainly as KOH or K_2CO_3) has been proved to beneficially produce carbons with very high surface areas, which could improve the performance of anaerobic digestion (Wang and Kaskel 2012).

2.1.2 Municipal solid waste

Municipal solid waste, commonly known as rubbish discarded by the humans, is a kind of waste including all solid or semi-solid materials disposed by citizens or businesses (Vergara and Tchobanoglous 2012). As a challenge all over the world, treatment and disposal of municipal solid waste have been concerned particularly. The conventional disposal methods of municipal solid waste, including landfill, combustion, and utilization as raw construction material or fertilizer, will more or less cause pollution to the environment (Chen et al. 2014).

As an economically and environmentally acceptable treatment, municipal solid waste may be used to obtain bio-based carbon materials as a main product by thermochemical conversion process. For example, owing to containing hydrocarbons and inorganic matter, municipal solid waste has been proved to be a good raw material in pyrolysis, which is a cost-effective and clean technology (Luque et al. 2012, Nipattummakul et al. 2010, Pedroza et al. 2014). During pyrolysis process, under the inert gas flow, the municipal solid waste is heated to a high temperature. After a short time, syngas, bio-oil, and biochar are generated, and the biochar, as the solid by-product of pyrolysis, can be obtained conveniently.

2.1.3 Animal excretion

With speedy expansion of livestock farms, the amount of animal excretion is increasing in China (Li et al. 2018). For example, the National Bureau of Statistics of People's Republic of China estimated that more than 100 million swine were raised, producing over 200 million tons of excretion in 2014 (Wu et al. 2017). Over the past decades, livestock farms of many countries are developing towards a large-scale industry of concentrated animal feeding operations. This fewer but larger model dramatically increases the number of animals on each farm as well as the amount of excretion, and has drawn attention to animal excretion recycling and disposal (Wang et al. 2018a).

As traditional excretion management practices, animal excretion could be used as fertilizer for crops. Unfortunately, due to large-scale centralization of animal farms, excretion production now far exceeds the nutritional demands of local crops, the excessive application of which to land causes both air and water pollution (Ro et al. 2007). In addition, arbitrary release of estrogens and pharmaceutically active ingredients may damage the ecosystems and the environment (Soupir et al. 2006).

This situation is likely to be mitigated by thermochemical conversion process synergistically (Ro et al. 2010), which can convert the excess excretion into products with energy added value, including gases, bio-oils, and bio-based carbon materials. For example, gasification and pyrolysis, only taking several minutes to hours, can also quickly destroy pathogens, and the final product composition is dependent on heating rate, reactor temperature, and residence time. In fact, as low cost, high carbon, and abundant residuals, animal excretion could be converted into biochar used as additives to enhance the performance of anaerobic digestion process, which can reduce the adverse environmental effects (Sohi et al. 2010).

2.1.4 Food waste

Every year, one-third of the worldwide food production originated from the whole food supply chain is wasted, causing significant environmental and economic problems (Browne and Murphy 2013). For example, in China, more than 90 million tons of food waste needs to be disposed of annually (Zhang et al. 2014), which is increasing as the population continues to grow and urbanization accelerates. As compared to the production and processing stages, the food waste generated from the consumer stage has the higher percentages of dairy and meat and the lower recycling rate, which may lead to a higher protein content (Li et al. 2013).

For the traditional disposing method, composting food waste causes emissions of greenhouse gases into the atmosphere, and landfills cause anaerobic decomposition of food waste that releases methane (Paritosh et al. 2017). As a more sustainable treatment, bio-based carbon materials fabricated from food waste have shown promising applications. Many researchers (Gupta et al. 2018, Kaushik et al. 2014, Saqib et al. 2018) have used hydrothermal carbonization or pyrolysis method to turn food waste into value added products, including hydrochar or pyrolytic carbon (solid product), bio-oil, and gases. So food waste is a qualified raw material for the synthesis of bio-based carbon materials.

2.1.5 Forest residue

As the rich biomass resource, forestry residue, including kinds of the unused forest biomass, almost counts for half of the total forest biomass, generally which is left in the forest to decompose (Kaygusuz 1997), and either used as boiler fuel or burnt in uncontrolled condition. Obviously, the biomass resource utilization efficiency is low and the environmental pollution will follow. Therefore, we urgently need to develop effective and clean methods to deal with forest residues (Chen et al. 2008).

As mentioned above, wood waste consisting of cellulose, lignin, and hemicellulose, is the main component of forest residue, which can be used to produce bio-based carbon materials by pyrolysis or activation process (Li et al. 2014, Salema and Ani 2011). A typical analysis of dry wood waste includes carbon (52%), hydrogen (6.3%), oxygen (40.5%), and nitrogen (0.4%) (Demirbas 2001). According to the reaction conditions, it is possible to produce carbon fibers, carbon particles, carbon nanoparticles, carbon nanofibers, and carbon nanotubes in various configurations (Omoriyekomwan et al. 2017).

2.2 Nonrenewable source

2.2.1 Industrial waste

Industrial waste, consisting of discarded materials during a manufacturing process of factories, industries, and mills, is the waste produced by industrial activity. Pollutants from unmanaged industrial wastes can damage the environment, and methane, which is released from rubbish dumps, also contributes to global climate change. It is estimated that the BOD_5 (5-day biochemical oxygen demand) of raw wastes from all industries is more than three times that of all urban sewers (Ghosh et al. 1985).

The BOD indicates the content of carbon organics. This means that the carbon content of industrial wastes is also much higher and it could be a more qualified source of bio-based carbon materials than sewage sludge. For example, the Kraft lignin, coming from the Kraft pulping process, which is characterized by low cost, high carbon content (60 wt%), and phenolic content, has been used to yield activated carbon by chemical activation (Kijima et al. 2011, Li et al. 2014). Similarly, the tires, as extremely complex items, containing carbon black up to 30 wt%, have been used to make biochar by microwave pyrolysis (Undri et al. 2013).

3. Methods used to Prepare Bio-based Carbon Materials

As we have already mentioned, bio-based carbon materials derived from biomass waste could be prepared by thermochemical conversion process. There are two simple and cost-effective thermochemical conversion methods: (1) hydrothermal carbonization, and (2) microwave pyrolysis.

3.1 Hydrothermal carbonization

Hydrothermal carbonization (HTC), as a sustainable technology which requires low energy consumption and can utilize waste biomass as raw materials, is a thermochemical conversion process to yield carbonaceous solid product (Funke and Ziegler 2010). The hydrothermal carbonization lowers the molecular ratio (O/C and H/C) of the feed by dehydration and decarboxylation (Funke and Ziegler 2010). The hydrothermal carbonization is achieved by implementing a suspension of biomass and water in temperature range of 150–350°C under autogenous pressure for several hours, and the detailed hydrothermal carbonization process is shown in Figure 5.3 (Bai et al. 2017).

The raw materials are crushed into powders, and then screened by the sieve to make the particle size uniform. Subsequently, the powders are mixed with deionized water and stirred for several hours. After that, the mixture is transferred into a stainless steel autoclave, which is heated to 150–350°C, and kept for several hours under autogenous pressure. Finally, by filtering, washing, and drying the resulting solid materials, hydrothermal carbonization chars are obtained (Wang et al. 2018b).

Generally, the hydrothermal carbonization has been used in the preparation of functional carbonaceous materials. Theoretically, any kind of biomass (cellulose, lignin, resin, peat, etc.) can be carbonized hydrothermally (Sevilla and Fuertes 2009, Mochidzuki et al. 2005, Titirici et al. 2007).

Many mechanisms that may occur during hydrothermal carbonization have been investigated. These chemical reactions include hydrolysis, dehydration, decarboxylation, condensation polymerization, and aromatization (Funke and Ziegler 2010), which do not happen step by step, but go through parallel paths. It is generally believed that the specific nature and importance of these reactions in hydrothermal carbonization will depend largely on the type of material used.

Hydrolytic reactions cause the disruption of mainly ester and ether bonds of the cellulose and lignin, the products of which include saccharides and phenolic fragments. Cellulose is hydrolyzed mainly under hydrothermal conditions higher than around 200°C (Peterson et al. 2008); due to its high amount of ether bonds, lignin is hydrolyzed significantly at approximately 200°C (Masselter et al. 1995).

Generally explained by elimination of hydroxyl groups, dehydration during hydrothermal carbonization involves both physical and chemical processes, removing water from the biomass molecules without changing its chemical structure (Funke and Ziegler 2010), which could lower the H/C and O/C ratios of biomass.

After the specific amount of water is obtained, significant decarboxylation occurs. Partial carboxyl and carbonyl groups are eliminated by hydrothermal treatment, which rapidly decompose above 150°C, yielding CO_2 and CO, respectively (Funke and Ziegler 2010). Due to the elimination

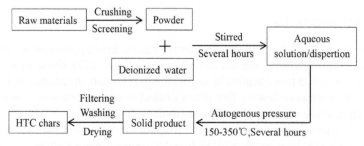

Figure 5.3: Schematic diagram of hydrothermal carbonization process.

of carboxyl and hydroxyl groups, easily polymerized unsaturated compounds are being generated. Therefore, it is understandable that the formation of hydrothermal carbon during the hydrothermal carbonization is mainly characterized by condensation polymerization. The aromatization is the aromaticity of carbon structures, which increases with the increase of reaction severity. The aromatic structure may be considered as a foundational structure of the prepared HTC-chars due to its high stability during the hydrothermal carbonization process.

The chemical elemental contents of the raw switchgrass and HTC biochar are shown in Table 5.1 (Regmi et al. 2012). As compared to feedstock, carbon percentage is much higher and inorganic content is comparatively lower in the HTC biochar. In addition, many micro-fissures (see the arrows in Figure 5.4) are generated in the biochar structures by hydrothermal carbonization process, as shown in scanning electron microscope (SEM) photography of cellulose and its biochar in Figure 5.4, which leads to higher surface areas (Kim et al. 2015).

Without using toxic chemicals, the hydrothermal carbonization is a low temperature and promising route, offering inexpensive, environment-friendly bio-based carbon materials (Wei et al. 2011). The aqueous solution or dispersion of organic materials is heated to temperatures as low as 150–350°C under autogeneous pressure. Generally, uniform chemical and structural properties are presented in the resulting carbon materials. In consequence, the hydrothermal carbonization process may have the opportunity to turn into a powerful technique used to synthesize valuable carbon materials from biomass waste.

Table 5.1: Composition of switchgrass and biochar derived from switchgrass via HTC. Data taken from Regmi et al. 2012.

Elements	C (wt%)	ash	Ca (mg kg^{-1})	K (mg kg^{-1})	Mg (mg kg^{-1})	P (mg kg^{-1})	Fe (mg kg^{-1})	Na (mg kg^{-1})
Switchgrass	44.6	1.3	2105	4082	4515	941	115	701
HTC biochar	70.5	3.2	2029	665	2215	481	258	395

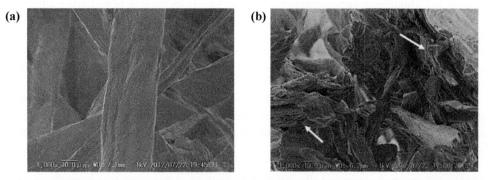

Figure 5.4: SEM photography of cellulose and its biochar; (a) Raw cellulose; (b) HTC biochar, scale bar: 10 μm. Reproduced with permission from Kim et al. 2015, copyright 2015 the Multidisciplinary Digital Publishing Institute.

3.2 Microwave pyrolysis

Due to its ability of quick and direct heating, microwave is considered a promising and practical heating technology for pyrolysis (Salema and Ani 2011, Yun et al. 2014, 2013). Compared to conventional heating methods, such as external heating by conduction, convection, or radiation (Chen et al. 2008), the integration of microwave heating that offers distinct heating mechanism can potentially and efficiently pyrolyze waste or biomass feedstocks (Budarin et al. 2011).

Schematic diagram of a typical experimental setup for the microwave pyrolysis (MP) process is as shown in Figure 5.5 (Salema and Ani 2011). In order to ensure an inert environment, nitrogen gas

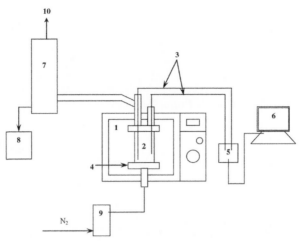

Figure 5.5: Schematic diagram of microwave pyrolysis system: (1) microwave cavity, (2) quartz glass fluidized bed reactor, (3) thermocouples, (4) distributor plate, (5) temperature data acquisition system, (6) personal computer, (7) condensing unit, (8) bio-oil collector, (9) rotameter, and (10) flue gas. Reproduced with permission from Salema and Ani 2011, copyright 2011 the Elsevier.

is supplied at a certain flow rate after the biomass is put in the fluidized bed quartz reactor. Nitrogen gas is continuously provided to maintain the inert environment while blowing out the vapor from the reactor during the experimental run. The inert gas also prevents explosion during the experiments. With this setup, in relatively short periods of time (typically minutes), the raw materials are heated to high temperature, and turned into various products including gases (e.g., CO, H_2, CO_2, CH_4), pyrolysis oil, and biochar.

During the microwave pyrolysis process, main advantages include that: (1) a wide range of materials can be used as a feedstock, and (2) selectable high-quality biochar, oil, or gas can be obtained according to the utilized conditions (Luque et al. 2012). Due to rapid and convenient start-up and shut-down causing its instantaneous heating, microwave heating offers a significantly better management of process and parameter for biomass pyrolysis, compared to conventional processes (Kappe 2004).

Unlike the conventional heating with temperature gradient from the hot surface to the interior, microwave heating offers the conversion of electromagnetic energy without differences in temperatures of the outer and inner surface. The quality of biochar produced under the microwave pyrolysis method has also been reported to be higher than that prepared by conventional heating (Salema and Ani 2011). As shown in Figure 5.6, significant cracks and fissures are found in conventional pyrolyzed oil palm shell char. On the contrary, uniform pores are observed in microwave pyrolyzed char. In addition, as compared to conventional pyrolysis processes, (1) the raw materials in microwave pyrolysis process require little pretreatment (Luque et al. 2012), which saves processing time and energy consumption, and results in cleaner products; (2) in the microwave pyrolysis process without oxygen, the formation of oxides is minimized under usual working conditions. It is clear that the most important advantage of the microwave pyrolysis method, as compared to conventional heating methods, is the significant reduction in temperature for pyrolysis reaction (minimum 200–380°C or less) (Zhao et al. 2011).

In general, the hydrothermal carbonization and microwave pyrolysis, as two convenient and effective techniques, have been widely used to prepare the bio-based carbon materials. As shown in Table 5.2, the biochar obtained by hydrothermal carbonization or microwave pyrolysis is usually produced at a low temperature. Therefore, the cost of biochar is low, while the surface area and pore volume are adequate for application. When the bio-based carbon materials are added in the anaerobic digestion process, the residual biochar can be directly left in digestates that can be used as soil fertilizer conveniently and efficiently (Shen et al. 2017).

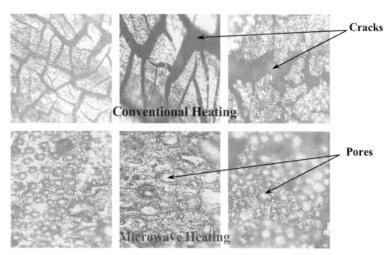

Figure 5.6: Surface SEM image of oil palm shell chars after conventional heating and microwave heating. Reproduced with permission from Salema and Ani 2011, copyright 2011 the Elsevier.

Table 5.2: Temperature conditions for different methods and properties of representative samples.

Thermochemical conversion process	Bio-based carbon material	Experimental temperature (°C)	BET surface area (m²g⁻¹)	Pore volume (cm³g⁻¹)	Reference
Hydrothermal carbonization	biochar	150–350	75.4	0.070	(Bai et al. 2017)
Microwave pyrolysis	biochar	200–380	38.06	0.096	(Antunes et al. 2017)

Table 5.3: Bio-based carbon materials derived from different wastes by thermochemical conversion process.

Raw material	Preparation method	BET surface area (m²g⁻¹)	Pore volume (cm³g⁻¹)	Reference
Wood sawdust	Hydrothermal + activation	2331	1.03	(Wei et al. 2011)
Food waste	hydrothermal carbonization	7.14	0.05	(Saqib et al. 2018)
Corn cob	Physical activation	980	0.52	(Song et al. 2013)
Orange skin	Physical activation	795	0.39	(Rosas et al. 2010)
Kraft lignin	Chemical activation	1479	0.69	(Li et al. 2014)
Municipal sludge	pyrolysis	67.60	0.099	(Chen et al. 2014)
Swine manure	pyrolysis	10.56	0.044	(Wang et al. 2018a)
Rice straw	pyrolysis	21.69	0.054	(Wang et al. 2018a)
Pine needle	pyrolysis	112.4	0.044	(Chen et al. 2008)
Corn cob	microwave pyrolysis	580	0.54	(Yun et al. 2018)
Waste carton	microwave pyrolysis	824	0.62	(Yun et al. 2018)
Walnut shell	microwave pyrolysis	818	0.64	(Yun et al. 2018)

Table 5.3 lists the synthesized bio-based carbon materials, which are derived from common biomass waste through different methods. Clearly, the thermochemical conversion (TCC) process is the common method used to prepare bio-based carbon materials. The bio-based carbon materials with high BET specific surface area and large pore volume, as accelerants in anaerobic digestion systems, may be beneficial to the enrichment of the anaerobic bacteria, which can provide a place of

growth and breeding for anaerobic bacteria, and responsible for improving the efficiency of anaerobic digestion systems (Li et al. 2018, Yun et al. 2018).

4. Characterization of Bio-based Carbon Materials

The morphologies and structures of bio-based carbon materials are generally characterized by using X-ray diffraction (XRD), Raman spectroscopy, X-ray photoelectron spectroscopy (XPS), Fourier transform infrared spectroscopy (FTIR), and scanning electron microscopy (SEM).

The structural parameters and features of the bio-based carbon materials can be obtained from the XRD measurements. The XRD patterns of different carbon materials are shown in Figure 5.7a (Yun et al. 2018) and Figure 5.7b (Wang et al. 2018b). For these bio-based carbons, the diffraction

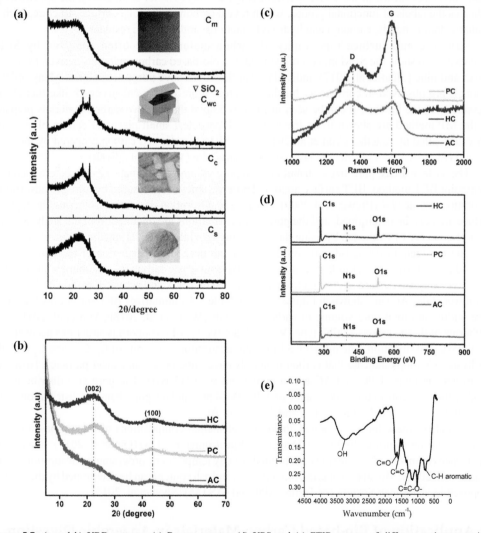

Figure 5.7: (a and b) XRD patterns, (c) Raman spectra, (d) XPS and (e) FTIR spectra of different carbon materials. C_m: mesoporous carbon; C_{wc}, C_c, and C_s derived from waste biomass resources such as waste carbon, corn cob, and sawdust, reproduced with permission from Yun et al. 2018, copyright 2018 the Elsevier. Hydrochar (HC), pyrolytic carbon (PC) and activated carbon (AC) derived from sunflower stalk, reproduced with permission from Wang et al. 2018b, copyright 2018 the American Chemical Society; (e) FTIR spectra of hydrothermal carbon, reproduced with permission from Titirici et al. 2007, copyright 2007 the American Chemical Society.

profiles exhibit two broad peaks at approximately 22° and 43°, which are assigned to the reflection from (002) and (100) planes, respectively. In addition, the XRD patterns reveal the presence of disordered carbon.

Raman spectroscopy and XPS of hydrochar (HC), pyrolytic carbon (PC), and activated carbon (AC) derived from the sunflower stalk are shown in Figures 5.7c and d (Wang et al. 2018b), respectively. From the Raman spectra survey (Figure 5.7c), the peak at 1345 cm^{-1} (D-band) corresponds to the amorphous carbon, and the other at around 1590 cm^{-1} (G-band) represents the hybridized graphite carbon, which shows the typical in-plane vibrations of amorphous carbons (Titirici et al. 2007). The compositions of samples are analyzed by XPS (Figure 5.7d), and elements of C, O, and N are found in the studied materials.

The surface functionality is analyzed using FTIR. To prevent the interference of water and CO_2, the bio-based carbon materials are purged with helium for several minutes prior to analysis (Brewer et al. 2009). As shown in the FTIR spectrum (Figure 5.7e) (Titirici et al. 2007), the spectroscopy indicates the labels of functional groups, which reveal carbons with polar surface structure, such as aliphatic double bonds, carbonyl and hydroxyl functions, and phenolic residues.

Structure and surface topography of carbon materials are often analyzed by SEM. Figures 5.8a–b show the SEM images of the prepared bio-based carbon materials derived from corn stover and pine (Shen et al. 2017), indicating that they both have porous structures. Many ordered channels are observed in corn stover biochar, whereas more heterogeneous crack-like tunnels are presented in pine biochar. It can be seen in Figures 5.8c–d that the two carbon materials prepared by the MP method are shown to be irregular block particles, small pieces of carbon blocks, and intricately piled up together (Yun et al. 2018). The SEM of activated carbon prepared by the HTC method (Figures 5.8e–f) shows the layer-stacking structure (Wang et al. 2018b).

The surface areas, pore size distribution, and pore volume of the materials are determined by a multipoint BET method. BET surface areas and pore volumes are measured by nitrogen adsorption-desorption tests at 77 K (Brewer et al. 2009). From the N_2 adsorption-desorption isotherms, the apparent surface area can be determined, and the pore size distributions are obtained from the analysis of the absorption curve of isotherms by using the Barret-Joyner-Halenda (BJH) method (Yun et al. 2018).

The nitrogen adsorption-desorption isotherms and pore size distribution of bio-based carbon derived from sawdust (C_s), bio-based carbon derived from waste carbon (C_{wc}), commercial activated carbon (C_a), pyrolytic carbon derived from sunflower stalk (PC), and activated carbon derived from sunflower stalk (AC) are shown in Figure 5.9, and Table 5.4 summarizes BET surface area, average pore width, and total pore volume of carbon materials (Wang et al. 2018b, Yun et al. 2018). The isotherms of these carbons exhibit type IV, which are typical of mesoporous structures according to the IUPAC classification. The nitrogen adsorption−desorption isotherms of C_s, C_{wc}, and C_a with the H3 hysteresis loops indicate that carbon materials have slits, pores, and sheet particles. However, the hysteresis loop of PC and AC are in agreement with H2 type. The pore size distribution of C_s, C_{wc}, and C_a is concentrated on a range of 2.49–28.54 nm, and the pore diameter distribution peaks are centered at 2.2 and 3.6 nm for PC, and 4.2 nm for AC.

Except for these tools mentioned above, the elemental compositions of the bio-based carbon materials can be identified by Energy Dispersive Spectrometer (EDS) (Qin et al. 2017). In addition, the surface chemistry of the samples can be analyzed by temperature-programmed desorption (TPD) (Rosas et al. 2010). Moisture, volatiles, fixed carbon, and ash content can be determined by the method proposed by American Society for Testing Material (ASTM) (Shen et al. 2015).

5. Application of Bio-based Carbon Materials in Anaerobic Digestion

5.1 Overview of Anaerobic Digestion (AD): History and current status

In 1896, biogas from sewage sludge digestion was used to fuel street lamps in England, which is the first practical application of AD for energy production (Marsh 2008). Later, the AD process received

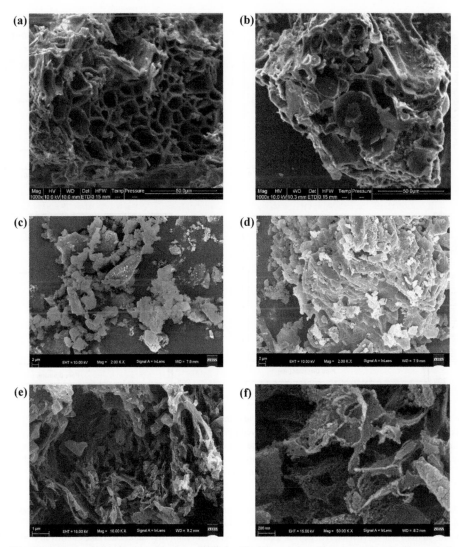

Figure 5.8: SEM images of (a) corn stover biochar and (b) pine biochar, scale bar: 50 μm, reproduced with permission from Shen et al. 2017, copyright 2017 the Elsevier. SEM images of C_{wc} (c) and C_{ws} (d), C_{wc}: bio-based carbon derived from waste carton; C_{ws}: bio-based carbon derived from walnut shell, scale bar: 2 μm, reproduced with permission from Yun et al. 2018, copyright 2018 the Elsevier. SEM images of activated carbon (e and f) derived from sunflower stalk, scale bar: (e) 1 μm, (f) 200 nm, reproduced with permission from Wang et al. 2018b, copyright 2018 the American Chemical Society.

plenty of attention in terms of developing its energy generation capacity. In the 1920s, Buswell established fermentation stoichiometry, and developed farm-scale digestion systems (Lusk et al. 1996). After that, numerous studies and demonstration projects have been attempted with the widest application of anaerobic bioconversion for disposing municipal sewage.

The AD process now can be more fully exploited to its maximum potential. Efficient systems, such as sequencing batch reactors, high-solids anaerobic digesters, or multiple-phase digester systems, have been devised to increase the solids' retention time and hydrolytic microbial contact, or optimize the environmental conditions for the hydrolytic process (Mao et al. 2015). Bio-based carbon materials derived from biomass waste as accelerants have been confirmed to facilitate the improvement of fermentation environment and performance in anaerobic digestion systems (Lu et al. 2016, Luo et al. 2015, Yun et al. 2018).

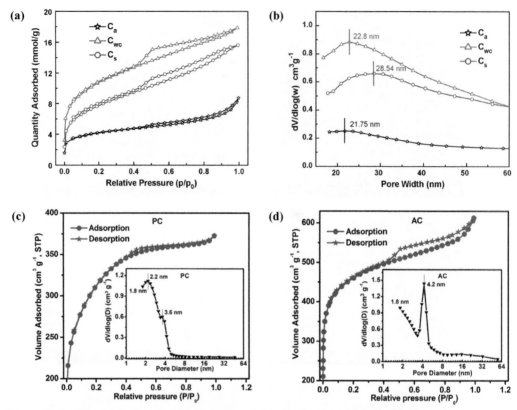

Figure 5.9: Nitrogen adsorption-desorption isotherms and the pore size distribution of carbon materials. C_s: bio-based carbon derived from sawdust; C_{wc}: bio-based carbon derived from waste carbon; C_a: commercial activated carbon, reproduced with permission from Yun et al. 2018, copyright 2018 the Elsevier. PC: pyrolytic carbon derived from sunflower stalk; AC: activated carbon derived from sunflower stalk, reproduced with permission from Wang et al. 2018b, copyright 2018 the American Chemical Society.

Table 5.4: The BET specific surface area, average pore width, and total pore volume of carbon materials. C_s: bio-based carbon derived from sawdust; C_{wc}: bio-based carbon derived from waste carbon; C_a: commercial activated carbon; PC: pyrolytic carbon derived from sunflower stalk; AC: activated carbon derived from sunflower stalk. Data taken from Yun et al. 2018 and Wang et al. 2018b.

Carbon samples	C_a	C_{wc}	C_s	PC	AC
BET specific surface area (m^2g^{-1})	314	824	604	1044	1505
Adsorption average pore width (nm)	36.1	30.1	35.9	2.6	3.6
Total pore volume (cm^3g^{-1})	0.25	0.62	0.54	0.58	0.94

Today, the AD as a waste management system not only provides pollution prevention of unmanaged pollutants from farm, industrial, and municipal wastes, but can also convert rubbish disposal into energy production (Yun 2016). Anaerobic digesters are commonly found alongside farms to recycle manure, or wastewater treatment facilities to reduce the costs of sludge disposal. In Austria and Germany, manure, food waste, and energy crops are digested in thousands of AD plants, producing biogas that is then used for electricity generation (Marsh 2008). In England, there were 259 facilities by 2014, and 500 projects planned to become operational by 2019 (Coker 2017). In China, 26.5 million biogas plants were built by 2007 with an output of 10.5 billion m^3, and it was increased to 248 billion m^3 (annually) by 2010 (Deng et al. 2014).

5.2 Use of activated carbon in AD system with poultry blood

Poultry abattoirs generate high amounts of organic wastes, such as poultry manure, feathers, blood, and intestinal wastes, traditionally, which have been considered a suitable co-substrate in AD systems (Cuetos et al. 2013, Yoon et al. 2014). However, as a complex substrate, residual blood, with high nitrogen content, can lead to intense inhibition due to the accumulation of ammonium in the digester. Nitrogen is an essential nutrient for microbes, but when total ammonia nitrogen is around 4–6 g N/L, ammonia inhibition occurs (Rajagopal et al. 2013).

To avoid inhibitory stages, the use of adsorbents in AD systems has been widely studied (Lu et al. 2016, Milan et al. 2001). For poultry blood as the sole substrate, the effect of activated carbon on performance of AD has been evaluated under batch digestion experiments and semi-continuous conditions (Cuetos et al. 2017). As shown in Figure 5.10a, compared to biogas yield of poultry blood digestion, the addition of granular activated carbon results in successful digestion of the residual blood, which indicates that granular activated carbon materials can reduce the ammonia inhibitory. The biogas yield at the initial stage of the experiments is different and the final cumulative production is similar, but obvious improvements of biogas production are observed, as compared to the control group. During the experiments, the average ammonia value of B_1.5 system is much lower, indicating that the addition of activated carbon with higher concentration is helpful to reduce the inhibition stage (Figure 5.10b), which is consistent with the higher methane production rate at the initial stage. Therefore, the increase in the amount of activated carbon added can prevent inhibitory condition. The results under semi-continuous conditions are shown in Figure 5.10c, which demonstrate biogas production is low at the start of the experiment and increases gradually over the first 8 days. After the equivalent time as hydraulic retention time, the biogas yield tends to be stable, with an average specific methane production (SMP) value of 216 ± 12 mL CH_4/g VS. Although there is a large gap

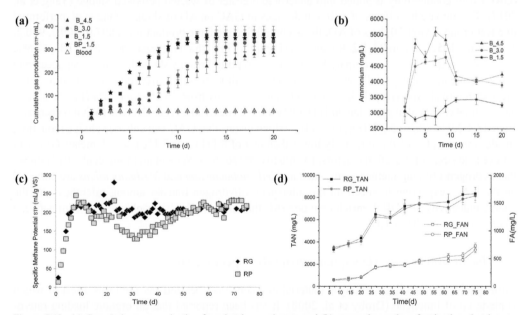

Figure 5.10: (a) Cumulative gas production from batch experiments and (b) ammonium values for the three batch tests: B_4.5, B_3.0, B_1.5, and Bp_1.5 represent the mixtures that are prepared by using ratios of blood (total solids (TS)) to mass of activated carbon added, of 4.5, 3.0, 1.5, and using powdered activated carbon as an adsorbent at a ratio of 1.5, respectively; (c) Specific methane production data obtained from Semi-continuous digestion at Standard Temperature and Pressure (STP), and (d) Total ammonia (TAN) and Free ammonia (FA) measurements for Semi-continuous digestion: RG and RP represent that semi-continuous digestion is carried out by using granular and powdered activated carbon, respectively, in a ratio of poultry blood (TS content) and mass of activated carbon of 3.0. Reproduced with permission from Cuetos et al. 2017, copyright 2017 the Elsevier.

between SMP and the values obtained in the batch experiments, this result is already prominent when considering the harsh conditions shown in Figure 5.10d.

The effect of activated carbon with different physical properties on the evolution of ammonia is not significantly different; the ammonia content in both reactors RG and RP is around 8,000, obviously reaching the inhibition level (Moestedt et al. 2016), which indicates that the improvement of digestion is not completely explained by the adsorption phenomenon. This means that aggregation of cells may play a key role for efficient methane production by facilitating direct interspecies electron transfer (DIET) between fermentation bacteria and methanogens (Luo et al. 2015, Zhao et al. 2016).

For AD systems with poultry blood as single substrate, it is feasible to use activated carbon to prevent inhibition conditions. In batch experiments, reactors containing activated carbon present much higher methane yield than the control group. In the semi-continuous experiment, although the ammonia concentrations (6–8 g/L) reach the inhibitory level, the methane yield of the system remains stable, which may be explained by the conductivity of activated carbon and the growth of microorganisms.

5.3 Use of granule activated carbon in AD system with wasted activated sludge

As an energy-saving and promising method with high value-added products, the AD has been used to dispose wasted activated sludge produced from municipal wastewater treatment plant (Feng et al. 2015, Garrido-Baserba et al. 2015, Larsson et al. 2015). High-strength macromolecule organics are the main components of waste activated sludge, and their decomposition is quite slow (Bougrier et al. 2006), which results in the low digestion efficiency and long retention time of anaerobic sludge digestion. On the other hand, for the low efficiency of AD of wasted activated sludge, the slow electron exchange rate among anaerobes is an important reason (Stams et al. 2006).

In order to enhance the sludge decomposition and methane production, granule activated carbon (GAC) with conductivity is added into anaerobic digester of wasted activated sludge (Yang et al. 2017). To compare the effects of different dosages of GAC on AD of sludge, Yang et al. performed five test groups. After 20 days of AD, the accumulative methane production is 897.0, 909.3, 920.2, 1028.2, and 1052.7 mL in R1, R2, R3, and R4 (Figure 5.11a), which increases by 1.4%, 2.6%, 14.6%, and 17.4%, respectively, as compared to R0. These reported results show that GAC can accelerate sludge fermentation, especially in the initial stage. With increase of GAC dosage from 0 to 5.0 g, the TSS removal ratio (i.e., sludge reduction rate) increases from 39.1% to 45.2% (Figure 5.11b), and the total chemical oxygen demand (TCOD)/soluble chemical oxygen demand (SCOD) decreases from 48.4/2.9 to 28.1/0.5 g/L (Figure 5.11c), which means that GAC can enhance sludge reduction. In addition, the pH of R0 is obviously lower than that of R4 (Figure 5.11d), which means GAC can relieve acid stress of digestion reactors. Eventually, the study of Yang et al. (Yang et al. 2017) shows that hydrogen-utilizing methanogens, such as *Methanobacterium* and *Methanoculleus* are enriched on the GAC, expecting to speed up the interspecies hydrogen transfer among microbes, and the good conductivity and DIET enrichment on GAC that improve the syntrophic metabolism between *Geobacter* and methanogens.

5.4 Use of biochar in AD system with citrus peel waste

Citrus peel waste is a lignocellulosic material that contains fiber and essential oils, most of which is made up of limonene (Droby et al. 2008). It has been reported that an organic loading rate of 2–3.5 g VS^{-1} day^{-1} of citrus peel can inhibit microbial activity in AD systems. Considering the inhibitory of the resulting metabolites during assimilation, physical removal of limonene from the digestion system is a preferable choice (Droby et al. 2008, Martin et al. 2013). To remove potentially inhibitory chemicals, biochar can be used as an additive in AD systems (Zhang et al. 2012).

The application of biochar in the AD of citrus peel waste, digested sewage sludge used as inoculum, has been investigated (Fagbohungbe et al. 2016), to evaluate the effect of different types of

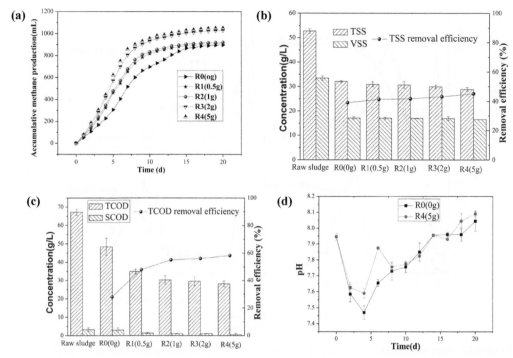

Figure 5.11: (a) Accumulative methane production, (b) total suspended solids (TSS)/volatile suspended solids (VSS) and TSS removal efficiency, (c) total chemical oxygen demand (TCOD)/soluble chemical oxygen demand (SCOD) and TCOD removal efficiency, and the (d) pH (d) in different groups. R0, R1, R2, R3, and R4 represent five test groups of mixtures containing diluted wasted activated sludge and the cultured inoculant sludge with a volume ratio of 9:1 by adding 0, 0.5, 1.0, 2.0, 5.0 g granule activated carbon (GAC), respectively. Reproduced with permission from Yang et al. 2017, copyright 2017 the Elsevier.

biochar and biochar ratios on the AD of citrus peel waste. As shown in Figure 5.12, over 30 days of digestion, the digestion system with citrus peel only experiences a maximum of 14 days of inhibition, which also proves that adding biochar could improve the recovery rate of anaerobic bacteria. This can be explained by the fact that biochar absorbs limonene, thereby reducing the concentration of the compounds in the medium. Moreover, the results are consistent with the report that adding biochar can shorten microbial lag phase, causing the rapid initiation of the growth phase (Mumme et al. 2014). Eventually, coconut shell biochar (CSB) digestion achieves the highest methane conversion efficiency, as compared to wood biochar (WB) and rice husk biochar (RHB) digestions, which also indicates that it is the most effective type of biochar (Fagbohungbe et al. 2016).

Further, in their studies, the effect of different ratios of WB on the CH_4 production performance of AD system over a 30-day incubation period was investigated (Figure 5.13). As the ratio of orange peel to WB decreases, the inhibition of methanogenesis also decreases, indicating that the bioavailability of limonene is reduced due to the presence of WB. In addition, the biochar provides a surface for immobilization of microbial community, which reduces the distance among microbes; this promotes the oxidation of volatile fatty acids and the production of hydrogen. Clearly, adding biochar to the AD system of citrus peel waste has two main effects: shortening the lag phase and increasing methane yield. The cumulative methane production of digestion systems with biochar ranges from 163.9 to 186.8 ml CH_4 g VS^{-1}, while system containing citrus peel only produces 165.9 ml CH_4 g VS^{-1}. In addition, the lag phase of the AD process decreases from 9.4 to 7.5 days as the content of biochar increases. These reported results illustrated that the biochar adsorbed limonene and immobilized microorganisms, which could improve the performance of anaerobic digestion.

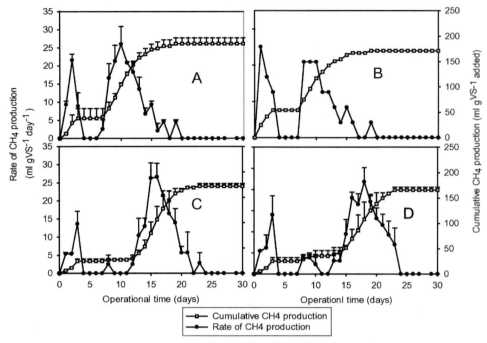

Figure 5.12: Rates and cumulative methane production in each reactor during the anaerobic digestion of different biochar with citrus peel at a mixing ratio of 1:1 based on the dry weight of the total solid, digested sewage sludge used as the source of microbial inoculum. In the second test, the substrate is made up of citrus peel and biochar at ratios of 1:3, 1:2, 1:1, and 2:1. All ratios are based on the dry weight of the total solid. (A) coconut shell biochar + citrus peel. (B) Wood biochar + citrus peel. (C) rice husk biochar + citrus peel. (D) citrus peel only. Reproduced with permission from Fagbohungbe et al. 2016, copyright 2016 the Elsevier.

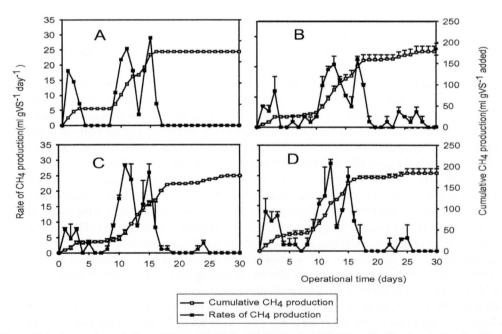

Figure 5.13: Rates and cumulative methane production in each reactor during the anaerobic digestion citrus peel to Wood biochar (WB) ratio, digested sewage sludge used as the source of microbial inoculum. (A) 2:1. (B) 1:1. (C) 1:2. (D) 1:3. Reproduced with permission from Fagbohungbe et al. 2016, copyright 2016 the Elsevier.

5.5 Use of biochar in AD system with dairy manure

Very recently, Yun et al. investigated the effect of seven bio-based carbon materials, prepared through a simple microwave pyrolytic carbonization method, as additives on the biogas yield (Yun et al. 2018). These carbon materials can significantly improve the biogas yield (380–502 mL/g TS) by 30–70%, as compared to the reference system (294 mL/g TS and 29.55%), as shown in Figure 5.14. Significant improvement in the AD efficiency can be microscopically attributed to the methanogenesis promoted by the conductive carbon that can facilitate direct interspecies electron transfer between fermenting bacteria and methanogens, accelerating syntrophic acetate metabolism and biogas yield. The result confirmed that these carbon materials developed were excellent additives in AD for improving the biogas yield and digestate stability. More importantly, the bio-based carbon additives in AD can promote the fertilizer utilization of the digestate. That is, the total weight loss of the digestate with bio-based carbon was 55.28%–65.38%, as compared to the CK group (70.54%). These increased properties lead to more efficient use of the waste feedstock in biogas systems, and promoting the resource utilization of waste biomass.

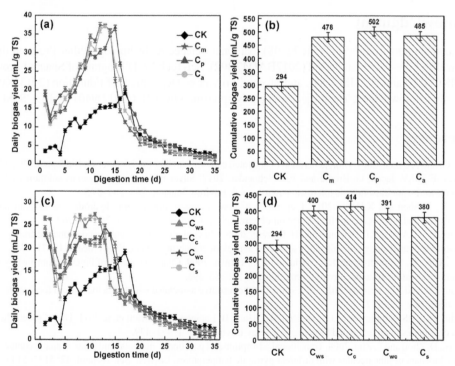

Figure 5.14: Daily biogas yield and cumulative biogas yield of bio-based carbon materials. C_a: activated carbon; C_m: mesoporous carbon; C_p: discarded toner of a printer; C_c, C_s, C_{wc}, and C_{ws}: bio-based carbon derived from waste biomass, such as corn cob, sawdust, waste carton, and walnut shell. Reproduced with permission from Yun et al. 2018, copyright 2018 the Elsevier.

6. Conclusion and Future Perspectives

In conclusion, due to the use of bio-based carbon materials in the anaerobic digestion process, the digestion performance is greatly enhanced. Bio-based carbon materials, derived from kinds of biomass wastes, can be synthesized by simple, low cost, and environment-friendly methods (such as hydrothermal carbonization and microwave pyrolysis). The as-prepared carbon materials, with high BET surface area (up to 1479 m²/g), have been proved to reduce ammonia and acid inhibition (inhibitory levels of VFA reaching values around 6 g/L and ammonium concentration is in the range of 6–8 g/L, biogas production is maintained with low variations), enhance biogas production rate (the increase is up to 70%) (Yun et al. 2018), increase the sludge reduction rate, and shorten the microbial lag phase (decreasing several days), as compared to the reference system or control group.

The significant improvements can be attributed to the overall coarseness of the surface of the biochars that offer a conductive environment for the colonization and growth of microbial cells, and the conductivity promotes direct interspecies electron transfer between fermenting bacteria and methanogens, but full understanding and description on the electrical and adsorption characteristics of bio-based carbon materials in anaerobic digestion system is still inadequate. Therefore, more comprehensive and thorough research, including first-principles density functional theory calculations and *ab initio* Car-Parrinello molecular dynamics simulations performed on the interface of solid/liquid in anaerobic digestion system (Yun et al. 2018), electrical properties, adsorption properties, etc., need to be implemented in the future, which will be extremely important for identifying and confirming the specific mechanisms of carbon materials as additives for improving the anaerobic digestion performance.

Acknowledgment

Financial support from NSFC (51672208), National Science & Technology Pillar Program during the Twelfth Five-year Plan Period (2012BAD47B02), Sci-Tech R&D Program of Shaanxi Province (2010K01-120 and 2015JM5183), and Shaanxi Provincial Department of Education (2013JK0927) is greatly acknowledged. The project was partly sponsored by SRF ((2012)940) for ROCS, SEM.

References

Aktas, O. and F. Cecen. 2007. Bioregeneration of activated carbon: a review. Int. Biodeterior. Biodegrad. 59: 257–272.

Antunes, E., M.V. Jacob, G. Brodie and P.A. Schneider. 2017. Silver removal from aqueous solution by biochar produced from biosolids via microwave pyrolysis. J. Environ. Manage. 203: 264–272.

Bai, C.X., F. Shen and X.H. Qi. 2017. Preparation of porous carbon directly from hydrothermal carbonization of fructose and phloroglucinol for adsorption of tetracycline. Chinese Chem. Lett. 28: 960–962.

Bougrier, C., C. Albasi, J.P. Delgenès and H. Carrère. 2006. Effect of ultrasonic, thermal and ozone pre-treatments on waste activated sludge solubilisation and anaerobic biodegradability. Chem. Eng. Process. 45: 711–718.

Brewer, C.E., K. Schmidt-Rohr, J.A. Satrio and R.C. Brown. 2009. Characterization of biochar from fast pyrolysis and gasification systems. Environ. Prog. Sustain. 28: 386–396.

Browne, J.D. and J.D. Murphy. 2013. Assessment of the resource associated with biomethane from food waste. Appl. Energ. 104: 170–177.

Budarin, V.L., P.S. Shuttleworth, J.R. Dodson, A.J. Hunt, B. Lanigan, R. Marriott et al. 2011. Use of green chemical technologies in an integrated biorefinery. Energy Environ. Sci. 4: 471–479.

Chen, B., D. Zhou and L. Zhu. 2008. Transitional adsorption and partition of nonpolar and polar aromatic contaminants by biochars of pine needles with different pyrolytic temperatures. Environ. Sci. Technol. 42: 5137–5143.

Chen, M., J. Wang, M. Zhang, M. Chen, X. Zhu, F. Min et al. 2008a. Catalytic effects of eight inorganic additives on pyrolysis of pine wood sawdust by microwave heating. J. Anal. Appl. Pyrol. 82: 145–150.

Chen, T., Y. Zhang, H. Wang, W. Lu, Z. Zhou, Y. Zhang et al. 2014. Influence of pyrolysis temperature on characteristics and heavy metal adsorptive performance of biochar derived from municipal sewage sludge. Bioresour. Technol. 164: 47–54.

Chen, Y., J.J. Cheng and K.S. Creamer. 2008b. Inhibition of anaerobic digestion process: a review. Bioresour. Technol. 99: 4044–64.

Coker, C. 2017. Status of organics recycling in the UK. Biocycle 58: 33–34.

Cuetos, M.J., X. Gomez, E.J. Martinez, J. Fierro and M. Otero. 2013. Feasibility of anaerobic co-digestion of poultry blood with maize residues. Bioresour. Technol. 144: 513–20.

Cuetos, M.J., E.J. Martinez, R. Moreno, R. Gonzalez, M. Otero and X. Gomez. 2017. Enhancing anaerobic digestion of poultry blood using activated carbon. J. Adv. Res. 8: 297–307.

Deal, C., C.E. Brewer, R.C. Brown, M.A.E. Okure and A. Amoding. 2012. Comparison of kiln-derived and gasifier-derived biochars as soil amendments in the humid tropics. Biomass Bioenerg. 37: 161–168.

Demirbas, A. 2001. Biomass resource facilities and biomass conversion processing for fuels and chemicals. Energ. Convers. Manage. 42: 1357–1378.

Deng, Y., J. Xu, Y. Liu and K. Mancl. 2014. Biogas as a sustainable energy source in China: Regional development strategy application and decision making. Renew. Sust. Energ. Rev. 35: 294–303.

Droby, S., A. Eick, D. Macarisin, L. Cohen, G. Rafael, R. Stange et al. 2008. Role of citrus volatiles in host recognition, germination and growth of Penicillium digitatum and Penicillium italicum. Postharvest Biol. Tec. 49: 386–396.

Fagbohungbe, M.O., B.M. Herbert, L. Hurst, H. Li, S.Q. Usmani and K.T. Semple. 2016. Impact of biochar on the anaerobic digestion of citrus peel waste. Bioresour. Technol. 216: 142–149.

Feng, Y., Y. Zhang, S. Chen and X. Quan. 2015. Enhanced production of methane from waste activated sludge by the combination of high-solid anaerobic digestion and microbial electrolysis cell with iron-graphite electrode. Chem. Eng. J. 259: 787–794.

Funke, A. and F. Ziegler. 2010. Hydrothermal carbonization of biomass: A summary and discussion of chemical mechanisms for process engineering. Biofue. Bioprod. Bior. 4: 160–177.

Garrido-Baserba, M., M. Molinos-Senante, J.M. Abelleira-Pereira, L.A. Fdez-Guelfo, M. Poch and F. Hernandez-Sancho. 2015. Selecting sewage sludge treatment alternatives in modern wastewater treatment plants using environmental decision support systems. J. Clean. Prod. 107: 410–419.

Ghosh, S., J.P. Ombregt and P. Pipyn. 1985. Methane production from industrial wastes by two-phase anaerobic digestion. Water Res. 19: 1083–88.

Gupta, S., H.W. Kua and H.J. Koh. 2018. Application of biochar from food and wood waste as green admixture for cement mortar. Sci. Total Environ. 619-620: 419–435.

Huang, X., S. Yun, J. Zhu, T. Du, C. Zhang and X. Li. 2016. Mesophilic anaerobic co-digestion of aloe peel waste with dairy manure in the batch digester: Focusing on mixing ratios and digestate stability. Bioresour. Technol. 218: 62–68.

Jindo, K., M.A. Sanchez-Monedero, T. Hernandez, C. Garcia, T. Furukawa, K. Matsumoto et al. 2012. Biochar influences the microbial community structure during manure composting with agricultural wastes. Sci. Total Environ. 416: 476–481.

Kappe, C.O. 2004. Controlled microwave heating in modern organic synthesis. Angew. Chem. Int. Ed. 43: 6250–6284.

Kaushik, R., G.K. Parshetti, Z. Liu and R. Balasubramanian. 2014. Enzyme-assisted hydrothermal treatment of food waste for co-production of hydrochar and bio-oil. Bioresour. Technol. 168: 267–74.

Kaygusuz, K. 1997. Rural energy resources applications and consumption in Turkey. Energ. Source 19: 549–557.

Kijima, M., T. Hirukawa, F. Hanawa and T. Hata. 2011. Thermal conversion of alkaline lignin and its structured derivatives to porous carbonized materials. Bioresour. Technol. 102: 6279–85.

Kim, D., K. Yoshikawa and K. Park. 2015. Characteristics of biochar obtained by hydrothermal carbonization of cellulose for renewable energy. Energies 8: 14040–14048.

Larsson, M., M. Jansson, S. Gronkvist and P. Alvfors. 2015. Techno-economic assessment of anaerobic digestion in a typical Kraft pulp mill to produce biomethane for the road transport sector. J. Clean. Prod. 104: 460–467.

Lee, S.H., H.J. Kang, Y.H. Lee, T.J. Lee, K. Han, Y. Choi et al. 2012. Monitoring bacterial community structure and variability in time scale in full-scale anaerobic digesters. J. Environ. Monit. 14: 1893–1905.

Lehmann, J., M.C. Rillig, J. Thies, C.A. Masiello, W.C. Hockaday and D. Crowley. 2011. Biochar effects on soil biota—A review. Soil Biol. Biochem. 43: 1812–1836.

Li, X., Q. Xu, Y. Fu and Q. Guo, 2014. Preparation and characterization of activated carbon from Kraft lignin via KOH activation. Environ. Prog. Sustain. 33: 519–526.

Li, X., S. Yun, C. Zhang, W. Fang, X. Huang and T. Du. 2018. Application of nano-scale transition metal carbides as accelerants in anaerobic digestion. Int. J. Hydrogen Energ. 43: 1926–1936.

Li, Y., R. Zhang, X. Liu, C. Chen, X. Xiao, L. Feng et al. 2013. Evaluating methane production from anaerobic mono- and co-digestion of kitchen waste, corn stover, and chicken manure. Energ. Fuel. 27: 2085–2091.

Lu, F., A. Bize, A. Guillot, V. Monnet, C. Madigou, O. Chapleur et al. 2014. Metaproteomics of cellulose methanisation under thermophilic conditions reveals a surprisingly high proteolytic activity. ISME J. 8: 88–102.

Lu, F., C. Luo, L. Shao and P. He. 2016. Biochar alleviates combined stress of ammonium and acids by firstly enriching Methanosaeta and then Methanosarcina. Water Res. 90: 34–43.

Luo, C., F. Lu, L. Shao and P. He. 2015. Application of eco-compatible biochar in anaerobic digestion to relieve acid stress and promote the selective colonization of functional microbes. Water Res. 68: 710–718.

Luque, R., J.A. Menéndez, A. Arenillas and J. Cot. 2012. Microwave-assisted pyrolysis of biomass feedstocks: the way forward? Energy Environ. Sci. 5: 5481–5488.

Lusk, P., P. Wheeler and C. Rivard. 1996. Deploying Anaerobic Digesters Current Status and Future Possibilities. NREL/TP-427-20558· UC Category 1503· DE96000508.

Mao, C., Y. Feng, X. Wang and G. Ren. 2015. Review on research achievements of biogas from anaerobic digestion. Renew. Sust. Energ. Rev. 45: 540–555.

Marsh, G. 2008. Rise of the anaerobic digester. Renewable Energy Focus 9: 28–30.

Martin, M.A., R. Fernandez, A. Serrano and J.A. Siles. 2013. Semi-continuous anaerobic co-digestion of orange peel waste and residual glycerol derived from biodiesel manufacturing. Waste Manage 33: 1633–1639.

Masselter, S.M., A.J. Zemann and O. Bobleter. 1995. Analysis of lignin degradation products by capillary electrophoresis. Chromatographia. 40: 51–57.

Milan, Z., E. Sanchez, P. Weiland, R. Borja, A. Martin and K. Ilangovan. 2001. Influence of different natural zeolite concentrations on the anaerobic digestion of piggery waste. Bioresour. Technol. 80: 37–43.

Mochidzuki, K., N. Sato and A. Sakoda. 2005. Production and characterization of carbonaceous adsorbents from biomass wastes by aqueous phase carbonization. Adsorption 11: 669–673.

Moestedt, J., B. Muller, M. Westerholm and A. Schnurer. 2016. Ammonia threshold for inhibition of anaerobic digestion of thin stillage and the importance of organic loading rate. Microb. Biotechnol. 9: 180–194.

Mumme, J., F. Srocke, K. Heeg and M. Werner. 2014. Use of biochars in anaerobic digestion. Bioresour. Technol. 164: 189–197.

Nipattummakul, N., I. Ahmed, S. Kerdsuwan and A.K. Gasco. 2010. High temperature steam gasification of wastewater sludge. Appl. Energ. 87: 3729–3734.

Nowak, O., S. Keil and C. Fimml. 2011. Examples of energy self-sufficient municipal nutrient removal plants. Water Sci. Technol. 64: 1–6.

Omoriyekomwan, J.E., A. Tahmasebi, J. Zhang and J. Yu. 2017. Formation of hollow carbon nanofibers on bio-char during microwave pyrolysis of palm kernel shell. Energ. Convers. Manage. 148: 583–592.

Paritosh, K., S.K. Kushwaha, M. Yadav, N. Pareek, A. Chawade and V. Vivekanand. 2017. Food waste to energy: an overview of sustainable approaches for food waste management and nutrient recycling. Biomed. Res. Int. 2017: 2370927.

Pedroza, M.M., J.F. Sousa, G.E.G. Vieira and M.B.D. Bezerra. 2014. Characterization of the products from the pyrolysis of sewage sludge in 1 kg/h rotating cylinder reactor. J. Anal. Appl. Pyrol. 105: 108–115.

Peterson, A.A., F. Vogel, R.P. Lachance, M. Fröling, J. Michael, J. Antal et al. 2008. Thermochemical biofuel production in hydrothermal media: A review of sub and supercritical water technologies. Energy Environ. Sci. 1: 32–65.

Qin, Y., H. Wang, X. Li, J.J. Cheng and W. Wu. 2017. Improving methane yield from organic fraction of municipal solid waste (OFMSW) with magnetic rice-straw biochar. Bioresour. Technol. 245: 1058–1066.

Rajagopal, R., D.I. Masse and G. Singh. 2013. A critical review on inhibition of anaerobic digestion process by excess ammonia. Bioresour. Technol. 143: 632–641.

Regmi, P., J.L. Garcia Moscoso, S. Kumar, X. Cao, J. Mao and G. Schafran. 2012. Removal of copper and cadmium from aqueous solution using switchgrass biochar produced via hydrothermal carbonization process. J. Environ. Manage. 109: 61–69.

Ro, K.S., K. Cantrell, Douglas Elliott and P.G. Hunt. 2007. Catalytic wet gasification of municipal and animal wastes. Ind. Eng. Chem. Res. 46: 8839–8845.

Ro, K.S., B.K. Cantrell and P.G. Hunt. 2010. High-temperature pyrolysis of blended animal manures for producing renewable energy and value-added biochar. Ind. Eng. Chem. Res. 49: 10125–10131.

Rosas, J.M., J. Bedia, J. Rodríguez-Mirasol and T. Cordero. 2010. On the preparation and characterization of chars and activated carbons from orange skin. Fuel Process. Technol. 91: 1345–1354.

Ryckebosch, E., M. Drouillon and H. Vervaeren. 2011. Techniques for transformation of biogas to biomethane. Biomass Bioenerg. 35: 1633–1645.

Salema, A.A. and F.N. Ani. 2011. Microwave induced pyrolysis of oil palm biomass. Bioresour. Technol. 102: 3388–95.

Saqib, N.U., S. Baroutian and A.K. Sarmah. 2018. Physicochemical, structural and combustion characterization of food waste hydrochar obtained by hydrothermal carbonization. Bioresour. Technol. 266: 357–363.

Schulz, H. and B. Glaser. 2012. Effects of biochar compared to organic and inorganic fertilizers on soil quality and plant growth in a greenhouse experiment. J. Plant Nutr. Soil Sci. 175: 410–422.

Sevilla, M. and A.B. Fuertes. 2009. The production of carbon materials by hydrothermal carbonization of cellulose. Carbon 47: 2281–2289.

Shanmugam, S.R., S. Adhikari, H. Nam and S. Kar Sajib. 2018. Effect of bio-char on methane generation from glucose and aqueous phase of algae liquefaction using mixed anaerobic cultures. Biomass Bioenerg. 108: 479–486.

Shen, Y., J.L. Linville, M. Urgun-Demirtas, R.P. Schoene and S.W. Snyder. 2015. Producing pipeline-quality biomethane via anaerobic digestion of sludge amended with corn stover biochar with *in situ* CO_2 removal. Appl. Energ. 158: 300–309.

Shen, Y., J.L. Linville, P.A.A. Ignacio-de Leon, R.P. Schoene and M. Urgun-Demirtas. 2016. Towards a sustainable paradigm of waste-to-energy process: Enhanced anaerobic digestion of sludge with woody biochar. J. Clean. Prod. 135: 1054–1064.

Shen, Y., S. Forrester, J. Koval and M. Urgun-Demirtas. 2017. Yearlong semi-continuous operation of thermophilic two-stage anaerobic digesters amended with biochar for enhanced biomethane production. J. Clean. Prod. 167: 863–874.

Sohi, S.P., E. Krull, E. Lopez-Capel and R. Bol. 2010. A review of biochar and its use and function in soil. Adv. Agron. 105: 47–82.

Song, M., B. Jin, R. Xiao, L. Yang, Y. Wu, Z. Zhong et al. 2013. The comparison of two activation techniques to prepare activated carbon from corn cob. Biomass Bioenerg. 48: 250–256.

Soupir, M.L., S. Mostaghimi, E.R. Yagow, C. Hagedorn and D.H. Vaughan. 2006. Transport of fecal bacteria from poultry litter and cattle manures applied to pasture land. Water Air Soil Poll. 169: 125–136.

Stams, A.J., F.A. de Bok, C.M. Plugge, M.H. van Eekert, J. Dolfing and G. Schraa. 2006. Exocellular electron transfer in anaerobic microbial communities. Environ. Microbiol. 8: 371–382.

Sun, Y. and J. Cheng. 2002. Hydrolysis of lignocellulosic materials for ethanol production: a review. Bioresour. Technol. 83: 1–11.

Titirici, M.M., A. Thomas, S.H. Yu, Jens-O. Muller and M. Antonietti. 2007. A direct synthesis of mesoporous carbons with bicontinuous pore morphology from crude plant material by hydrothermal carbonization. Chem. Mater. 19: 4205–4212.

Undri, A., S. Meini, L. Rosi, M. Frediani and P. Frediani. 2013. Microwave pyrolysis of polymeric materials: Waste tires treatment and characterization of the value-added products. J. Anal. Appl. Pyrol. 103: 149–158.

United States Biochar Market Report. 2017. https://www.sohu.com/a/225439949_763925.

Vergara, S.E. and G. Tchobanoglous. 2012. Municipal solid waste and the environment: a global perspective. Annu. Rev. Env. Resour. 37: 277–309.

Wang, H., C. Fang, Q. Wang, Y. Chu, Y. Song, Y. Chen et al. 2018a. Sorption of tetracycline on biochar derived from rice straw and swine manure. RSC Adv. 8: 16260–16268.

Wang, J. and S. Kaskel. 2012. KOH activation of carbon-based materials for energy storage. J. Mater. Chem. 22: 23710.

Wang, X., S. Yun, W. Fang, C. Zhang, X. Liang, Z. Lei et al. 2018b. Layer-stacking activated carbon derived from sunflower stalk as electrode materials for high-performance supercapacitors. ACS Sustain. Chem. Eng. 6: 11397–11407.

Wei, L., M. Sevilla, A.B. Fuertes, R. Mokaya and G. Yushin. 2011. Hydrothermal carbonization of abundant renewable natural organic chemicals for high-performance supercapacitor electrodes. Adv. Energy Mater. 1: 356–361.

Wu, J., Y.Y. Hu, S.F. Wang, Z.P. Cao, H.Z. Li, X.M. Fu et al. 2017. Effects of thermal treatment on high solid anaerobic digestion of swine manure: Enhancement assessment and kinetic analysis. Waste Manage. 62: 69–75.

Yang, Y., Y. Zhang, Z. Li, Z. Zhao, X. Quan and Z. Zhao. 2017. Adding granular activated carbon into anaerobic sludge digestion to promote methane production and sludge decomposition. J. Clean. Prod. 149: 1101–1108.

Yoon, Y.M., S.H. Kim, S.Y. Oh and C.H. Kim. 2014. Potential of anaerobic digestion for material recovery and energy production in waste biomass from a poultry slaughterhouse. Waste Manage. 34: 204–209.

Yuan, J.H., R.K. Xu and H. Zhang. 2011. The forms of alkalis in the biochar produced from crop residues at different temperatures. Bioresour. Technol. 102: 3488–97.

Yun, S., L. Wang, C. Zhao, Y. Wang and T. Ma. 2013. A new type of low-cost counter electrode catalyst based on platinum nanoparticles loaded onto silicon carbide (Pt/SiC) for dye-sensitized solar cells. Phys. Chem. Chem. Phys. 15: 4286–90.

Yun, S., A. Hagfeldt and T. Ma. 2014. Superior catalytic activity of sub-5 μm-thick Pt/SiC films as counter electrodes for dye-sensitized solar cells. ChemCatChem. 6: 1584–1588.

Yun, S. 2016. Use of transition metal compounds in solar and biomass energy. Nano Energy Systems 1: 1–4. http://www.onecentralpress.com/use-of-transition-metalcompounds-in-solar-and-biomass-energy/.

Yun, S., W. Fang, T. Du, X. Hu, X. Huang, X. Li et al. 2018. Use of bio-based carbon materials for improving biogas yield and digestate stability. Energy. 164: 898–909.

Yun, S., C. Zhang, Y. Wang, J. Zhu, X. Huang, T. Du et al. 2019. Synergistic effects of Fe salts and composite additives on anaerobic digestion of dairy manure. Int. Biodeter. Biodegr. 136: 82–92.

Zhang, C., H. Su, J. Baeyens and T. Tan. 2014. Reviewing the anaerobic digestion of food waste for biogas production. Renew. Sust. Energ. Rev. 38: 383–392.

Zhang, C., S. Yun, X. Li, Z. Wang, H. Xu and T. Du. 2018. Low-cost composited accelerants for anaerobic digestion of dairy manure: Focusing on methane yield, digestate utilization and energy evaluation. Bioresour. Technol. 263: 517–524.

Zhang, J., L. Zhang, K.C. Loh, Y. Dai and Y.W. Tong. 2017a. Enhanced anaerobic digestion of food waste by adding activated carbon: Fate of bacterial pathogens and antibiotic resistance genes. Biochem. Eng. J. 128: 19–25.

Zhang, T., S. Yun, X. Li, X. Huang, Y. Hou, Y. Liu et al. 2017b. Fabrication of niobium-based oxides/oxynitrides/ nitrides and their applications in dye-sensitized solar cells and anaerobic digestion. J. Power Sources 340: 325–336.

Zhang, Y., C.J. Banks and S. Heaven. 2012. Co-digestion of source segregated domestic food waste to improve process stability. Bioresour. Technol. 114: 168–78.

Zhao, X., J. Zhang, Z. Song, H. Liu, L. Li and C. Ma. 2011. Microwave pyrolysis of straw bale and energy balance analysis. J. Anal. Appl. Pyrol. 92: 43–49.

Zhao, Z., Y. Zhang, D.E. Holmes, Y. Dang, T.L. Woodard, K.P. Nevin et al. 2016. Potential enhancement of direct interspecies electron transfer for syntrophic metabolism of propionate and butyrate with biochar in up-flow anaerobic sludge blanket reactors. Bioresour. Technol. 209: 148–156.

Zhou, G., X. Xu, X. Qiu and J. Zhang. 2019. Biochar influences the succession of microbial communities and the metabolic functions during rice straw composting with pig manure. Bioresour. Technol. 272: 10–18.

CHAPTER 6

Potential Applications of Nanobiocatalysis for Sustainable Biofuels Production

Madan L. Verma,[1,2,]* *Shivali Sahota*[3] and *Asim K. Jana*[4]

1. Introduction

Nanotechnology is a fast-growing domain involving the fabrication and use of nanomaterials (Park et al. 2018, Xie and Huang 2018, Lima et al. 2017). Nanomaterials are the advanced materials that have at least one dimension between 1 and 100 nanometers. Nanomaterials can be synthesised by physio-chemical and biological methods. Although production of nanomaterials by physicochemical methods have employed harsh chemical and environmental conditions, green methods of production of nanomaterials by biological methods are highly preferred (Xie and Huang 2018, Verma et al. 2019). Synthesis of nanomaterials is done by either 'bottom-up' or 'top-down' processes (Qi et al. 2018, Verma et al. 2016). The properties of nanomaterials differ significantly from those of other materials because of their increased surface area and quantum effects. Recent advances in nanofabrication techniques have made many tailored nanomaterials with unique properties, such as optical, electrical, chemical, and mechanical, etc. (Gao et al. 2018, Han et al. 2018, Li et al. 2017, Verma et al. 2012).

Nanobiotechnology, a combination of two technologies, namely biotechnology and nanotechnology, has had a great growth in multidisciplinary fields that have potential applications in food, pharmaceutical, and bioenergy sectors. Various forms of nanomaterials, such as nanoparticles, nanotubes, and nanocomposites, etc., have recently employed enzyme for bioenergy production as a robust nanocarrier (Dutta and Saha 2018, Rastian et al. 2016, Grewal et al. 2017, Verma et al. 2013a, b, Kalia et al. 2009, Shah et al. 2007).

Biofuels are the need of the hour in the current scenario due to depleting fossil fuels and hike in fuel prices. There is an urgent need of cost-effective biofuels production. Cost-effective production

[1] Centre for Chemistry and Biotechnology, Deakin University, Victoria-3216, Australia.
[2] Department of Biotechnology, Dr YS Parmar University of Horticulture and Forestry, Himachal Pradesh-177001, India.
[3] Centre for Rural Development and Technology, Indian Institute of Technology, Delhi, India.
[4] Department of Biotechnology, National Institute of Technology, Jalandhar, Punjab-144011, India.
* Corresponding author: madanverma@gmail.com

of biofuels, either in gaseous form (biohydrogen and biomethane) or in the liquid form, such as biodiesel and bioethanol, need recyclable biocatalysts (Fan et al. 2017, Roth et al. 2016, Verma et al. 2016, Puri et al. 2013). Recyclable biocatalysts can be reused many times, which reduces the harmful emissions and hike in biofuels price. Enzymes, such as lipase, cellulase, glucosidase, play a vital role in biofuel production. Enzyme cost for large scale biofuel production can be minimized by enzyme immobilization techniques. Enzyme immobilization technique basically needs a support on which enzyme binds covalently or non-covalently. Prior to nanotechnology era, a variety of support, such as inorganic or organic, synthetic and natural polymers, have been employed for enzyme immobilization (Mehrasbi et al. 2017, Kanwar et al. 2015, Verma and Barrow 2015, Kanwar and Verma 2010, Verma and Kanwar 2008, Kanwar et al. 2008a, b, Verma et al. 2008a, b, Kanwar et al. 2007a, b, Kanwar et al. 2006, Kanwar et al. 2005). However, this support material has many issues related to biocatalytic potential of the enzyme, such as mass transfer and diffusion, etc. Such issues can be easily addressed with the advancement of the materials, such as nanomaterials (Lotti et al. 2018, Verma 2018, Rastian et al. 2017, Verma 2017, Verma 2016, El-Batal et al. 2016).

The present write up discusses types of carbon nanomaterials employed for the biofuel producing enzymes, an important step towards biofuel production. The prime enzyme lipase and cellulase are discussed at the interface of carbon nanomaterial and enzyme interaction. Recent studies employed for biofuels produced using carbon nanomaterial bound enzyme is also discussed.

2. Contributions of Carbon Nanomaterials in the Bioenergy Sector

Recently many research papers have appeared in many journals of repute that clearly demonstrate the importance of nanomaterial-based enzyme immobilization for potential applications in different sectors ranging from biomedical to industrial applications (Verma 2018, Verma 2017a–c). Nanomaterials possess unique extra properties that differ from bulk materials, including micromaterial and macromaterial (Puri et al. 2013). For example, nanomaterials show exceptional properties beneficial to an enzyme carrier, such as higher surface area to volume ratios, high amenability of surface modification, high mass transfer rate, low steric hinderance, and quick enzyme separation from product (Verma et al. 2016).

The unique and sonorous properties of carbon nanomaterials enable them to become a good carrier material in the field of bioenergy and environmental challenges (Mauter and Elimelech 2008). High electronic conductivity is one of the most significant characteristics of carbon-based nanomaterials, by the virtue of which they have become a significant choice of carrier materials in the field of bioenergy, especially fuel-cell applications (Lee et al. 2009). Along with high conductivity, they also have very large surface area and high porosity, which contribute to the utilization of carbon-based nanomaterials in various applications in the field of bioenergy (Su and Centi 2013). For instance, in case of fuel cells, when carbon nanotubes (CNTs) were utilized as electrode materials, they enhanced the active surface area, fuel mass transport, conductivity, and subsequently high resistance to corrosion. CNTs also showed catalytic mechanism with faster reaction kinetics, thus causing higher power density and energy conversion efficiency (Hoa et al. 2017). In case of many catalytic and biodegradation applications (such as absorption of air and water-pollutant gases, decomposition of various organic and inorganic substances, etc.), higher surface area and porosity of carbon-based nanomaterials (such as graphene-doped/co-doped with various heteroatoms, their composites with metal/metal oxides and/or conducting polymers) have shown remarkable results in removing harmful substances from soil and water both (Hoa et al. 2017).

3. Classification of Carbon-based Nanomaterials (NM)

Carbon-based nanomaterials (NM) consist of carbon nanotubes (CNTs), fullerenes, nanoonions, nanodiamonds, single-walled carbon nanotube, multi-walled carbon nanotube, and graphene

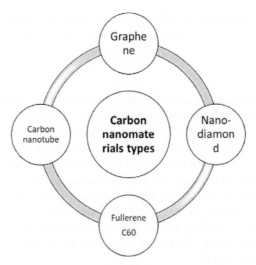

Figure 6.1: Various types of carbon nanomaterials.

(Figure 6.1). Due to their extraordinary optical, electrical, mechanical, and thermal properties, they cover a broader range of applications in the field of environment, biomedicine, and bioenergy (Fan et al. 2017). These are a class of engineered nanomaterials which can be used in various applications, such as adsorbents, high-flux membranes, depth filters, antimicrobial agents, environmental sensors, renewable energy technologies, and pollution prevention strategies (Mauter et al. 2008).

From mid-80s, the era of carbon nanomaterials started with fullerenes and related compounds. The attraction of carbon-based nanomaterials has increased in the scientific community with a great interest due to the revelation of several allotropes, such as fullerenes, carbon nanotubes, and graphenes (Gogotsi and Presser 2013). Particles that are present in nanometer range for at least one dimension with some special and novel properties have several advantages, such as high surface to volume ratio with unique thermal, mechanical, and electronic properties. Nowadays, a number of carbon-based nanomaterials have been evaluated as notable sorbent materials in sample preparation, including graphene, single-walled carbon nanotube, multi-walled carbon nanotube, graphene, C_{60}, nanodiamonds, and fullerenes, as well as their functionalized forms (Georgakilas et al. 2015). The unique structures of carbon nanomaterials allow them to interact with organic molecules via non-covalent forces, such as hydrogen bonding, p–p stacking, electrostatic forces, van der Waals forces, and hydrophobic interactions. The hollow, layered, and nanosized structures of carbon nanomaterials make them good candidates for use as adsorbents (Zhang et al. 2013). Nanocarbon materials play a critical role in the development of new or improved technologies and devices for sustainable production and use of renewable energy (Su et al. 2013).

3.1 Fullerene (C_{60})

The basic structure of C_{60} consists of 20 hexagonal and 12 pentagonal rings arranged in a systematic arrangement to form the buckyball structure. The chemical reactivity of C_{60} is equivalent to a strained electron deficient polyalkene, with localized double bonds. Fullerenes have proven to be potential nanomaterials in various applications, and their efficiency could be enhanced using functionalization methods (Mauter et al. 2008). In cycloaddition technique, addition of functional group occurs across ring junctions where electron density is much higher than at the free atoms (Georgakilas et al. 2015). The reactions which take place on the spherically-shaped carbon core of fullerenes involve additions of nucleophiles, free radicals, and carbenes, as well as π^2-complexation with various transition metal elements. The functionalized carbon atoms change their hybridization from a trigonal sp^2 to a less strained tetrahedral sp$_3$ configuration in fullerene adducts; thus, the primary driving force for addition reactions is the relief of strain in the fullerene cage (Gogotsi et al. 2013).

3.2 Graphene

Graphene has become an essential material of interest and has utterly alerted the scientific curiosity of many research communities around the world. Graphene structure can be configured to different dimensional forms, such as zero-dimensional for buckyball, to first-dimensional nanotubes. Nowadays, graphene has become one of the researchers' interesting, contemplated, and promising nanomaterials (Ghany et al. 2017). This is due to the reason of its exclusive combination of magnificent properties, which are inconceivable to normal materials and have paved the way for its exploitation in a large variety of applications (Li et al. 2013a). Many important carbon-based nanomaterials contain graphene as the primary building block of their structure. It has been practically proven that the quality of produced graphene directly affects their electronic and optical properties. The presence of defects, impurities, structural disorders, and wrinkles in the graphene sheet can adversely affect those properties (Nasir et al. 2018).

Graphene possesses many functional groups, such as carboxyl group and hydroxyl group, etc. This natural functionalized nanosupport is an ideal carrier for a variety of bioactive compounds, and hence has a plethora of applications, ranging from bioremediation to bioenergy sectors, due to excellent functional group and biocompatibility properties (Hermanová et al. 2015).

3.3 Carbon nanotubes (CNTs)

Carbon nanotubes were discovered in 1991 as a minor by-product of fullerene synthesis. Till date, remarkable progress has been achieved, including the discovery of two basic types of nanotubes (single-wall and multi-wall). Great achievements have been recorded in their synthesis and purification, elucidation of their fundamental physical properties, and many important steps are still being taken towards realistic practical applications (Mauter et al. 2008). Carbon nanotubes are composed of long cylindrical structures and pyramidalized by curvature from the pure sp^2 hybridization of graphene, towards the diamond-like sp^3. Ideally, it is believed that a perfect tube is capped at both ends by hemi-fullerenes, leaving no dangling bonds. A single-walled carbon nanotube (SWNT) is made up of a single layer of carbon-sheet, while multi-wall tubes (MWNT) consist of many nested cylinders whose successive radii differ by the interlayer spacing of graphite, and lengths of CNT may vary up to 3–4 mm (Gogotsi et al. 2013).

Carbon nanotubes (CNTs) have become a center of interest due to their unique properties, such as high specific surface area, highly polar functional groups, high porosity efficient heat conductivity, and good chemical stability. All the above properties make them good candidates for application in the fields of food, pharmaceutical, drug delivery, biosensor, and biofuel production (Ji et al. 2010, Pavlidis et al. 2012, Verma et al. 2013a–d, Li et al. 2013).

3.4 Nanodiamonds

Nanodiamonds (lengths in the scale of ~ 1 to 100 nm) have broader applications, such as pure-phase diamond films, diamond particles, and their structural assemblies (Lai and Barnard 2012). Nanodiamonds have unique properties, such as excellent mechanical and optical properties, high surface areas, and tunable surface structures. The recently fabricated one-dimensional (1-D) diamond nanorods (DNR) and 2-D diamond nanoplates complete the set of possible dimensionalities of these structures (Georgakilas et al. 2015). There is a special class of nanodiamond material whose length is of the order of a few nanometers, and they are called "ultra-nanocrystalline". Doped diamond nanoparticles exhibited the highest impact advanced material due to high mechanical performance, high chemical resistance, excellent biocompatibility, unique magneto-optical and electronic properties (Zhang et al. 2018b).

Due to all available unique existing properties of nanodiamonds, they have potential applications in drug delivery, bioimaging and tissue engineering, tribology, and also as protein mimics and a filler material for nanocomposites (Mochalin et al. 2012).

It is pertinent from the above discussed studies that unique properties of various forms of carbon nanomaterials exhibit a plethora of applications, ranging from bioenergy to environmental sectors.

4. Enzyme (Lipase/cellulase/glucosidase) Immobilizing Methods to Carbon Nanomaterials

The term "immobilized enzyme" was first accepted in the Enzyme Engineering Conference in 1971. The meaning of immobilization as a word is movement limiting because binding free enzymes to supports limits their mobility (Verma et al. 2011, Verma and Kanwar 2010, Kalia et al. 2009, Verma et al. 2009, Kanwar et al. 2008a, b). Enzymes can be isolated from diverse sources, such as animals, plants, and microorganisms. However, enzymes sourced from microbes are the most sought choice due to ease of production and easy scale up at industrial setting. Enzymes are easily denatured outside the living cell, and such ultrasensitive biocatalyst moiety can be protected inside the organisms. However, enzyme can be protected from external harsh environment conditions by enzyme immobilization (Figure 6.2). Enzyme immobilization provides operational stability and reusability that makes any bioprocess cost effective.

Enzyme immobilization techniques is classify further nature of interaction between enzyme and substrate and carrier-free approach (Verma 2017a–d, Verma et al. 2016, Abraham et al. 2014). Enzyme is immobilized by physical adsorption, covalent binding, cross-linking, and entrapment methods (Figure 6.3). However, each and every technique has its advantages and disadvantages. Physical adsorption is the most simple and cost-effective method, where enzyme and support interact through weak forces. Enzyme leaking is the problem associated with this method. The covalent immobilization methods involve strong bonding between enzyme and support that can lead to recycling of enzymes several times without loss of biocatalytic function. Cross-linking is a form of covalent binding where enzymes are linked to other enzymes by covalent bonding without using supports. Entrapment involves trapping the enzyme inside the support; however, enzyme leakage and substrate diffusion is the limitations of this method. Based on the pros and cons of each enzyme immobilization method,

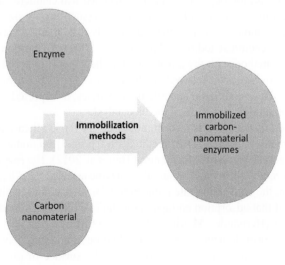

Figure 6.2: Schematic representations of developing an enzyme-carbon nanomaterials bioconjugate.

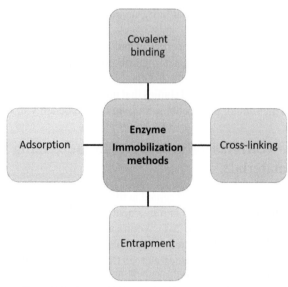

Figure 6.3: Types of immobilization methods employed for binding of enzyme-carbon nanomaterials.

enzyme immobilization is done using two approaches, e.g., physical adsorption followed by covalent binding (Puri et al. 2013, Verma et al. 2013a–d, Verma et al. 2012, Verma and Kanwar 2012).

4.1 Types of nano-immobilization methods

4.1.1 Adsorption method

Adsorption is a simple, reversible, and mild process. It binds (such as hydrophobic and van der Waals interactions) the enzyme to the support material (Jegannathan et al. 2008). The strength of these bonds is relatively low (weak forces) and the enzymes can be displaced from the surface upon application of physical stimuli (ionic strength, temperature, and pH). Immobilisation via adsorption is achieved simply by mixing aqueous solution of the enzyme with the support material and washing away the excess enzyme from the immobilised enzyme on the support material (Costa et al. 2005). The efficiency of this immobilisation depends on the physicochemical characteristics of the procedures (such as the pH and the ionic strength) and support materials (such as pore size and shape, along with hydrophilicity and hydrophobicity). This is the most widely used method because of its simplicity, low cost, and no need for chemical additives (Fukuda et al. 2001). Adsorption is the only reversible enzyme immobilisation method with the advantage that the support material can be recovered for repeated use. Its major drawback is the relatively high enzyme usage. *Candida rugosa* was immobilised onto alkyl trimethoxy silane modified magnetite nanoparticles, as reported by Wang et al. (2012), which provided easy separation and reuse of the biocatalyst.

Lipase from *Burkholderia* sp. was also immobilised on ferric silica nanocomposite through physical binding by dimethyl octadecyl [3-(trimethoxysilyl) propyl] ammonium chloride. As a result, only 29.45 mg g^{-1} lipase was bound to the membrane (Tran et al. 2012). The enzyme nanobiocomposite exhibited improved thermal stability and remarkable enzyme activity, i.e., 78 percent. However, due to enzyme deactivation, the conversion rate reduced to 58% after eight repetitive runs (Wang et al. 2011). Authors reported that adsorption entrapment technique has been found to be more effective than surface adsorption (Hernandez-Martin and Otero 2008) *Candida rugosa* lipase, CRL, was also encapsulated inside peptide nanotubes, through hydrogen bonding between amide groups of the nanotube and complementary functional groups on the surface of proteins. As a result, the catalytic activity of nanotube-bound lipases achieved was 33% higher as compared to free-standing lipases at room temperature (Yu et al. 2005). In general, this physical binding method is known to achieve weak interactions between the enzyme and the support, even though the ionic binding method is generally stronger than other interactions (Veum and Hanefeld 2006). The other major drawback of the adsorption process is enzyme leaching, which makes both the support and enzyme unusable.

4.1.2 *Covalent binding*

Covalent binding is a typical and effective immobilisation method, which involves formation of covalent bonds between a support surface and the active amino acid residues of the surface of the enzyme (Klibanov et al. 1983, Mano et al. 2002, 2003, Soukharev et al. 2004, Costa et al. 2005, Asuri et al. 2007). Several types of supports have been used, such as nanopores (Mano et al. 2002) nanoparticles (Soukharev et al. 2004) or nanofibers (Asuri et al. 2007) to attach the bioactive entities through covalent links. The main advantage of the covalent binding is the interactions between lipase and support prevent enzyme leaching during use (Tan et al. 2010). Recently, Xie and Wang (2012) reported a technique for immobilization of *Candida rugosa* lipase on magnetic chitosan microspheres for transesterification of soybean oil. As a result, the immobilised enzyme was determined as an effective catalyst for the conversion of soybean oil and was shown to retain its activity during four repetitive cycles by transesterification (Xie and Wang 2012).

Zhang et al. (2012) used two different types of lipase immobilization to effectively synthesize the biodiesel production (Zhang et al. 2012). According to Cho et al. (2012), β-glucosidase was immobilised on polymer nanofibers by glutaraldehyde treatment onto covalently attached β-G molecules to achieve high stability and enzyme activity. Lipases from *Rhizopus orizae* and *Candida rugosa* were covalently bound to silica, which was used for the conversion of biodiesel from crude canola oil. Under optimized conditions, the conversion rate of degummed crude canola oil to fatty acid methyl esters reached 88%, which is higher compared to the conversion obtained by free enzyme (84.25%) (Jang et al. 2012). The limitation of covalently bound enzymes is that they have a lower activity than the native enzymes. This is due to the limited conformational transitions that are available on the supports for the chemical interaction. Galliker et al. (2010) utilised silica nanoparticles as support material for laccase immobilisation. They concluded the easy recovery of product with 91.5% loss in enzyme activity.

4.1.3 *Entrapment or encapsulation of enzymes*

Entrapment is a method to entrap an enzyme into polymeric or inorganic frameworks, such as an organic polymer and sol-gel matrices, which retains the enzyme but allows the substrate and products to pass through (O'driscoll 1976). This method can be simply defined as the mixing of enzyme and polymer solution and then crosslinking the polymer (entrap enzyme) to form the lattice structure (O'driscoll 1976). The entrapment method is widely used for industrial applications as it is fast, cheap, easy, and usually involves mild conditions (Trevan 1988, Ikeda et al. 2002). In addition, entrapment is further divided into three categories, gel entrapping, fiber entrapping, and microencapsulation. For example, in sol-gel matrices, a wide variety of enzymes has been used for enzyme immobilisation by hydrolytic polymerization of metal alkoxides or tetraethoxysilane (Trevan 1998, Ikeda et al. 2002). The sol-gel-entrapped lipase was found to have improved operational stability than when the lipase was immobilised by adsorption on the same material (Reetz et al. 1996). It is probably due to leakage of lipase from the adsorbed preparation under the conditions used. In another method, CALB were entrapped into thermosensitive polymer, such as poly-N-isopropylacrylamide at high temperature (typically to 50ºC), causing shrinkage of the polymer. The shrinkage was reversible, so enzyme leaked out of the particles. Considerable inactivation of enzyme was observed as compared to a procedure involving only low temperature (Gawlitza et al. 2012). The limitation of entrapment methods is confined to mass transfer and low enzyme loading due to the act of the support as barrier (Zhang et al. 2012). According to Coradin et al. (2003), the agonic acid-based microencapsulation creates an insoluble polymeric matrix for enzyme entrapment. Alginate micro beads or microspheres have been used in the enzymatic reactor process. As a result, mechanical stability limits their use to few applications (Coradin et al. 2003).

4.1.4 Cross linking method

This technique involves covalent interaction of one enzyme molecule to other enzyme molecules with the aid of a cross-linker (Yucel et al. 2013). The advantage of cross-linking enzyme is obtaining a stable lipase with better reusability due to the strong interactions between the support material and the lipase. This method is based on the formation of covalent bonds between the enzyme molecules, by means of multifunctional reagents, leading to three-dimensional cross-linked enzyme aggregates, and of covalent bonds onto a substrate. The most widely used chemical compound for cross-linking is glutaraldehyde. *Candida rugosa* lipase was also immobilised on glutaraldehyde-activated Poly (vinyl alcohol-coethylene) (PVA-co-PE) nanofibrous membranes via covalent bonds with the aldehyde groups. This method gave 60% enzyme activity and 39.52 mg lipase/g membrane (Zhu and Sun 2012). Lipase immobilisation on nanoparticles by covalent attachment using glutaraldehyde as a cross-linker is a common method (Wang et al. 2009). It provides good enzyme activity (70%) and enzyme loading efficiency (25 mg lipase/g particles). It is noted in all studies that the increment of GA initial concentration induced the increase in enzyme loading, but also a decrease in the activity of the immobilized enzyme (Zhu and Sun 2012). In addition, glutaraldehyde and its derivatives cause environmental concerns (Wissink et al. 2001, Yurekli and Altinkaya 2011). Other immobilisation chemistries have been developed. In one approach, the nitrile groups of the polyacrylonitrile (PAN) nanofibers were activated by an amidination reaction, followed by reaction with lipase in phosphate buffer solution (Li et al. 2011). Activated nanofibers showed good enzyme loading efficiency (43 mg lipase/g fibers) and retained 80% activity of lipase compared to free enzyme, which is much higher than is the typical case for glutaraldehyde.

Lipase can also be covalently attached to multi-walled carbon nanotubes (MWNTs) through carbodiimide activation. It resulted in improved stability at high temperatures (Ji et al. 2010). Authors studied the preparation of *Pseudomonas cepacia* lipase cross-linked enzyme aggregates (Kumari et al. 2007). It has been reported that the cross-linked lipases have a greater stability than the free enzymes in denaturing conditions. The enzyme was found to catalyse the conversion of fatty acid methyl esters from Butter tree (*Madhuca indica*) oil, where transesterification is difficult by chemical routes due to its high free fatty acid content. The conversion obtained after 2.5 hours was 92 percent. Yan et al. (2011) investigated a modification procedure for the preparation of cross-linked *Geotrichum* sp. The obtained cross-linked lipase revealed improved pH and thermal stability as compared to free lipase. The relative biodiesel yield obtained was 85% through transesterification of waste cooking oil with methanol (Yan et al. 2011). However, enzyme immobilization approach may cause changes in enzyme structure, in particular to active site of enzyme that lead to poor interaction of enzyme-substrate molecules, rendering loss of poor enzyme activity.

5. Factors Affecting Biodiesel Production

Transesterification is a well-known method for biodiesel production from vegetable oils and animal fats (Mukhtar et al. 2018, Arumugam and Ponnusami 2017, Yan et al. 2016, Rodrigues et al. 2016, Meher et al. 2006). This process involves stoichiometric interaction of three moles of alcohols, e.g., methanol and one mole of triglyceride to produce three moles of fatty acid methyl esters along with one mole of by-product glycerol (Shimada et al. 1999). In transesterification or alcoholysis, fats or oils react with alcohol in presence of a catalyst to form alkyl esters and glycerol (Leung et al. 2010).

Selecting a suitable alcohol and catalyst are important factors for transesterification (Noureddini et al. 2005, Salis et al. 2008, Sakai et al. 2010, Sawangpanya et al. 2010, Moreno-Pirajan and Giraldo 2011). Various alcohols, such as methanol, butanol, ethanol, propanol, and amyl alcohol are feasible for transesterification (Desai et al. 2006, Devanesan et al. 2007, Da Ros et al. 2010, Li et al. 2011). Among these, methanol is used widely because it is relatively cheaper than other alcohols. Though high concentrations of methanol inhibit the enzyme, Shimada et al. (2002) used an alternative technique of stepwise addition of methanol. The other critical factor is glycerol, a by-product of alcoholysis, which is physically adsorbed on the surface of immobilized lipase. This attenuates the mass transfer of oil to lipase active sites and accordingly lowers the rate of biodiesel production. In addition, a good catalyst is also needed to obtain high conversion rate (Lotero et al. 2005). Interestingly, utilisation of lipases as a catalyst for biodiesel production has a great potential, compared to chemical and acid catalysts. This is because no complex operations are required for the recovery of glycerol. The amount of water in the reaction mixture is another critical parameter during transesterification process. High water and free fatty acid in oil reduce the effectiveness of catalysts, produce soap, and require considerable amount of catalysts, because higher water content results in a lower reaction rate because hydrolysis of oil to fatty acids is reduced when water is abundant (Al-Zuhair 2007) due to the increase in the available interfacial area. Moreover, the ratio of the two reactants of transesterification, namely methanol and oil, is also recognized as a crucial factor that needs to be optimized.

Surface functionalisation allows great flexibility in the modification of support materials to achieve effective enzyme immobilisation. Specifically, immobilisation provides a good catalytic environment (better resistance to extreme pH, high temperature), improved stability, and re-use advantages (Li et al. 2013, Kuo et al. 2013). So far, various materials have been employed as supports for enzyme immobilisation, such as natural or synthetic polymers (e.g., acrylic resins, celite, silica, polystyrene, ceramic beads, chitosan, alginate, agarose and startch), nanomaterials (e.g., nanoparticles, nanofibers, nanotubes, nanocomposites, and nanoporous multi-walled carbon nanotubes). Nanomaterials typically have a higher specific surface area, allowing higher enzyme loading of enzyme per unit of mass than nanostructured counterparts. In addition, it has been reported that proteins adsorbed onto nanomaterials are more stable in strongly denaturing environment conditions of high temperatures and organic solvents (Hanefeld et al. 2009). Enzyme immobilization involves covalent or non-covalent interaction with the supporting matrix, such as hydrophobic/hydrophilic polymeric and inorganic material (Lupoi and Smith 2011, Liu et al. 2012). Such interaction between the biomolecules and materials is categorised into primarily four types, namely, adsorption, covalent, cross-linking, and entrapment (Cao 2005, Mateo et al. 2007, Chen et al. 2009). The selection of immobilisation method and support material is critical to achieve high enzyme efficiencies. For example, the catalytic potential of the same biocatalyst exhibits very different behavior of enzyme activities after immobilization onto different carriers (bulk vs nanomaterials). For instance, Iso et al. (2001) reported higher conversion yield (> 85%) with *Pseudomonas fluorogens* lipase using porous kaolinite particle as support, whereas Salis et al. (2008) showed a relatively low conversion efficiency (58%) after using propylene powder as support material for biodiesel production.

It is very obvious from the above cited recent studies that applications of various forms of carbon nanomaterials as supports for enzyme immobilization improve the stability and activity of industrial enzymes that has further strengthened the bioprocessing at the industrial setting.

6. Impact of Nanobiotechnology for Biodiesel Production

Microbial lipase has been playing a crucial role in biodiesel production (Nehdi et al. 2018, He et al. 2018, Law et al. 2018, Lotti et al. 2018, Wang et al. 2018, Verma and Barrow 2015, Kanwar and Verma 2010, Verma and Kanwar 2010, Verma et al. 2008a, b). Efficiency of enzyme has further improved with the intervention of nanotechnology (Verma 2018, Verma 2017, Verma et al. 2016). Various lipases sourced from diverse microorganism have been employed for improved biocatalytic potential with respect to biodiesel production, as listed in Table 6.1.

Table 6.1: List of various carbon-based nanomaterials used for lipases immobilization for various biotechnological applications.

Name of the carbon nanomaterials	Enzymes (Lipase sources)	Immobilization method	Features	Application	References
Carbon nanotube (CNT)	*C. rugosa*	Covalent	Excellent solubility in organic solvents	Biodiesel production	Shi et al. 2007
Carbon nanotube (CNT)	*C. rugosa*	Adsorption	High retention of lipase activity	Construction of miniaturized biosensor devices	Shah et al. 2007
Carbon nanotube (CNT)	*C. Antarctica*	Adsorption	High catalytic activity, excellent storage, and operational stability	To catalyze esterification and transesterification in food and pharmaceutical industries	Pavlidis et al. 2010
SWNT	*C. rugosa, C. antarctica B, T. lanuginosus*	Adsorption	Protected lipase inactivation, enhanced stability	Biosensors and nanobiocatalysts	Lee et al. 2010a
MWNT	*T. lanuginosus*	Adsorption, covalent	High synthetic activity	Biosensor applications	Lee et al. 2010b
MWNT	*Rhizopus arrhizus*	Covalent	Improved resolution efficiency	Ethanol production	Ji et al. 2010
MWNTs	*Thermomyces lanuginosus* lipase	covalent	Significantly improved thermal stability	Biosensor applications	Verma et al. 2013
Graphene oxide/ MWNTs	*Rhizopus oryzae Candida rugosa Penicillium camemberti Alcaligenes* sp. *Candida antartica*	Covalent Adsorption Ionic Cross linking	Efficient catalyst in organic medium	Biocatalyst	Li et al. 2013, Pavlidis et al. 2012
MWCNTs	*P. fluorescens*	Covalent	Significant improvement of thermal stability	Kinetic resolution of racemic compounds (Nanobiocatalyst)	Dwivedee et al. 2017
MWCNTs	*Chromobacterium viscosum Pseudomonas cepacian*	Covalent	Enhanced interface and hyper activation	Biodiesel production	Su et al. 2016
Magnetic m-MWCNTs	*Rhizomucor miehei lipase*	Covalent	Enhanced esterification activity	Biodiesel production	Fan et al. 2016
Magnetic m-MWCNTs	*Burkholderia cepacia* lipase	Covalent	Catalyze transesterification	Biodiesel production	Fan et al. 2017
MWCNTs	*Candida Antarctica*	Covalent	Catalyzing methanolysis	Biodiesel production	Rastian et al. 2016
Graphene oxide	*Candida rugosa*	Covalent	High catalytic activity of esterification	Biodiesel production	Lau et al. 2014
Graphene oxide	*Alcaligenes* sp.	Non-covalent hydrophobic interactions	Hydrolyzing carboxylic ester bonds	Excellent biocatalytic potential	Mathesh et al. 2016
Graphene oxide (GO)	*Rhizopus oryzae*	Covalent	Better resistance to heat inactivation	Industrial scale uses of enzymes for the conversion of lipids into fuels	Hermanová et al. 2015

Magnetic nanocomposite was synthesized by encapsulating magnetic nanoparticle into the graphene oxides (Xie and Huang 2018). Lipase sourced from *Candida rugosa* was immobilized on the nanobiocomposite by hydrophobic interaction via interfacial activation of the enzyme. The immobilized enzyme showed ca. 65% activity recovery and ca. 86% immobilization efficiency. The nanosupport immobilized enzyme and native enzyme was characterized by analytical techniques, such as transmission electron microscopy, x-ray powder diffraction, x-ray photoelectron spectroscopy, Fourier transform infrared spectroscopy, vibrating-sample magnetometer, and nitrogen adsorption-desorption techniques. Nanobiocatalysts showed excellent transesterification process by high yield of biodiesel production. Higher yield of biodiesel production was obtained (93%) at ambient temperature. Recovery of the immobilized biocatalyst was quite fast and achieved by applying an external magnetic field. Nanobiocatalysts showed excellent recyclability without the loss of catalytic potential of enzyme.

Jafarian et al. (2018) studied the immobilization of lipases sourced from *Candida rugosa* on graphene and its modified graphene oxide nanosheets. Nanosheet was modified using 2,4,6-tricholor-1.3.5-triazine and 3-aminopropyltriethoxysilane. The modified nanosheet and native nanosheet was evaluated for the immobilization and biocatalytic performance of the enzyme. The kinetic studies of the immobilized enzyme were evaluated at the structural level using several techniques, such as x-ray diffraction, Scanning electron microscopy, Energy dispersive x-ray spectroscopy, Uv-Vis spectroscopy, Fourier Transform infrared spectroscopy, and Atomic force microscopy. Authors investigated modified as well as native nanosheet for maximum binding, immobilization yield, and leakage study of the *Candida rugosa* lipase. The efficient biocatalytic potential of the lipase was exhibited by the surface modified nanosheet with these two functionalization moieties.

Dutta and Saha (2018) developed a graphene oxide nanosheet based on a simple immobilization methodology for the mesophilic lipase sourced from *Brevibacillus borstelensis*. The activity and stability of the enzyme was enhanced after immobilization on the graphene surface. Thermal stability of the mesophilic enzyme was enhanced up to 95°C. Higher activity of the mesophilic lipase was due to the structural stability of the enzyme. The immobilized lipase showed broader pH stability in the range of 7 to 11. Kinetic parameters of the immobilized enzyme were improved by lowering the activation energy value and Michaelis constant (Km) value. Enzyme loading on the surface of the graphene nanosheet was quite higher than the other carbonaceous materials.

Magnetic nanoparticle was functionalized with lipase from *Candida antarctica* (Mehrasbi et al. 2017). Magnetic nanoparticle was modified with silica coating, followed by functionalization with silanization agent. The immobilization of lipase on magnetic nanoparticle was confirmed by various analytical techniques. Thermal stabililty of the immobilized lipase was quite higher than the free enzyme. Moreover, the immobilized lipase showed organic solvent tolerance. The waste cooking oil and methanol were used as substrates for the transesterification process of biodiesel production. Various parameters, such as waste oil to alcohol, tert-butanol, and molecular sieve were optimized to get higher activities of the enzyme. Robust nanobiocatalysts showed 100% efficient transesterification process. Nanobiocatalysts showed high recyclability without loss of biocatalytic potential.

Robust magnetic nanocarrier was developed using surface modification of magnetic nanoparticle with the functionalizing agents, such as 2,3-epoxypropyltrimethylammonium chloride (Li et al. 2017). Functionalized aminated magnetic nanoparticle was treated with glutaraldehyde. The lipase sourced from *Burkholderia cepacia* was immobilized on the surface of the functionalized nanoparticle by anion exchange and covalent binding methods. The immobilized biocatalyst was employed for biodiesel production using substrates of soybean oil and methanol. Nanobiocatalysts showed excellent transesterification process for biodiesel production at 12 hours; it was approximately two-fold to free enzyme. Nanobiocatalysts showed excellent performance in terms of high yield of transesterification process. The immobilized biocatalysts demonstrated high operational stability and recyclability.

Thangaraj et al. (2017) developed nanobiocatalysts for biodiesel production using modified magnetic nanoparticle. The magnetic nanoparticle core was coated by silica shell. It was functionalized by silanization process. Lipase was covalently bound to functionalized nanoparticle using a glutaraldehyde crosslinker. Nanobiocatalyst showed a maximum yield of biodiesel production using methanol and oil.

Wan et al. (2017) studied the effects of different imidazole-based ionic liquids with different functional groups, namely hydroxyl, carboxyl, amino, and alkyl to modify the multi-walled carbon nanotubes. The modified nanocarrier was employed to immobilize the lipase sourced from *Candida antarctic*. The influence of ionic liquid-treated carbon nanotubes was correlated with respect to biocatalytic potential of the immobilized lipase. The ionic liquid-treated multi-walled carbon nanotubes improved the biocatalytic potential of the enzyme. Authors demonstrated that ionic liquid containing hydroxyl functional group-treated multiwalled carbon nanotubes was the most suitable nanocarrier amongst the other functional groups. The immobilized lipase showed excellent thermal stability and reusability of the biocatalyst.

Fan et al. (2017) developed a methodology for lipase sourced from *Burkholderia cepacia* to covalently immobilize on the surface of modified carbon nanotubes. In order for quick separation under the influence of magnetic field, carbon nanotube was made into superparamagnetic nanotubes by iron oxide filling. The magnetic carbon nanotube was surface-functionalized using polyamidoamine dendrimers. The amino-functionalized nanocarrier was employed for enzyme immobilization using a glutaraldehyde cross-linker. The immobilized enzyme showed 17-fold higher activity as compared to free lipase. The lipase in the bound form exhibited improved thermal stability as well as broader pH tolerance. This hyperactive immobilized enzyme was employed for biodiesel production. The maximum yield of biodiesel production was obtained (ca. 93%). Quick recyclability of the immobilized enzyme under the influence of external magnetic device provided retention of 90% esterification activity of biodiesel production even at the 20th cycles. Authors demonstrated the excellent bioprocess for biodiesel production with the incorporation of iron oxide filling to the multi-walled carbon nanotubes.

The magnetic barium ferrite nanoparticle was employed for immobilizing lipase sourced from *Aspergillus niger* (El-Batal et al. 2016). Nanobiocatalysts were used for biodiesel production using waste cooking oil and methyl alcohol. Nanobiocatalysts exhibited excellent performance by retaining approximately 90% activity even after the many recyclability studies. The biodiesel showed properties fit as per the standard parameters. For example, the measured flash point, calorific value, and cetane number of biodiesels were 188°C, 43.1 MJ/Kg, and 59.5, respectively.

Magnetic carbon nanotubes-based nanocomposite was employed for covalent binding of a fungal lipase sourced from *Rhizomucor miehei* (Fan et al. 2016). Magnetic multi-walled carbon nanotubes were amino-functionalized by grafting polyamidoamine dendrimer. The interaction of enzyme at the interface of nanocarrier was studied using three-dimensional analysis that exhibited an oriented immobilization of enzyme on the surface of the carbon nanotubes. The magnetic nanocomposite was employed for biodiesel production using waste cooking oil in a butanol solvent. The rate of esterification was quite high and it was 27-fold higher as compared to free enzyme. The magnetic nanocomposite was quickly separated from the product by external magnetic field and reused up to the 10th cycles without any significant loss of biocatalytic potential.

Fan et al. (2017) developed multi-wall carbon nanotubes (MWCNTs) immobilized lipase methodology for the production of biodiesel. Hydrophobic enzyme lipase sourced from *Burkholderia cepacia* bound in polyethyleneimine microcapsules was modified with oxidized multi-wall carbon nanotubes. The immobilized lipase exhibited higher catalytic stability and excellent stability; it was 22-fold higher than the free enzyme. The immobilized enzymes exhibited stronger endurance against seven organic solvents, namely, methanol, ethanol, propyl alcohol, isopropanol, butanol, isobutanol, and acetone, as compared to free enzyme. Authors claimed the enhancement in the activity of these interfacial enzymes was due to numerous polycondensation interfaces. The immobilized enzyme was characterized by scanning electron microscopy and zeta-potential studies. The same behavior was reported using confocal laser scanning microscopy. Multi-walled carbon nanotubes-modified microcapsules demonstrated an enhancement of biocatalytic potential. The immobilized enzyme showed higher yield of biodiesel production (97%) from soybean oil in 12 hours as compared to free enzyme. The bacterial enzyme was highly stable in this immobilized preparation, as it retained nearly full activities even up to the 10th cycles.

Prlainovic et al. (2013) developed a facile immobilization method for the immobilization of *Candida rugosa* lipase on the pristine multi-walled carbon nanotubes. Various parameters employed for the immobilization of the lipase were optimized with respect to non-covalent binding of the enzyme on the surface of the supporting nanomaterial. Enzyme loading, time of optimal enzyme-nanomaterial interaction, and suitable buffer for optimal enzyme activity were evaluated to get the higher biocatalytic potential of the enzyme. Biochemical studies were also investigated and compared to the structural studies using various techniques, such as Fourier Transform Infrared Spectroscopy, Atomic Force Microscopy, and Cyclic voltammetry. The non-covalent bound enzyme on nanomaterials showed higher enzyme loading (85%) at high ionic strength of the buffer.

Verma et al. (2013d) developed a facile approach to immobilized lipase on carbon nanotubes in order to correlate the structural and functional aspects. The multi-walled carbon nanotubes were treated with ethylenediamine to amino-functionalize the nanotubes. The coupling agent, O-(7-Azabenzotriazol-1-yl)-N,N,N9,N9-tetramethyluroniumhexafluorophosphate, was employed to increase the yield of reaction. Amino-functionalized carbon nanotubes were treated with lipase enzyme sourced from *Thermomyces lanuginosus* with the aid of a glutaraldehyde cross-linker. The covalently-bound enzyme was compared to a free enzyme with respect to structural changes by transmission electron microscopy, and spectroscopy techniques, including x-ray photoelectron, Fourier-transform infrared, and circular dichroism. The size and morphology of the multi-walled carbon nanotube and nanotubes-bound enzyme was studied using transmission electron microscopy (TEM). TEM studies confirmed the slight increment in the size of the nanotube after covalent binding of the enzyme. The chemical composition of the functionalized carbon nanotubes-bound lipase was studied using x-ray photoelectron spectroscopy. The atomic composition of oxygen and nitrogen in the enzyme-bound multi-walled carbon nanotubes was compared to the pristine multi-walled carbon nanotubes. The higher concentrations of the nitrogen atoms on the surface of functionalized carbon nanotubes confirmed the presence of the amide group as compared to native multi-walled carbon nanotubes. The considerable increase in the contents of the surface oxygen and nitrogen atoms indicated the covalent bonding at the interphase between the amino-functionalized multi-walled carbon nanotubes and enzyme. The secondary changes in the structure of the protein after immobilization onto the multi-walled carbon nanotubes were investigated using circular dichroism spectroscopy. The α-helix content of the free enzyme and carbon nanotube immobilized enzyme was compared, and the content of total α-helix remained unaltered, confirming that the covalent binding between the carbon nanotube and enzyme remained unaltered in the secondary structure of the enzyme. The kinetic properties of the multi-walled carbon nanotubes-bound enzymes were slightly changed as compared to free form of enzyme. This parameter clearly indicated that the biocatalytic potential of the enzyme remained functional. The immobilized lipase retained good biocatalytic potential, and biochemical properties of the immobilized enzyme were shown at the same temperature and pH optima. The immobilized lipase was successfully reused up to ten cycles. Multi-walled carbon nanotube immobilized lipase demonstrated excellent thermal stabilities at extreme temperature conditions. Authors claimed the functionalization agents as well as coupling agents play a considerable role in the degree of the surface functionalization, as well as higher enzyme loading on the surface of the nanocarrier employed for enzyme immobilization for potential applications in bioenergy production.

Pavlidis et al. (2010) developed a simple methodology for lipase immobilization on to the carbon nanotubes using physical adsorption method. The surface of multi-walled carbon nanotubes were modified with different functional groups, such as amine-, carboxyl-. and ester-. The functionalized nanotube was led to higher lipase loading to the maximum of 25% of the weight of nanocarrier. The immobilized enzyme showed higher catalytic activity as compared to the free enzyme. The immobilized enzyme demonstrated increased operational stability even after six months of storage.

The immobilized lipase was quite stable in organic solvent that retained one-fourth activity even after one month time period.

Bencze et al. (2016) developed nanobioconjugates using single-walled carbon nanotubes and *Candida antarctica* lipase B. The native carbon nanotube was functionalized with carboxyl group. The functionalized carbon nanotube was employed for covalent immobilization of lipase with the aid of variable crosslinkers. The best nanobiocatalyst in terms of low diffusional limitation was selected for biodiesel production. The biodiesel was produced using co-substrates of sunflower oil and ethanol at ambient reaction conditions, i.e., 35°C. The higher yield of biodiesel was produced using single-walled carbon nanotube immobilized lipase in acetonitrile. The recyclable enzyme retained almost 90% activity even after the use of the 10th cycles.

Multi-walled carbon nanotubes were modified with carboxyl functional group. It was further surface functionalized with amidation process using either butylamine or octadecylamine (Rastian et al. 2016). Lipase sourced from *Candida antarctica* B was employed for immobilization studies. The immobilized lipase was employed for biodiesel production using rapeseed oil and methanol. The loading of immobilized enzyme was higher on the nanotube at the amount of 20 mg enzyme per gram of the nanocarrier. Higher yield of biodiesel production was achieved at the yield of 92 percent. The immobilized enzyme retained higher catalytic activity and demonstrated exceptional thermal stability.

Guan et al. (2017) employed sulfonated multi-walled carbon nanotubes for biodiesel production using triglycerides transesterification method. Authors demonstrated a facile approach for sulfonated multi-walled carbon nanotubes for biodiesel production. Higher yield of biodiesel production was achieved at the rate of 98% in 1 hour at higher temperature. Authors demonstrated regeneration methods for the effective transesterification of triglyceride for biodiesel production.

It is very obvious from the above-cited studies that biocatalytic potential of lipase has significantly improved at the interface of carbon nanomaterial and enzyme. Thus, carbon nanomaterial is the ideal nanocarrier for the enzyme immobilization studies with respect to potential biodiesel production.

7. Impact of Nanobiotechnology for Bioethanol Production

Cellulase is one of the prime candidates in the bioethanol production, as it converts lignocellulosic biomass to fermentable sugars (Qi et al. 2018, Yu et al. 2018, Han et al. 2018, Gomes et al. 2018b). However, high cost of enzyme and ease of denaturation has hampered the industrial setting. Here, the immobilization approach plays a vital role in making the robust recyclable biocatalysts with improved biocatalytic potential that make bioprocess cost effective, and is the most sought technology (Prakash et al. 2018, Zhang et al. 2018a, Lima et al. 2017, Roth et al. 2016, Nguyen et al. 2016). Here, carbon-based nanomaterials are discussed to improve the biocatalytic potential with respect to biofuel production. Various cellulases and glucosidases sourced from diverse microorganisms have been employed for improved biocatalytic potential with respect to bioethanol production, as listed in Table 6.2.

Magnetic nanoparticles were surface modified and immobilized β-glucosidase using a glutaraldehyde with different molecular weights of polyethylene glycol space-linkers (Park et al. 2018). Nanobiocatalysts demonstrated improved thermal activity and retained high biocatalytic activity even after ten cycles.

Nanobiocatalysts were developed by entrapment of magnetic nanoparticle into the cross-linked ionic liquid and epoxy polymer (Hosseini et al. 2018). Cellulase was covalently immobilised onto the surface via activation of epoxy group. Higher amount of enzyme was loaded on the surface of the nanocomposite. Nanobiocatalysts retained higher activity and improved thermal stability.

Nanobiocatalysts were developed by covalently immobilizing cellulase from *Trichoderma reesei* onto the chitosan modified magnetic nanoparticle (Sánchez-Ramírez et al. 2017). Nanobiocatalysts showed higher biocatalytic activity using carboxymethylcellulose and retained higher activity.

Table 6.2: List of various carbon-based nanomaterials used for cellulases immobilization for various biotechnological applications.

Enzymes and Source	Nanomaterials	Immobilization methods	Performance or Improved properties	References
Cellulase	Graphene	Hydrophobic interaction	Higher biocatalytic activity	Gao et al. 2018
Cellulase	Multi-walled carbon nanotube	Covalent bonding	Four-fold increased half live	Ahmad and Khare 2018
Cellulase	Graphene oxide	Cross linking	Reusability and storage stability	Dutta et al. 2016
Cellulase	PEGylated graphene oxide nanosheets	Covalent bonding	Enhances stability	Xu et al. 2016
Cellulase	$Fe_3O_4@SiO_2$– graphene oxide	Covalent binding	High thermal stability and reusability	Li et al. 2015
Cellulase	Multi-walled carbon nanotube	Physical adsorption	Retained 52% of its activity after 6th cycle of reactions	Mubarak et al. 2014
Cellulase	Graphene	Covalent bonding	Improvement in the bio-receptivity	Gokhale et al. 2013
Cellulase	Graphene	Covalent binding	Improved biocatalytic potential	Dutta et al. 2016
Glucosidase	Graphene	Covalent binding	Higher biocatalytic activity	Gomes et al. 2018

Nanobiocatalysts showed high activity towards hydrolysis of real biomass, i.e., *Agave atrovirens* leaves. Gao et al. (2018) modified the graphene oxide using a hydrophobic spacer, and cellulase was quickly immobilized onto the modified surface of graphene sheet. Hydrophobic moiety, namely P-β-sulfuric acid ester ethyl sulfone aniline, was grafted onto the functionalized graphene oxide. Industrially important cellulase was immobilized onto the functionalized carbon nanomaterial by covalent binding method. An efficient immobilization methodology was achieved within ten minutes, and this yielded high immobilization and efficiency over 90 percent. As compared to free enzyme, immobilized biocatalyst improved high thermal and operational stabilities. Immobilized cellulase has a six-fold higher half-life as compared to free enzyme. Kinetic studies of immobilized enzymes showed better affinity by lower Km value as compared to free enzyme. Authors demonstrated that this immobilization protocol on activated graphene oxide nanocarrier is an ideal support for the other industrially important enzymes.

The cellulase sourced from *Aspergillus niger* was immobilized on multi-wall carbon nanotubes by covalent binding method (Ahmad and Khare 2018). The enzyme immobilization on the nanomaterial was structurally confirmed by Fourier Transform Infrared Spectroscopy and Scanning electron microscopy. Immobilized cellulase showed improved thermal stabilities that led to four-fold better half-life of immobilized biocatalyst as compared to free enzyme. Michalis Menten parameter Km values of the immobilized enzyme were two-fold higher than the free enzyme and immobilized enzyme retained biocatalyst potential even at the 10th recycles. Authors claimed the efficient nanocarrier support for cellulose hydrolysis.

Nanobiocatalysts were developed using two nanoparticles, namely magnetic nanoparticle and silica nanoparticle (Grewal et al. 2017). The cellulase immobilized on magnetic nanoparticle showed higher catalytic activity as compared to silica nanoparticle. Magnetic nanoparticle was further employed for saccharification of ionic liquid pre-treated sugarcane bagasse and wheat straw. Magnetic nanoparticles demonstrated high hydrolysis yield. Nanobiocatalysts were developed by immobilizing cellulase from *Aspergillus fumigatus* onto MnO_2 nanoparticles (Cherian et al. 2015). Nanobiocatalysts showed improved stability and reusability.

Robust nanobiocatalysts were developed by covalent binding of cellulase onto the surface of superparamagnetic nanoparticle (Abraham et al. 2014). Enzyme loading on the surface of the nanomaterials was optimized using different concentrations of the protein loading on the fixed quantity of the nanomaterials. The weight ratio of enzyme and nanomaterials was found optimal at equal concentrations of the proteins and nanomaterials. Robust nanomaterial-bound cellulase was characterized and compared to enzyme kinetics of the free enzyme. The kinetic parameters of the nanomaterial-bound cellulase enzyme was found to be quite comparable to free enzyme. The K_M and Vmax value of the nanomaterial-bound enzyme was found to be 2.6 mg/mL and 2.0 mg/mL/min, respectively. Thermostabilities of the nanomaterial-bound enzyme was significantly higher than the free enzyme, as free enzyme lost its fully activity at 80°C in first two hours, however, immobilized enzyme retained more than 80% activity even after six hours. Nanobiocatalysts showed high hydrolysis of microcrystalline cellulose and hemp hurds (natural cellulosic substrate). Robust nanobiocatalysts showed better storage stability, recyclability, and improved thermal stability.

Zhao et al. (2014) investigated the hydrolysis of cellulose to glucose using graphene oxide nanosupport. Authors claimed the functional groups, namely phenolic and carboxylic and its layered, soft structure, rendered graphene oxide with superior hydrolytic potential. Mubarak et al. (2014) developed a simple protocol for cellulase immobilization on the carbon nanotubes by physical adsorption method. Pristine multi-walled carbon nanotubes were functionalized using acid-oxidation method. The functionalized carbon nanotubes and cellulase enzyme were dispersed in aqueous phosphate buffer. Enzyme was non-covalently bound to functionalized carbon nanotubes. The physical adsorption of cellulase on carbon nanotube was confirmed using Fourier transform infrared (FTIR) spectroscopy and field emission scanning electron microscopy (FESEM). The carbon nanotubes hold the highest enzyme loading. The immobilized biocatalyst retained high biocatalytic potential even at the 6th cycle of the reusability.

Nanobiocatalysts were developed by covalent immobilization of β-glucosidase onto the surface modified magnetic nanoparticle (Verma et al. 2013). Robust showed improved catalytic activity towards synthetic substrate, and the same studies were compared to microparticle immobilized enzyme (Verma et al. 2013). Robust magnetic nanoparticles demonstrated high stability and fast recovery as compared to microparticles.

It is very pertinent from the above-cited studies that biocatalytic potential of lignocellulosic biomass converting enzymes, such as cellulase and glucosidase, has considerably improved at the interface of carbon nanomaterial and enzyme. Thus, carbon nanomaterial is the ideal nanocarrier for the study of enzyme immobilization with respect to potential bioethanol production.

8. Conclusion

The present article critically discusses the recent technological advances made in the production of biofuels through nanobiotechnology intervention. Robust nanobiocatalysts were successfully developed and employed for a plethora of applications, ranging from biomedical to bioenergy sector. Nanobiocatalysts were fully characterized by spectroscopy and microscopy techniques to understand the structure-function relationship at the interface of enzyme and nanomaterial.

Recyclable robust nanobiocatalysts can be reused, thus economising the process of biofuel production. The results of the various studies encompassing use of various novel nanomaterials for immobilizing enzymes involved in oil transesterification/biomass hydrolysis that demonstrate a sustainable biofuel production is very promising.

References

Abraham, R.E., M.L. Verma, C.J. Barrow and M. Puri. 2014. Suitability of magnetic nanoparticle immobilised cellulases in enhancing enzymatic saccharification of pretreated hemp biomass. Biotechnol. Biofuels 7: 90.

Ahmad, R. and S.K. Khare. 2018. Immobilization of *Aspergillus niger* cellulase on multiwall carbon nanotubes for cellulose hydrolysis. Bioresour. Technol. 252: 72–75.

Al-Zuhair, S. 2007. Production of biodiesel: possibilities and challenges. Biofuels Bioprod. Bioref. 1(1): 57–66.

Arumugam, A. and V. Ponnusami. 2017. Production of biodiesel by enzymatic transesterification of waste sardine oil and evaluation of its engine performance. Heliyon. 3(12): e00486.

Asuri, P., S.S. Bale, R.C. Pangule, D.A. Shah, R.S. Kane and J.S. Dordick. 2007. Structure, function, and stability of enzymes covalently attached to single-walled carbon nanotubes. Langmuir. 23(24): 12318–12321.

Bencze, L.C., J.H. Bartha-Vari, G. Katona, M.I. Tosa, C. Paizs and F.D. Irimie. 2016. Nanobioconjugates of *Candida antarctica* lipase B and single-walled carbon nanotubes in biodiesel production. Bioresour. Technol. 200: 853–860.

Cao, L. 2005. Immobilised enzymes: science or art? Curr. Opin. Chem. Biol. 9(2): 217–226.

Chen, Y.Z., C.B. Ching and R. Xu. 2009. Lipase immobilization on modified zirconia nanoparticles: Studies on the effects of modifiers. Process Biochem. 44(11): 1245–1251.

Cherian, E., M. Dharmendira kumar and G. Baskar. 2015. Immobilization of cellulase onto MnO_2 nanoparticles for bioethanol production by enhanced hydrolysis of agricultural waste. Chin. J. Cat. 36: 1223–1229.

Cho, E.J., S. Jung, H.J. Kim, Y.G. Lee, K.C. Nam, H.-J. Lee et al. 2012. Co-immobilization of three cellulases on Au-doped magnetic silica nanoparticles for the degradation of cellulose. Chem. Commun. (Camb.) 48(6): 886–888.

Coradin, T., N. Nassif and J. Livage. 2003. Silica-alginate composites for microencapsulation. Appl. Microbiol. Biotechnol. 61(5-6): 429–434.

Costa, S.A., H.S. Azevedo and R.L. Reis. 2005. Enzyme immobilization in biodegradable polymers for biomedical applications. pp. 301–323. *In*: R.L. Reis and J.S. Roman (eds.). Biodegradable Systems in Tissue Engineering and Regenerative Medicine. London: CRC Press LLC.

Da Ros, P.C.M., G.A.M. Silva, A.A. Mendes, J.C. Santos and H.F. de Castro. 2010. Evaluation of the catalytic properties of *Burkholderia cepacia* lipase immobilized on non-commercial matrices to be used in biodiesel synthesis from different feedstocks. Bioresour. Technol. 101(14): 5508–5516.

Desai, P., A. Dave and S. Devi. 2006. Alcoholysis of salicornia oil using free and covalently bound lipase onto chitosan beads. Food Chem. 95(2): 193–199.

Devanesan, M., T. Viruthagiri and N. Sugumar. 2007. Transesterification of Jatropha oil using immobilized *Pseudomonas fluorescens*. African J. Biotechnol. 6(21).

Dutta, N., S. Biswas and M.K. Saha. 2016. Biophysical characterization and activity analysis of nano-magnesium supplemented cellulase obtained from a psychrobacterium following graphene oxide immobilization. Enzyme Microb. Technol. 95: 248–258.

Dutta, N. and S. Saha. 2018. Immobilization of a mesophilic lipase on graphene oxide: stability, activity and reusability insights. Meth. Enzymol. 609: 247–272.

Dwivedee, B.P., J. Bhaumik, S.K. Rai, J.K. Laha and U.C. Banerjee. 2017. Development of nanobiocatalysts through the immobilization of *Pseudomonas fluorescens* lipase for applications in efficient kinetic resolution of racemic compounds. Bioresour. Technol. 239: 464–471.

El-Batal, A.I., A.A. Farrag, M.A. Elsayed and A.M. El-Khawaga. 2016. Biodiesel production by *Aspergillus niger* lipase immobilized on barium ferrite magnetic nanoparticles. Bioeng. 3: 14.

Fan, Y., G. Wu, F. Su, K. Li, L. Xu, X. Han et al. 2016. Lipase oriented-immobilized on dendrimer-coated magnetic multi-walled carbon nanotubes toward catalyzing biodiesel production from waste vegetable oil. Fuel. 178: 172–178.

Fan, Y., F. Su, K. Li, C. Ke and Y. Yan. 2017. Carbon nanotube filled with magnetic iron oxide and modified with polyamidoamine dendrimers for immobilizing lipase toward application in biodiesel production. Sci. Rep. 7: 45643.

Fukuda, H., A. Kondo and H. Noda. 2001. Biodiesel fuel production by transesterification of oils. J. Biosci. Bioeng. 92(5): 405–416.

Galliker, P., G. Hommes, D. Schlosser, P.F. Corvini and P. Shahgaldian. 2010. Laccase-modified silica nanoparticles efficiently catalyze the transformation of phenolic compounds. J. Colloid Interface Sci. 349(1): 98–105.

Gao, J., C.L. Lu, Y. Wang, S.S. Wang, J.J. Shen, J.X. Zhang et al. 2018. Rapid immobilization of cellulase onto graphene oxide with a hydrophobic spacer. Catal. 8: 2073–4344.

Gawlitza, K., C. Wu, R. Georgieva, D. Wang, M.B. Ansorge-Schumacher and R. von Klitzing. 2012. Immobilization of lipase B within micron-sized poly-N-isopropylacrylamide hydrogel particles by solvent exchange. Phys. Chem. Chem. Phys. 14(27): 9594–9600.

Georgakilas, V., J.A. Perman, J. Tucek and R. Zboril. 2015. Broad family of carbon nanoallotropes: classification, chemistry, and applications of fullerenes, carbon dots, nanotubes, graphene, nanodiamonds, and combined superstructures. Chem. Rev. 115: 4744–4822.

Ghany, N.A.A., S.A. Elsherif and H.T. Handal. 2017. Revolution of graphene for different applications: State-of-the-art. Surf. Interfaces 9: 93–106.

Gogotsi, Y. and V. Presser. 2013. Carbon Nanomaterials. CRC Press.

Gokhale, A.A., J. Lu and I. Lee. 2013. Immobilization of cellulase on magnetoresponsive graphene nano-supports. J. Mol. Catal. B Enzym. 90: 76–86.

Gomes, A.A., E.P. Telli, L.C. Miletti, E. Skoronski, M.G. Ghislandi, G. Felippe da Silva et al. 2018a. Improved enzymatic performance of graphene-immobilized β-glucosidase A in the presence of glucose-6-phosphate. Biotechnol. Appl. Biochem. 65: 246–254.

Gomes, D., M. Gama and L. Domingues. 2018b. Determinants on an efficient cellulase recycling process for the production of bioethanol from recycled paper sludge under high solid loadings. Biotechnol. Biofuels 11: 111.

Grewal, J., R. Ahmad and S.K. Khare. 2017. Development of cellulase-nanoconjugates with enhanced ionic liquid and thermal stability for *in situ* lignocellulose saccharification. Bioresour. Technol. 242: 236–243.

Guan, Q., Y. Li, Y. Chen, Y. Shi, J. Gu, B. Li et al. 2017. Sulfonated multi-walled carbon nanotubes for biodiesel production through triglycerides transesterification. RSC Adv. 7: 7250–7258.

Han, J., P. Luo, Y. Wang, L. Wang, C. Li, W. Zhang et al. 2018. The development of nanobiocatalysis via the immobilization of cellulase on composite magnetic nanomaterial for enhanced loading capacity and catalytic activity. Int. J. Biol. Macromol. 119: 692–700.

Hanefeld, U., L. Gardossi and E. Magner. 2009. Understanding enzyme immobilisation. Chem. Soc. Rev. 38(2): 453–468.

He, Y., T. Wu, X. Wang, B. Chen and F. Chen. 2018. Cost-effective biodiesel production from wet microalgal biomass by a novel two-step enzymatic process. Bioresour. Technol. 268: 583–591.

Hermanová, S., M. Zarevúcká, D. Bouša, M. Pumera and Z. Sofer. 2015. Graphene oxide immobilized enzymes show high thermal and solvent stability. Nanoscale 7: 5852–5858.

Hernandez-Martin, E. and C. Otero. 2008. Different enzyme requirements for the synthesis of biodiesel: Novozym 435 and Lipozyme TL IM. Bioresour. Technol. 99(2): 277–286.

Hoa, L.Q., M.D.C. Vestergaard and E. Tamiya. 2017. Carbon-based nanomaterials in biomass-based fuel-fed fuel cells. Sens. 17: 2587.

Hosseini, S.H., S.A. Hosseini, N. Zohreh, M. Yaghoubi and A. Pourjavadi. 2018. Covalent immobilization of cellulase using magnetic poly (ionic liquid) support: improvement of the enzyme activity and stability. J. Agric. Food Chem. 66: 789–798.

Ikeda, Y., Y. Kurokawa, K. Nakane and N. Ogata. 2002. Entrap-immobilization of biocatalysts on cellulose acetate-inorganic composite gel fiber using a gel formation of cellulose acetate-metal (Ti, Zr) alkoxide. Cellulose 9(3-4): 369–379.

Iso, M., B. Chen, M. Eguchi, T. Kudo and S. Shrestha. 2001. Production of biodiesel fuel from triglycerides and alcohol using immobilized lipase. J. Mol. Catal. B Enzym. 16(1): 53–58.

Jafarian, F., A.K. Bordbar, A. Zare and A. Khosropour. 2018. The performance of immobilized *Candida rugosa* lipase on various surface modified graphene oxide nanosheets. Int. J. Biol. Macromol. 111: 1166–1174.

Jang, M.G., D.K. Kim, S.C. Park, J.S. Lee and S.W. Kim. 2012. Biodiesel production from crude canola oil by two-step enzymatic processes. Renew. Energ. 42: 99–104.

Jegannathan, K.R., S. Abang, D. Poncelet, E.S. Chan and P. Ravindra. 2008. Production of biodiesel using immobilized lipase—a critical review. Crit. Rev. Biotechnol. 28(4): 253–264.

Ji, P., H. Tan, X. Xu and W. Feng. 2010. Lipase covalently attached to multiwalled carbon nanotubes as an efficient catalyst in organic solvent. AIChE J. 56: 3005–3011.

Kalia, S.B., G. Kaushal, M. Kumar, S.S. Cameotra, A. Sharma, M.L. Verma et al. 2009. Antimicrobial and toxicological studies of some metal complexes of 4-methylpiperazine-1-carbodithioate and phenanthroline mixed ligands. Brazilian J. Microbiol. 40: 916–922.

Kanwar, S.S., R.K. Kaushal, M.L. Verma, Y. Kumar, G.S. Chauhan, R. Gupta et al. 2005. Synthesis of ethyl laurate by hydrogel immobilized lipase of *Bacillus coagulans* MTCC-6375. Indian J. Microbiol. 45: 187–193.

Kanwar, S.S., H.K. Verma, S. Pathak, R.K. Kaushal, Y. Kumar, M.L. Verma et al. 2006. Enhancement of ethyl propionate synthesis by poly (AAc-co-HPMA-cl-MBAm)-immobilized *Pseudomonas aeruginosa* MTCC-4713, exposed to Hg^{2+} and NH^{4+} ions. Acta Microbiol. Immunol. Hung. 53: 195–207.

Kanwar, S.S., R.K. Kaushal, M.L. Verma, Y. Kumar, W. Azmi, R. Gupta et al. 2007a. Synthesis of ethyl oleate employing synthetic hydrogel-immobilized lipase of *Bacillus coagulans* MTCC-6375. Indian J. Biotechnol. 6: 68–73.

Kanwar, S.S., M.L. Verma, M.C. Maheshwari, S. Chauhan, S.S. Chimni and G.S. Chauhan. 2007b. Properties of poly (AAc-co-HPMA-cl-EGDMA) hydrogel-bound lipase of *Pseudomonas aeruginosa* MTCC-4713 and its use in synthesis of methyl acrylate. J. Appl. Polym. Sci. 104(1): 183–191.

Kanwar, S.S., C. Sharma, M.L. Verma, S. Chauhan, S.S. Chimni and G.S. Chauhan. 2008a. Short-chain ester synthesis by transesterification employing poly (MAc-co-DMA-cl-MBAm) hydrogel-bound lipase of *Bacillus coagulans* MTCC-6375. J. Appl. Polym. Sci. 109: 1063–1071.

Kanwar, S.S., S. Gehlot, M.L. Verma, R. Gupta, Y. Kumar and G.S. Chauhan. 2008b. Synthesis of geranyl butyrate with the poly (acrylic acid-co-hydroxy propyl methacrylate-cl-ethylene glycol dimethacrylate) hydrogel immobilized lipase of *Pseudomonas aeruginosa* MTCC-4713. J. Appl. Polym. Sci. 110: 2681–2692.

Kanwar, S.S. and M.L. Verma. 2010. Lipases. pp. 1–16. *In*: Encyclopaedia of Industrial Biotechnology. Wiley Publishers, USA.

Kanwar, S.S., M.L. Verma, S. Puri and G.S. Chauhan. 2015. Synthetic hydrogel: Characteristics and applications. pp. 173–212. *In*: Emerging Areas in Biotechnology. Publisher: Nirmal Book Agency, Kurukshetra, India.

Klibanov, A.M., T.-M. Tu and K.P. Scott. 1983. Peroxidase-catalyzed removal of phenols from coal conversion waste waters. Science 221(4607): 259–261.

Kumari, V., S. Shah and M.N. Gupta. 2007. Preparation of biodiesel by lipase-catalyzed transesterification of high free fatty acid containing oil from *Madhuca indica*. Energy Fuels 21(1): 368–372.

Kuo, C.-H., L.-T. Peng, S-C. Kan, Y.-C. Liu and C.-J. Shieh. 2013. Lipase-immobilized biocatalytic membranes for biodiesel production. Bioresour. Technol. 145: 229–232.

Lai, L. and A.S. Barnard. 2012. Surface phase diagram and thermodynamic stability of functionalisation of nanodiamonds. J. Math. Chem. 22: 16774–16780.

Lau, S.C., H.N. Lim, M. Basri, H.R.F. Masoumi, A.A. Tajudin, N.M. Huang et al. 2014. Enhanced biocatalytic esterification with lipase-immobilized chitosan/graphene oxide beads. PLoS One 9: 104695.

Law, S.Q.K., R. Halim, P.J. Scales and G.J.O. Martin. 2018. Conversion and recovery of saponifiable lipids from microalgae using a nonpolar solvent via lipase-assisted extraction. Bioresour. Technol. 260: 338–347.

Lee, C.H., T.S. Lin and C.Y. Mou. 2009. Mesoporous materials for encapsulating enzymes. Nano Today 4: 165–179.

Lee, H.K., J.K. Lee, M.J. Kim and C.J. Lee. 2010a. Immobilization of lipase on single walled carbon nanotubes in ionic liquid. Bull. Korean Chem. Soc. 31: 650–652.

Lee, S.H., T.T.N. Doan, K. Won, S.H. Ha and Y.M. Koo. 2010b. Immobilization of lipase within carbon nanotube-silica composites for non-aqueous reaction systems. J. Mol. Catal. B: Enzym. 62: 169–172.

Leung, D.Y., X. Wu and M. Leung. 2010. A review on biodiesel production using catalyzed transesterification. App. Energy 87(4): 1083–1095.

Li, K., Y. Fan, Y. He, L. Zeng, X. Han and Y. Yan. 2017. *Burkholderia cepacia* lipase immobilized on heterofunctional magnetic nanoparticles and its application in biodiesel synthesis. Sci. Rep. 71: 6473.

Li, Q., F. Fan, Y. Wang, W. Feng and P. Ji. 2013a. Enzyme immobilization on carboxyl-functionalized graphene oxide for catalysis in organic solvent. Ind. Eng. Chem. Res. 52: 6343–6348.

Li, S.-F., Y.-H. Fan, R.-F. Hu and W.-T. Wu. 2011. *Pseudomonas cepacia* lipase immobilized onto the electrospun PAN nanofibrous membranes for biodiesel production from soybean oil. J. Mol. Catal. B Enzym. 72: 40–45.

Li, X., H. Zhu, J. Feng, J. Zhang, X. Deng, B. Zhou et al. 2013b. One-pot polylol synthesis of graphene decorated with size-and density-tunable Fe_3O_4 nanoparticles for porcine pancreatic lipase immobilization. Carbon 60: 488–497.

Li, Y., X.Y. Wang, X.P. Jiang, J.J. Ye, Y.W. Zhang and X.Y. Zhang. 2015. Fabrication of graphene oxide decorated with $Fe_3O_4@SiO_2$ for immobilization of cellulase. J. Nanopart. Res. 17: 8.

Lima, J.S., P.H. Araújo, C. Sayer, A.A. Souza, A.C. Viegas and D. de Oliveira. 2017. Cellulase immobilization on magnetic nanoparticles encapsulated in polymer nanospheres. Bioprocess Biosyst. Eng. 40(4): 511–518.

Liu, C.-H., C.-C. Huang, Y.-W. Wang, D.-J. Lee and J.-S. Chang. 2012. Biodiesel production by enzymatic transesterification catalyzed by *Burkholderia* lipase immobilized on hydrophobic magnetic particles. Appl. Energy. 100: 41–46.

Lotero, E., Y. Liu, D.E. Lopez, K. Suwannakarn, D.A. Bruce and J.G. Goodwin. 2005. Synthesis of biodiesel via acid catalysis. Ind. Eng. Chem. Res. 44(14): 5353–5363.

Lotti, M., J. Pleiss, F. Valero and P. Ferrer. 2018. Enzymatic production of biodiesel: strategies to overcome methanol inactivation. Biotechnol. J. 13(5): e1700155.

Lupoi, J.S. and E.A. Smith. 2011. Evaluation of nanoparticle immobilized cellulase for improved ethanol yield in simultaneous saccharification and fermentation reactions. Biotechnol. Bioeng. 108(12): 2835–2843.

Mano, N., F. Mao and A. Heller. 2002. A miniature biofuel cell operating in a physiological buffer. J. Am. Chem. Soc. 124(44): 12962–12963.

Mano, N., F. Mao and A. Heller. 2003. Characteristics of a miniature compartment-less glucose-O_2 biofuel cell and its operation in a living plant. J. Am. Chem. Soc. 125(21): 6588–6594.

Mateo, C., J.M. Palomo, G. Fernandez-Lorente, J.M. Guisan and R. Fernandez-Lafuente. 2007. Improvement of enzyme activity, stability and selectivity via immobilization techniques. Enzyme Microb. Technol. 40(6): 1451–1463.

Mathesh, M., B. Luan, T.O. Akanbi, J.K. Weber, J. Liu, C.J. Barrow et al. 2016. Opening lids: modulation of lipase immobilization by graphene oxides. ACS Catal. 6: 4760–4768.

Mauter, M.S. and M. Elimelech. 2008. Environmental applications of carbon-based nanomaterials. Environ. Sci. Technol. 42: 5843–5859.

Meher, L., D.V. Sagar and S. Naik. 2006. Technical aspects of biodiesel production by transesterification—a review. Renew. Sust. Ener. Rev. 10(3): 248–268.

Mehrasbi, M.R., J. Mohammadi, M. Peyda and M. Mohammadi. 2017. Covalent immobilization of *Candida antarctica* lipase on core-shell magnetic nanoparticles for production of biodiesel from waste cooking oil. Renew. Energ. 101: 593–602.

Mochalin, V.N., O. Shenderova, D. Hoand and Y. Gogotsi. 2012. The properties and applications of nanodiamonds. Nat. Nanotech. 7: 11.

Moreno-Pirajan, J. and L. Giraldo. 2011. Study of immobilized *Candida rugosa* lipase for biodiesel fuel production from palm oil by flow microcalorimetry. Arabian J. Chem. 4(1): 55–62.

Mubarak, N.M., J.R. Wong, K.W. Tan, J.N. Sahu, E.C. Abdullah, N.S. Jayakumar et al. 2014. Immobilization of cellulase enzyme on functionalized multiwall carbon nanotubes. J. Mol. Catal. B: Enzym. 107: 124–131.

Mukhtar, H., S.M. Suliman, A. Shabbir, M.W. Mumtaz, U. Rashid and S.A. Rahimuddin. 2018. Evaluating the potential of oleaginous yeasts as feedstock for biodiesel production. Protein Pept. Lett. 25(2): 195–201.

Nasir, S., M.Z. Hussein, Z. Zainal and N.A. Yusof. 2018. Carbon-based nanomaterials/allotropes: a glimpse of their synthesis, properties and some applications. Materials 11(2): e295.

Nehdi, I.A., H.M. Sbihi, L.E. Blidi, U. Rashid, C.P. Tan and S.I. Al-Resayes. 2018. Biodiesel production from *Citrillus colocynthis* oil using enzymatic based catalytic reaction and characterization studies. Protein Pept. Lett. 25(2): 164–170.

Nguyen, L.T., Y.S. Lau and K.L. Yang. 2016. Entrapment of cross-linked cellulase colloids in alginate beads for hydrolysis of cellulose. Colloids Surf. B Biointerfaces 145: 862–869.

Noureddini, H., X. Gao and R. Philkana. 2005. Immobilized *Pseudomonas cepacia* lipase for biodiesel fuel production from soybean oil. Bioresour. Technol. 96(7): 769–777.

O'driscoll, K.F. 1976. Techniques of enzyme entrapment in gels. Methods Enzymol. 44: 169–183.

Park, H.J., A.J. Driscoll and P.A. Johnson. 2018. The development and evaluation of β-glucosidase immobilized magnetic nanoparticles as recoverable biocatalysts. Biochem. Eng. J. 133: 66–73.

Pavlidis, I.V., T. Tsoufis, A. Enotiadis, D. Gournis and H. Stamatis. 2010. Functionalized multi-wall carbon nanotubes for lipase immobilization. Adv. Eng. Mater. 12(5): B179–B183.

Pavlidis, I.V., T. Vorhaben, T. Tsoufis, P. Rudolf, U.T. Bornscheuer, D. Gournis et al. 2012. Development of effective nanobiocatalytic systems through the immobilization of hydrolases on functionalized carbon-based nanomaterials. Bioresour. Technol. 115: 164–171.

Prakash, H., P.S. Chauhan, T. General and A.K. Sharma. 2018. Development of eco-friendly process for the production of bioethanol from banana peel using inhouse developed cocktail of thermo-alkali-stable depolymerizing enzymes. Bioprocess Biosyst. Eng. 41(7): 1003–1016.

Prlainovic, N.Z., D.I. Bezradica, Z.D. Knezevic-Jugovic, S.I. Stevanovic, M.L. Avramov Ivic, P.S. Uskokovic et al. 2013. Adsorption of lipase from Candida rugosa on multi walled carbon nanotubes. J. Ind. Eng. Chem. 19(1): 279–285.

Puri, M., C.J. Barrow and M.L. Verma. 2013. Enzyme immobilization on nanomaterials for biofuel production. Trends Biotechnol. 31(4): 215–216.

Qi, B., J. Luo and Y. Wan. 2018. Immobilization of cellulase on a core-shell structured metal-organic framework composites: Better inhibitors tolerance and easier recycling. Bioresour. Technol. 268: 577–582.

Rastian, Z., A.A. Khodadadi, Z. Guo, F. Vahabzadeh and Y. Mortazavi. 2016. Plasma functionalized multiwalled carbon nanotubes for immobilization of *Candida antarctica* lipase B: Production of biodiesel from methanolysis of rapeseed oil. Appl. Biochem. Biotechnol. 178(5): 974–989.

Reetz, M.T., A. Zonta and J. Simpelkamp. 1996. Efficient immobilization of lipases by entrapment in hydrophobic sol gel materials. Biotechnol. Bioeng. 49(5): 527–534.

Rodrigues, J., V. Perrier, J. Lecomte, E. Dubreucq and S. Ferreira-Dias. 2016. Biodiesel production from crude jatropha oil catalyzed by immobilized lipase/acyltransferase from *Candida parapsilosis* in aqueous medium. Bioresour. Technol. 218: 1224–1229.

Roth, H.C., S.P. Schwaminger, F. Peng and S. Berensmeier. 2016. Immobilization of cellulase on magnetic nanocarriers. Chem. Open. 5(3): 183–187.

Sakai, S., Y. Liu, T. Yamaguchi, R. Watanabe, M. Kawabe and K. Kawakami. 2010. Production of butyl-biodiesel using lipase physically-adsorbed onto electrospun polyacrylonitrile fibers. Bioresour. Technol. 101(19): 7344–7349.

Salis, A., M. Pinna, M. Monduzzi and V. Solinas. 2008. Comparison among immobilised lipases on macroporous polypropylene toward biodiesel synthesis. J. Mol. Catal. B Enzym. 54(1): 19–26.

Sánchez-Ramírez, J., J.L. Martínez-Hernández, P. Segura-Ceniceros, G. López, H. Saade, M.A. Medina-Morales et al. 2017. Cellulases immobilization on chitosan-coated magnetic nanoparticles: application for *Agave Atrovirens* lignocellulosic biomass hydrolysis. Bioprocess Biosyst. Eng. 40: 9–22.

Sawangpanya, N., C. Muangchim and M. Phisalaphong. 2010. Immobilization of lipase on $CaCO_3$ and entrapment in calcium alginate bead for biodiesel production. Sci. J. UBU 2: 46–51.

Shah, S., K. Solanki and M.N. Gupta. 2007. Enhancement of lipase activity in non-aqueous media upon immobilization on multi-walled carbon nanotubes. Chem. Cent. J. 1: 30.

Shi, Q., D. Yang, Y. Su, J. Li, Z. Jiang, Y. Jiang et al. 2007. Covalent functionalization of multi-walled carbon nanotubes by lipase. J. Nanopart. Res. 9: 1205–1210.

Shimada, Y., Y. Watanabe, T. Samukawa, A. Sugihara, H. Noda, H. Fukuda et al. 1999. Conversion of vegetable oil to biodiesel using immobilized *Candida antarctica* lipase. J. Am. Oil Chem. Soc. 76(7): 789–793.

Shimada, Y., Y. Watanabe, A. Sugihara and Y. Tominaga. 2002. Enzymatic alcoholysis for biodiesel fuel production and application of the reaction to oil processing. J. Mol. Catal. B Enzym. 17(3): 133–142.

Soukharev, V., N. Mano and A. Heller. 2004. A four-electron O_2-electroreduction biocatalyst superior to platinum and a biofuel cell operating at 0.88 V. J. Am. Chem. Soc. 126(27): 8368–8369.

Su, D.S. and G. Centi. 2013. A perspective on carbon materials for future energy application. J. Energy Chem. 22: 151–173.

Su, F., G. Li, Y. Fan and Y. Yan. 2016. Enhanced performance of lipase via microcapsulation and its application in biodiesel preparation. Sci. Rep. 6: 29670.

Tan, T., J. Lu, K. Nie, L. Deng and F. Wang. 2010. Biodiesel production with immobilized lipase: a review. Biotechnol. Adv. 28(5): 628–634.

Thangaraj, B., Z. Jia, L. Dai, D. Liu and W. Du. 2017. Lipase NS81006 immobilized on functionalized ferric-silica magnetic nanoparticles for biodiesel production. Biofuels 1: 9.

Tran, D.T., C.L. Chen and J.S. Chang. 2012. Immobilization of *Burkholderia* sp. lipase on a ferric silica nanocomposite for biodiesel production. J. Biotechnol. 158(3): 112–119.

Trevan, M. 1988. Enzyme immobilization by entrapment. pp. 491–494. *In*: New Protein Techniques. Springer.

Verma, M.L. and S.S. Kanwar. 2008. Properties and application of Poly (MAc-2011co-DMA-cl-MBAm) hydrogel immobilized *Bacillus cereus* MTCC 8372 lipase for synthesis of geranyl acetate. J. Appl. Polym. Sci. 110: 837–846.

Verma, M.L., G.S. Chauhan and S.S. Kanwar. 2008a. Enzymatic synthesis of isopropyl myristate using immobilized lipase from *Bacillus cereus* MTCC 8372. Acta Microbiol. Immunol. Hung. 55: 327–342.

Verma, M., W. Azmi and S.S. Kanwar. 2008b. Microbial lipases: at the interface of aqueous and non-aqueous media: a review. Acta Microbiol. Immunol. Hung. 55: 265–294.

Verma, M., W. Azmi and S.S. Kanwar. 2009. Synthesis of ethyl acetate employing celite-immobilized lipase of *Bacillus cereus* MTCC 8372. Acta Microbiol. Immunol. Hung. 56: 229–242.

Verma, M.L. and S.S. Kanwar. 2010. Lipases. Wiley Encyclopedia of Industrial Biotechnology. Bioprocess Bioseparation and Cell Technology, Hoboken, New Jersey, USA 5: 3550–3565.

Verma, M.L. and S.S. Kanwar. 2010. Purification and characterization of a low molecular mass alkaliphilic lipase of *Bacillus cereus* MTCC 8372. Acta Microbiol. Immunol. Hung. 57: 191–207.

Verma, M.L., W. Azmi and S.S. Kanwar. 2011. Enzymatic synthesis of isopropyl acetate by immobilized *Bacillus cereus* lipase in organic medium. Enzyme Res. (Article ID 919386): 7 pages.

Verma, M.L. and S.S. Kanwar. 2012. Harnessing the potential of thermophiles: the variants of extremophiles. Dyn. Biochem. Process Biotechnol. Mol. Biol. 6: 28–39.

Verma, M.L., C.J. Barrow, J.F. Kennedy and M. Puri. 2012. Immobilization of β-d-galactosidase from *Kluyveromyces lactis* on functionalized silicon dioxide nanoparticles: characterization and lactose hydrolysis. Int. J. Biol. Macromol. 50: 432–437.

Verma, M.L., C.J. Barrow and M. Puri. 2013a. Nanobiotechnology as a novel paradigm for enzyme immobilisation and stabilisation with potential applications in biodiesel production. Appl. Microbiol. Biotechnol. 97: 23–39.

Verma, M.L., R. Chaudhary, T. Tsuzuki, C.J. Barrow and M. Puri. 2013b. Immobilization of β-glucosidase on a magnetic nanoparticle improves thermostability: application in cellobiose hydrolysis. Bioresour. Technol. 135: 2–6.

Verma, M.L., R. Rajkhowa, X. Wang, C.J. Barrow and M. Puri. 2013c. Exploring novel ultrafine Eri silk bioscaffold for enzyme stabilisation in cellobiose hydrolysis. Bioresour. Technol. 145: 302–306.

Verma, M.L., M. Naebe, C.J. Barrow and M. Puri. 2013d. Enzyme immobilisation on amino-functionalised multi-walled carbon nanotubes: structural and biocatalytic characterisation. PloS One 8: 73642.

Verma, M.L. and C.J. Barrow. 2015. Recent advances in feedstocks and enzyme-immobilised technology for effective transesterification of lipids into biodiesel. pp. 87–103. *In*: Microbial Factories. Springer, New Delhi.

Verma, M.L., M. Puri and C.J. Barrow. 2016. Recent trends in nanomaterials immobilised enzymes for biofuel production. Crit. Rev. Biotechnol. 36: 108–119.

Verma, M.L. 2017a. Enzymatic nanobiosensors in the agricultural and food industry. *In*: Nanoscience in Food and Agriculture 4: 229–245. Springer, Cham.

Verma, M.L. 2017b. Fungus-mediated bioleaching of metallic nanoparticles from agro-industrial by-products. pp. 89–102. *In*: Fungal Nanotechnology. Springer, Cham.

Verma, M.L. 2017c. Nanobiotechnology advances in enzymatic biosensors for the agri-food industry. Environ. Chem. Lett. 15: 555–560.

Verma, M.L., N.M. Rao, T. Tsuzuki, C.J. Barrow and M. Puri. 2019. Suitability of recombinant lipase immobilised on functionalised magnetic nanoparticles for fish oil hydrolysis. Catalysts 9: 420.

Veum, L. and U. Hanefeld. 2006. Carrier enabled catalytic reaction cascades. Chem. Commun. (8): 825–831.

Wan, X., X. Xiang, S. Tang, D. Yu, H. Huang and Y. Hu. 2017. Immobilization of *Candida antarctic* lipase B on MWNTs modified by ionic liquids with different functional groups. Colloids Surf. B 160: 416–422.

Wang, J., G. Meng, K. Tao, M. Feng, X. Zhao, Z. Li et al. 2012. Immobilization of lipases on alkyl silane modified magnetic nanoparticles: effect of alkyl chain length on enzyme activity. PloS One 7(8): e43478.

Wang, M., M. Chen, Y. Fang and T. Tan. 2018. Highly efficient conversion of plant oil to bio-aviation fuel and valuable chemicals by combination of enzymatic transesterification, olefin cross-metathesis, and hydrotreating. Biotechnol. Biofuels 11:30.

Wang, X., P. Dou, P. Zhao, C. Zhao, Y. Ding and P. Xu. 2009. Immobilization of lipases onto magnetic Fe_3O_4 nanoparticles for application in biodiesel production. ChemSusChem. 2(10): 947–950.

Wang, X., X. Liu, X. Yan, P. Zhao, Y. Ding and P. Xu. 2011. Enzyme-nanoporous gold biocomposite: excellent biocatalyst with improved biocatalytic performance and stability. PLoS One 6(9): e24207.

Wissink, M., R. Beernink, J.S. Pieper, A.A. Poot, G.H. Engbers, T. Beugeling et al. 2001. Binding and release of basic fibroblast growth factor from heparinized collagen matrices. Biomaterials 22(16): 2291–2299.

Xie, W. and J. Wang. 2012. Immobilized lipase on magnetic chitosan microspheres for transesterification of soybean oil. Biomass Bioenergy 36: 373–380.

Xie, W. and M. Huang. 2018. Immobilization of *Candida rugosa* lipase onto graphene oxide Fe_3O_4 nanocomposite: Characterization and application for biodiesel production. Energy Convers. Manag. 159: 42–53.

Xu, J., Z. Sheng, X. Wang, X. Liu, J. Xia, P. Xiong et al. 2016. Enhancement in ionic liquid tolerance of cellulase immobilized on PEGylated graphene oxide nanosheets: Application in saccharification of lignocellulose. Bioresour. Technol. 200: 1060–1064.

Yan, J., Y. Yan, S. Liu, J. Hu and G. Wang. 2011. Preparation of cross-linked lipase-coated micro-crystals for biodiesel production from waste cooking oil. Bioresour. Technol. 102(7): 4755–4758.

Yan, W., F. Li, L. Wang, Y. Zhu, Z. Dong and L. Bai. 2016. Discovery and characterization of a novel lipase with transesterification activity from hot spring metagenomic library. Biotechnol. Rep. (Amst.) 14: 27–33.

Yu, L., I.A. Banerjee, X. Gao, N. Nuraje and H. Matsui. 2005. Fabrication and application of enzyme-incorporated peptide nanotubes. Bioconjugate Chem. 16(6): 1484–1487.

Yu, N., L. Tan, Z.Y. Sun, H. Nishimura, S. Takei, Y.Q. Tang et al. 2018. Bioethanol from sugarcane bagasse: Focused on optimum of lignin content and reduction of enzyme addition. Waste Manag. 76: 404–413.

Yucel, S., P. Terzioğlu and D. Özçimen. 2013. Lipase applications in biodiesel production. Biodiesel-Feedstocks, Production and Applications 8: 209–249.

Yurekli, Y. and S.A. Altinkaya. 2011. Catalytic performances of chemically immobilized urease under static and dynamic conditions: A comparative study. J. Mol. Catal. B Enzym. 71(1): 36–44.

Zhang, B.T., X. Zheng, H.F. Li and J.M. Lin. 2013. Application of carbon-based nanomaterials in sample preparation: a review. Anal. Chim. Acta 784: 1–17.

Zhang, B., Y. Weng, H. Xu and Z. Mao. 2012. Enzyme immobilization for biodiesel production. Appl. Microbiol. Biotechnol. 93: 61–70.

Zhang, Q., Y. Wei, H. Han and C. Weng. 2018a. Enhancing bioethanol production from water hyacinth by new combined pretreatment methods. Bioresour. Technol. 251: 358–363.

Zhang, Y., K.Y. Rhee, D. Hui and S.J. Park. 2018b. A critical review of nanodiamond based nanocomposites: Synthesis, properties and applications. Compos. Part B-Eng. 143: 19–27.

Zhao, X., J. Wang, C. Chen, Y. Huang, A. Wang and T. Zhang. 2014. Graphene oxide for cellulose hydrolysis: how it works as a highly active catalyst? Chem. Commun. 50: 3439–3442.

Zhu, J. and G. Sun. 2012. Lipase immobilization on glutaraldehyde-activated nanofibrous membranes for improved enzyme stabilities and activities. Reactive and Functional Polymers 72(11): 839–845.

CHAPTER *7*

Carbon-based Nanomaterials for Energy Storage and Sensing Applications

Elochukwu Stephen Agudosi,[1] *Ezzat Chan Abdullah,*[1,*]
Nabisab Mujawar Mubarak[2,*] *and Mohammad Khalid*[3,*]

1. Introduction

Nowadays, the science and technology of developing nanomaterials have been in top gear due to the dire need of effective alternative sources of materials in the field of materials science and engineering. The primary reason for this paradigm shift is for cost reduction, reliability, and durability of nanomaterials, while making their production sustainable. In the electronics industry, for instance, the quest for renewable and effective alternative energy storage systems has increased over the years. The utilization of nanomaterials in developing different kinds of products is attributed to their properties and potential to further enhance properties. The field of nanoscience has seen some breakthroughs in the applications of nanomaterials in several areas of need, including energy storage and sensing (Lee et al. 2008).

In this chapter, carbon-based nanomaterials and their potential applications in energy storage and sensing have been reviewed. So far, a significant progress has been made in the use of carbon-based nanomaterials due to their intriguing and excellent properties. Carbon-based nanomaterials are also adjudged to have little or no adverse human and environmental impacts associated with their usage. Globally, the huge research interest from scientists in developing novel carbon nanomaterials is mainly due to their properties, such as excellent electrical and thermal conductivity (Balandin et al. 2008), large surface area (Chae et al. 2004), high electron mobility (Novoselov et al. 2005), and superior mechanical strength (Rao et al. 2010). When carbon-based nanomaterials are functionalized

[1] Department of Chemical Process Engineering, Malaysia-Japan International Institute of Technology (MJIIT) Universiti Teknologi Malaysia (UTM), Jalan Sultan Yahya Petra, 54100 Kuala Lumpur, Malaysia.
[2] Department of Chemical Engineering, Faculty of Engineering and Science, Curtin University, 98009 Sarawak, Malaysia.
[3] Graphene & Advanced 2D Materials Research Group (GAMRG), School of Science and Technology, Sunway University, 47500 Subang Jaya, Selangor, Malaysia.
Emails: elochukwu@live.utm.my, elochukwuagudosi@gmail.com
* Corresponding authors: ezzatc@utm.my, ezzatchan@gmail.com; mubarak.mujawar@curtin.edu.my, mubarak.yaseen@gmail.com; khalids@sunway.edu.my

with other nanomaterials or nanoparticles or functional species, their overall properties are further enhanced for specific applications. In the electronics industry, the flexibility and ease of manipulation of the properties of carbon-based nanomaterials with their functional species through functionalization enable the development of nano-electronics devices with improved qualities at a lower cost. Carbon-based nanomaterials have been employed in so many applications and in developing several devices, which include sensors (chemical, bio-, and gas) (Shan et al. 2009, Choi et al. 2010, Ge et al. 2012, Sun et al. 2011), lithium-ion batteries (Sun et al. 2011), supercapacitors (SCs) (Pendashteh et al. 2013, Nguyen and Zhao 2014), transparent electrodes (Sun et al. 2011, Hone et al. 2001), composites (Anwar et al. 2015), transistors (Haddad et al. 2018, Park et al. 2018), hole transporting layers (Iakobson et al. 2018), green tires (Seo et al. 2018), and adsorbents (Lingamdinne et al. 2018, 2019).

Several methods of synthesizing carbon nanomaterials have been developed over the years. They include exfoliation, thermal decomposition, chemical vapor deposition, chemical-based techniques (including Hummer's method), laser abrasion, and arc-discharge method (Choi et al. 2010, Guermoune et al. 2011, Wu et al. 2009, Cano-Márquez et al. 2009). The techno-economic analysis demonstrated that certain synthesis routes may not be viable due to high cost (occasioned by the energy requirements), and purity of the products. The chemical vapor deposition has been adjudged the most economical means of synthesizing graphene in large quantities. However, when the quality or purity of carbon-based nanomaterials becomes important for a specific purpose/application, certain synthesis methods are inevitable. In the synthesis of carbon-based nanomaterials, certain factors are critical in the process. Such factors as the carbon precursor, cost of the carbon precursor, the nature of the carbon precursor, the synthesis routes to be used, and the intended applications for the generated carbon nanomaterials are considered. Known sources of carbon precursors are hydrocarbons, highly oriented pyrolytic graphite (HOPG), which can either be in the form of solid or gas. Examples of carbon-based nanomaterials are carbon nanotubes (CNTs), carbon black (CB), carbon nanofibers, carbon nanoribbons, fullerenes, and graphene (Liu et al. 2018). So far, graphene, which is a two-dimensional (2-D) carbon material, has proved to be the most promising candidate among others due to its excellent conductivity (electrical and thermal), ultra-thinness, large surface area, excellent mechanical strength, and low cost of production. The great potential of graphene has been demonstrated in the area of conductor-switching, bio-imaging, gas separation, photo-catalysis, supercapacitors, lithium-ion batteries, electro-catalysis, and water desalination (Liu et al. 2018). Another form of novel carbon nanomaterial with interesting 3-D-helical or spiral morphology and excellent properties are the helical carbon fibers (HCFs). The morphology is wholly dependent on their fiber diameter, coil diameter, and coil pitch. HCFs have either solid or hollow structures and are useful in reinforcing composites due to their helical morphologies which is advantageous in achieving better anchorage in the embedding matrix. Higher infiltration and better load transfer are achieved with HCFs.

Figure 7.1 shows the various synthesis routes for carbon nanomaterial with special reference to graphene production.

Catalyzed thermal chemical vapor deposition (CTCVD) is the common synthesis technique used for the production of HCFs (Raghubanshi et al. 2016). Regarding the feedstock, acetylene is used as the carbon source while transition metal oxides are used as the catalysts. HCFs have such properties as high structural stability, high Young's modulus, super-elasticity, good electrical conductivity, torsion emission, magnetization, EM wave absorption, field emission, magnetoresistances (MR), etc. These factors present them as suitable candidates in developing catalysts for hydrogen sorption, supercapacitors, sensors, fuel cell catalyst support, EM wave absorber, etc. Due to the specific requirements of several applications, modifying carbon-based nanomaterials for specific applications has been made possible. For instance, different kinds of doping of graphene for different purposes can come in the form of nitrogen-doped, boron-doped, sulfur-doped, etc. In the modification of carbon-based nanomaterials, different dimensionalities are achieved to perform specific functions, namely; 0-D, 1-D, 2-D, and 3-D.

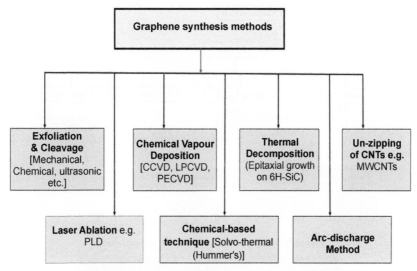

Figure 7.1: Synthesis methods for carbon nanomaterials.

2. Types of Nanomaterials and Classification of CNMs

Generally, nanomaterials are categorized into two groups, namely, natural and artificial types. The natural type of nanomaterials includes those that exist in biological systems, such as viruses (e.g., capsid) and substances in the human bone matrix. On the other hand, artificial nanomaterials are those synthesized through laboratory experiments. These artificial nanomaterials are further classified into four (4) groups, viz., carbon-based nanomaterials, metal-based nanomaterials, dendrimers, and composites (Figure 7.2; Table 7.2). Carbon-based nanomaterials are those which are made up of carbon materials, e.g., fullerenes, CNTs, HCFs, carbon nanofibers, etc. The metal-based nanomaterials are those of metallic materials, such as gold nanoparticles, silver nanoparticles, zinc nanomaterials, etc. Dendrimers are nanomaterials which have very useful applications in drug delivery. Composites are nanomaterials which are combined to form different structures and shapes with specific morphologies for specific applications, e.g., polymer nanocomposites, ceramics nanocomposites, metal nanocomposites, etc. For the purposes of this chapter, only carbon nanomaterials (CNMs) are studied alongside their energy storage and sensing applications.

3. Synthesis of Carbon Nanomaterials (CNMs)

There are several synthesis methods developed over the years for carbon nanomaterials. This study focused on the production of novel CNMs, such as CNTs and graphene. It is important to note that certain applications of CNMs will require specific synthesis routes. The structure of CNMs and their properties are very critical in the synthesis process. For CNTs, the arc discharge, laser deposition, and chemical vapor deposition methods are mainly employed. However, in the synthesis of graphene, more methodological approaches are being used due to the high commercial demand for graphene occasioned by its extraordinary properties. Graphene has been successfully synthesized through exfoliation, thermal decomposition (Epitaxial growth), arc discharge, pulse laser deposition, chemical-based techniques, unzipping of CNTs, and chemical vapor deposition processes. There are two basic synthesis approaches adopted for carbon nanomaterials, namely, top down and bottom up (Table 7.1) (Mitra et al. 2016). Figure 7.1 shows the typical synthesis methods for the production of carbon nanomaterials (graphene).

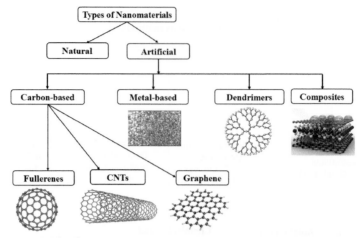

Figure 7.2: Types and classification of nanomaterials.

Color version at the end of the book

Table 7.1: Basic approaches adopted in graphene synthesis.

Approach	Method	Brief description	Advantages/disadvantages
Top-down	Mechanical exfoliation	Exfoliation of layers from graphite to form by scotch tape	Low output, High quality graphene
	Chemical reduction of graphite derivatives	Exfoliation of graphite derivative in suspension with chemical reducing agents	Large scale production, Significant defects
	Longitudinal unzipping of carbon nanotubes (CNTs)	CNTs are cut longitudinally with Ar or Ne plasma	Controlled edge graphene, Low scalability
Bottom-up	Chemical vapor Deposition	Decomposition of hydrocarbon on metal substrate at high temperature with low pressure or low temperature	Large area graphene, Poor yield
	Thermal decomposition (Epitaxial growth)	Evaporation of SiC on Si-wafer at high temperatures	Micron length graphene with few defects, but poor scalability and costly method due to high energy requirements
	Organic synthesis	Stepwise organic synthesis to produce polycyclic aromatic hydrocarbon	Large scale graphene, Few defects, High cost

3.1 Exfoliation and cleavage

This method has been used extensively for the production of graphene in large quantities. It involves the use of mechanical and chemical energy to collapse or break down the weak van der Waal forces in a pure graphite sheet. It is the repeated peeling of an HOPG sheet to form a graphene sheet. This method has been used to synthesize high quality graphene sheets (both single layer graphene (SLG) and few layer graphene (FLG)) (Lotya et al. 2009). Another approach in this method is to use a potassium metal to intercalate a graphite sheet and exfoliate with ethanol to form graphene sheets (Viculis et al. 2003). In a chemical approach, sonication of carbon nanosheets is done, which gives rise to nanoscrolls (Niyogi et al. 2006, Velicky et al. 2017, Novoselov et al. 2012), by using HOPG of 1 mm thickness in the oxygen plasma and then subjecting it to dry etching to produce 5-micrometer mesas, which are then baked. A scotch tape is used to peel off single-few layer graphene sheets from the graphite sheet. Silicon substrate has been used for this process and it was found to be a sustainable

Table 7.2: Categories of artificial nanomaterials.

Artificial nanomaterials	Brief description	Examples
Carbon-based	Carbon taking the form of hollow spheres, tubes, ellipses or cylindrical shapes	Fullerenes, carbon nanotubes (CNTs), graphene
Metal-based	Nanosized metals and metal oxides	Quantum dots, nanogold, nanosilver, Titanium oxide, Zinc oxide, etc.
Dendrimers	Nanosized polymers build from branched units. Useful in the process of catalysis and those with 3-D can be utilized in drug delivery	Poly-amidoamines (PAMAMs)
Composites	Formed as a result of combining nanoparticles with nanoparticles, or with larger/bulky type materials which may be carbon-based, metal-based, or polymer based	(a) Metal-Ni/Al$_2$O$_3$, Mg/CNT (b) Ceramic-Al$_2$O$_3$/SiO$_2$, Al$_2$O$_3$/TiO$_2$ (c) Polymer-Polymer/CNT, polyester/TiO$_2$, etc.

Nanoscale range for carbon-based nanomaterials (1 to 100 nanometer)
0-D–0-dimensional; 3 dimensions in nanoscale range
1-D–1-dimensional; 1 out of 3 dimensions in nanoscale range
2-D–2-dimensional; 2 out of 3 dimensions in nanoscale range
3-D–3-dimensional; Formed by the arrangement of multiple 0-D, 1-D, or 2-D materials in forming 3-D structure

approach. So many techniques have been adopted to exfoliate a high purity graphite sheet in order to produce graphene, including a green approach where citric acid was employed as the solvent to produce graphene nanosheets (Tung et al. 2009). With some adjustments or modification of the techniques, large volume graphene can be synthesized in different quality for electrical purposes. The main limitations of this technique are the structural defects of graphene in the liquid exfoliation phase due to the oxidation and reduction process, which leads to poor electrical properties of graphene.

3.2 Chemical vapor deposition

So far, the chemical vapor deposition (CVD) is adjudged a powerful and effective method for the production of CNMs. It has a great advantage of commercial production of large surface area graphene for so many industrial applications. This process employs high temperature for the decomposition of carbonaceous precursors or carbon-containing materials on the substrate surfaces for the formation of high quality graphene films (Dathbun and Chaisitsak 2013). Here, precursors are typically absorbed on the crystalline substrate surface, followed by their thermal decomposition, which leads to the generation of sites of absorption on the substrate surface. These sites of absorption are regarded as the nucleus, and then through nucleation, the continuous growth of thin films is promoted (Reinke et al. 2015). Methane (CH$_4$), ethylene (C$_2$H$_2$), polycyclic aromatic hydrocarbons (PAHs), and benzene are commonly known sources of carbon for the CVD of graphene. CVD process was first used to synthesize planar FLG (Somani et al. 2006).

There are mainly three different approaches to the CVD technique, namely, atmospheric pressure CVD, low pressure CVD, and plasma enhanced CVD (e.g., microwave plasma enhanced CVD) (Chae et al. 2009, Yuan et al. 2009, Li et al. 2009, Dervishi et al. 2009a, b). So many other CVD processes have been developed, as a solution for high temperature requirements is being sought. A system of CVD that will drastically reduce the activation energy during thermal cracking (to achieve high cracking efficiencies) will be the most novel CVD process. During thermal cracking, the carbonaceous precursors decompose to form carbon atoms, which are then nucleated to form a graphitic nanostructure from the dissociated carbon atoms. Higher energy requirements discourage a CVD process that is free of a metal catalyst. Without a metal catalyst, about 2500°C is required for a graphitic structure to be formed. Transition metals are used to reduce the reaction temperatures, as they work as an effective catalyst to quickly transform hydrocarbons into graphitic materials (Lin et al. 2012). The CVD method of the synthesis of CNMs is reliable, feasible, and cost-effective for so many industrial applications. Meanwhile, recent research focuses on finding a more sustainable approach to produce CNMs, such as graphene and CNTs in large volume and high quality through the CVD process at temperatures lower than 1000°C (Li et al. 2011).

Factors such as the size and type of the catalyst, hydrocarbon source, and the CVD approach will influence the growth temperature of the CVD process. Catalysts are critical in the CVD process. Catalysts such as Cu, Ni, and Pt, which are transition metals, are the most widely used in the CVD of graphene. Moreover, the flow rate of the precursor gas, the ratio of the gases (in the case of gas mixture), the reaction time, and the temperature of the furnace are prime factors in the CVD process (Gallo et al. 2012, Jang et al. 2017, Naghdi et al. 2018, Zhang et al. 2013, Reina and Kong 2012, Wu et al. 2015, Kataria et al. 2014). SLG is preferred to multi-layer graphene (MLG) as they possess the best graphene properties. Hydrocarbon sources with specific chemical structures and low activation energies deliver production at desirable low temperatures (Guermoune et al. 2011). Nevertheless, the use of PAHs and benzene are encouraged due to the ring shapes for carbon atoms they possess, which is a better way of reducing the activation energy for the production of CNMs using the CVD process. Thus, the defect density is drastically reduced.

3.3 *Chemical-based techniques*

This method is the chemical extraction of graphene films from graphite, unlike the liquid phase exfoliation technique. The key processes in chemical-based technique are oxidation, purification, dilution, and centrifugation. Carbon nanofilms (CNF) are synthesized from natural graphite through oxidation and purification processes (Horiuchi et al. 2004). Immediately after synthesis of CNF, dilution is done in methanol (for dispersion), and centrifugation steps help in the extraction of the thinnest sheets from the dispersion. The thickness of CNF is dependent on the dilution factor. FLG of 1–6 layers have been reported with this technique. A different chemical approach uses sulfuric and nitric acids in a first step to initiate intercalation between the layers of graphite, followed by thermal expansion (fast heating to 1000°C) to produce thin graphite sheets, while the explosive evaporation of the acid molecules goes on. In the second step of intercalation, sonication is carried out using a surfactant soon after tetrabutylammonium hydroxide and oleum mixture is employed to produce single-FLG (Park and Ruoff 2009). This method is suitable for the commercial production of graphene. The crystallinity of the source material (input graphite) and the lateral size will be a measure of the graphene layer number after synthesis. The main disadvantage of this technique is the high temperature requirements for thermal expansion. This would mean high cost and high carbon footprint; hence, it is not suitable for large volume graphene production. Furthermore, despite the high temperature requirements, the excessive use of chemicals in the interaction process usually damages the final product and creates undesirable defects due to aggressive redox reaction. Another chemical approach for large volume graphene production is the solvothermal technique (the Modified Hummer's process). Modified Hummer's technique employs graphite powder to synthesize the powdered form of graphene oxide, which is then chemically reduced to graphene (Solvothermal method). The experimental procedure for this approach is illustrated in the research study conducted by Zaaba et al. (Zaaba et al. 2017). The highs of this technique are not only that it is eco-friendly since it involves no toxic gas emission, but that the synthesis method to produce graphene in commercial quantities is viable.

3.4 *Thermal decomposition*

This method is mostly known as epitaxial growth of CNMs on an electrically insulated surface, such as 6H-SiC. Graphene sheets are produced through thermal decomposition of 6H-SiC at temperatures of about 1250–1450°C for 20 minutes with H_2 as the etching gas (Wu et al. 2009). Graphene is epitaxially grown on the H_2 etched surface of 6H-SiC. The number of layers generated depends on the decomposition temperature (Berger et al. 2004). This method is most suitable for the commercial production of wafer-size graphene films. Low atom thickness (as low as one) of graphene has been produced through this method, and this production route is viable in the semiconductor industry for applications mostly in post-CMOS age (Berger et al. 2004, Hass et al. 2008, Rollings et al. 2006, Ohta et al. 2008, Cambaz et al. 2008). A much lower temperature of decomposition is required when Ni substrate is used to catalyze the process (Ohta et al. 2008, Juang et al. 2009). The challenge of this method is the control of graphene thickness and repeated production of multilayer (up to 60 layers which is undesirable) graphene. The mechanism of the growth process and effect of the interface need to be further studied for a better understanding towards the application in developing future electronics devices/systems.

3.5 Unzipping CNTs

This method is a new synthesis method for graphene and it utilizes multi-wall nanotubes (MWNTs) as the material source. When the unzipping of multi-wall carbon nanotubes (MWNTs) through chemical means is done, graphene flakes and nanoribbons are formed. The very first successful step came about by the longitudinal opening up of MWNT through the intercalation of lithium and ammonia and then exfoliating in acid medium, followed by thermal expansion. The initial products are basically nanoribbons or partially opened MWNTs and graphene flakes (Cano-Márquez et al. 2009). Another approach adopts etching treatment to open up the MWNTs to form graphene. A chemical treatment approach is a multi-step procedure where MWNTs are unzipped by chemical exfoliation using concentrated H_2SO_4 through a stepwise oxidation process and then using $KMnO_4$ reduction in NH_4OH and surfactant (Kosynkin et al. 2009). This method of unzipping MWNTs enables graphene synthesis in a substrate free manner. Another current approach of producing multilayer graphene (MLG) is through calcination of Al_2S_3 under CO and argon combined gas environment. CO reduces aluminum sulfide to gaseous carbon and alpha alumina, thereby precipitating graphene sheets on the alumina particles (Kim et al. 2009). Even though this approach is simple and fascinating, there are certain limitations in understanding the process mechanism in terms of the control of process parameters, bulk production of graphene, and control of the number of graphene layers. The main limitation of this technique is the inadequate knowledge of the mechanism of transformation.

3.6 Arc discharge method

High quality B- and N-doped graphene have been synthesized using the arc discharge method. The nature of graphene synthesized is a function of the environment in which the synthesis occurs. Krätschmer et al. (Krätschmer et al. 1990) were the first scientists to synthesize graphene through this means. FLG (2–4 layers) has been synthesized using this technique (Chen et al. 2012). Moreover, several carbon-based nanomaterials have been successfully synthesized using the arc discharge technique, e.g., fullerenes and carbon nanotubes (CNTs) (Xu et al. 2005, Li et al. 2010a, Antisari et al. 2010, Kim et al. 2016, Tseng et al. 2018, Mubarak et al. 2014). In 2010, FLG sheets having 2–4 layers were successfully synthesized under H_2-He gas mixture (Rao et al. 2010). It should be noted that different discharge atmospheres may induce different bonds, which can result in different properties exhibited by the as-produced graphene. The electric arc discharge system is comprised mainly of two electrodes and a chamber made of steel, and is systematically cooled by water while in operation. Typically, a steady current of 100–150 A is required for the discharge process after the arc is generated and becomes crucial for the smooth running of the process.

The atmosphere of arc graphite rod evaporation for the production of graphene determines the type of graphene to be generated. Air, H_2, and $(NH_3 + He)$ are the common environments in which graphene has been synthesized using this process. N-doped graphene sheets are produced using ammonia-helium (NH_3 + He) mixture as the discharge atmosphere. For graphene sheet synthesis, with high electrical conductivity and dispersivity, C-O and O-C=O bonds are injected in the CO_2-He gas mixture. The discharge occurs when the graphite rods are brought very close together and plasma is generated. By continuously rotating the cathode and keeping a constant close distance of about 1–2 mm, the anode is constantly consumed and graphene is generated. Collection of the soot generated is done after the discharge process under ambient conditions. B-doped graphene is synthesized when the arc discharge is done in the atmosphere mixture of H_2 and ammonia B_2H_6 (Subrahmanyam et al. 2009), while N-doped is achieved in combined environment of He and NH_3 (Li et al. 2010b).

Many of the synthesis processes that adopt the arc discharge method have demonstrated that H_2-Inert gas mixtures (atmospheres) can produce highly-crystalline and high-purity FLG sheets. In this process, H_2 plays the role of inhibiting the rolling and closing of graphene sheets by disallowing the loosely attached carbon bonds to yield FLG sheets. Particularly, graphene sheets produced by the arc discharge technique can be used to enhance the electrical and electrochemical performances of LIBs, which are ubiquitous power sources for portable electronics systems/devices. The property improvement of LIBs comes in handy when the graphene sheets are functionalized or fused with other NPs for use as candidates for the anode and cathode electrodes. Direct current (DC) hydrogen arc discharge of pure graphite electrodes in different gas mixtures can be used to synthesize high-crystalline FLG sheets in commercial quantities. Studies show that the initial discharge capacity of LIB when graphene sheets are used to improve the electrodes can reach as high as 1332 mAh/g at a current density of 50 mA/g. In this situation, after 300 cycles, the capacitance retention stands at 323 mAh/g, showing excellent cyclic stability (Chen et al. 2012).

3.7 Laser ablation

This method relies on the laser exfoliation or ablation of amorphous graphite, and is sometimes called pulsed laser deposition (Kazemizadeh and Malekfar 2018). The technique is relatively new and has similar working principles as the arc discharge method for graphene synthesis. It keeps attracting research interest due to the fact that it utilizes solid carbon sources and requires lower temperatures. Laser ablation has been used to fabricate porous graphene. Porous graphene (PG) is a new carbon-based nanomaterial and this allotrope is currently gaining huge attention in the field of nanomaterial due to its crystallinity, excellent porosity properties, and light weight; thus making it suitable for some applications in biotechnology and nanoelectronics. Here, PG is synthesized (direct deposition on a substrate) when graphite is vaporized through laser irradiation and under a high temperature in an inert atmosphere. This process utilizes less energy due to lower temperature demands during a reaction, and gives excellent crystallinity of graphene since the quality of the synthesized graphene is a function of the process temperature.

In the past, PG had been synthesized through physical removal of carbon atoms from graphene sheets by means of mechanical exfoliation and using electron beam (Fischbein and Drndić 2008), helium bombardment (Bell et al. 2009), etching with nitrogen (Fox et al. 2011), and reduction with silver NPs (Liu et al. 2015). The key parameters in the laser ablation method of synthesizing CNMs include laser intensity, beam wavelength, pulse duration, type of pressure, and temperature of the carrier gas. A good control of these parameters will generate a high quality PG and several kinds of PG can equally be made possible by manipulating them. The interesting thing with this method is that the reactor/equipment generates a laser, which produces PGs after exfoliating graphene sheets from graphite in a one-step simultaneous process. Production of PG from HOPG as the solid source of carbon using the laser ablation method for the first time was reported by Russo (Russo et al. 2014). Laser ablation is a one-step process and takes place in vacuum and under an inert gas environment. Other techniques use CVD process to create the porous substrate before the PG synthesis, and this method is known as the 'template method' (Ning et al. 2011).

The period of laser irradiation determines the number of layers of graphene synthesized and the laser intensities determine the structural crystallinity of the PGs. With high laser intensities, highly crystalline PGs are produced. The control of the gas flow rate is also a critical factor in this process, since the gas flow rate greatly influences the formation of PGs. Powdered metal catalysts, such as Ni, enhance the synthesis process and determine the rate of graphitic nanostructure formation. However, not all metal catalysts will form Sp^2 structure in the laser ablation. Thus, metal catalysts, such as Fe, Co, Pt, are not suitable for the formation of Sp^2 structure in porous graphene structures in the laser ablation method, but rather will lead to the formation of carbon nanoparticles (CNPs) independent of laser intensity. Studies have shown that few-layered graphene can be synthesized using laser deposition on catalytic Ni thin film (Cappelli et al. 2004, Bertke et al. 2009). Factors such as the period of irradiation (ablation time) and substrate temperature determine the number of layer of graphene to be synthesized. The basic steps in the process include carbon atom adsorption, crystallization, segregation, and re-crystallisation, which happen as a result of the interaction between carbon atoms and the metal substrates. For high growth model of CNMs, the laser ablation method offers a one-step process for the formation of lightweight, porosity, and crystalline properties of porous graphene (Kazemizadeh and Malekfar 2018). Only a few studies on the use of this method for the synthesis of CNMs in large volume have been recorded.

4. Carbon Nanotubes (CNTs)

One of the most researched carbon materials in the 21st century are carbon nanotubes, and they are one allotrope of carbon discovered by Iijima (a Japanese physicist). The international attention grown for CNTs is due to its superior properties, such as high melting point, a wide band gap, high tensile strength, and excellent thermal conductivity (Mubarak et al. 2014, Singh and Song 2012). CNTs possess these qualities due to their small size and the carbon atoms strongly bonded together covalently. The three main types of CNTs are SWCNT, DWCNT, and MWCNT. Typically, CNTs are synthesized from either solid or gaseous carbonaceous materials, such as HOPG and hydrocarbons. Three main synthesis processes for the production of CNTs are the arc discharge method, CVD process, and laser deposition method. Others are the arc-jet plasma, floating catalyst, simplified carbon arc, and fluidized-bed CVD processes (Ying et al. 2011). Some progress has been made in the synthesis of CNTs for commercial purposes through the continuous processes that ensure sustainability. Continuous production of CNTs has been achieved through the arc discharge and CVD methods. To maintain a reliable continuous synthesis process for CNTs, scientists use the idea of feeding the reactor with the carbon materials, catalysts, and discharging the final products continuously at a constant rate. The main challenge of the continuous production process for CNTs with catalysts is the preparation of the catalysts and purification of the final products (synthesized CNTs). So far, catalysts are prepared separately with no purification step, and the synthesized CNTs are purified after production. An integrated system will be a novel approach, where all the activities ranging from feeding the reactor with carbon materials, catalyst treatment, and product purification are done in one continuous step. With this kind of system in place, purified CNTs can be produced in large quantities in a sustainable manner. Scientists are encouraged to look at the yield of CNTs in terms of g_{CNT}/mol_{carbon} source instead of $g_{CNT}/g_{catalyst}$. Lastly, the issue of handling bulk particles and the interaction with the reactor needs to be studied further in the continuous CVD of CNTs (Ying et al. 2011).

5. Graphene as a Superlative Carbon-based Nanomaterial

Graphene is an allotrope of carbon with a single Sp^2 atomic carbon layer, and has gained unprecedented attention in the field of nanoscience and nanotechnology since its discovery in 2004 (Chae et al. 2004) due to its excellent electrical properties (Novoselov et al. 2004, Ferrari et al. 2015). A plane layer of graphite of a single atom graphene can be produced from a variety of materials, which include hydrocarbon and other carbonaceous gases (Geim 2009). Graphene possesses such properties as high electron mobility at room temperature ($\sim 250,000$ cm^{-3}g^{-1}) (Novoselov et al. 2005, Geim 2009), and large active surface area (2,630 m^{2}g^{-1}) (Chae et al. 2004, Stoller et al. 2008), excellent thermal conductivity (5000 Wm^{-1}K^{-1}) (Balandin et al. 2008), thermal resistance of $\sim 4 \times 10^{-8}$ Km^{2}W^{-1} (Freitag et al. 2009), and superior mechanical properties with Young's modulus of 1 TPa (Rao et al. 2010, Taha-Tijerina et al. 2013). These tremendous properties make graphene a good candidate to be employed in electrochemical applications. Studies have shown that carbon nanomaterials, such as graphene, fullerenes, generate electrocatalytic effects in amperometric detective systems (Pérez-López and Merkoçi 2012). These qualities present these carbon nanomaterials as the ultimate electrode materials both in energy storage and biosensing applications (Zhou et al. 2009).

The synthesis methods for graphene include exfoliation (Niyogi et al. 2006, Novoselov et al. 2012, Li et al. 2008), chemical vapor deposition (CVD) (Reina and Kong 2012, Wu et al. 2015, Kataria et al. 2014), thermal decomposition-epitaxial growth on electrically insulated surfaces (Wu et al. 2009, Juang et al. 2009), unzipping of carbon nanotubes-CNTs (Cano-Márquez et al. 2009), chemical-based techniques (Horiuchi et al. 2004), Hummer's method (Zaaba et al. 2017), arc-discharge method (Antisari et al. 2010), and laser ablation (Pulsed laser deposition) (Yang and Hao 2016). The techno-economic viability of these synthesis routes enables scientists to determine the most viable synthesis routes for the commercial production of graphene in term of cost, suitability, and sustainability. So many potential applications of graphene have been recorded, which include single molecule detection (e.g., hydrogen storage), transparent conducting electrodes, composites, transistors, sensors, solar cells, photo-detectors, and energy storage devices (supercapacitors and Lithium-ion batteries) (Lee et al. 2008). Future electronics, semiconductor and composite industries have continued to see graphene as a breakthrough in the quest for new materials for nanoelectronics devices. It has also been recorded that when graphene is used as filler with the insulating polymer matrix, it enhances the electrical conductivity of the composites. Excellent properties of graphene include electrical, mechanical, anomalous quantum Hall effect, thermal, and optical strengths. Graphene-based nanoparticles (NPs) have unique electric transport and structural properties, including high carrier mobility and saturation velocities, excellent thinness, chemical inertness, high mechanical strength, high current carrying capacity, and high thermal/electrical conductivity. Huge research interest in graphene and graphene-based NPs has continued globally.

In the process of graphene synthesis, the operating parameters are statistically optimized to produce high graphene yield in terms of the mass and purity. Graphene and graphene-based NPs have wide industrial applications, which include flat panel field emission displays (Ruoff and Lorents 1995, Eda et al. 2008), nanoelectronics devices (Hone et al. 2001), chemical sensors (Shan et al. 2009), hydrogen storage (Moon et al. 2001, Kaiser 2001) and scanning probe tip (Das et al. 2014), transistors (Haddad et al. 2018, Park et al. 2018), energy storage (Sun et al. 2011), gas sensors (Choi et al. 2010, Toda et al. 2015), bio-sensors (Ge et al. 2012, Lu et al. 2007), and supercapacitors (Pendashteh et al. 2013, Nguyen and Nguyen 2016). Graphene has also been found to be an excellent material in the fabrication of several nanocomposites. When transition metal oxides are used to functionalize graphene/graphene sheet, they enhance specific capacity and improve thermal conductivity and cyclic stability in LIBs. This process of developing nanocomposites using graphene for high-end electronic applications ensures sustainability in terms of cost-effectiveness and improvements in LIBs battery's charge and discharge process.

However, carbon nanomaterials are not accurately defined in their chemical structure and physical credentials, e.g., size, shape, and number of layers (Kochmann et al. 2012). It is therefore important to control the synthesis process in order to achieve a relatively uniform structure with a slight difference in overall properties. Up until now, graphene used in the electrochemical analysis is mainly synthesized through redox means, which introduces a lot of defects to its structure and generates hetero-atoms (Hummers and Offeman 1958, Marcano et al. 2010). The defects create room for oxygen-containing groups, which negatively influence the electrochemical properties of graphene. More so, graphene derived from other synthesis methods, such as CVD, also contains some form of defects in its structure, and some impurities are generated in the course of transferring graphene from metallic support to an insulating substrate (Kim et al. 2009, Shang et al. 2008). Other methods

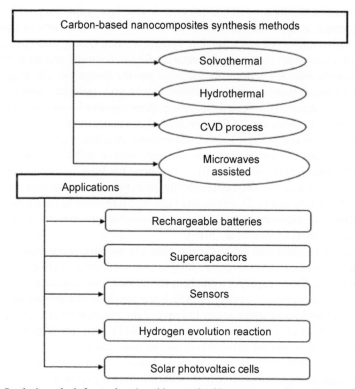

Figure 7.3: Synthetic methods for graphene/transition metal oxides nanocomposites and potential applications.

of graphene production have been reported in fabricating graphene in different sizes, shapes, and quality. The most commonly used techniques are exfoliation; mechanical exfoliation (Novoselov et al. 2012), and chemical exfoliation (Park and Ruoff 2009). The mechanical exfoliation method appears to be suitable for large volume graphene production. All of the synthesis routes provide some advantages in some ways. It is important to know that the defects in carbon nanomaterials do not only influence the conductance (Tang et al. 2009), but offer the possibility of functionalizing them with biomolecules and/or metal/metal oxide nanoparticles (usually transition metal oxides). This process provides the opportunity to enhance the sensitivity of these materials so as to make them useful as sensors in great variability. The act of designing the size and the morphology of graphene and combining it with other nanomaterials gives rise to composite materials which have improved sensing capabilities (Ambrosi et al. 2014). Figure 7.3 shows the synthetic methods for graphene/ transition metal oxides nanocomposites and potential applications.

6. Energy Storage Applications of CNMs

In the field of materials science, electrochemical energy storage has become a big challenge due to the rising need for portable electronic devices and systems. Traditionally, batteries are routinely used to store electrical energy in chemical form due to high energy density that they possess. Even though batteries possess high energy density, they have low power output, unlike the conventional di-electric capacitors that have high power output and low energy density instead. High power density, which is a characteristic feature of the supercapacitors or ultra-capacitors, has made them become an intermediate between batteries and dielectric capacitors (Zhao et al. 2011). Due to the high power density of supercapacitors, they can charge/discharge quickly with long cycle life (> 100,000 cycles). These properties make them the best energy storage devices (Xiong et al. 2014). Supercapacitors can be manufactured in the form of EDLC (electric double-layer capacitor), which stores energy through ion adsorption at the electrode/electrolyte boundary or in the form of pseudo-capacitor, where energy is stored due to fast and reversible Faradic reaction (Zhang and Zhao 2009, Wasiński et al. 2015). EDLC chiefly uses carbon-based materials for the electrons, and the power electrostatic charge is built up at the interface between the electrode and the electrolyte. The large surface area of graphene in EDLCs makes them a suitable candidate for electrode materials. Several applications of graphene have been recorded in this regard. Specific graphene synthesis methods will inform specific applications. Graphene has been used to functionalize other nanomaterials to enhance their electrical conductivity by making its composites. The two-dimensional structure of graphene, high surface area, great optical magnetic and electronic properties, and high electron mobility attracted more scientists to it (Rakhi and Alshareef 2011, Wu et al. 2012). Graphene has been applied as conducive support and building block to enhance the electrical properties of transition metal oxides, such as SnO, NiO, ZnO, MnO_2, CO_3O_4, RuO_2, and Fe_2O_3 (Yang and Hao 2016). In order to tackle the issue of a poor electrical conductivity of transition metal oxides, their nanostructures are spread over the extremely conductive graphene matrix (Anwar et al. 2015).

Nanostructure formation can be made possible via so many techniques, which include co-precipitation, solvothermal, hydrothermal, sol-gel, and microwave micro-emulsion (Xiao et al. 2013). In the fabrication of supercapacitors, graphene has replaced other carbon matrices, such as CNF, mesoporous carbon as the conductive materials with better capabilities. Carbon-based nanomaterials have been supplemented into a carbon matrix to improve the capacitance of CNF (Kafy et al. 2017, Zhou and Wu 2013, Dong et al. 2013, Selvakumar et al. 2017). A variety of carbon materials other than graphene have been employed as reliable candidate materials for EDLC electrode (Zhang and Zhao 2009). Supercapacitors are seen as the new generation of renewable energy storage devices since they can be regarded as alternatives to batteries due to their high power density ($\sim 1,000–10,000$ W/kg), outstanding cycle performance, no memory effect, and low maintenance cost (Zhang and Zhao 2009, Winter and Brodd 2004, He et al. 2017). Functionalizing graphene with other nanomaterials or functional species enhances the electrical conductivity and electrochemical performance of the materials for so many energy storage applications.

6.1 Supercapacitors

With amazing properties, such as high power density, ultra-thinness, energy density, and long cycle life, supercapacitors are considered to be one of the most promising candidates in the electrochemical renewable energy storage systems (Borenstein et al. 2017, Chen et al. 2017). Typically, two kinds of energy storage mechanisms exist for supercapacitors–the electrical double-layer capacitance (EDLC) and pseudocapacitance. The carbon material is a type of EDLC due to its low cost and excellent stability (Chen et al. 2010, Chua et al. 2011). On the other hand, transition metal oxides and conducting polymers are considered pseudocapacitors due to the reversible Faradic oxidation-reduction reaction. In an effort to enhance the electrochemical properties of the capacitors, several attempts have been made for the functionalization of different components and nanostructured optimization of materials (Aricò et al. 2010). To this end, graphene has been synthesized by some different methods and investigated for electrode materials for supercapacitors. A lot of research has seen graphene manifest great stability and enhance specific capacitance in synergy with other nanoparticles or nanomaterials.

The addition of functional species into graphene matrix improves the rate performance and the electrochemical stability of supercapacitors. The shape and size of the functional species on graphene or graphene oxide influence the electrochemical performance of the supercapacitors. A statistically optimized morphology of nanomaterials during synthesis of the functional species will increase the performance of the supercapacitor electrochemically. Many efforts have been made in the fabrication of graphene-metal oxides hybrid material, such as transition metal oxides (MnO_2, RuO_2) which offer a high specific capacitance and superior cyclability (Chua et al. 2011, Chen et al. 2016). Increase in the EDLC and pseudocapacitance will influence the performance of carbon-based supercapacitors.

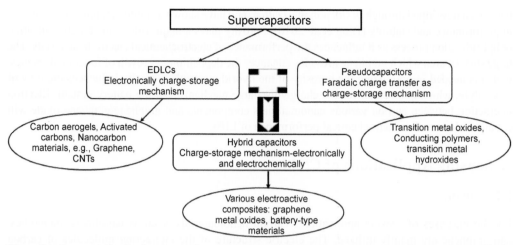

Figure 7.4: Representation for three types of supercapacitors and the most common corresponding materials used in the device fabrication.

For the EDLC, the most effective approach is to improve the surface area and optimize the porosity of the material. Then the introduction of functional species with redox capability will drastically enhance the pseudocapacitance, and therefore produce a synergistic effect on the electrochemical performance. Figure 7.4 represents the three types of supercapacitors and the most common corresponding materials used in the device fabrication.

6.2 Batteries

Today, portable mobile devices and systems have batteries as their major component. Graphene has conveniently and completely replaced graphite as anode material. Lithium-ion and sodium-ion batteries combine with metal transition oxides for improvement of electrochemical properties. This is due to the durability, high energy density, excellent reversibility, specific capacity, and the good cyclic ability of the lithium-ion batteries occasioned by the synergistic effects that exist between them and the metal transition oxides. Basically, lithium-ion batteries (LIBs) are fabricated by the assemblage of an anode electrode, cathode electrode, and a separator. Batteries store an electrochemical energy based on intercalation and de-intercalation of lithium ions (Wang and Cao 2008). The typical performance of lithium-ion or sodium-ion batteries depends on the physical and chemical properties of the electrode materials (Brownson and Banks 2014). In recent times, metals, metal oxides (containing lithium), and carbon materials are widely investigated as LIB electrode materials (Chen et al. 2010, Chua et al. 2011).

Currently, graphene and graphene-based materials have been investigated as reliable and effective electrode materials (Roy and Srivastava 2015). Some breakthroughs are already recorded in terms of their capabilities to improve the specific capacitance, power density, and cyclic stability of lithium-ion and sodium-ion batteries. The two main areas where graphene application is crucial in the performance of the LIB electrode are in the battery's specific capacitance and stability. A lot of graphene-based functionalized nanomaterials have been developed in the past through frantic efforts by scientists. Lithium metal oxides ($LiMn_2O_4$, $LiFePO_4$) have been widely employed as cathode material for LIBs. Despite their high theoretical capacitance, the main limitation with lithium metal oxides for battery energy application is their poor conductivity and poor lithium ion diffusion. In solving this problem, foreign atoms are doped with the materials to further decrease their particle size, and then electronically conductive materials are used to form a cover of protective materials through coating (Wang and Cao 2008). Graphene sheets are now used to modify the lithium metal oxides due to their excellent conductivity, high electron mobility, and large surface area. Nanostructured graphene-based composites have been developed through *in situ* polymerization, and have shown a significant improvement both in performance and stability (Ding et al. 2010). Porosity plays a major role in the functionalization or hybridization process as it influences the performance of electrochemical electrode materials. The large surface area which graphene provides enhances the deposition or growth of functional species, which is needed in optimizing the porosity of the hybrid materials. Porosity influences the flow of electrolyte, which is essential for the charge/discharge mechanism of the battery system. Electron interaction between and on various nanomaterials components and graphene/graphene oxide will equally influence the electrochemical performance of LIBs.

7. Sensing Applications of CNMs

7.1 Sensors

For the purposes of sensing applications, carbon nanomaterials, such as nanofibers, nanotubes, and graphene are mainly utilized. The electric structure of the interacting molecules of carbon nanomaterials makes the conventional semiconductors less attractive in the sensing applications (Elhaes et al. 2016). Carbon nanomaterials can be used directly or in conjunction with other materials to make or develop great sensing devices or systems. Low-cost sensors with high sensitivity, low power consumption are made possible using carbon-based nanomaterials. Consequent upon their high surface-to-volume ratio, carbon-based nanomaterials have opened a door of opportunities to develop more advanced sensitive sensor platforms. A significant progress has been made even though a major challenge is their vulnerability against contaminations, which affects their ability to achieve intrinsic sensitivities when ultra-low level detections are needed. To achieve high sensitivity during gas sensing, continuous *in-situ* cleaning of CNTs is done using ultraviolet light to enhance performance. This procedure was applied in nitric oxide detection under a controlled atmosphere, and a detection limit of 590 ppq was achieved at room temperature (Chen et al. 2012).

Recently, in immunosensing, some studies have developed and demonstrated the electrochemical detection of antigens and biomarkers necessary for cancer diagnosis using graphene-based immune-devices (Hasanzadeh et al. 2016). These devices come in handy for delivering clinical information by the simple way of using the POC testing devices. One major drawback of using microfluidic bioelectronics immune devices is the non-specific adsorption, which gives false errors, resulting in a decrease in sensitivity. A detection limit of 0.1 μM for hydrogen peroxide has been reported with graphene-based electrodes for an enzymatic system immobilized with horseradish peroxidase (Lu et al. 2013). However, this process of functionalization has recorded some challenges, such as immobilization protocol of horseradish peroxidase on an enzymatic system, low temporal stability, pH influence, temperature, and humidity on the enzyme activity. Metal oxides (Cui et al. 2015), noble metals (Chen et al. 2013), metal sulfides (Li et al. 2014), and composites materials with other carbon materials (Chen et al. 2014), have been used to develop non-enzymatic peroxide detective devices. The decision to use these materials is due to their ability to enhance electron transfer at faster rates and their catalytic activity which results in high sensitivity. In this light, the influence of graphene as a choice material on the amperometric properties in hydrogen peroxide detection is always a crucial factor. Different characterization techniques are employed in validating the electrochemical performance of graphene in amperometric detective systems. They are Raman spectrometry, cyclic voltammetry, electrochemical impedance spectroscopy, and amperometry. When sensitivity in direct detection of H_2O_2 is investigated, the influence of defects in the nanomaterials, such as graphene, can be determined by the sensitivity so obtained. In a non-enzymatic detective system for H_2O_2, non-functionalized graphene fabricated from mechanical exfoliation and CVD process demonstrates the best performance in exhibiting sensitivity towards the catalytic reduction of H_2O_2 and can detect H_2O_2 to the limit of 0.1 micro M (Li et al. 2010). This shows a fast amperometric response upon successive addition of H_2O_2. Graphene fabricated through mechanical exfoliation and CVD process produces higher sensitivity than that of reduced graphene oxide (rGO) produced through chemical reduction of graphite oxide.

7.2 CNTs and graphene in biosensors

In the world today, the use of biosensors has become critical and an important aspect of security due to their roles, which range from diagnostics of life-threatening diseases to the detection of biological warfare agents and terrorists attacks. Carbon nanotubes (CNTs) have been incorporated as sensing elements in several newly developed biosensors. CNTs were the first carbon-based nanomaterials used in the fabrication of biosensors before the novel application of graphene for the same purpose. Generally, the use of carbon nanomaterials has contributed immensely to the science of technology of developing biosensors in so many ways. Typically, electrochemical, electrical, and optical biosensors have been fabricated using two main carbon allotropes, namely, CNTs and graphene. The bright future of graphene and CNTs as great elements in the fabrication of biosensors gives scientist the needed encouragement to concentrate efforts in the development of carbon-based nanomaterials for various sensing applications. There have been questions as to which of these two carbon allotropes is best suited for application in the fabrication of biosensors. This is because the development of graphene as a 2-D material has dwarfed interest in CNTs as per their application in biosensing.

An electrochemical biosensor is a device which produces a specific quantitative analytical data or information by the use of a biological element which recognizes spatial direct contact with a transducer (Thévenot et al. 2001). There are various sized domains in biosensors (macroscale or nanoscale sensing elements), but interestingly, they place no constraint on the performance of biosensing. Carbon nanomaterials play a significant role in the development of biosensors as they help in addressing key issues involved in the design of the biosensing interface. This role is critical since the interaction of biosensing surface depends on the selectivity of the analytes. More so, carbon nanomaterials assist efficient transduction, increase the selectivity and sensitivity of biosensors, and improve the response times in sensitive systems. Carbon nanomaterials are great in solving some specific challenges in the sensing application, which include achieving compatibility of biological matrices with biosensors in order to use them in *in vivo* applications, fabricating reliable biosensors that can function in a confined environment, for example inside cells, and using multiple biosensors for the detection of multiple analytes on one sensing device (Yang et al. 2010). Before the massive application of CNTs and graphene in biosensing, other nanomaterials with different dimensionalities (zero-, one-, two-, and three-) have been used in the fabrication of biosensors. They include semiconductor quantum dots, metallic semiconductor nanowires, metallic nanoparticles, CNTs, mesoporous materials, and nanostructured conductive polymers or nanocomposites (Yang et al. 2010). CNTs have high mechanical strength, high surface-to-volume ratios, high aspect ratios, excellent chemical and thermal stability, and rich optical and electronic characteristics. These properties of CNTs make them potential candidates to be used in developing biosensors. The success recorded by CNTs in developing biosensors further fueled huge research interest in graphene to push further the boundaries of this science. Electrochemical sensors are typically macroscale biosensors manufactured with nanomaterials. FET-based devices are nanoscale transducers, while CNTs are used in the manufacture or fabrication of nanoscale optical biosensors.

The rising need and use of graphene have overtaken CNTs in various applications of biosensing. Certain factors influence the properties of graphene. They include the number of layers, the substrate used in the synthesis process, adsorbed impurities, flatness and defects, size of the sheet, and edge type and functionalization. The main objective of modern science is the real-time monitoring of biomolecular interactions with sensitivity in order to detect single-molecule activities in natural samples. Nowadays, several bio-detectors can be used to sense DNA and protein molecules. The extraordinary properties of CNTs and graphene, such as large surface area and high conductivity, allow for the development of electrochemical, optical, and FET-based biosensors. These devices are useful in single molecule and single cell detection.

7.3 Recent advancements in carbon-based sensors

Biosensors and chemical sensors for the detection of glucose and H_2O_2, respectively can be fabricated with carbon-based nanomaterials. In the past half a decade, the outstanding optical and electrical properties of graphene quantum dots (GQDs) have made them the new type electrochemiluminescence (ECL) luminophores, which are useful in sensing applications. GQDs ECL has obvious advantages of facile preparation, easy labeling, low cost, and excellent ECL activity (Li et al. 2012). The addition of strong oxidizing agents, such as $S_2O_8^{2-}$ as the co-reactant (e.g., GQDs/$S_2O_8^{2-}$), generates better properties for DNA detection (Lu et al. 2014), metal ion sensing (Chen et al. 2015), and immunoassay (Zhang et al. 2015). GQDs are seen as future candidates for ECL nanomaterials. Some study has found that sulfide (SO_3^{2-}) as a reducing agent is a promising co-reactant of GQDs ECL, and makes the ECL system four times stronger than carbon quantum dots due to the enriched surface-states of GQDs (Zhou et al. 2017). With this ECL system of sensing, the LOD was measured as 5.0 μM for glucose at S/N of 3. In the absence of carbon nanotubes and gold nanoparticles, carbon-based NPs, such as graphene and graphene oxide, have the capabilities to be used (as biosensors) in biomedical diagnosis and therapeutic conditions due to the possibilities of varying their properties to meet specific needs. The fluorescence-quenching characteristics of graphene and GO have made them good candidates for signal detection since they can generate great fluorescence that can be utilized as signals for DNA, metal ions, and RNA detection. A mono-dispersed acid catalyzed carbon dots (of 4.8 nm) synthesized through hydrothermal method could be suitable for bioimaging due to the bright fluorescence characteristic that it possesses. When ACDs is functionalized with DNA (MUC-1 aptamer), its specificity is strengthened to detect MCF-7 cancer cells in the human breast (Wang et al. 2016).

In water purification technology, the hybrid of two carbon nanomaterials enhances the selectivity and selective detection of heavy metals. The combination of graphene and graphene quantum dots generates synergistic effects that enable them to act as sensors for the detection of Cu(II) in drinking water samples using DPASV. The hybridization enhanced the sensing performance of the nanomaterials to yield a detection limit as low as 1.34 nM for Cu(II) (Wang et al. 2017). A design of a composite by thermally initiating free radicals of polymeric ionic liquids (PIL) on multi-walled carbon nanotube (MWCNT) serves as an efficient electrode in the simultaneous determination of ascorbic acid (AA), uric acid (UA), and dopamine (DA) (Wang et al. 2015). The imidazole functional groups within the PIL chains are responsible for the separation by oxidation of the biomolecules, whereas the differential pulse voltammetry enables the simultaneous selectivity and sensitivity of the biomolecules. It has been demonstrated that the determination or detection of these compounds in urine samples is feasible. Flexible multi-layer graphene film (FGF) has been employed as an antenna based wearable sensor due to its excellent mechanical flexibility, structural stability, and reversible deformability. FGF has exhibited a tremendous strain sensitivity of 9.8 and 9.36 for compressive and tensile bending, respectively (Tang et al. 2018). These results are superior to the copper antenna sensor with a compressive and tensile bending strain sensitivity of 5.39 and 4.05, respectively. A recent study has developed a sensing platform which is based on Nafion-Gr/GCE for caffeine determination. This sensor exhibited great electrocatalytic activity in caffeine oxidation without co-existence interference. Nafion enrichment effect enhances the sensitivity of caffeine determination with a low limit of detection (1.2×10^{-7} μM at S/N = 3) (Sun et al. 2011). A high sensitivity sensor (LOD of 1.1 μM for TNrGO) has been fabricated using modified GC electrodes, which were prepared with reduced graphene oxide doped with titanium nitride for Gallic acid quantification in sweet wines (Stanković et al. 2017). A nano-sized fluorescence chemosensor (signal probe) fabricated with graphene quantum dots (GQDs) has demonstrated great potential to be utilized in the detection of highly toxic pollutants, such as Hg^{2+} in an aqueous solution with a pH of 7.0 (Chakraborti et al. 2013).

A non-enzymatic glucose sensing was carried out using hybrid nanostructured materials (carbon nanotubes, activated carbon matrices, and nanofibers) onto glassy carbon electrodes, and a rapid time of 11 seconds and linear range of 0–2 mM glucose were recorded (Rathod et al. 2010). A biosensor fabricated using 3DG-GOD as nano-biocomposite exhibited great sensitivity for glucose sensing, and has a linear dynamic range of about 6 mM (Mansouri et al. 2017). Here, GOD was immobilized on 3DG film. This biosensor has demonstrated a great capability for clinical applications, and can be used in home care for rapid glucose monitoring. Further application of hybrid nano-biocomposite (GCE/3DG-GOD) can be readily utilized in fabricating other oxidase-based biosensors for lactate, alcohol, cholesterol, xanthine, hypoxanthine, and acetylcholine. A recent study has shown that drop-coating carbon black onto a glassy CE will produce a sensor used in the detection of phenolic compounds (catechol, p-cresol, and p-nitrophenol) electrochemically (Lounasvuori et al. 2018). The carbon black-glassy carbon electrode (CB-GCE) has higher current density compared to bare GCE and less positive oxidation potential for phenolic compounds. A carbon cloth decorated with TiO_2 nanosheets arrays have been used for biosensing applications. An efficient DET for glucose oxidase supported on the carbon cloth/TiO_2 was achieved, and it exhibited a stable, high sensitivity (52 μAm M^{-1} Cm^{-2}), good selectivity, low limit of detection (23.4 μM, S/N = 3), and low response time (0–4s) (Liu et al. 2016). Another study demonstrated that by electrochemically depositing Prussian blue/graphene composite film on GCE, a high sensitivity sensor (1.6 μA μM^{-1} H_2O_2 per cm^2) was fabricated and exhibited a great electrocatalytic activity and low detection limit (20 nM) for the detection of H_2O_2 (Li et al. 2012). These properties were a combination of the catalytic nature of PB and the large surface area of graphene, thereby producing a good electrochemical synergistic effect between them.

Another graphene-based sensor was fabricated by functionalizing glassy carbon electrode with reduced graphene oxide, and this sensor exhibited high sensitivity (LOD: 0.8 μM) and selectivity for hydroquinone determination (Li et al. 2012). The oxidation potential of HQ was lowered while the redox peak currents of HQ were enhanced in the electrochemical response of HQ. *In vivo* applications, a flexible single-walled CNT ISFET sensor was fabricated for sensing. Suitable for chemical and biological sensing applications, SWCNT ISFET sensors exhibited a sensitivity of 18–45 μA/mM, and indicated high selectivity for glucose oxidase (Lee and Cui 2010). An electrochemical sensor which is based on graphene has demonstrated tremendous electrocatalytic activity in the oxidation and reduction of paracetamol. By functionalizing GCE with graphene, a high sensitivity is achieved, and there is a clearly promoted paracetamol determination with a low detection limit of 3.2×10^{-8} M (Kang et al. 2010). This graphene/GCE sensor, therefore, has a promising future in the preparation of pharmaceutical tablets. When a glassy carbon electrode is functionalized with SnO_2 nanofibers, a biosensor is fabricated, which is useful in the amperometric hydrogen peroxide determination. The GCE/SnO_2 sensor exhibited high catalytic activity towards hydrogen peroxide (2–160 μM H_2O_2) with a low limit of detection (0.5 μM) (Kafi et al. 2017). From the amperometric analysis, GCE/SnO_2 has great storage capacity, repeatability, and reproducibility. Carbon dots (CNDs) functionalized with multicolor fluorescence sulfur has demonstrated an ultrasensitive sensing ability for neurotransmitter dopamine detection. The CNDs exhibited a detection limit of 47 pM for dopamine (DA) and a detection limit of 92 pM for DA in plasma sample and in PC-12 live cell under hypoxic condition (Gupta and Nandi 2017). Detection of nitro-aromatic explosives has been demonstrated using graphite micro-particle-based electrodes decorated with graphene nanoribbons (singe-, few-, and multilayer). This biosensor exhibited high sensitivity for bioanalytes, and showed a great electrochemical reduction of 2,4,6-trinitrotoluene (Goh and Pumera 2011). Carbon dots (CDs) functionalized with Arg (Arg-CDs) has been utilized for simultaneous sensing and cell imaging. As a fluorescence probe, it has a great sensitivity and emission ability for the detection of cancer cells (MCF-7 cells) due to its biocompatibility on cells (Fu et al. 2017). A disposable pencil graphite electrode (PGE) has been fabricated using silver nanoparticles (AuNPs), single-walled carbon nanotubes (SWCNTs), and polyaniline for DNA detection (Eksin et al. 2015).

An electrochemiluminescence sensor was fabricated with nitrogen-doped graphene quantum dots and chitosan film to generate electroluminescence signal for nitroaniline (NA) sensing with the detection limit of 0.005 μmol L^{-1} (Chen et al. 2016). The electrocatalytic response and the stable photoluminescence (PL) exhibited by N-GQDs/Chitosan ECL sensor enabled linear ECL intensity activity of 0.01–1 µML^{-1}. This has found potential sensing applications in fuel cells and real water sampling. The amperometric xanthine determination at a nanomolar level has been demonstrated by an amperometric enzyme biosensor through layer-by-layer architecture by hybridizing carboxymethylcellulose-modified GO with platinum nanoparticles and PAMAM-MNP NPs. This enzyme electrode exhibited high sensitivity up to 140 mA/M cm^2 and LOD of 13 nM in the range of 50 nM–12 µM (Borisova et al. 2016). Zinc oxide-MWNTs composite electrode sensor has shown glucose sensing applications through enzyme immobilization (Bai et al. 2010).

Table 7.3 represents the non-enzymatic detection of hydrogen peroxide study on graphene-based electrode materials made with graphene production from different synthesis routes.

Table 7.3: A non-enzymatic detection of H$_2$O$_2$ study on graphene-based electrode materials.

Electrode material	Potential [V]	Sensitivity [mA·M^{-1}·cm^{-2}]	LOD [µM]	Reference
Polydopamine-rGO/Ag	−0.5 (vs Ag/AgCl)	355.8	2.1	Fu et al. 2014
rGO-PMS/AuNPs	−0.75 (vs Ag/AgCl)	39.2	0.06	Maji et al. 2014
rGO	−0.3 (vs SCE)	25	9.2	Zöpfl et al. 2016
CVDG	−0.3 (vs SCE)	173	15.1	Zöpfl et al. 2016
SG	−0.3 (vs SCE)	202	651.5	Zöpfl et al. 2016
MnO$_2$/GO	−0.3 (vs SCE)	8.2	0.8	Li et al. 2010
PB/Graphene	−0.05 (vs Pt)	196.6	1.9	Jin et al. 2010
CoOxNPs/ERG	0.75 (vs SCE)	148.6	0.2	Li et al. 2014
AgNPs/PQ11/Graphene	−0.4 (vs Ag/AgCl)	56.6	28	Liu et al. 2010

8. Conclusion and Future Prospects

In conclusion, nanomaterials are of tremendous importance in the field of nanoscience and nanotechnology due to their properties and ease of applications in several areas. Basically, nanomaterials are of two types, namely, natural and artificial nanomaterials. The growing need for effective alternative yet sustainable energy storage devices and the desire to keep the environment safe and free from pollutants have necessitated huge research interest in carbon-based nanomaterials. Rapid industrialization has resulted in the generation of toxic chemicals released into the environment in the form of airborne, aquatic, and solid wastes. CNMs have exceptional properties, such as excellent thermal conductivity, large surface area, high electron mobility, and superior mechanical strength. Several synthesis methods are used for the production of CNMs, such as exfoliation, thermal decomposition, chemical vapor deposition, chemical-based method, laser ablation, arc discharge, and unzipping of CNTs. Although some breakthroughs have been recorded in this regard, some issues associated with the synthesis operations need to be studied for better understanding. Several industrial applications of novel CNMs (e.g., HCFs, CNTs, and graphene) other than energy storage (LIBs and supercapacitors) and sensing have been recorded, which include flat panel field emission displays, nanoelectronics devices, hydrogen storage, scanning probe tip, transistors, gas sensor, bio-sensor, hole transporting layers (HTLs), green tires, and adsorbents. CVD process of CNMs promises to be the most economical synthesis method because of high yield and high quality product achieved in a sustainable manner. In the CVD method, the issues with the transfer of graphene onto the dielectric substrate require attention and solution as the graphene structure should be maintained in order to exhibit excellent properties. This is particularly true when a specific application requires a high purity graphene since the structural modification of graphene may be desirable for other specialized applications. Also, the procedure of transfer from the surface of the metal catalyst to an insulating surface (dielectric substrates) contaminates the thin film and causes wrinkles or serious vacancy defects (damage to the graphene sheets). This is undesirable. The high energy requirements (high carbon footprint) for some of the synthesis processes, synthesis of CNMs in a substrate free manner, size control of the CNMs during synthesis, and cooling methods used are serious issues that need solutions.

Generally, carbon-based nanomaterials are considered safer than other nanomaterials (such as metal nanomaterials, dendrimers) on humans and the environment owing to the compatibility of carbon with that of the living systems. However, the practical applications of graphene in sensing and energy storage need to be extensively demonstrated for novel and effective sensor, battery, and supercapacitor development. In recent times, a reliable, sensitive, and cost-effective sensor platform is sought for proper environmental assessment and remediation. A considerable progress has been made in developing reliable batteries, supercapacitors, electrochemical and biosensor systems based on carbon nanomaterials. Novel examples of carbon nanomaterials are graphene, nanofibers, carbon nanotubes (CNTs), and helical carbon fibers (HCFs). The unique electrical and physicochemical properties of carbon nanomaterials have made the development of new functionality, high specificity, and sensitivity of sensor platforms possible. For an enhanced performance of sensor platforms, some critical factors, such as size, shape, composition, architecture, and functionality of the materials are basic.

On the other hand, high energy density, high power density, excellent conductivity, and stability of lithium-ion batteries and supercapacitors have been developed to solve the energy-related issues in the electronics industry. Some of the existing and current synthesis methods for the production of carbon nanomaterials are the CVD, exfoliation and cleavage, thermal decomposition, laser ablation, arc discharge method, and chemical-based technique. Carbon-based nanocomposites have been developed or fabricated through solvothermal, hydrothermal, CVD and microwave assisted CVD processes for energy storage and sensing applications. Carbon-based nanomaterials have been employed in developing several devices, which include sensors (bio-, gas-, and chemical-sensors), lithium-ion batteries, sodium-ion batteries, photo-catalysts, supercapacitors, single molecule detectors, transparent electrodes, composites, transistors, hydrogen storage, scanning probe tips, solar cells, photo-detectors, hole transporting layers, adsorbents, green tires, and other nanoelectronics devices. Besides all these applications, the utilization of graphene as the building block to enhance specific capacitance, electrical and thermal conductivity, energy density, power density, and cyclic stability in lithium-ion batteries and supercapacitors needs to be explored more. Nature/structure and properties of graphene determine how it influences the overall performance of the materials which it combines or synergizes with. Although detection in living cells has been achieved, a majority of the biosensors only employed fluorescence-labeled probes which are meant to absorb unto the graphene/graphene oxide surface. This mechanism is undesirable, as it influences the stability of the platform and raises issues of absorption of other substrates that are non-specific. Therefore, it is imperative that the cytotoxicity of carbon-based nanoparticles be carefully considered and thoroughly investigated in terms of the oxygen content, lateral size, and the dose utilized in the fabrication of sensors (Zhang et al. 2013, Hu et al. 2011).

Despite these exciting advancements, there are certain issues associated with the applications of carbon-based nanomaterials in developing novel nanoelectric devices. In the energy storage and sensing applications, the following issues should be addressed to achieve the best in the wake of the current developments.

i. Control of the shape and size of nanomaterials during synthesis to attain optimal electric, thermal, electrochemical active sites.

ii. Optimization and control of architectures of bi- or trimetallic nanomaterials to enhance stability and specificity.

iii. Proper functional molecule discovery to achieve high specificity with the analytes.

iv. Exceptional selectivity design requires a good correlation between structure, composition, and reactivity of nanomaterials.

v. Improvement of electrochemical properties through the discovery of highly conductive, stable, and great surface area of substrate materials, both chemically and mechanically.

vi. Proper design of the nanometric interfaces to enhance electrocatalytic properties, sensitivity, and stability of the sensor device.

Conflicts of Interest

The authors declare no conflict of interest.

Acknowledgments

The authors would like to acknowledge Malaysia-Japan International Institute of Technology, Universiti Teknologi Malaysia (MJIIT-UTM), Malaysia for funding this research.

Nomenclature

CNM	Carbon nanomaterials
NPs	Nanoparticles
CCVD	Catalytic Chemical Vapor Deposition
LPCVD	Low Pressure Chemical Vapor Deposition
PECVD	Plasma Enhanced Chemical Vapor Deposition
PLD	Pulse Laser Deposition
H_2O_2	Hydrogen peroxide
GQDs ECL	Graphene quantum dots electrochemiluminiscence
$SO_3^{2-}/S_2O_8^{2-}$	Sulfides
DNA	Deoxyribonucleic acid
RNA	Ribonucleic acid
S/N	Signal to error ratio
LOD	Limit of detection
MCF-7	Human breast cancer cells (Michigan Cancer Foundation-7)
DPASV	Differential pulse anodic stripping voltammetry
AA	Ascorbic acid
UA	Uric acid
DA	Dopamine
NA	Nitroaniline
GC	Glassy carbon
PIL	Polymeric ionic liquids
MWCNT	Multi-walled carbon nanotube
MUC-1	Mucin 1Aptamer
FGF	Flexible multilayer graphene films
Gr/GCE	Glassy carbon electrode
TNrGO	Titanium reduced graphene oxide
N-GQDs	Nitrogen-doped graphene quantum dots
ACDs	Acid catalyzed carbon dots
GOD	Glucose oxidase
CE	Carbon electrode
GCE	Glassy carbon electrode
3DG-GOD	3-dimensional graphene glucose oxidase
GCE/3DG-GOD	Glassy carbon electrode/3-dimensional graphene glucose oxidase
CB-GCE	Carbon black/glassy carbon electrode

TiO$_2$	Titanium oxide
NSA	Nanosheet arrays
POC	Point of care
DET	Direct electron transfer
PB	Prussian blue
HQ	Hydroquinone
GCE/SnO$_2$	Glassy carbon electrode/Tin oxide
CNDs	Carbon dots
ppq	Part per quadrillion
EM	Electromagnetic wave absorption
MR	Magnetoresistances
CTCVD	Catalyzed thermal chemical vapor deposition
TNT	2, 4, 6-trinitrotoluene
Arg-CDs	Arginine-modified carbon dots
AuNPs	Gold nanoparticles
SWCNTs	Single-walled carbon nanotubes
PANI	Polyaniline
PGEs	Pen graphite electrode
MNP	Magnetic nanoparticles
NCDs	N-doped carbon dots
EDLC	Electrical double-layer capacitance
HOPG	Highly oriented pyrolytic graphite
CNMs	Carbon nanomaterials
CNFs	Carbon nanofilms/Carbon nanofibers
SWCNT	Single-walled carbon nanotubes
DWCNT	Double-walled carbon nanotubes
MWCNT	Multi-walled carbon nanotubes
SLG	Single layer graphene
FLG	Few layer graphene
NPs	Nanoparticles
LIBs	Lithium-ion batteries
SCs	Supercapacitors
DC	Direct current
CMOS	Complementary metal-oxide-semiconductor
CNT ISFET	Carbon nanotubes/Ion Selective Field Effect Transistor
SWCNT ISFET	Single Walled Carbon nanotubes/Ion Selective Field Effect Transistor
PAMAM-MNP	Polyamidoamine Magnetite nanoparticles

References

Ambrosi, A., C.K. Chua, A. Bonanni and M. Pumera. 2014. Electrochemistry of graphene and related materials. Chem. Rev. 114(14): 7150–7188. Doi: 10.1021/cr500023c.

Antisari, M.V., D.M. Gattia, L. Brandão, R. Marazzi and A. Montone. 2010. Carbon nanostructures produced by an AC arc discharge. pp. 1766–1771. *In*: Mater. Sci. Forum Vol. 638-642.

Anwar, A.W., A. Majeed, N. Iqbal, W. Ullah, A. Shuaib, U. Ilyas et al. 2015. Specific capacitance and cyclic stability of graphene based metal/metal oxide nanocomposites: A review. J. Mater. Sci. Technol. 31(7): 699–707. Doi: 10.1016/j.jmst.2014.12.012.

Aricò, A.S., P. Bruce, B. Scrosati, J.M. Tarascon and W. Van Schalkwijk. 2010. Nanostructured materials for advanced energy conversion and storage devices. pp. 148–159. *In*: Mater. Sustainable Energy: A Collection of Peer-Reviewed Research and Review Articles from Nature Publishing Group.

Bai, D., Z. Zhang and K. Yu. 2010. Synthesis, field emission and glucose-sensing characteristics of nanostructural ZnO on free-standing carbon nanotubes films. Appl. Surf. Sci. 256(8): 2643–2648. Doi: 10.1016/j.apsusc.2009.11.011.

Balandin, A.A., S. Ghosh, W. Bao, I. Calizo, D. Teweldebrhan, F. Miao et al. 2008. Superior thermal conductivity of single-layer graphene. Nano Lett. 8(3): 902–907. Doi: 10.1021/nl0731872.

Bell, D.C., M.C. Lemme, L.A. Stern, J.R. Williams and C.M. Marcus. 2009. Precision cutting and patterning of graphene with helium ions. Nanotechnol. 20(45). Doi: 10.1088/0957-4484/20/45/455301.

Berger, C., Z. Song, T. Li, X. Li, A.Y. Ogbazghi, R. Feng et al. 2004. Ultrathin epitaxial graphite: 2D electron gas properties and a route toward graphene-based nanoelectronics. J. Phys. Chem. B 108(52): 19912–19916. Doi: 10.1021/jp040650f.

Bertke, S.J., D.H. Tomich, J.E. Hoelscher and R.L. Jacobsen. 2009. Laser precision-based graphene growth processes. Paper presented at the ICALEO 2009—28th Int. Congr. Appl. Lasers Electro-Opt. Congr. Proc.

Borenstein, A., O. Hanna, R. Attias, S. Luski, T. Brousse and D. Aurbach. 2017. Carbon-based composite materials for supercapacitor electrodes: A review. J. Mater. Chem. A 5(25): 12653–12672. Doi: 10.1039/c7ta00863e.

Borisova, B., A. Sánchez, S. Jiménez-Falcao, M. Martín, P. Salazar, C. Parrado et al. 2016. Reduced graphene oxide-carboxymethylcellulose layered with platinum nanoparticles/PAMAM dendrimer/magnetic nanoparticles hybrids. Application to the preparation of enzyme electrochemical biosensors. Sens. Actuators B 232: 84–90. Doi: 10.1016/j.snb.2016.02.106.

Brownson, D.A.C. and C.E. Banks. 2014. The Handbook of Graphene Electrochemistry.

Cambaz, Z.G., G. Yushin, S. Osswald, V. Mochalin and Y. Gogotsi. 2008. Noncatalytic synthesis of carbon nanotubes, graphene and graphite on SiC. Carbon 46(6): 841–849. Doi: 10.1016/j.carbon.2008.02.013.

Cano-Márquez, A.G., F.J. Rodríguez-Macías, J. Campos-Delgado, C.G. Espinosa-González, F. Tristán-López, D. Ramírez-González et al. 2009. Ex-MWNTs: Graphene sheets and ribbons produced by lithium intercalation and exfoliation of carbon nanotubes. Nano Lett. 9(4): 1527–1533. Doi: 10.1021/nl803585s.

Cappelli, E., S. Orlando, G. Mattei, C. Scilletta, F. Corticelli and P. Ascarelli. 2004. Nano-structured oriented carbon films grown by PLD and CVD methods. Appl. Phys. A: Mater. Sci. Process. 79(8): 2063–2068. Doi: 10.1007/s00339-004-2862-0.

Chae, H.K., D.Y. Siberio-Pérez, J. Kim, Y. Go, M. Eddaoudi, A.J. Matzger et al. 2004. A route to high surface area, porosity and inclusion of large molecules in crystals. Nature 427(6974): 523–527. Doi: 10.1038/nature02311.

Chae, S.J., F. Günes, K.K. Kim, E.S. Kim, G.H. Han, S.M. Kim et al. 2009. Synthesis of large-area graphene layers on nickel film by chemical vapor deposition: Wrinkle formation. Paper presented at the Proceedings of SPIE—The International Society for Optical Engineering.

Chakraborti, H., S. Sinha, S. Ghosh and S.K. Pal. 2013. Interfacing water soluble nanomaterials with fluorescence chemosensing: Graphene quantum dot to detect Hg^{2+} in 100% aqueous solution. Mater. Lett. 97: 78–80. Doi: 10.1016/j.matlet.2013.01.094.

Chen, D., L. Tang and J. Li. 2010. Graphene-based materials in electrochemistry. Chem. Soc. Rev. 39(8): 3157–3180. Doi: 10.1039/b923596e.

Chen, G., T.M. Paronyan, E.M. Pigos and A.R. Harutyunyan. 2012. Enhanced gas sensing in pristine carbon nanotubes under continuous ultraviolet light illumination. Sci. Rep. 2. Doi: 10.1038/srep00343.

Chen, S., R. Yuan, Y. Chai and F. Hu. 2013. Electrochemical sensing of hydrogen peroxide using metal nanoparticles: A review. Microchim. Acta 180(1-2): 15–32. Doi: 10.1007/s00604-012-0904-4.

Chen, S., X. Chen, T. Xia and Q. Ma. 2016. A novel electrochemiluminescence sensor for the detection of nitroaniline based on the nitrogen-doped graphene quantum dots. Biosens. Bioelectron. 85: 903–908. Doi: 10.1016/j.bios.2016.06.010.

Chen, X., G. Wu, Z. Cai, M. Oyama and X. Chen. 2014. Advances in enzyme-free electrochemical sensors for hydrogen peroxide, glucose, and uric acid. Microchim. Acta 181(7-8): 689–705. Doi: 10.1007/s00604-013-1098-0.

Chen, X., R. Paul and L. Dai. 2017. Carbon-based supercapacitors for efficient energy storage. Natl. Sci. Rev. 4(3): 453–489. Doi: 10.1093/nsr/nwx009.

Chen, Y., H. Zhao, L. Sheng, L. Yu, K. An, J. Xu et al. 2012. Mass-production of highly-crystalline few-layer graphene sheets by arc discharge in various H_2-inert gas mixtures. Chem. Phys. Lett. 538: 72–76. Doi: 10.1016/j.cplett.2012.04.020.

Chen, Y., Y. Dong, H. Wu, C. Chen, Y. Chi and G. Chen. 2015. Electrochemiluminescence sensor for hexavalent chromium based on the graphene quantum dots/peroxodisulfate system. Electrochim. Acta 151: 552–557. Doi: 10.1016/j.electacta.2014.11.068.

Chen, Y., J. Xu, Y. Yang, Y. Zhao, W. Yang, X. He et al. 2016. Enhanced electrochemical performance of laser scribed graphene films decorated with manganese dioxide nanoparticles. J. Mater. Sci.: Mater. Electron 27(3): 2564–2573. Doi: 10.1007/s10854-015-4059-z.

Choi, B.G., H. Park, M.H. Yang, Y.M. Jung, S.Y. Lee, W.H. Hong et al. 2010. Microwave-assisted synthesis of highly water-soluble graphene towards electrical DNA sensor. Nanoscale 2(12): 2692–2697. Doi: 10.1039/c0nr00562b.

Choi, W., I. Lahiri, R. Seelaboyina and Y.S. Kang. 2010. Synthesis of graphene and its applications: A review. Crit. Rev. Solid State Mater. Sci. 35(1): 52–71. Doi: 10.1080/10408430903505036.

Chua, C.K., A. Ambrosi and M. Pumera. 2011. Graphene based nanomaterials as electrochemical detectors in Lab-on-a-chip devices. Electrochem. Commun. 13(5): 517–519. Doi: 10.1016/j.elecom.2011.03.001.

Cui, X., S. Wu, Y. Li and G.G. Wan. 2015. Sensing hydrogen peroxide using a glassy carbon electrode modified with *in-situ* electrodeposited platinum-gold bimetallic nanoclusters on a graphene surface. Microchim. Acta 182(1-2): 265–272. Doi: 10.1007/s00604-014-1321-7.

Das, S., P. Sudhagar, Y.S. Kang and W. Choi. 2014. Graphene synthesis and application for solar cells. J. Mater. Res. 29(3): 299–319. Doi: 10.1557/jmr.2013.297.

Dathbun, A. and S. Chaisitsak. 2013. Effects of three parameters on graphene synthesis by chemical vapor deposition. Paper presented at the 8th Annual IEEE International Conference on Nano/Micro Engineered and Molecular Systems, IEEE NEMS 2013.

Dervishi, E., Z. Li, F. Watanabe, A. Biswas, Y. Xu, A.R. Biris et al. 2009. Large-scale graphene production by RF-cCVD method. Chem. Commun. (27): 4061–4063. Doi: 10.1039/b906323d.

Dervishi, E., Z. Li, F. Watanabe, A. Courte, A. Biswas, A.R. Biris et al. 2009. Versatile catalytic system for the large-scale and controlled synthesis of single-wall, double-wall, multi-wall, and graphene carbon nanostructures. Chem. Mater. 21(22): 5491–5498. Doi: 10.1021/cm902502c.

Ding, Y., Y. Jiang, F. Xu, J. Yin, H. Ren, Q. Zhuo et al. 2010. Preparation of nano-structured LiFePO$_4$/graphene composites by co-precipitation method. Electrochem. Commun. 12(1): 10–13. Doi: 10.1016/j.elecom.2009.10.023.

Dong, Q., G. Wang, H. Hu, J. Yang, B. Qian, Z. Ling et al. 2013. Ultrasound-assisted preparation of electrospun carbon nanofiber/graphene composite electrode for supercapacitors. J. Power Sources 243: 350–353. Doi: 10.1016/j.jpowsour.2013.06.060.

Eda, G., H. Emrah Unalan, N. Rupesinghe, G.A.J. Amaratunga and M. Chhowalla. 2008. Field emission from graphene based composite thin films. Appl. Phys. Lett. 93(23). Doi: 10.1063/1.3028339.

Eksin, E., G. Bolat, F. Kuralay, A. Erdem and S. Abaci. 2015. Preparation of gold nanoparticles/single-walled carbon nanotubes/polyaniline composite-coated electrode developed for DNA detection. Polym. Bull. 72(12): 3135–3146. Doi: 10.1007/s00289-015-1457-6.

Elhaes, H., A. Fakhry and M. Ibrahim. 2016. Carbon nano materials as gas sensors. Paper presented at the Mater. Today: Proceedings.

Ferrari, A.C., F. Bonaccorso, V. Fal'ko, K.S. Novoselov, S. Roche, P. Bøggild et al. 2015. Science and technology roadmap for graphene, related two-dimensional crystals, and hybrid systems. Nanoscale 7(11): 4598–4810. Doi: 10.1039/c4nr01600a.

Fischbein, M.D. and M. Drndić. 2008. Electron beam nanosculpting of suspended graphene sheets. Appl. Phys. Lett. 93(11). Doi: 10.1063/1.2980518.

Fox, D., A. O'Neill, D. Zhou, M. Boese, J.N. Coleman and H.Z. Zhang. 2011. Nitrogen assisted etching of graphene layers in a scanning electron microscope. Appl. Phys. Lett. 98(24). Doi: 10.1063/1.3601467.

Freitag, M., M. Steiner, Y. Martin, V. Perebeinos, Z. Chen, J.C. Tsang et al. 2009. Energy dissipation in graphene field-effect transistors. Nano Lett. 9(5): 1883–1888. Doi: 10.1021/nl803883h.

Fu, C., K. Qian and Fu, A. 2017. Arginine-modified carbon dots probe for live cell imaging and sensing by increasing cellular uptake efficiency. Mater. Sci. Eng. C 76: 350–355. Doi: 10.1016/j.msec.2017.03.084.

Fu, L., G. Lai, B. Jia and A. Yu. 2014. Preparation and electrocatalytic properties of polydopamine functionalized reduced graphene oxide-silver nanocomposites. Electrocatalysis 6(1): 72–76. Doi: 10.1007/s12678-014-0219-9.

Gallo, E.M., B.I. Willner, J. Hwang, S. Sun, M. Spencer, T. Salgaj et al. 2012. Chemical vapor deposition of graphene on copper at reduced temperatures. Paper presented at the Proceedings of SPIE—The International Society for Optical Engineering.

Ge, S., M. Yan, J. Lu, M. Zhang, F. Yu, J. Yu et al. 2012. Electrochemical biosensor based on graphene oxide-Au nanoclusters composites for l-cysteine analysis. Biosens. Bioelectron. 31(1): 49–54. Doi: 10.1016/j.bios.2011.09.038.

Geim, A.K. 2009. Graphene: Status and prospects. Science 324(5934): 1530–1534. Doi: 10.1126/science.1158877.

Goh, M.S. and M. Pumera. 2011. Graphene-based electrochemical sensor for detection of 2,4,6-trinitrotoluene (TNT) in seawater: The comparison of single-, few-, and multilayer graphene nanoribbons and graphite microparticles. Anal. Bioanal. Chem. 399(1): 127–131. Doi: 10.1007/s00216-010-4338-8.

Guermoune, A., T. Chari, F. Popescu, S.S. Sabri, J. Guillemette, H.S. Skulason et al. 2011. Chemical vapor deposition synthesis of graphene on copper with methanol, ethanol, and propanol precursors. Carbon 49(13): 4204–4210. Doi: 10.1016/j.carbon.2011.05.054.

Gupta, A. and C.K. Nandi. 2017. PC12 live cell ultrasensitive neurotransmitter signaling using high quantum yield sulphur doped carbon dots and its extracellular Ca^{2+} ion dependence. Sens. Actuators B 245: 137–145. Doi: 10.1016/j.snb.2017.01.145.

Haddad, P.A., D. Flandre and J.P Raskin. 2018. Intrinsic rectification in common-gated graphene field-effect transistors. Nano Energy 43: 37–46. Doi: 10.1016/j.nanoen.2017.10.049.

Hasanzadeh, M., N. Shadjou, A. Mokhtarzadeh and M. Ramezani. 2016. Two dimension (2-D) graphene-based nanomaterials as signal amplification elements in electrochemical microfluidic immune-devices: Recent advances. Mater. Sci. Eng. C 68: 482–493. Doi: 10.1016/j.msec.2016.06.023.

Hass, J., W.A. De Heer and E.H. Conrad. 2008. The growth and morphology of epitaxial multilayer graphene. J. Phys. Condens. Matter. 20(32). Doi: 10.1088/0953-8984/20/32/323202.

He, X., Q. Liu, J. Liu, R. Li, H. Zhang, R. Chen et al. 2017. Hierarchical $NiCo_2O_4$@NiCoAl-layered double hydroxide core/shell nanoforest arrays as advanced electrodes for high-performance asymmetric supercapacitors. J. Alloys Compd. 724: 130–138. Doi: 10.1016/j.jallcom.2017.06.256.

Hone, J., B. Batlogg, Z. Benes, M.C. Llaguno, N.M. Nemes, A.T. Johnson et al. 2001. Thermal properties of single-walled carbon nanotubes. Materials Research Society Symposium—Proceedings 633: A1711–A17112.

Horiuchi, S., T. Gotou, M. Fujiwara, T. Asaka, T. Yokosawa and Y. Matsui 2004. Single graphene sheet detected in a carbon nanofilm. Appl. Phys. Lett. 84(13): 2403–2405. Doi: 10.1063/1.1689746.

Hu, W., C. Peng, M. Lv, X. Li, Y. Zhang, N. Chen et al. 2011. Protein corona-mediated mitigation of cytotoxicity of graphene oxide. ACS Nano. 5(5): 3693–3700. Doi: 10.1021/nn200021j.

Hummers, W.S. and R.E. Offeman. 1958. Preparation of graphitic oxide. J. Am. Chem. Soc. 80(6): 1339. Doi: 10.1021/ja01539a017.

Iakobson, O.D., O.L. Gribkova, A.R. Tameev, A.A. Nekrasov, D.S. Saranin and A. Di Carlo. 2018. Graphene nanosheet/polyaniline composite for transparent hole transporting layer. J. Ind. Eng. Chem. Doi: 10.1016/j.jiec.2018.04.042.

Jang, B., C.H. Kim, S.T. Choi, K.S. Kim, K.S. Kim, H.J. Lee et al. 2017. Damage mitigation in roll-to-roll transfer of CVD-graphene to flexible substrates. 2D Mater. 4(2). Doi: 10.1088/2053-1583/aa57fa.

Jin, E., X. Lu, L. Cui, D. Chao and C. Wang. 2010. Fabrication of graphene/prussian blue composite nanosheets and their electrocatalytic reduction of H_2O_2. Electrochim. Acta 55(24): 7230–7234. Doi: 10.1016/j.electacta.2010.07.029.

Juang, Z.Y., C.Y. Wu, C.W. Lo, W.Y. Chen, C.F. Huang, J.C. Hwang et al. 2009. Synthesis of graphene on silicon carbide substrates at low temperature. Carbon 47(8): 2026–2031. Doi: 10.1016/j.carbon.2009.03.051.

Kafi, A.K.M., Q. Wali, R. Jose, T.K. Biswas and M.M. Yusoff. 2017. A glassy carbon electrode modified with SnO_2/nanofibers, polyaniline and hemoglobin for improved amperometric sensing of hydrogen peroxide. Microchim. Acta 184(11): 4443–4450. Doi: 10.1007/s00604-017-2479-6.

Kafy, A., A. Akther, L. Zhai, H.C. Kim and J. Kim. 2017. Porous cellulose/graphene oxide nanocomposite as flexible and renewable electrode material for supercapacitor. Synth. Met. 223: 94–100. Doi: 10.1016/j.synthmet.2016.12.010.

Kaiser, A.B. 2001. Electronic transport properties of conducting polymers and carbon nanotubes. Rep. Prog. Phys. 64(1): 1–49. Doi: 10.1088/0034-4885/64/1/201.

Kang, X., J. Wang, H. Wu, J. Liu, I.A. Aksay and Y. Lin. 2010. A graphene-based electrochemical sensor for sensitive detection of paracetamol. Talanta 81(3): 754–759. Doi: 10.1016/j.talanta.2010.01.009.

Kataria, S., S. Wagner, J. Ruhkopf, A. Gahoi, H. Pandey, R. Bornemann et al. 2014. Chemical vapor deposited graphene: From synthesis to applications. Physica Status Solidi (A) Applications and Materials Science 211(11): 2439–2449. Doi: 10.1002/pssa.201400049.

Kazemizadeh, F. and R. Malekfar. 2018. One step synthesis of porous graphene by laser ablation: A new and facile approach. Phys. B Condens. Mater. 530: 236–241. Doi: 10.1016/j.physb.2017.11.052.

Kim, C.D., B.K. Min and W.S. Jung. 2009. Preparation of graphene sheets by the reduction of carbon monoxide. Carbon 47(6): 1610–1612. Doi: 10.1016/j.carbon.2009.02.025.

Kim, K.S., Y. Zhao, H. Jang, S.Y. Lee, J.M. Kim, K.S. Kim et al. 2009. Large-scale pattern growth of graphene films for stretchable transparent electrodes. Nature 457(7230): 706–710. Doi: 10.1038/nature07719.

Kim, S., Y. Song, J. Wright and M.J. Heller. 2016. Graphene bi- and trilayers produced by a novel aqueous arc discharge process. Carbon 102: 339–345. Doi: 10.1016/j.carbon.2016.02.049.

Kochmann, S., T. Hirsch and O.S. Wolfbeis. 2012. Graphenes in chemical sensors and biosensors. TrAC, Trends Anal. Chem. 39: 87–113. Doi: 10.1016/j.trac.2012.06.004.

Kosynkin, D.V., A.L. Higginbotham, A. Sinitskii, J.R. Lomeda, A. Dimiev, B.K. Price et al. 2009. Longitudinal unzipping of carbon nanotubes to form graphene nanoribbons. Nature 458(7240): 872–876. Doi: 10.1038/nature07872.

Krätschmer, W., L.D. Lamb, K. Fostiropoulos and D.R. Huffman. 1990. Solid C60: a new form of carbon. Nature 347(6291): 354–358.

Lee, C., X. Wei, J.W. Kysar and J. Hone. 2008. Measurement of the elastic properties and intrinsic strength of monolayer graphene. Science 321(5887): 385–388. Doi: 10.1126/science.1157996.

Lee, D. and T. Cui. 2010. Low-cost, transparent, and flexible single-walled carbon nanotube nanocomposite based ion-sensitive field-effect transistors for pH/glucose sensing. Biosens. Bioelectron. 25(10): 2259–2264. Doi: 10.1016/j.bios.2010.03.003.

Li, L., Z. Du, S. Liu, Q. Hao, Y. Wang, Q. Li et al. 2010. A novel nonenzymatic hydrogen peroxide sensor based on MnO_2/graphene oxide nanocomposite. Talanta 82(5): 1637–1641. Doi: 10.1016/j.talanta.2010.07.020.

Li, L.L., J. Ji, R. Fei, C.Z. Wang, Q. Lu, J.R. Zhang et al. 2012. A facile microwave avenue to electrochemiluminescent two-color graphene quantum dots. Adv. Funct. Mater. 22(14): 2971–2979. Doi: 10.1002/adfm.201200166.

Li, N., Z. Wang, K. Zhao, Z. Shi, Z. Gu and S. Xu. 2010a. Large scale synthesis of N-doped multi-layered graphene sheets by simple arc-discharge method. Carbon 48(1): 255–259. Doi: 10.1016/j.carbon.2009.09.013.

Li, N., Z. Wang, K. Zhao, Z. Shi, Z. Gu and S. Xu. 2010b. Synthesis of single-wall carbon nanohorns by arc-discharge in air and their formation mechanism. Carbon 48(5): 1580–1585. doi:10.1016/j.carbon.2009.12.055.

Li, S.J., J.M. Du, Y.F. Shi, W.J. Li and S.R. Liu. 2012. Functionalization of graphene with Prussian blue and its application for amperometric sensing of H_2O_2. J. Solid State Electrochem. 16(6): 2235–2241. Doi: 10.1007/s10008-012-1653-3.

Li, S.J., Y. Xing and G.F. Wang. 2012. A graphene-based electrochemical sensor for sensitive and selective determination of hydroquinone. Microchim. Acta 176(1-2): 163–168. Doi: 10.1007/s00604-011-0709-x.

Li, S.J., J.M. Du, J.P. Zhang, M.J. Zhang and J. Chen. 2014. A glassy carbon electrode modified with a film composed of cobalt oxide nanoparticles and graphene for electrochemical sensing of H_2O_2. Microchim. Acta 181(5-6): 631–638. Doi: 10.1007/s00604-014-1164-2.

Li, X., X. Wang, L. Zhang, S. Lee and H. Dai. 2008. Chemically derived, ultrasmooth graphene nanoribbon semiconductors. Science 319(5867): 1229–1232. Doi: 10.1126/science.1150878.

Li, X., W. Cai, J. An, S. Kim, J. Nah, D. Yang et al. 2009. Large-area synthesis of high-quality and uniform graphene films on copper foils. Science 324(5932): 1312–1314. Doi: 10.1126/science.1171245.

Li, Z., P. Wu, C. Wang, X. Fan, W. Zhang, X. Zhai et al. 2011. Low-temperature growth of graphene by chemical vapor deposition using solid and liquid carbon sources. ACS Nano. 5(4): 3385–3390. Doi: 10.1021/nn200854p.

Lin, M.Y., W.C. Guo, M.H. Wu, P.Y. Wang, T.H. Liu, C.W. Pao et al. 2012. Low-temperature grown graphene films by using molecular beam epitaxy. Appl. Phys. Lett. 101(22). Doi: 10.1063/1.4768948.

Liu, J., Z. He, S.Y. Khoo and T.T.Y. Tan. 2016. A new strategy for achieving vertically-erected and hierarchical TiO_2/ nanosheets array/carbon cloth as a binder-free electrode for protein impregnation, direct electrochemistry and mediator-free glucose sensing. Biosens. Bioelectron. 77: 942–949. Doi: 10.1016/j.bios.2015.10.070.

Liu, J., H.J. Choi and L.Y. Meng. 2018. A review of approaches for the design of high-performance metal/graphene electrocatalysts for fuel cell applications. J. Ind. Eng. Chem. 64: 1–15. Doi: 10.1016/j.jiec.2018.02.021.

Liu, S., J. Tian, L. Wang, H. Li, Y. Zhang and X. Sun. 2010. Stable aqueous dispersion of graphene nanosheets: Noncovalent functionalization by a polymeric reducing agent and their subsequent decoration with Ag nanoparticles for enzymeless hydrogen peroxide detection. Macromolecules 43(23): 10078–10083. Doi: 10.1021/ma102230m.

Liu, Y., X. Liu, M. Li, Y. Liu, Z. Guo, Z. Xue et al. 2015. A novel synthesis of porous graphene nanoarchitectures using silver nanoparticles for fabricating enzyme sensor. RSC Adv. 5(121): 100268–100271. Doi: 10.1039/c5ra17717k.

Lotya, M., Y. Hernandez, P.J. King, R.J. Smith, V. Nicolosi, L.S. Karlsson et al. 2009. Liquid phase production of graphene by exfoliation of graphite in surfactant/water solutions. J. Am. Chem. Soc. 131(10): 3611–3620. Doi: 10.1021/ja807449u.

Lounasvuori, M.M., D. Kelly and J.S. Foord. 2018. Carbon black as low-cost alternative for electrochemical sensing of phenolic compounds. Carbon 129: 252–257. Doi: 10.1016/j.carbon.2017.12.020.

Lu, J., L.T. Drzal, R.M. Worden and I. Lee. 2007. Simple fabrication of a highly sensitive glucose biosensor using enzymes immobilized in exfoliated graphite nanoplatelets nafion membrane. Chem. Mater. 19(25): 6240–6246. Doi: 10.1021/cm702133u.

Lu, L.M., X.L. Qiu, X.B. Zhang, G.L. Shen, W. Tan and R.Q. Yu. 2013. Supramolecular assembly of enzyme on functionalized graphene for electrochemical biosensing. Biosens. Bioelectron. 45(1): 102–107. Doi: 10.1016/j.bios.2013.01.065.

Lu, Q., W. Wei, Z. Zhou, Z. Zhou, Y. Zhang and S. Liu. 2014. Electrochemiluminescence resonance energy transfer between graphene quantum dots and gold nanoparticles for DNA damage detection. Analyst 139(10): 2404–2410. Doi: 10.1039/c4an00020j.

Maji, S.K., S. Sreejith, A.K. Mandal, X. Ma and Y. Zhao. 2014. Immobilizing gold nanoparticles in mesoporous silica covered reduced graphene oxide: A hybrid material for cancer cell detection through hydrogen peroxide sensing. ACS Appl. Mater. Interfaces 6(16): 13648–13656. Doi: 10.1021/am503110s.

Mansouri, N., A.A. Babadi, S. Bagheri and S.B.A. Hamid. 2017. Immobilization of glucose oxidase on 3D graphene thin film: Novel glucose bioanalytical sensing platform. Int. J. Hydrogen Energy 42(2): 1337–1343. Doi: 10.1016/j.ijhydene.2016.10.002.

Marcano, D.C., D.V. Kosynkin, J.M. Berlin, A. Sinitskii, Z. Sun, A. Slesarev et al. 2010. Improved synthesis of graphene oxide. ACS Nano. 4(8): 4806–4814. Doi: 10.1021/nn1006368.

Moon, J.M., S.C. Lim, Y.H. Jeong, Y.H. Lee, D.J. Bae, E.K. Suh et al. 2001. Transport phenomena in an anisotropically aligned single-wall carbon nanotube film. Phys. Rev. B 64(23). Doi: 10.1103/PhysRevB.64.233401.

Mubarak, N.M., E.C. Abdullah, N.S. Jayakumar and J.N. Sahu. 2014. An overview on methods for the production of carbon nanotubes. J. Ind. Eng. Chem. 20(4): 1186–1197. Doi: 10.1016/j.jiec.2013.09.001.

Naghdi, S., K.Y. Rhee and S.J. Park. 2018. A catalytic, catalyst-free, and roll-to-roll production of graphene via chemical vapor deposition: Low temperature growth. Carbon 127: 1–12. Doi: 10.1016/j.carbon.2017.10.065.

Nguyen, B.H. and V.H. Nguyen. 2016. Promising applications of graphene and graphene-based nanostructures. Advances in Natural Sciences: Nanosci. Nanotechnol. 7(2): 023002. Doi: 10.1088/2043-6262/7/2/023002.

Nguyen, K.T. and Y. Zhao. 2014. Integrated graphene/nanoparticle hybrids for biological and electronic applications. Nanoscale 6(12): 6245–6266. Doi: 10.1039/c4nr00612g.

Ning, G., Z. Fan, G. Wang, J. Gao, W. Qian and F. Wei. 2011. Gram-scale synthesis of nanomesh graphene with high surface area and its application in supercapacitor electrodes. Chem. Commun. 47(21): 5976–5978. Doi: 10.1039/c1cc11159k.

Niyogi, S., E. Bekyarova, M.E. Itkis, J.L. McWilliams, M.A. Hamon and R.C. Haddon. 2006. Solution properties of graphite and graphene. J. Am. Chem. Soc. 128(24): 7720–7721. Doi: 10.1021/ja060680r.

Novoselov, K.S., A.K. Geim, S.V. Morozov, D. Jiang, Y. Zhang, S.V. Dubonos et al. 2004. Electric field in atomically thin carbon films. Science 306(5696): 666–669. Doi: 10.1126/science.1102896.

Novoselov, K.S., A.K. Geim, S.V. Morozov, D. Jiang, M.I. Katsnelson, I.V. Grigorieva et al. 2005. Two-dimensional gas of massless Dirac fermions in graphene. Nature 438(7065): 197–200. Doi: 10.1038/nature04233.

Novoselov, K.S., V.I. Fal'Ko, L. Colombo, P.R. Gellert, M.G. Schwab and K. Kim. 2012. A roadmap for graphene. Nature 490(7419): 192–200. Doi: 10.1038/nature11458.

Ohta, T., F. El Gabaly, A. Bostwick, J.L. McChesney, K.V. Emtsev, A.K. Schmid et al. 2008. Morphology of graphene thin film growth on SiC(0001). New J. Phys. 10. Doi: 10.1088/1367-2630/10/2/023034.

Park, I.J., T.I. Kim, I.T. Cho, C.W. Song, J.W. Yang, H. Park et al. 2018. Graphene electrode with tunable charge transport in thin-film transistors. Nano Res. 11(1): 274–286. Doi: 10.1007/s12274-017-1630-3.

Park, S. and R.S. Ruoff. 2009. Chemical methods for the production of graphenes. Nature Nanotechnol. 4(4): 217–224. Doi: 10.1038/nnano.2009.58.

Pendashteh, A., M.F. Mousavi and M.S. Rahmanifar. 2013. Fabrication of anchored copper oxide nanoparticles on graphene oxide nanosheets via an electrostatic coprecipitation and its application as supercapacitor. Electrochim. Acta 88: 347–357. Doi: 10.1016/j.electacta.2012.10.088.

Pérez-López, B. and A. Merkoçi. 2012. Carbon nanotubes and graphene in analytical sciences. Microchim. Acta 179(1-2): 1–16. Doi: 10.1007/s00604-012-0871-9.

Raghubanshi, H., E.D. Dikio and E.B. Naidoo. 2016. The properties and applications of helical carbon fibers and related materials: A review. J. Ind. Eng. Chem. 44: 23–42. Doi: 10.1016/j.jiec.2016.08.023.

Rakhi, R.B. and H.N Alshareef. 2011. Enhancement of the energy storage properties of supercapacitors using graphene nanosheets dispersed with metal oxide-loaded carbon nanotubes. J. Power Sources 196(20): 8858–8865. Doi: 10.1016/j.jpowsour.2011.06.038.

Rao, C.N.R., K.S. Subrahmanyam, H.S.S.R. Matte, B. Abdulhakeem, A. Govindaraj, B. Das et al. 2010. A study of the synthetic methods and properties of graphenes. Sci. Technol. Adv. Mater. 11(5). Doi: 10.1088/1468-6996/11/5/054502.

Rathod, D., C. Dickinson, D. Egan and E. Dempsey. 2010. Platinum nanoparticle decoration of carbon materials with applications in non-enzymatic glucose sensing. Sens. Actuators B 143(2): 547–554. Doi: 10.1016/j.snb.2009.09.064.

Reina, A. and J. Kong. 2012. Graphene growth by CVD methods. pp. 167–203. *In*: Graphene Nanoelectronics: From Materials to Circuits (Vol. 9781461405481).

Reinke, M., Y. Kuzminykh and P. Hoffmann. 2015. Low temperature chemical vapor deposition using atomic layer deposition chemistry. Chem. Mater. 27(5): 1604–1611. Doi: 10.1021/cm504216p.

Rollings, E., G.H. Gweon, S.Y. Zhou, B.S. Mun, J.L. McChesney, B.S. Hussain et al. 2006. Synthesis and characterization of atomically thin graphite films on a silicon carbide substrate. J. Phys. Chem. Solids 67(9-10): 2172–2177. Doi: 10.1016/j.jpcs.2006.05.010.

Roy, P. and S.K. Srivastava. 2015. Nanostructured anode materials for lithium ion batteries. J. Mater. Chem. A 3(6): 2454–2484. Doi: 10.1039/c4ta04980b.

Ruoff, R.S. and D.C. Lorents. 1995. Mechanical and thermal properties of carbon nanotubes. Carbon 33(7): 925–930. Doi: 10.1016/0008-6223(95)00021-5.

Russo, P., A. Hu, G. Compagnini, W.W. Duley and N.Y. Zhou. 2014. Femtosecond laser ablation of highly oriented pyrolytic graphite: A green route for large-scale production of porous graphene and graphene quantum dots. Nanoscale 6(4): 2381–2389. Doi: 10.1039/c3nr05572h.

Selvakumar, D., A. Alsalme, A. Alswieleh and R. Jayavel. 2017. Freestanding flexible nitrogen doped-reduced graphene oxide film as an efficient electrode material for solid-state supercapacitors. J. Alloys Compd. 723: 995–1000. Doi: 10.1016/j.jallcom.2017.06.333.

Seo, J.G., C.K. Lee, D. Lee and S.H. Song. 2018. High-performance tires based on graphene coated with Zn-free coupling agents. J. Ind. Eng. Chem. Doi: 10.1016/j.jiec.2018.04.015.

Shan, C., H. Yang, J. Song, D. Han, A. Ivaska and L. Niu. 2009. Direct electrochemistry of glucose oxidase and biosensing for glucose based on graphene. Anal. Chem. 81(6): 2378–2382. Doi: 10.1021/ac802193c.

Shang, N.G., P. Papakonstantinou, M. McMullan, M. Chu, A. Stamboulis, A. Potenza et al. 2008. Catalyst-free efficient growth, orientation and biosensing properties of multilayer graphene nanoflake films with sharp edge planes. Adv. Funct. Mater. 18(21): 3506–3514. Doi: 10.1002/adfm.200800951.

Singh, C. and W. Song. 2012. Carbon nanotube structure, synthesis, and applications. pp. 1–37. *In*: The Toxicology of Carbon Nanotubes (Vol. 9781107008373).

Somani, P.R., S.P. Somani and M. Umeno. 2006. Planer nano-graphenes from camphor by CVD. Chem. Phys. Lett. 430(1-3): 56–59. Doi: 10.1016/j.cplett.2006.06.081.

Stanković, D.M., M. Ognjanović, M. Fabian, L. Švorc, J.F.M.L. Mariano and B. Antić. 2017. Design of titanium nitride- and wolfram carbide-doped RGO/GC electrodes for determination of gallic acid. Anal. Biochem. 539: 104–112. Doi: 10.1016/j.ab.2017.10.018.

Stoller, M.D., S. Park, Z. Yanwu, J. An and R.S. Ruoff. 2008. Graphene-based ultracapacitors. Nano Lett. 8(10): 3498–3502. Doi: 10.1021/nl802558y.

Subrahmanyam, K.S., L.S. Panchakarla, A. Govindaraj and C.N.R. Rao. 2009. Simple method of preparing graphene flakes by an arc-discharge method. J. Phys. Chem. C 113(11): 4257–4259. Doi: 10.1021/jp900791y.

Sun, J.Y., K.J. Huang, S.Y. Wei, Z.W. Wu and F.P. Ren. 2011. A graphene-based electrochemical sensor for sensitive determination of caffeine. Colloids Surf. B 84(2): 421–426. Doi: 10.1016/j.colsurfb.2011.01.036.

Sun, Y., Q. Wu and G. Shi. 2011. Graphene based new energy materials. Energy Environ. Sci. 4(4): 1113–1132. Doi: 10.1039/c0ee00683a.

Taha-Tijerina, J., L. Peña-Paras, T.N. Narayanan, L. Garza, C. Lapray, J. Gonzalez et al. 2013. Multifunctional nanofluids with 2D nanosheets for thermal and tribological management. Wear 302(1-2): 1241–1248. Doi: 10.1016/j.wear.2012.12.010.

Tang, D., Q. Wang, Z. Wang, Q. Liu, B. Zhang, D. He et al. 2018. Highly sensitive wearable sensor based on a flexible multi-layer graphene film antenna. Sci. Bull. Doi: 10.1016/j.scib.2018.03.014.

Tang, L., Y. Wang, Y. Li, H. Feng, J. Lu and J. Li. 2009. Preparation, structure, and electrochemical properties of reduced graphene sheet films. Adv. Funct. Mater. 19(17): 2782–2789. Doi: 10.1002/adfm.200900377.

Thévenot, D.R., K. Toth, R.A. Durst and G.S. Wilson. 2001. Electrochemical biosensors: Recommended definitions and classification. Biosens. Bioelectron. 16(1-2): 121–131. Doi: 10.1016/S0956-5663(01)00115-4.

Toda, K., R. Furue and S. Hayami. 2015. Recent progress in applications of graphene oxide for gas sensing: A review. Anal. Chim. Acta 878: 43–53. Doi: 10.1016/j.aca.2015.02.002.

Tseng, K.H., C.J. Chou, S.H. Shih, D.C. Tien, H.C. Ku and L. Stobinsk. 2018. Submerged arc discharge for producing nanoscale graphene in deionised water. Micro and Nano Lett. 13(1): 31–34. Doi: 10.1049/mnl.2017.0387.

Tung, V.C., M.J. Allen, Y. Yang and R.B. Kaner. 2009. High-throughput solution processing of large-scale graphene. Nat. Nanotechnol. 4(1): 25–29. Doi: 10.1038/nnano.2008.329.

Velický, M., P.S. Toth, A.M. Rakowski, A.P. Rooney, A. Kozikov, C.R. Woods et al. 2017. Exfoliation of natural van der Waals heterostructures to a single unit cell thickness. Nat. Commun. 8. Doi: 10.1038/ncomms14410.

Viculis, L.H., J.J. Mack and R.B. Kaner. 2003. A chemical route to carbon nanoscrolls. Science 299(5611): 1361. Doi: 10.1126/science.1078842.

Wang, X., K. Zheng, X. Feng, C. Xu and W. Song. 2015. Polymeric ionic liquid functionalized MWCNTs as efficient electrochemical interface for biomolecules simultaneous determination. Sens. Actuators B 219: 361–369. Doi: 10.1016/j.snb.2015.04.128.

Wang, Y. and G. Cao. 2008. Developments in nanostructured cathode materials for high-performance lithium-ion batteries. Adv. Mater. 20(12): 2251–2269. Doi: 10.1002/adma.200702242.

Wang, Y., S. Zhao, M. Li, W. Li, Y. Zhao, J. Qi et al. 2017. Graphene quantum dots decorated graphene as an enhanced sensing platform for sensitive and selective detection of copper(II). J. Electroanal. Chem. 797: 113–120. Doi: 10.1016/j.jelechem.2017.05.031.

Wang, Z., B. Fu, S. Zou, B. Duan, C. Chang, B. Yang et al. 2016. Facile construction of carbon dots via acid catalytic hydrothermal method and their application for target imaging of cancer cells. Nano Res. 9(1): 214–223. Doi: 10.1007/s12274-016-0992-2.

Wasiński, K., M. Walkowiak, P. Półrolniczak and G. Lota. 2015. Capacitance of Fe_3O_4/rGO nanocomposites in an aqueous hybrid electrochemical storage device. J. of Power Sources 293: 42–50. Doi: 10.1016/j.jpowsour.2015.05.064.

Winter, M. and R.J. Brodd. 2004. What are batteries, fuel cells, and supercapacitors? Chem. Rev. 104(10): 4245–4269. Doi: 10.1021/cr020730k.

Wu, J., Q. Liu, K. Wang and J. Cai. 2012. Enhanced direct electrochemistry of glucose oxidase and glucose biosensing based on TiO_2-Decorated graphene nanohybrids. pp. 507–510. *In*: Adv. Mater. Res. Vol. 496.

Wu, T., Y. Jiang and X. Zhang. 2015. The synthesis of CVD single crystal graphene growth on copper substrate. Gongneng Cailiao. J. Funct. Mater. 46(16): 16037–16043 and 11651. Doi: 10.3969/j.issn.1001-9731.2015.16.006.

Wu, Z.S., W. Ren, L. Gao, B. Liu, C. Jiang and H.M. Cheng. 2009. Synthesis of high-quality graphene with a pre-determined number of layers. Carbon 47(2): 493–499. Doi: 10.1016/j.carbon.2008.10.031.

Xiao, W., Z. Wang, H. Guo, Y. Zhang, Q. Zhang and L. Gan. 2013. A facile PVP-assisted hydrothermal fabrication of Fe_2O_3/Graphene composite as high performance anode material for lithium ion batteries. J. Alloys Compd. 560: 208–214. Doi: 10.1016/j.jallcom.2012.12.166.

Xiong, P., C. Hu, Y. Fan, W. Zhang, J. Zhu and X. Wang. 2014. Ternary manganese ferrite/graphene/polyaniline nanostructure with enhanced electrochemical capacitance performance. J. Power Sources 266: 384–392. Doi: 10.1016/j.jpowsour.2014.05.048.

Xu, J., Z. Wang, Z. Shi and Z. Gu. 2005. Synthesis, isolation and spectroscopic characterization of Yb-containing high metallofullerenes. Chem. Phys. Lett. 409(4-6): 192–196. Doi: 10.1016/j.cplett.2005.05.034.

Yang, W., K.R. Ratinac, S.R. Ringer, P. Thordarson, J.J. Gooding and F. Braet. 2010. Carbon nanomaterials in biosensors: Should you use nanotubes or graphene. Angewandte Chemie—International Edition 49(12): 2114–2138. Doi: 10.1002/anie.200903463.

Yang, Z. and J. Hao. 2016. Progress in pulsed laser deposited two-dimensional layered materials for device applications. J. Mater. Chem. C 4(38): 8859–8878. Doi: 10.1039/c6tc01602b.

Ying, L.S., M.A. Bin Mohd Salleh, H.B. Mohamed Yusoff, S.B. Abdul Rashid and J.B. Abd Razak. 2011. Continuous production of carbon nanotubes—A review. J. Ind. Eng. Chem. 17(3): 367–376. Doi: 10.1016/j.jiec.2011.05.007.

Yuan, G.D., W.J. Zhang, Y. Yang, Y.B. Tang, Y.Q. Li, J.X. Wang et al. 2009. Graphene sheets via microwave chemical vapor deposition. Chem. Phys. Lett. 467(4-6): 361–364. Doi: 10.1016/j.cplett.2008.11.059.

Zaaba, N.I., K.L. Foo, U. Hashim, S.J. Tan, W.W. Liu and C.H. Voon. 2017. Synthesis of Graphene Oxide using Modified Hummers Method: Solvent Influence. Paper presented at the Procedia Engineering.

Zhang, H., C. Peng, J. Yang, M. Lv, R. Liu, D. He et al. 2013. Uniform ultrasmall graphene oxide nanosheets with low cytotoxicity and high cellular uptake. ACS Appl. Mater. Interfaces 5(5): 1761–1767. Doi: 10.1021/am303005j.

Zhang, L.L. and X.S. Zhao. 2009. Carbon-based materials as supercapacitor electrodes. Chem. Soc. Rev. 38(9): 2520–2531. Doi: 10.1039/b813846j.

Zhang, T.T., H.M. Zhao, X.F. Fan, S. Chen and X. Quan. 2015. Electrochemiluminescence immunosensor for highly sensitive detection of 8-hydroxy-2'-deoxyguanosine based on carbon quantum dot coated Au/SiO_2 core-shell nanoparticles. Talanta 131: 379–385. Doi: 10.1016/j.talanta.2014.08.024.

Zhang, Y., L. Zhang and C. Zhou. 2013. Review of chemical vapor deposition of graphene and related applications. Acc. Chem. Res. 46(10): 2329–2339. Doi: 10.1021/ar300203n.

Zhao, X., B.M. Sánchez, P.J. Dobson and P.S. Grant. 2011. The role of nanomaterials in redox-based supercapacitors for next generation energy storage devices. Nanoscale 3(3): 839–855. Doi: 10.1039/c0nr00594k.

Zhou, C., Y. Chen, X. You, Y. Dong and Y. Chi. 2017. An electrochemiluminescent biosensor based on interactions between a graphene quantum dot–sulfite co-reactant system and hydrogen peroxide. ChemElectroChem. 4(7): 1783–1789. Doi: 10.1002/celc.201600921.

Zhou, M., Y. Zhai and S. Dong. 2009. Electrochemical sensing and biosensing platform based on chemically reduced graphene oxide. Anal. Chem. 81(14): 5603–5613. Doi: 10.1021/ac900136z.

Zhou, Z. and X.F. Wu. 2013. Graphene-beaded carbon nanofibers for use in supercapacitor electrodes: Synthesis and electrochemical characterization. J. Power Sources 222: 410–416. Doi: 10.1016/j.jpowsour.2012.09.004.

Zöpfl, A., M. Sisakthi, J. Eroms, F.M. Matysik, C. Strunk and T. Hirsch. 2016. Signal enhancement in amperometric peroxide detection by using graphene materials with low number of defects. Microchim. Acta 183(1): 83–90. Doi: 10.1007/s00604-015-1600-y.

CHAPTER **8**

Carbon Nanomaterials in Biosensors for Biotechnological Applications

Madan L. Verma,[1,2,]* *Meenu Thakur,*[3] *Pankaj Kumar,*[4]
Irfan Ahmad[5] and *Krishan D. Sharma*[6]

1. Introduction

Nanotechnology is a multi-disciplinary domain that has unlimited applications, ranging from sustainable agriculture to health sectors (Liu et al. 2017, Eguilaz et al. 2016, Hou et al. 2015, Lim et al. 2014, Verma et al. 2013a–d, Verma et al. 2012, Verma and Kanwar 2012, Verma and Kanwar 2010). Nanotechnology involves the synthesis and applications of the nanomaterials (Zhu 2017, Inbaraj and Chen 2016, Verma et al. 2016). Nanomaterial is defined as any advanced material of size in the range of 1–100 nanometer (nm). Various forms of nanomaterials, such as nanoparticles, nanofibers, nanocomposite, nanotubes (single-walled and multi-walled), and nanosheet have been synthesized by various physicochemical and biological routes. Although each and every production method has pros and cons, biological methods of nanomaterials production, an environment-friendly biogenic synthesis, provides an upper edge due to adherence to green chemistry rules (Pelle and Compagnone 2017, Verma 2017c, Yu et al. 2015). Commercially, various forms of mono-dispersed nanomaterials have been synthesized in the powder form (nanopowder), as well as stable suspension form. Nanomaterials can be produced from the bulk material as well as atomic materials. Two approaches, namely top-down and bottom up, are employed for the production of the nanomaterials. The stability of the nanomaterials is a more important aspect, and hence a very sophisticated method/ instrument is to be chosen in order to avoid any surface defect on the nanomaterials. Recently,

[1] Centre for Chemistry and Biotechnology, Deakin University, Victoria-3216, Australia.
[2] Department of Biotechnology, Dr YSP University of Horticulture and Forestry, Himachal Pradesh-177001, India.
[3] Department of Biotechnology, Shoolini Institute of Life Sciences and Business Management, Solan, Himachal Pradesh-173212, India.
[4] Science and Engineering Research Board, Department of Science and Technology, Government of India, New Delhi-110070, India.
[5] ABPG College Ranapar, DDU Gorakhpur University, Gorakhpur-273405, India.
[6] Department of Food Science and Technology, Dr YSP University of Horticulture and Forestry, Solan, Himachal Pradesh-173230, India.
* Corresponding author: madanverma@gmail.com

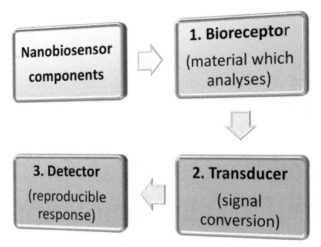

Figure 8.1: Biosensor composed of three components in the following order: bioreceptor, transducer, and detector.

nanobiotechnology, a portable biosensor technology, is getting more attention due to wide arrays of biotechnological applications.

The biosensor can be characterized as an investigative device which fuses a biologically active component with a suitable physical transducer to produce a quantifiable signal corresponding to the concentration of chemical species in any type of sample (Tauhami 2014). Biosensor assembly is composed of three major parts, namely bioreceptor, transducer, and detector (Figure 8.1). The main purpose of a biosensor is to detect biological material, such as antibodies, proteins, enzymes, nucleic acids, bacteria, viruses (Lee et al. 2012). The bioreceptor perceives the target and reacts, while transducer changes over this response into a noticeable signal which can be measured and correlated with the analyte concentration. Biosensor was first reported in 1962 by Clark and Lyons (Clark and Lyons 1962). Since then, biosensors were studied intensively and used in various types of applications, such as public health, environmental monitoring, food and agriculture sector (Pelle and Compagnone 2017, Verma 2017a–c, Sonawane et al. 2014, Tauhami 2014, Verma et al. 2010, Jain et al. 2010). The biosensor is different from chemical sensor in two ways: (i) sensing element consists of biological material, such as proteins, oligo- or polynucleotide, microorganisms, and in some cases, whole biological tissues, and (ii) sensor is utilized for monitoring biological processes or to recognize biomolecules (Balasubramanian and Burghard 2006). For biosensing *in vitro*, the sample solution, such as blood serum, urine, milk products, etc., is dropped at the top of the biosensor, and the output signal gives information regarding the composition of the solution taken. On the other hand, *in vivo* biosensing represents a system which aims, for instance, to measure rate of uptake or efflux of relevant species or to detect the distribution of concentration of analyte in a living organism (Lee et al. 2010, Wilson and Gifford 2005).

This chapter discusses the various types of nanomaterials employed for biosensing applications. Contribution of the carbon nanomaterials in improving biosensing technology is discussed in detail. The impact of carbon nanomaterials-based biosensor applications, especially in agriculture and food industry, is critically discussed.

2. Contribution of the Carbon Nanomaterial in Biosensing Applications

A considerable measure of research activity has been engaged in creating biosensors for different purposes. The time has come to convey this innovation to the front line and make it financially accessible (Nayak et al. 2009, Tothill 2009, Wang 2008). Efforts and funds should be mobilized to make biosensors on a large scale to profit and be useful to the overall population. With introduction to the commercial market, the uses of this technology would be significantly upgraded. A couple of such applications could be the detection of virulence of a vaccine just before it is injected in order to prevent the coincidental procurement of a sickness, bandages recognizing a septic injury, deadly viruses in the environment or from the patient sample (quick and early recognition), and so forth in the medical field. Constant observation of dairy items and distilleries may help encourage a cleaner and sterile condition, and experiment with different tastes imparted by specific microorganisms in specific concentrations, giving rise to new products. A far-fetched and plausible utilization of this innovation could be in space investigation, where the concentration of the living beings would be exceptionally low and may prompt noting a large number of the long-standing inquiries with respect to the presence of life in space. Many research studies have been accomplished in the past couple of years in the overall range on creating novel biosensors with high sensitivity and selectivity.

The on-going, quick advancement of nanomaterials has made a significant impact on the improvement of biosensors. The use of nanomaterials has been given to every single specialized segment of biosensors—from recognition components to signal processors. At the point when the material's size is reduced to the nanoscale, the fascinating changes in substance and physical properties occur because of two primary elements: surface effect and quantum effect. Due to the increase in surface to volume ratio of nanomaterials as compared to their bulk form, the sensitivity of biosensors increases through expanding the interface for recognition of the element allocated. The quantum confinement phenomenon can prompt an expansion in the band-gap energy and a blue shift in light emission with decreasing size. Therefore, the electrical and optical properties of nanomaterials become size and shape dependent. These fundamental highlights of nanomaterials make it possible to turn chemical and physical properties to particular biosensor applications by controlling their size, shape, and chemical composition (Tothil 2011). Nanomaterials consist of various materials having different sizes, shape, composition, and chemistry. They can play a very efficient role in the detection mechanism of the biosensor technology. Nanoelectromechanical system consists of nanomaterials and an electrical system, which are integrated in such a way that it can make a very good electrical transduction system. Based on their electrical and mechanical properties, various nanomaterials are examined for their application in enhancing biological signaling and transduction mechanism. Materials, such as nanotubes, nanowires, nanorods, nanoparticles, and thin films that can be utilized for increasing biological signal, are made up of nanocrystalline matter (Jianrong et al. 2004).

To manufacture biosensors, carbon-nanomaterials such as carbon nanotube, single-walled carbon nanotubes, multi-walled carbon nanotubes, graphene, C_{60}, and nanodiamond are very popular because of their properties, for example, high surface-to-volume ratio, very good electrical conductivity, and great biocompatibility. Carbon nanomaterials can altogether enhance the performance of biosensors in the perspective of sensitivity and stability, which have been widely utilized as the transducer or nanocarrier in food and agriculture sector (Figure 8.2). Carbon nanomaterials, including graphene, carbon nanotubes, carbon spots, and other carbon nanomaterials, have unique physical, chemical, and electrical properties (Xia and Gao 2015, Marin and Merkoci 2012, Musameh et al. 2012), making them suitable for the fields of energy storage and conversion, high-frequency electronics, biological and chemical sensors (Wang and Dai 2015). As indicated by their diverse structures, carbon nanomaterials can be separated into zero-dimensional (i.e., carbon spots and fullerene), one-dimensional (i.e., carbon nanotubes), and two-dimensional (i.e., graphene) nanomaterials. Among them, graphene (including its derivatives), carbon nanotubes, and carbon dabs are the most broadly utilized carbon nanomaterials for designing biosensors. It was two centuries back that carbon first appeared to be available in organic molecules and biomolecules, and also as natural carbon materials, for example, the different kinds of amorphous carbon, diamond, and graphite. The extraordinary capacity of carbon particles to partake in strong covalent bonds with other carbon atoms in differing hybridization states (sp, sp^2, sp^3) or with non-metallic components empower them to form a wide range of structures, from small molecules to long chains. Diamond comprises tetrahedral sp^3 carbon atoms that form a unique large crystal. Conversely, graphite is comprised of stacked graphene monolayers that are held

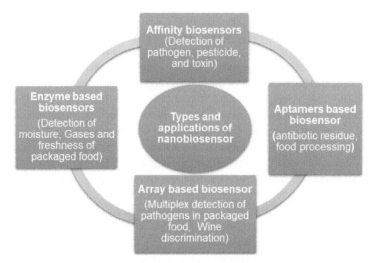

Figure 8.2: Types and applications of nanobiosensors.

together by van der Waals interactions. These graphene monolayers comprise sp^2 carbon particles that are packed densely in a two-dimensional hexagonal lattice (Georgakilas et al. 2015). Over the past few years, new materials are identified as having unique properties. The first of these carbon nanostructures to be found was the C_{60} particle, which is known as fullerene (the smallest and stable nanostructure), and was at first detailed by Kroto et al. in 1985 and Kratschmer et al. in 1990. Each C_{60} particle comprises 60 sp^2 carbon atoms arranged in a series of hexagons and pentagons to form a spherical (truncated icosahedral) structure. Fullerenes exist between the boundary of molecules and nanomaterials. Another significant advance in the improvement of carbon nanomaterials happened six years after the fact, with the discovery of carbon nanotubes (CNTs) by Iijima. Multi-walled carbon nanotubes (MWNTs) and single-walled nanotubes (SWNTs) were reported by Iijima (1991) and Iijima and Ichihashi (1993), respectively, after which CNTs have risen as a standout amongst the most intensively examined nanostructured materials (Journet et al. 1997, Balasubramanian and Burghard 2006). The most recently isolated carbon nanostructure is graphene, the building block of graphite. While its existence had been anticipated decades back (Wallace 1947), and tentatively recognized in 1962 by Boehm et al., it was first discovered and characterized in 2004 by Nobel Prize winner Andre Geim and Novoselov (Boehm et al. 1962, Novoselov et al. 2004).

Since effectively stripped from graphite by Novoselov et al. in 2004, graphene (GR) has been a constant research topic of high importance (Cai et al. 2018, Stankovich et al. 2006). Formed by hexagonal lattice through sp^2 hybridized carbon atoms, this two-dimensional (2-D) carbon nanomaterial has a stable and solid structure, unique thermal conductivity, great surface area, electron transfer capacity, and desired flexibility and impermeability (Wei et al. 2016). GR can be converted to graphene oxide (GO) by chemical oxidation, and further responded to form reduced GO (rGO). Compared to pristine GR, these functionalized materials have plentiful carboxy, epoxide, and hydroxy groups, making them scatter in polar solvents effortlessly and additionally alter with biomolecules by covalent or non-covalent bonding (Zhou et al. 2018). Moreover, the imperfections caused by chemical reactions give these materials diverse and even desired processing features (Zhou et al. 2017, Liu et al. 2011).

Carbon nanotubes (CNTs) are one-dimensional (1-D) carbon nanomaterials, which are viewed as the processed results of GR after being rolled up. As a sort of macromolecular system, CNTs open a new field in the zone of science, material science, materials, and life advancements since the early 1990s (Wang and Lin 2008). In view of the quantity of layers, CNTs can be categorised into single-walled CNTs (SWCNTs) and multi-walled CNTs (MWCNTs). SWNTs are one rolled up graphene sheet, while MWNTs are concentric tubes of at least two rolled up graphene sheets, separated by around 0.34 nm. SWNTs have exceptionally remarkable electrical properties, depending upon the chirality of the wrap, and they can carry on as either metals or semiconductors (Schnorr and Swager 2011, Eder 2010). Recent investigations have set up the way that few captivating properties of CNTs, for example, their nanodimensions and surface chemistry (Balasubramanian and Burghard 2006), make them more versatile materials for new kinds of electrochemical, electric, and optical biosensors (Yang et al. 2010). At the point when they are utilized in biosensors, CNTs can serve either as transducers or nanocarriers because of their predominant electrochemical properties or numerous other properties (Wang and Dai 2015). It has been discovered that the usage of CNTs can enhance the recognition ability for surface adsorbates (Chiashi et al. 2015). In the previous decades, the advancement of CNTs-based biosensors is one of the most captivating research fields among these applications, in view of its special properties, for example, incredible electrical conductivity, favorable chemical stability, and high mechanical quality.

Carbon dots (CDs) are small-sized, zero-dimensional (0D) carbon nanoparticles. Since being revealed by the group of Scrivens in 2004 (Xu et al. 2004), CDs have raised an on-going consideration in different fields. Due to the quantum confinement and edge effects, CDs have a close resemblance with normal quantum spots (QDs), for example, the remarkable optical and electro-optical properties (Wang and Dai 2015). Besides, the exceptional customizable luminous range, astounding light stability, low toxicity, and prevalent biocompatibility make CDs applications in the fields of bio-imaging, biosensor, catalysis, and photoelectric conversion (Liu et al. 2017, Lin et al. 2017, Lv et al. 2017). Presently, carbon quantum dots (CQDs) and graphene quantum dots (GRQDs) are the critical parts of CDs. Unlike CQDs, GRQDs are solely formed by GR-based or class-GR materials, for example, polycyclic aromatic hydrocarbon molecules (Li et al. 2017). Due to the assorted variety of compositions, CQDs and GRQDs have widened the range of applications.

Carbon black (CB), a sort of amorphous carbon, is a fine powder with a huge surface area. As one of the industrial raw materials, CB has been broadly utilized in the industry of rubber, plastic, coating, cable, printing ink, and dry battery. Furthermore, CB assumes a vital part in the electrocatalysis of oxidation/reduction processes because of its electrically conductive properties and capacity. Recently, a few scientists have utilized CB as the electrochemical substrate to change terminals by drop-casting technique (Arduini et al. 2017, Talarico et al. 2016, Lee et al. 2010), which can cause a speedier electron transfer rate compared to bare electrodes.

Porous carbon materials are the new non-silicon-based materials (Lee et al. 2006), which resulted in the development of new supporting material for biomolecules carrier. As compared to pure mesoporous silica materials, porous carbon materials have some unique properties, for example, high specific surface area, large pore volume, open pore structure, proficient mass transportation, high conductivity, and fantastic chemical stability (Chen et al. 2012, Stein et al. 2009, Lee et al. 2006). Ordered mesoporous carbon (OMC), as one of the porous carbon materials, has ideal biocompatibility and porousness (Zhang et al. 2016, Zhang et al. 2015b), which could be utilized as a compelling substrate to immobilized catalyst and different biomolecules without the decrease of their unique stability and activity.

Nanodiamonds are sp^3 carbon nanoparticles that comprise crystal domain with a diamondoid-like topology, and diameters are in the range of 1–2 nm, but under 20 nm (Williams 2011). Such nanostructures are not dispersible and are generally synthesized by top-down techniques, for example, as jet milling or abrasion of microdiamonds (Boudou et al. 2009, Niwase et al. 1995). Nanostructures of this sort have diameters more than 20 nm and act like bulk diamonds. While, sp^3 carbon nanostructures having diameters less than 1 nm are called diamondoids, and can be found in petroleum deposits (Dahl et al. 2003). The sp^3 hybridized surface-bound carbon atoms of these little diamondoids are found to be attached to hydrogen or other noncarbon atoms. As a result, characteristics of nanodiamonds are similar to organic molecules rather than bulk diamonds.

Nanohorns are tubule-like/cone-like structures produced using a solitary graphenic layer. They generally exist in vast spherical aggregates that are around 80–100 nm in diameter and found to be similar to dahlia flowers. Diameters of individual nanohorn at the tip and base of the cone are 1–2 nm and 4–5 nm, respectively. The wall-to-wall distance between adjacent single-walled carbon nanohorn (SWNH) is around 0.4 nm. Different kinds of SWNH exhibit similar features of buds and seeds (Yudasaka et al. 2008).

Dong et al. (2013) designed a nanobiosensor by immobilizing acetylcholinesterase enzyme on multi-walled carbon nanotubes chitosan nanocomposites modified glassy carbon electrode. Nanobiosensor exhibited detection of minute level of pesticides in water and soil in a concentration range between 1.0×10^{-12} and 5.0×10^{-7} M, and a detection limit of 7.5×10^{-13} M. For soil nitrate detection, Pan et al. (2013) built up a solid-state sensor by utilizing polypyrrole doped with nitrate, which exhibited the capacity of graphene layer to limit the water layer formation and successfully elevate ion-to-electron transition, thus significantly upgrading the stability and response rate. A biosensor was designed by Ali et al. (2017a) for the detection of nitrate in the soil. Graphene oxide (GO) nanosheets and poly (3,4-ethylenedioxythiophene) nanofibers (PEDOT-NFs) were utilized to make this microfluidic impedimetric sensor. The capacity of nanocomposite of hosting nitrate reductase enzyme and quantification of nitrate ions was shown in samples extracted from the soil of a *Zea mays* field, with a wide concentration range of 0.44–442 mg/L, and a detection limit of 0.135 mg/L.

Zhao et al. (2011) developed a humidity sensor utilizing distinctive dispersion concentrations, acquiring high detecting capacitance, and additionally quick reaction and great repeatability of graphene oxide films. A nanocomposite material constituted of multi-walled carbon nanotubes and Nafion was deposited by drop-casting on the surface of an acoustic wave resonator, giving enhanced sensitivity and dynamic trademark because of its large specific area and special ionic conductivity (Lei et al. 2011). Without a doubt, this sensor demonstrated a high sensitivity up to 260 kHz/% RH, good linearity with $R^2 > 0.99$, high accuracy of 0.3% RH at low humidity level below 10% RH. A graphene oxide film was utilized to coat a SU8 polymer channel waveguide with a linear response in less than 1 second (Lim et al. 2014).

Wu et al. (2014) have utilized graphene oxide (GO) as an acceptor and multicolor upconversion fluorescent nanoparticle as a benefactor for multiplexed fluorescence reverberation energy transfer-based aptasensor for mycotoxin determination. For detection of ochratoxin A (OTA), this aptasensor was utilized with a linear range from 0.05 to 100 ng/mL and a detection limit of 0.02 ng/mL. Whereas, for detection of aflatoxin B1, an immunosensor with an improved electrochemical performance utilizing graphene or conducting polymer or gold nanoparticles or the ionic fluid nanocomposite film on altered gold cathode, has been used (Linting et al. 2012). Nanocomposite demonstrated a magnificent sensitivity, long-term stability, and can be utilized to distinguish 1-fm concentration of aflatoxin B1. Srivastava et al. (2013) have used antibody functionalized reduced graphene oxide for aflatoxin B1, and acquired a detection limit of 0.12 ng/mL and a high sensitivity of 68 µA/(ng/mL)/cm².

Wu et al. (2013) designed an amperometric electrochemical biosensor by using graphene for detection of dichlorvos in water and agricultural products. Modified nanocomposite of rGO and Nafion on the electrode surface was employed for AChE loading. Biosensor could sensitively detect the target with the LOD of 2.0 µg L⁻¹. Similarly, Ding et al. (2014) developed a MWCNTs-based sensing platform on a nanoporous gold (NPG) electrode for the detection of malathion. Nanobiosensor showed a LOD of OPS (organophosphorous pesticide) down to 0.5 ng L⁻¹.

Mogha et al. (2016) fabricated an AChE-based amperometric biosensor for sensitive detection of chlorpyrifos in water samples. The rGO (reduced graphene oxide) supporting zirconium oxide nanoparticles (rGO-ZrO₂NPs) as nanocomposite was sprayed to deposit on indium tin oxide (ITO) glass, followed by enzyme immobilization via adsorption method. Nanobiosensor could easily detect chlorpyrifos in ultra-low concentration with a limit of detection (LOD) of 0.1 pM. Bao et al. (2016) developed an OPH (organophosphorus hydrolase)-based biosensor for determination of methyl parathion and parathion using titanium dioxide nanofibers (TiO₂NFs)-decorated carboxylic functionalized MWCNTs (cMWCNTs-TiO$_2$NFs). Nanobiosensor exhibited detection time of less than 5 seconds and low LOD (12 nM and 10 nM, respectively).

Kaur et al. (2016) developed an amperometric biosensor for malathion detection based on the nanocomposite of MWCNTs-poly(3,4-ethylenedioxythiophene). The limit of detection of malathion was calculated to be 1.0 fM within the linear range from 1.0 fM to 1.0 μM. Cinti et al. (2016) designed an amperometric pesticide biosensor using cobalt phthalocyanine (CoPc)-loaded carbon black (CB-CoPc) as the transducing materials. Nanobiosensor showed a low LOD value of 18 nM. For the detection of phorate, Zheng et al. (2015) prepared an AChE biosensor using a composite of ionic liquid functionalized graphene (IL-GR) and polyvinyl alcohol (PVA), denoted as IL-GR-PVA. This amperometric biosensor exhibited a LOD of 8.0 fM.

An AChE-based electrochemical biosensor was prepared by Zhang et al. (2017), utilizing the nanocomposite of rGO and palladium nanoparticles (PdNPs). Biosensor could well detect chlorpyrifos with the LOD of 0.23 μM. Sun et al. (2013) proposed an immunosensor based on gold nanoparticles (AuNPs) and MWCNTs-PANI (MWCNTs-AuNPs-PANI) nanocomposite. The developed immunosensor could detect even very low concentration of chlorpyrifos with a LOD of 0.17 nM. Moreover, this nanobiosensor could be successfully applied in vegetable samples for the detection of chlorpyrifos. Yu et al. (2015) designed an electrochemical biosensor for the determination of paraoxon. The amino functionalized MWCNTs (aMWCNTs), with better immobilization efficiency for AChE, exhibited the LOD (limit of detection) of 0.08 nM, with two linear ranges of 0.2–1.0 nM and 1.0–30 nM, respectively. Similarly, Wang et al. (2014) reported an electrochemical biosensor for chlorpyrifos and carbofuran detection based on a composite of carboxylic functionalized GR (cGR) and zinc oxide nanoparticles (ZnONPs) (cGR-ZnONPs). Biosensor showed excellent detection of chlorpyrifos and carbofuran, with a LOD of 0.05 pM and 0.52 pM, respectively.

Similarly, Ye et al. (2017) prepared an electrochemical biosensor based on the MWCNTs and metal nanoparticles for higher sensitive and selective detection of methyl parathion. Nanocomposite was further used to immobilize MPD (methylparathion degrading) enzyme. Nanobiosensor could detect a target with a favorable LOD of 0.3 μg L^{-1}.

Arvand and Mirroshandel (2017) prepared a GO/aptamer-QDs-based biosensor for detection of edifenphos with a LOD of 0.13 μg L^{-1} and a wide linear range from 0.5 to 6.0 μg L^{-1}. Based on the quenching feature of SWCNTs, Abnous et al. (2017) invented a fluorescent biosensor for acetamiprid detection, with the LODs of 198 and 130 pM, respectively.

Li et al. (2016) constructed an ECL (Electrochemiluminescence) biosensor for the detection of carbofuran using CDs as the signal indicator. The CDs-tagged DNA aptamers acted as the receptor, while fullerene (C$_{60}$, another important carbon nanomaterial)-loaded AuNPs (C60-AuNPs) acted as the energy donor, leading to the occurrence of ECLET between the above two materials. This assay could be used to detect carbofuran with the LOD of 0.88 pM, which offered a strategy for the monitoring of pesticide residues in real fruit and vegetable samples.

A fluorescent sensor utilizing L-tyrosine methyl ester-altered CQDs (carbon quantum dots) was set up by Hou et al. (2015) to detect methyl parathion in vegetables and soft drinks. Subsequently, this CQDs-based fluorescent biosensor could identify methyl parathion in the range of 0.1–10^5 nM, with a LOD of 48 pM. Zhang et al. (2015) developed an ECL biosensor on OMC substrate doped with ZnS QDs for the detection of glyphosate with a wide linear range from 0.1 to 10 nM in vegetal juice samples. The group of MacFarlane in 2016 developed N and S co-doped CQDs (N, S-CQDs) for carbaryl detection with the LOD of 5.4 μg L^{-1} (Li et al. 2016). Pan et al. (2013) used poly(amidoamine) dendrimer (PAMAM) to hybridize with MWCNTs to obtain MWCNTs-PAMAM nanocomposite biosensor for metolcarb detection with a LOD of 0.019 mg L^{-1} and a linear range between 0.1–50.0 mg L^{-1}.

It is pertinent from the above discussed studies that various forms of carbon nanomaterials are the integral part of the biosensor technology. In addition to ease of portability, incorporation of the carbon nanomaterials improves the response time, reproducibility, and limit of detection as compared to the conventional biosensors.

3. Applications of Carbon Nanomaterials based Nanobiosensors in Food Industry

Nanobiotechnology plays a pivotal role in the food industry. Various forms of nanobiosensors are employed to check the safety and quality of the food, and hence nanobiosensors have been become an integral part of the food industry (Perez-Lopez and Merkoci 2011; Table 8.1). As the food chain is becoming more complex, powerful analytical methods are required to verify the performance of food safety and quality systems (Campuzano et al. 2017). Unique properties of nanomaterials enable the food products to be more innovative while ensuring food safety (Lim and Kim 2016). Nanomaterials, such as carbon nanotubes, metal nanoparticles, nanowires, nanocomposite, and nanostructured materials are playing an increasing role in the design of sensing and biosensing systems, with applications in food analysis (Perez-Lopez and Merkoci 2011).

In the food industry, nanosensor has been utilized for monitoring effective environmental conditions for food processing, food packaging and distribution, and food storage and transport. Nanosensors have also been used for traceability, food nutrition, and monitoring products which have a limited shelf-life. Nanocapsules incorporated into food for effective delivery of nutrients without affecting the food color, texture, taste, and odors, such as nanocochleates (50 nm coiled nanoparticles), have successfully delivered the nutrients, i.e., vitamins, lycopene, and omega fatty acids more efficiently to cells (Gomez-Arribas et al. 2018).

The biosensing applications of the carbon nanomaterials biosensors, particularly electrochemical sensors, are improved due to inherent properties of nanomaterials, such as low detection limit, quick response, and desired portability. Carbon nanotubes (CNT)-based electrochemical biosensors can be divided into biocatalytic sensors using enzymes and bioaffinity sensors (Zhu 2017). Moreover, food-borne diseases are caused by consuming food or beverages contaminated by pathogenic microorganisms, such as bacteria, viruses, and parasites (Inbaraj and Chen 2016). Biosensors address three broad categories of food analysis expectations: safety, quality, and authenticity (Rotariu et al. 2016). Detection of contaminants, such as pesticides and antibiotic residues, allergens, biological toxins, and pathogenic microbes is done for food safety screening (Ye et al. 2017, Ribeiro et al. 2014).

Table 8.1: Various types of nanobiosensors for applications in the food industry.

Application of Biosensor	Carbon nanomaterials used	Transducer	Applications	References
Pathogen Detection *Salmonella typhi, E. coli, Listeria monocytogenes, Campylobacter, Mycobacterium* sp.	CNT-based	Affinity-based, Enzyme-based	In freshness of packaged and processed foods	Inbaraj and Chen 2016, Arduini et al. 2016, Anik et al. 2007, Branen et al. 2007
Toxins detection (Bacterial and Fungal) Enterotoxins, Shiga toxin, Cholera toxin, Saxitoxin, Aflatoxin, Ochratoxin	CNT-based	Electochemical and Affinity-based	For detection of harmful toxins	Tothill 2011, Branen et al. 2007
Pesticides detection Parathion, Paraoxan	SWCNT MWCNT	Electrochemical	Detection of pesticide residues in food	Viswanathan et al. 2006, Prakash-Deo et al. 2005
Allergen Detection In Milk and nuts	CNT-based	Electrochemical and optical	Detection of allergens in foods	Ashley et al. 2017
Mixing of pork with other meat sausages Moisture content and freshness of food	CNT-based	Electrochemical Methylene blue-based	Detection of mixing of pork Moisture content in packaged food	Ali et al. 2011a

Applications of nanosensors are very diverse, ranging from detection of ions, organic compounds, heavy metals, microbial toxins, to pesticide residues in food (Omanovic-Miklicanina and Maksimovic 2016, Perez-Lopez and Merkoci 2011).

Different biological components have improved the detection and sensitivity that is real time analysis, such as integration of enzymes (Ozdemir et al. 2010, Li et al. 2009, Prakash-Deo et al. 2005), antibodies (Vinayaka et al. 2009, Gong-Jun et al. 2009, Zhang et al. 2009) or DNA sequences (Viswanathan et al. 2009, 2006) with nanomaterials, e.g., gold nanoparticles (AuNPs), single wall carbon nanotubes (SWCNTs), multi-walled carbon nanotubes (MWCNTs), cadmium telluride quantum dot nanoparticles (CdTe QDs), etc. essentially provide novel hybrid systems for food applications, combining the conductive or semi-conductive properties of the nanoscale materials with the specific properties of the biomolecules. In this section, the main focus will be on nanobiosensors with carbon nanotubes (CNTs), which may be single wall carbon nanotubes (SWCNTs), multi-walled carbon nanotubes (MWCNTs), and graphene, etc.

Pelle and Compagnone (2017) have reported nanomaterial-based sensing and biosensing of phenolic compounds and related antioxidant capacity in food. Total polyphenolic content in tea extracts was investigated using polytyrosine (GCE-SWCNTs/Polytyr), which were covalently functionalized with single-walled CNTs (Eguilaz et al. 2016). Efficient food quality assurance (consumer acceptable product quality, price, good shelf life, safety, etc.) and the improvement of automation, reduction of production time and production cost, production feasibility and quality sorting, are of utmost demand in the food industry. And, these factors act as driving force for the development of highly efficient sensing/sensor system (Gomez-Arribas et al. 2018). Recently, Gomez-Arribas et al. (2018) have also concluded efficient use of carbon nanomaterials biosensors for the analysis of food allergens and discussed biosensing based on nanoparticles for detection of food allergens.

Modified carbon nanotube was employed as supporting electrode for microbial cell (*Pseudomonas fluorescens*) immobilization for glucose detection (Anik et al. 2007). Sucrose is the most common food sweetener, followed by fructose. Carbon nanotube-based biosensor for fructose detection in honey has been reported by Antichoa et al. (2004). The biosensor was constructed by modifying the transducer of CNTs paste with an electropolymerized 3,4-dihydroxybenzaldehyde (3,4-DHB) film. Immobilized enzyme (D-fructose dehydrogenase) was placed on the top of electrode surface. The detection of fructose with this biosensor in real samples gives a high current sensitivity as compared to conventional solid and paste electrodes. Other than enzymatic biosensor, toxin detection is also very important in food analysis. Saxitoxin (STX), a poisonous tetrahydropurine molecule, blocks the sodium transport pump that causes paralytic shellfish poisoning (PSPs). However, ingestion of STX at the concentration of PSP-contaminated shellfish of 80 µg STX equivalents per 100 g shellfish has been regulated in most countries (Hou et al. 2016, Egmond and Marine 2004). Aptamers are the specific oligonucleotide or peptide molecules that bind to a specific target molecule. Aptamer binding to STX leads to quick conformation changes that cause poor signal of dye. The oxidation peak obtained recoveries ranging from 63% to 121%, which indicated an acceptable accuracy and reproducibility (Hou et al. 2016). However, detection of changes in the electron-transfer resistance enables cyanotoxin detection below the World Health Organization's (WHO) provisional concentration limit of 1 µg/L in drinking water (WHO guidelines 1998). Certain bacterial toxins can also be detected with the help of nanobiosensors. Bacterium (*Vibrio cholera*) secretes a protein complex known as cholera toxin, which causes the watery diarrhoea (Kaittanis et al. 2010). Viswanathan et al. (2006) developed a sensitive method for the detection of cholera toxin (CT) using an electrochemical immunosensor with liposomic magnification. Researchers developed a sandwich assay by bound toxin to anti-CT antibody, followed by conjugation with ganglioside functionalized liposome, and measured the release molecule by voltammetry (Zhao et al. 2004). The limit of detection was estimated to be 3 mg/mL. *Staphylococcal* enterotoxins (SEs), a thermostable toxin produced by *S. aureus*, is associated with food-borne diseases, resulting from consumption of contaminated foods (Sonawane et al. 2014, Yang et al. 2009, Goldman et al. 2004). However, ELISA (Enzyme Linked Immunosorbent Assay) has a limited detection range. To overcome this drawback, Yang et al. (2009) developed an optical immunosensor using CNTs for detection of SEs through binding with anti-SE primary antibody immobilized on CNT, followed by binding of HRP-labeled secondary antibody and detection of HRP fluorescence. In an attempt to simultaneously detect multiple toxins, multiplexed fluoroimmunoassays were developed by conjugating highly luminescent semiconductor nanocrystals (CdSeeZnS core-shell QDs) and antibodies for analyzing CT, ricin, Shiga-like toxin, and SE B (Goldman et al. 2004). Other electrochemical nanosensing systems based on carbon nanotubes include an immunosensor based on a cerium oxide nanoparticle and chitosan nanocomposite which detects staphylococcal enterotoxin B and cholera-toxin (Viswanathan et al. 2006). Shiga toxin secreted by *E. coli* O157:H7 can inhibit protein synthesis and activate apoptosis and necrosis, whereas listeriolysin O secreted by *Listeria monocytogenes* can create pores for subsequent disruption of the phospholipid bilayer and eventual cell lysis (Kayal et al. 2006). Although toxins are not usually transmitted through infected individuals, they are able to cause significant devastating effects on organs and tissues. Besides, the toxins can remain in the food, environmental, and clinical samples even after the death of their corresponding pathogens. Therefore, prompt screening of toxins is highly essential to minimize intoxication. However, these methods are time consuming and laborious, besides requiring homogeneous or purified samples.

Food allergies provoke a type 1 hypersensitivity response, which can even cause loss of life. The detectors are categorized according to their sensing mechanism, such as optical, electromechanical, and electrochemical biosensor (Neethirajan et al. 2018). Multiple food allergens can be detected by biosensors designed with a range of technological properties and detection limits. For example, ovalbumin contamination poses a serious health risk to individuals allergic to eggs. A magnetic bead-based nanoparticle technique combined with screen-printed transducers is an immunosensor used for ovalbumin detection (Cadkova et al. 2015). The immune sandwich complex formed between ovalbumin antigen-antibody-HRP conjugated secondary antibody interaction is detected by a standard ELISA procedure. Thionine is an electron mediator that reduces non-specific signals from the hydrogen peroxide, and generates highly sensitive quantitative capturing in linear sweep voltammetry. This approach offers simple and rapid analysis of complex food samples with a detection range between 11 and 222 nM, and a detection lower limit of 5 nM. Egg powder is used in wine processing to remove polyphenols and other undesired compounds. Trace ovalbumin in wine poses a threat to allergic individuals. One SPR biosensor can successfully identify ovalbumin in both white and red wine samples using a single step immunoassay. Samples undergo a two-step pre-treatment process: size exclusion chromatography combined with polyvinylpolypyrrolidone purification. Red wine samples that were artificially fortified with egg white powder were used to confirm the efficiency of the pre-treatment purification and verify that trace amounts of allergen were detected. Cow's milk contains more than 25 protein allergens in casein and whey fractions (Ashley et al. 2017).

Biosensors have been developed to detect milk allergens in food products, such as non-bovine milk and other dairy products. Casein is a frequently used marker because it is also used as an indicator in the final steps of milk manufacturing process. Bovine casein can be measured in goat milk using a specially designed, label-free optoelectronic biosensor that has a Mach-Zehnder interferometer with broadband and monolithical integration features (Angelopoulou et al. 2015). The biosensor was compared to conventional ELISA using the same antibodies. The assay time was configured to 10 minutes, with a recovery rate of 93 to 110%, detection ranges from 0.1 to 1.0%, and a limit of detection of 0.04 percent. Moreover, the milk allergen β-lactoglobulin can be detected by a magnetic electrochemical biosensor designed with disposable screen-printed electrodes and carboxylic magnetic beads. The accuracy was confirmed using a commercially available ELISA kit and the same immune reagents. Another casein-detecting biosensor was designed based on an SPR gold sensing chip with immobilized polyclonal anti-casein antibodies (Ashley et al. 2017). Meatball, a pulverized meat product, is popular in many Asian and European countries (Ali et al. 2014, 2011a, b). However, meatball has been found to be contaminated with pork (Ali et al. 2012). Moreover, from the religious and health point of view, the adulterant of pork-related products in food industries cause the violation of Kosher and Halal food laws (Ali et al. 2014, 2011). Moreover, it will also provoke life-threatening allergic reactions (Ali et al. 2014, Kayal et al. 2006), and consumption of these foods at high levels can cause accumulation of cholesterol in the human body, resulting in chronic diseases, such as cardiovascular disease (Ali et al. 2011b). Thus, quick detection of pork adulteration in meatball needs an ultrasensitive analytical method.

Owing to the high-water solubility of most pesticides, the high level of toxicity, and their overusage in agriculture, an ultrasensitive method for residue analysis of these pollutants is required (Valdes et al. 2009). Viswanathan et al. (2006) developed an enzyme "acetylcholinesterase" based electrochemical biosensor. Enzyme was immobilized on vertically assembled polyaniline modified single-walled carbon nanotubes wrapped with thiol terminated single strand oligonucleotide (ssDNA). It was used for the determination of organophosphorus insecticides such as methyl parathion and chlorpyrifos. The thiol terminated ssDNA SWCNTs has been immobilized onto gold surface via AueS chemical bonding. With the development of this device, they described that the incorporation of SWCNTs enhanced the flexible three-dimensional conductive supports for acetylcholinesterase enzyme. The methyl parathion and chlorpyrifos were determined through inhibition of enzyme reaction, so that the normal AChE activity in presence of an organophosphate or carbamate pesticide is altered, decreasing the response signal of the biosensor. This decrease can be related to the pesticide concentration. The many advantages that OP pesticides detection based on a CNTs-modified transducer have over the amperometric biosensing have also been used for developing biosensors using an organophosphorus hydrolase (OPH) biocatalyst for detection of paraoxon, as reported by Wang's group (Prakash-Deo et al. 2005). Nanobiosensor was developed using a nanotube-enzyme bilayer approach. The CNTs-modified electrodes greatly improve anodic detection of the p-nitrophenol (product of the OPH reaction), including higher sensitivity and stability compared to the bare electrode. The great properties of OPH-CNTs amperometric biosensor decrease its electrocatalytic activity and minimize the surface fouling associated with the oxidation process of the phenolic compounds (Wang et al. 2003), making CNTs a hopeful nanomaterial to control the working potential of the electrochemical measurements.

Food-borne illnesses associated with pathogens, toxins, and other contaminants pose a serious threat to human health. Besides, a large amount of money is spent on both analyses and control measures, which causes significant loss to the food industry (Inbaraj and Chen 2016). The worldwide statistics on food-borne diseases published for 2011–2012 by the Centers for Disease Control and Prevention reported a total of 1,632 outbreaks, 29,112 affected patients, 1,750 hospitalizations, and 68 deaths. Pathogenic microorganisms causing food-borne disease are *Salmonella* (31%), *Listeria* (28%), *Campylobacter* (5%), and *Escherichia coli* O157:H7 (3%) (Centers for disease control 2015, Valdes et al. 2009). Conventional methods for detecting pathogens involve many tedious techniques, and have been replaced by novel molecular techniques. However, most of the published studies have primarily focused on biomedical applications of nanomaterials for disease detection and treatment (Fan et al. 2014, Sonawane et al. 2014, Perez-Lopez and Merkoci 2011, Ali et al. 2011, Kaittanis et al. 2010). *E. coli* O157:H7 has drawn considerable attention owing to its toxin producing capability, thereby damaging the intestinal lining and causing anemia with an infective dose as low as 100 cells (Sonawane et al. 2014). For detection of *E. coli* O157:H7 in milk, a disposable immunosensing strip containing double antibodies for indirect sandwich enzyme-linked immunoassay was fabricated by attaching 13-nm GNPs onto screen-printed carbon electrodes (SPCEs) (Lin et al. 2008). In a similar study, Kalelle et al. (2006) used rabbit immunoglobulin G (IgG) antibody-conjugated silver nanoshells for rapid and highly selective detection of *E. coli* in the range of 5 to 10^9 cells by monitoring the change in the SPR band shift in the presence of *E. coli* cells. In addition, a rapid anodic stripping voltammetric detection of *E. coli* using coreshell Cu@GNPs as anti-*E. coli* sensors was also reported (Zhang et al. 2009). More recently, Maurer et al. (2012) developed a novel nanobiosensor platform through studding of CNTs with RNA-coated GNPs for the selective detection of *E. coli*.

A highly sensitive electrochemical immunoassay for determination of *S. typhimurium* was demonstrated by Dungchai et al. (2008). It has been based on using immobilized monoclonal antibodies on polystyrene for capturing bacteria, followed by adding polyclonal antibody followed by GNPs conjugate to bind the bacteria in the presence of copper-enhancer solution and ascorbic acid. The copper was deposited on the surface of gold nanoparticles after the reduction. The copper bound gold nanoparticle was employed for direct measurement of *S. typhimurium* concentration by anodic stripping voltammetry, with the limit of detection being 98.9 CFU/mL (Dungchai et al. 2008). A detection limit of 100 CFU/mL was obtained for *Salmonella* in milk (Joo et al. 2012). In an attempt to improve the performance of electrochemical biosensor by CNTs, Jain et al. (2012) immobilized CNTs functionalized monoclonal antibodies onto a glassy carbon electrode for *S. typhimurium* detection using electrochemical impedance spectroscopy as a function of change in charge transfer resistance and impedance. Recent advances in developing nanosensors provide attractive solutions for fast, sensitive, and high throughput analysis (Kaittanis et al. 2010, 2007, Yakes et al. 2008). Kaittanis et al. (2007) developed a one-step nanoparticle-mediated bacterial detection method in milk and blood by exploiting the magnetic relaxation property of superparamagnetic iron oxide nanoparticles (SPIONs). This method was employed for simultaneous detection of multiple pathogenic microorganisms. *E. coli* O157:H7, *S. typhimurium*, and *B. cereus* have been detected by Zhao et al. (2004). In addition, both Gram-negative and Gram-positive microorganisms could be quantified (Zhao et al. 2004). In a similar study, the gold nanoparticles were conjugated with p-conjugated polymer poly(para-phenylene ethynylene) for positive identification of 12 different bacterial strains within a few minutes (Phillips et al. 2008). Three different nanoparticle preparations were prepared by the researchers, and distinct fluorescence could be observed for each bacterium, including *Amycolatopsis azurea, Amycolatopsis orientalis, Bacillus licheniformis, Bacillus subtilis*, some *E. coli* strains (BL21DE3, DH5a, and XL1-Blue), *Lactobacillus lactis, Lactobacillus plantarum, Pseudomonas putida, Streptomyces coelicolor*, and *Streptomyces griseus* (Phillips et al. 2008). The salient feature of this method included minimum incubation period with no further modification of biomolecules on AF-MNPs. However, the main drawback is low specificity as compared to antibody conjugates methods. A portable nanobiosensor for detection of pathogens in milk has been developed by Biscotti et al. (2018). Nowadays, traceability of foodstuffs and monitoring with respect to food safety using suitable biosensing devices has become the prime objective in many food industries (Maksimovic et al. 2015).

It is pertinent from the above cited recent studies that incorporation of the various forms of carbon nanomaterials has improved the ultra-sensitivity as well as robustness of the biosensors.

4. Applications of Carbon Nanomaterials-based Nanobiosensors in Agriculture

Recent studies have shown the key attention for utilization of carbon nanomaterials (CNMs), such as carbon nanotubes, fullerenes, carbon nanoparticles, and carbon nanohorns, etc., in agricultural/plant sciences because of increased germination potential, root and shoot length, growth and biomass, and enhanced nutrient uptake (Pasinszki et al. 2017, Mukherjee et al. 2016, Srivastava et al. 2015, Rai et al. 2012, Khot et al. 2012, Petersen et al. 2011).

Due to small size and optimized properties of bionanomaterials, they are most suitable for sensing inter/intra cellular microenvironments for physiological and biological characteristics. Nanobiosensors are designed to detect contaminants, pests, nutrient contents, and plant stresses, such as biotic and abiotic (Prameela 2017, Pasinszki et al. 2017, Joyner and Kumar 2015). Carbon nanomaterials-based nanobiosensors are promising tools for harnessing agricultural applications, such as diagnostic tools for soil quality and disease assessment, as agents to promote sustainable agriculture, as a tool for effective detection of DNA and protein, and as a tool for analysis in food products and food production, etc. (Mukherjee et al. 2016).

Nanosensors have huge potential and applicability in agri-food sector, as discussed by Omanovic-Miklicanina and Maksimovic (2016). In agriculture, they are used for monitoring soil conditions (e.g., soil pH and moisture) and for monitoring range of fertilizers, insecticides, pesticides, herbicides, pathogens, and to improve crop growth as well. Nanosensors have been reported to detect food-borne contaminants for monitoring optimal environmental conditions at the farm (Prameela 2017). Nanosensors and nanobased smart delivery systems have been utilized for efficient use of agricultural natural resources, i.e., water, nutrients, and chemicals through precision farming and nanochips for identity preservation and tracking as well, and nanocapsules for controlled use and effective delivery of pesticides, herbicides, fertilizers, and vaccines have also been practiced in agricultural farms (Prameela 2017, Omanovic-Miklicanina and Maksimovic 2016). Moreover, nanomaterial-based biosensors, such as aptasensors, have been used for determination of pesticides and insecticides (phorate, acetamiprid, isocarbophos, etc.), antibiotics, drugs and their residues (cocaine, oxytetracycline, tetracycline, kanamycin, etc.) and for the determination of heavy metals (Hg^{2+}, As^{3+}, etc.) (Omanovic-Miklicanina and Maksimovic 2016). Recently, nanoparticles smart nanosensors have been utilized for early warning of changing environmental and climatic conditions for capable response to different conditions (Pelle and Compagnone 2018, Gomez-Arribas et al. 2018).

Application of CNT-based gas sensor provides sensitive, non-contact, and non-destructive process for food analysis and to maintain food quality, which is highly desirable. CNT-based biosensors, such as CNT/Polymer gas sensor and MWNT-based biosensor have been successfully utilized for evaluation of meat freshness and its monitoring, particularly during storage, shipment, and processing (Sinha et al. 2006). CNT-based biosensors (CNTs/poly methyl methacrylate gas sensor) are used to detect the concentration of certain chemicals, such as ethyl acetate (volatile component) released during initial bacterial putrefaction of meat, to assess and control the quality of meat. CNTs-based optical sensors provide practical applications for remote *in-situ* monitoring. CNMs-based electrochemical biosensors have been widely used in wastewater monitoring, and CNTs humidity sensors have a significant application to monitor humidity in protected horticulture or green house agriculture, because temperature and humidity directly impact the product quality and quantity. Sinha et al. (2006) have shown that MWNT-coated quartz crystal microbalance humidity sensor monitored the relative humidity of the range (5 to 97%) with response and recovery time of about 60 to 70 seconds, respectively.

Carbon nanomaterials-based pressure sensors can be successfully utilized for uniform and efficient spraying of insecticides, herbicides, and fertilizers (liquid) in the agricultural fields (Sinha et al. 2006). Farmers get benefited from this application of CNMs by uniform spraying and staying away from extra cost of fertilizers and pesticides, which resulted in higher crop yield. CNMs have also been used to inhibit the organic contaminant uptake by plants (Ghosh et al. 2015, Husen and Siddiqi 2014, Petersen et al. 2009). One of the promising research areas for CNMs-based biosensors consist of assessment and evaluation of the quality of fruits and vegetables and their damage. Carbon nanomaterials-based chemical sensor has been reported to identify undesired chemical residues/waste resulted from food and agriculture industries (such as environmental contaminants (in raw/processed foods) and from food additives, herbicides, and pesticides, etc.) (Gomez-Arribas et al. 2018, Sinha et al. 2006). CNTs biosensors help to maintain optimum, accurate pH balance of water quality, which is essential for the survival and growth of cultured fishes and shrimps in fishing industry. CNT-based pH sensors are also used to determine pH value when selecting fish farming location to avoid any abnormalities in fishing grounds and hatcheries (Ghosh et al. 2015, Sinha et al. 2006).

In protective agriculture, such as green house and controlled garden atmosphere, carbon dioxide is utilized by the plants for the photosynthesis process, which is the guideline procedure in plants on earth that convert light energy (in which CO_2 and water combine with the help of light energy) to chemical energy (to form sugar). In this process, some of the sugars molecules formed are converted into the complex compounds that enhance the sustained growth of the cultivated plant (Mukherjee et al. 2016). However, if the carbon dioxide concentration is too high because of CO_2 dissolution in water to make carbonic acid, then plants don't grow properly because of acidic soil and air. So, carbon nanomaterials biosensors, such as ($CNTs$-CO_2) sensors, have been used to monitor the concentration of carbon dioxide (CO_2) to provide optimal environments for the growth of plants inside the green house and controlled garden atmosphere (Sinha et al. 2006).

Plants are the fundamental components of the biological system and the key part of metabolic pathways. Plant metabolism is considered a complex mechanism of events (physical and chemical), i.e., photosynthetic process, respiration, and organic compounds (biosynthesis and degradation) that take place in the plant cell (Marslin et al. 2017). Improvement of plant growth, such as enhanced seed germination, root and shoot growth, and positive effect on physiological parameters of the plant, has been observed upon exposure to carbon nanomaterials. Plants growing in their natural ecological conditions are to a great extent affected by biotic (insect pest, bacterial, fungus, and viruses) and abiotic (heat, cold, salinity, drought, water logging, metal, UV radiation, etc.) stresses (Kumar and Srivastava 2016). Different studies demonstrated that NMs play an imperative role in alleviating abiotic stresses and stress-induced alterations in plants (Marslin et al. 2017, Torabian et al. 2016, Siddiqui et al. 2014). Interaction of nanomaterials with plants alters the morphological, physiological, and physicochemical characteristics due to changes in the plant metabolism, and they largely depend upon size-dependent properties and efficacy of nanoparticles. Mondal et al. (2011) reported improved plant growth characteristics in mustard (*B. juncea*) upon exposure of pristine and oxidized-multi-wall CNTs at exposure of 2.3–46.0 mg/L. Lin and Xing (2007) reported remarkable increase (17%) in root length in ryegrass (*Lolium perenne*) when exposed to 2000 mg/L multi-wall CNTs, as compared to untreated controls. Canas et al. (2008) assessed the toxicity of uncoated and coated single-walled CNTs poly-3-aminobenzene sulfonic acid in six vegetable crops, i.e., tomato (*Lycopersicon esculentum*), onion (*Allium cepa*), cucumber (*Cucumis sativus*), carrot (*Daucus carota*), cabbage (*Brassica oleracea*), and lettuce (*Lactuca sativa*). Upon exposure, uncoated-CNTs showed increased root length in cucumber and onion when contrasted with the coated-CNTs. Microscopic studies revealed only surface adsorption and no internalization of CNTs into the roots.

Exposure of 50 mg/mL single wall CNTs (SWCNT) and multi wall CNTs (MWCNT) to Murashige and Skoog (MS) growth medium resulted in enhanced (75 and 110%) total fresh biomass of tomato seeds, when compared to graphene and activated carbon (Khodakovskaya et al. 2011). Significant increase in root length (50% and 32%) was reported in wheat seedlings upon exposure to 40–160 mg/L multi wall CNTs for 3 and 7 days, respectively (Wang et al. 2012). Khodakovskaya et al. (2012) observed an increased growth rate of 55–64% in tobacco cells at exposure to 5–500 mg/mL MWCNT. However, activated carbon exposure at low concentrations (5 mg/mL) resulted in improved cell growth, but growth was suppressed at higher exposures (100–500 mg/mL) in tobacco. Tiwari et al. (2013) reported tremendous increase in plant fresh biomass (43%) and nutrient uptake (1.6x iron and 2x calcium) in corn, when exposed to 60 mg/L multi wall CNTs as contrasted to control. Upon exposure to multi wall CNTs, to study its effect on growth and germination in barley, soybean, and corn on agar medium, there was a remarkable increase in germination rate, i.e., 90% (in corn) and 50% (in barley and soybean) were observed when contrasted with untreated controls (Lahiani et al. 2013). Similarly, root and shoot lengths were enhanced by 40%, i.e., more than three-fold in corn, and in soybean, the root length was increased up to 26%, respectively. Sonkar et al. (2012) and Tripathi and Sarkar (2014) observed improved growth in gram and enhanced root growth in wheat upon exposure to water soluble carbon nanodots (ws-CNDs) when contrasted with controls, respectively.

In the CNT-plant interaction at the DNA (deoxyribonucleic acid) level, it has been observed that there was upregulation of genes, i.e., *NtLRX* and *CycB* and an aquaporin (*NtPIP1*) that plays an important role in cell division, cell wall formation, and in water transport (Khodakovskaya et al. 2012). Enhancement in plant growth characteristics was also studied using water-soluble carbon nanoparticle (ws-CNP), such as fullerols. Kole et al. (2013) reported that upon exposure of ws-CNP in bitter melon (*Momordica charantia*), there was increase in the biomass and phyto-medicinal content. Fullerol exposure in the fruit tissues remarkably enhances the contents of anticancerous compounds, i.e., cucurbitacin-B (74%) and lycopene (82%) and antidiabetic molecules, i.e., charantin (20%) and insulin (91%). Concentration-dependent effects of ws-CNPs were examined in wheat (Saxena et al. 2014). Treatment of 10–150 mg/L ws-CNPs was given to plants in the soil up to 20 days. Significant increase in root and shoot lengths was recorded, which was three times higher as compared to untreated controls with optimized 50 mg/L ws-CNPs treatment.

A positive impact on the growth of terrestrial plants has been reported using carbon nanohorns (CNHs), which are disordered single layered grapheme spherical structures sheets by Lahiani et al. (2015). At the variable concentration of low to high dose of CNHs, upon exposure to tobacco cells for 24 hours, there was notable increase, i.e., 78% in cultured tobacco cells at 100 mg/ml. However, Zhang et al. (2015) reported significant increase in germination rate of treated tomato seeds when exposed to 40 mg/ml graphene as compared to untreated controls.

Soil diversity and soil microbial biomass protection is the foremost challenge in agriculture. At present, very little literature is available on the interaction of carbon nanomaterials CNTs with soil microbial community (Mukherjee et al. 2016, Dinesh et al. 2012). Tong et al. (2007) studied the effectiveness of fullerenes (C_{60}) at the concentration of 1 and 1000 mg C_{60}/g on soil microbial populations for 180 days using total lipid derived phosphate. However, the fullerenes had no impact on the soil microbial population (structure and function), as well as on soil enzymatic activities (Tong et al. 2007).

Thus, carbon nanomaterials have improved the performance of biosensors in terms of sensitivity and robustness in agriculture sectors.

5. Conclusion

Even with these developments in biosensing field, biosensors are mainly still in laboratory research compared to the other biosensing fields, which means the potential markets need to be developed eagerly. Parameters for improving stability, reliability, and cost-effectiveness in research and development of biosensors are still big challenges. Although the application of carbon nanomaterials can solve the above problems well, the complex preparation and high cost of nanomaterials are key factors in the practical utilization and popularization of biosensors.

Carbon nanomaterial-based biosensors have revolutionized various analytical techniques important in food technology. All these nanobiosensors have pivotal roles in detection of pathogens, antibiotic residues, pesticide residues, and food allergens. However, real time analysis, small sample size, and portability are some issues which limit the wide applications of these sensors in the food industry. On the other hand, microfluidic and array-based biosensor holds great potential for developing future analytical techniques due to stringent conditions of food safety.

Carbon nanomaterials have potential applications in agriculture, such as to improve the seed germination, enhance nutrient uptake, plant growth, and fruit quality, etc. However, these positive impacts of CNMs largely depend upon the CNM type/concentration, growth conditions, growth medium in nature, and most importantly, on the selected plant species. Numerous studies which reported a positive impact upon exposure to CNTs, are from short-term studies and were done in artificial plant growth media. However, to assure its consistency with real agricultural conditions, further detailed investigation is mandatory to harness the actual potential, and to spread the application of CNMs in agriculture.

References

Abnous, K., N.M. Danesh, M. Ramezani, M. Alibolandi, P. Lavaee and S.M. Taghdisi. 2017. Aptamer based fluorometric acetamiprid assay using three kinds of nanoparticles for powerful signal amplification. Microchim. Acta 184: 81–90.

Ali, M.A., H. Jiang, N.K. Mahal, R.J. Weber, R. Kumar, M.J. Castellano et al. 2017. Microfluidic impedimetric sensor for soil nitrate detection using graphene oxide and conductive nanofibers enabled sensing interface. Sensors Actuators B Chem. 239: 1289–1299.

Ali, M.E., U. Hashim, S. Mustafa, Y.B. Che Man, M.H.M. Yusop, M. Kashif et al. 2011a. Nanobiosensor for detection and quantification of DNA sequences in degraded mixed meats. J. Nanomater. 2011: 781098.

Ali, M.E., U. Hashim, S. Mustafa, Y.B. Man, M.H. Yusop, M.F. Bari et al. 2011b. Nanoparticle sensor for label free detection of swine DNA in mixed biological samples. Nanotechnol. 22: 195503.

Ali, M.E., U. Hashim, S. Mustafa, Y.B. Che Man and K.N. Islam. 2012. Gold nanoparticle sensor for the visual detection of pork adulteration in meatball formulation. J. Nanomater. 2012: 103607.

Ali, M.E., U. Hashim, S. Mustafa, Y.B. Che Man, T. Adam and Q. Humayun. 2014. Nanobiosensor for the detection and quantification of pork adulteration in meatball formulation. J. Exp Nanosci. 9: 152–60.

Angelopoulou, M., A. Botsialas, A. Salapatas, P.S. Petrou, W. Haasnoot, E. Makarona et al. 2015. Assessment of goat milk adulteration with a label-free monolithically integrated optoelectronic biosensor. Anal. Bioanal. Chem. 407: 3995–4004.

Anik, U., S. Timur, D. Odaci, B. Perez, S. Alegret and A. Merkoci. 2007. Carbon nanotube composite as novel platform for microbial biosensor. Electroanalysis 19(7-8): 893–898.

Antichoa, R., I. Lavagnini and F. Magno. 2004. Amperometric mediated carbon nanotube paste biosensor for fructose determination. Analyt. Lett. 37(8): 1657–1669.

Arduini, F., S. Cinti, V. Scognamiglio and D. Moscone. 2016. Nanomaterials in electrochemical biosensors for pesticide detection: Advances and challenges in food analysis. Microchimica Acta 183(7): 2063–2083.

Arduini, F., M. Forchielli, V. Scognamiglio, K.A. Nikolaevna and D. Moscone. 2017. Organophosphorous pesticide detection in olive oil by using a miniaturized, easy-to-use, and cost-effective biosensor combined with QuEChERS for sample clean-up. Sensors 17: 1–9.

Arvand, M. and A.A. Mirroshandel. 2017. Highly-sensitive aptasensor based on fluorescence resonance energy transfer between L-cysteine capped ZnS quantum dots and graphene oxide sheets for the determination of edifenphos fungicide. Biosens. Bioelectron. 96: 324–331.

Ashley, J., M. Piekarska, C. Segers, L. Trinh, T. Rodgers, R. Wiley et al. 2017. AnSPR based sensor for allergens detection. Biosens. Bioelectron. 88: 109–113.

Balasubramanian, K. and M. Burghard. 2006. Biosensors based on carbon nanotubes. Anal. Bioanal. Chem. 385(3): 452–468.

Bao, J., C.J. Hou, Q.C. Dong, X.Y. Ma, J. Chen, D.Q. Huo et al. 2016. ELP-OPH/BSA/TiO$_2$ nanofibers/c-MWCNTs based biosensor for sensitive and selective determination of p-nitrophenyl substituted organophosphate pesticides in aqueous system. Biosens. Bioelectron. 85: 935–942.

Biscotti, A., R. Lazzarini, G. Virgilli, F. Ngatcha, A Valisi and M. Rossi. 2018. Optimizing a portable biosensor system for bacterial detection in milk-based mix for ice-cream. Sensing Biosens. Res. 18: 1–6.

Boehm, H.P., A. Clauss, G.O. Fischer, Hofmann and U.Z. Naturforsch. 1962. Dünnste Kohlenstoff-Folien. 17(3): 150–153.

Boudou, J.P., P.A. Curmi, F. Jelezko, J. Wrachtrup, P. Aubert, M. Sennour et al. 2009. High yield fabrication of fluorescent nanodiamonds. Nanotechnology 20(23): 235602.

Branen, J.R., M.J. Hass, W.C. Douthit, W.C. Maki and A.L. Branen. 2007. Detection of *Escherichia coli* O157, Salmonella enterica serovar typhimurium, and Staphylococcal enterotoxin B in a single sample using enzymatic bio-nanotransduction. J. Food Prot. 70: 841–50.

Cadkova, M., R. Metelka, L. Holubová, D. Horák, V. Dvořáková, Z. Bílková et al. 2015. Magnetic beads-based electrochemical immunosensor for monitoring allergenic food proteins. Anal. Biochem. 484: 4–8.

Cai, G.N., Z.Z. Yu, R.R. Ren and D.P. Tang. 2018. Exciton plasmon interaction between AuNPs/grapheme nanohybrids and CdS quantum dots/TiO$_2$ for photoelectrochemical aptasensing of prostate-specific antigen. ACS Sensors 3(3): 632–639.

Campuzano, S., P. Yáñez-Sedeño and J.M. Pingarrón. 2017. Electrochemical affinity biosensors in food safety. Chemosensors 5(1): 8.

Canas, J.E., M. Long, S. Nations, R. Vadan, L. Dai, M. Luo et al. 2008. Effects of functionalized and nonfunctionalized single-walled carbon nanotubes on root elongation of select crop species. Environ. Toxicol. Chem. 27: 1922–1931.

Centers for Disease Control and Prevention. Available from: http://www.cdc.gov/features/foodborne-diseases-data/ [Last accessed 16.02.15].

Chen, L.F., X.D. Zhang, H.W. Liang, M.G. Kong, Q.F. Guan, P. Chen et al. 2012. Synthesis of nitrogen-doped porous carbon nanofibers as an efficient electrode material for supercapacitors. ACS Nano. 6: 7092–7102.

Chiashi, S., K. Kono, D. Matsumoto, J. Shitaba, N. Homma, A. Beniya et al. 2015. Adsorption effects on radial breathing mode of single-walled carbon nanotubes. Phys. Rev. B 91: 155415.

Cinti, S., D. Neagu, M. Carbone, I. Cacciotti, D. Moscone and F. Arduini. 2016. Novel carbon black-cobalt phthalocyanine nanocomposite as sensing platform to detect organophosphorus pollutants at screen-printed electrode. Electrochim. Acta 188: 574–581.

Clark, L.C.J. and C. Lyons. 1962. Electrode systems for continuous monitoring in cardiovascular surgery. Ann. N. Y. Acad. Sci. 102: 29–45.

Dahl, J.E., S.G. Liu and R.M.K. Carlson. 2003. Isolation and structure of higher diamondoids, nanometer-sized diamond molecules. Science 299(5603): 96–99.

Dinesh, R., M. Anandaraj, V. Srinivasan and S. Hamza. 2012. Engineered nanoparticles in the soil and their potential implications to microbial activity. Geoderma 19: 173–174.

Ding, J.W., H.X., Zhang, F.L. Jia, W. Qin and D. Du. 2014. Assembly of carbon nanotubes on a nanoporous gold electrode for acetylcholinesterase biosensor design. Sens. Actuators B 199: 284–290.

Dong, J., X. Fan, F. Qiao, S. Ai and H. Xin. 2013. A novel protocol for ultra-trace detection of pesticides: combined electrochemical reduction of Ellman's reagent with acetylcholinesterase inhibition. Anal. Chim. Acta 761: 78–83.

Dungchai, W., W. Siangproh, W. Chaicumpa, P. Tongtawe and O. Chailapakul. 2008. Salmonella typhi determination using voltammetric amplification of nanoparticles: a highly sensitive strategy for metallo-immunoassay based on a copper-enhanced gold label. Talanta 77: 727–732.

Eder, D. 2010. Carbon nanotube-inorganic hybrids. Chem. Rev. 110(3): 1348–1385.

Egmond, V. and H. Marine. 2004. Biotoxins: food and nutrition paper. Food and Agricultural Organization of the United Nations (FAO): Rome, Italy.

Eguilaz, M., A. Gutierrez, F. Gutierrez, J.M. Gonzalez-Domínguez, A. Ansón-Casaos, J. Hernández-Ferrer et al. 2016. Covalent functionalization of single-walled carbon nanotubes with polytyrosine: Characterization and analytical applications for the sensitive quantification of polyphenols. Anal. Chim. Acta 909: 51–59.

Fan, Z., P.P. Fu, H. Yu and P.C. Ray. 2014. Theranostic nanomedicine for cancer detection and treatment. J. Food Drug Anal. 22: 3–17.

Georgakilas, V., J.A. Perman, J. Tucek and R. Zboril. 2015. Broad family of carbon nanoallotropes: classification, chemistry, and applications of fullerenes, carbon dots, nanotubes, graphene, nanodiamonds, and combined superstructures. Chem. Rev. 115: 4744–4822.

Ghosh, M., S. Bhadra, A. Adegoke, M. Bandyopadhyay and A. Mukherjee. 2015. MWCNT uptake in Allium cepa root cells induces cytotoxic and genotoxic responses and results in DNA hyper-methylation. Mutat. Res. 774: 49–58.

Goldman, E.R., A.R. Clapp, G.P. Anderson, H.T. Uyeda, J.M. Mauro, I.L. Medintz et al. 2004. Multiplexed toxin analysis using four colors of quantum dot fluororeagents. Anal. Chem. 76: 684–688.

Gomez-Arribas, L.N., E. Benito-Peña, M.C. Hurtado-Sánchez and M.C. Moreno-Bondi. 2018. Biosensing based on nanoparticles for food allergens detection. Sensors 18: 1087.

Gong-Jun, Y., H. Jin-Lin, M. Wen-Jing, S. Ming and J. Xin-An. 2009. A reusable capacitive immunosensor for detection of Salmonella spp. based on grafted ethylene diamine and self-assembled gold nanoparticle monolayers. Analytica Chimica Acta 647: 159–166.

Hou, J.Y., J. Dong, H.S. Zhu, X. Teng, S.Y. Ai and M.L. Mang. 2015. A simple and sensitive fluorescent sensor for methyl parathion based on L-tyrosine methyl ester functionalized carbon dots. Biosens. Bioelectron. 68: 20–26.

Hou, L., L. Jiang, Y. Song, Y. Ding, J. Zhang, X. Wu et al. 2016. Amperometric aptasensor for saxitoxin using a gold electrode modified with carbon nanotubes on a self-assembled monolayer, and methylene blue as an electrochemical indicator probe. Microchim. Acta. 183: 1971–1980.

Husen, A. and K.S. Siddiqi. 2014. Carbon and fullerene nanomaterials in plant system. J. Nanobiotechnol. 12: 16.

Iijima, S. 1991. Helical microtubules of graphitic carbon. Nature 354: 56–58.

Iijima, S. and T. Ichihashi. 1993. Single-shell carbon nanotubes of 1-nm diameter. Nature 363: 603–605.

Inbaraj, B.S. and B.H. Chen. 2016. Nanomaterial-based sensors for detection of foodborne bacterial pathogens and toxins as well as pork adulteration in meat products. J. Food Drug Anal. 24: 15–28.

Jain, S., S.R. Singh, D.W. Horn, V.A.R. Davis and S.R. Pillai. 2012. Development of an antibody functionalized carbon nanotube biosensor for foodborne bacterial pathogens. J. Biosens. Bioelectron. 11: 002.

Jain, Y., A. Goel, C. Rana, N. Sharma, M.L. Verma and A.K. Jana. 2010. Biosensors, types and applications. Int. Conf. Biomed. Eng. Assist. Technol. N. I. T. Jalandhar, India. December 17–19, 2010 pp. 1–6.

Jianrong, C., M. Yuqing, H. Nongyue, W. Xiaohua and L. Sijiao. 2004. Nanotechnology and biosensors. Biotechnol. Adv. 22(7): 505–518.

Joo, J., C. Yim, D. Kwon, J. Lee, H.H. Shin, H.J. Cha et al. 2012. A facile and sensitive detection of pathogenic bacteria using magnetic nanoparticles and optical nanocrystal probes. Analyst 137(16): 3609–3612.

Journet, C., W.K. Maser, P. Bernier, A. Loiseau, M. Lamy de la Chapelle, S. Lefrant et al. 1997. Large-scale production of single-walled carbon nanotubes by the electric arc technique. Nature 388: 756–758.

Joyner, J.R. and D.V. Kumar. 2015. Nanosensors and their applications in food analysis: a review. Int. J. Sci. Technol. 1: 80–90.

Kaittanis, C., S.A. Naser and J.M. Perez. 2007. One-step, nanoparticle mediated bacterial detection with magnetic relaxation. Nano Lett. 7: 380–383.

Kaittanis, C., S. Santra and J.M. Perez. 2010. Emerging nanotechnology based strategies for the identification of microbial pathogenesis. Adv. Drug Deliv. Rev. 62: 408–423.

Kalelle, S.A., A.A. Kundu, S.W. Gosavi, D.N. Deobagkar, D.D. Deobagkar and S.K. Kulkarni. 2006. Rapid detection of *Escherichia coli* by using antibody-conjugated silver nanoshells. Small 2: 335–338.

Kaur, N., H. Thakur and N. Prabhakar. 2016. Conducting polymer and multi-walled carbon nanotubes nanocomposites based amperometric biosensor for detection of organophosphate. J. Appl. Electrochem. 775: 121–128.

Kayal, S., A. Charbit and O. Listeriolysin. 2006. A key protein of Listeria monocytogenes with multiple functions. FEMS Microbiol. Rev. 9: 76–85.

Khodakovskaya, M.V., K. DeSilva, D.A. Nedosekin, E. Dervishi, A.S. Biris, E.V. Shashkov et al. 2011. Complex genetic, photothermal, and photoacoustic analysis of nanoparticle-plant interactions. Proc. Natl. Acad. Sci. U.S.A. 108: 1028–1033.

Khodakovskaya, M.V., K. de Silva, A.S. Biris, E. Dervishi and H. Villagarcia. 2012. Carbon nanotubes induce growth enhancement of tobacco cells. ACS Nano. 6: 2128–2135.

Khot, L.R., S. Sankaran, J.M. Maja, R. Ehsani and E.W. Schuster. 2012. Applications of nanomaterials in agricultural production and crop protection: a review. Crop. Protect. 35: 64–70.

Kole, C., P. Kole, K.M. Randunu, P. Choudhary, R. Podila, P.C. Ke et al. 2013. Nanobiotechnology can boost crop production and quality: first evidence from increased plant biomass, fruit yield and phytomedicine content in bitter melon (Momordica charantia). BMC Biotechnol. 13: 37.

Kratschmer, W., L.D. Lamb, K. Fostiropoulos and D.R. Huffman. 1990. Solid C60: a new form of carbon. Nature 347: 354–358.

Kroto, H.W., J.R. Heath, S.C. O'Brien, R.F. Curl and R.E. Smalley. 1985. C60: Buckminsterfullerene. Nature 318: 162–163.

Kumar, P. and D.K. Srivastava. 2016. Biotechnological advancement in genetic improvement of broccoli (Brassica oleracea L. var. italica), an important vegetable crop. Biotechnol. Lett. 38: 1049–1063.

Lahiani, M.H., E. Dervishi, J. Chen, Z. Nima, A. Gaume, A.S. Biris et al. 2013. Impact of carbon nanotubes exposure to seeds of valuable crops. ACS Appl. Mater. Interfaces 5: 7965–7973.

Lahiani, M.H., J. Chen, F. Irin, A.A. Puretzky, M.J. Green and M.V. Khodakovskaya. 2015. Interaction of carbon nanohorns with plants: uptake and biological effects. Carbon 81: 607–619.

Lee, J.H., J.Y. Park, K. Min, H.J. Cha, S.S. Choi and Y.J. Yoo. 2010. A novel organophosphorus hydrolase-based biosensor using mesoporous carbons and carbon black for the detection of organophosphate nerve agents. Biosens. Bioelectron. 25: 1566–1570.

Lee, J., J. Kim and T. Hyeon. 2006. Recent progress in the synthesis of porous carbon materials. Adv. Mater. 18: 2073–2094.

Lee, S.H., J.H. Sung and T.H. Park. 2012. Nanomaterial-based biosensor as an emerging tool for biomedical applications. Ann. Biomed. Eng. 40(6): 1384–1397.

Lei, S., D. Chen, W. Chen and Y. Chen. 2011. A promising surface acoustic wave humidity sensor based on multi-walled carbon nanotubes/nafion composite film for low humidity detection. Sens. Lett. 9: 1606–1611.

Li, H.T., C.H. Sun, R. Vijayaraghavan, F.L. Zhou, X.Y. Zhang and D.R. MacFarlane. 2016. Long lifetime photoluminescence in N, S co-doped carbon quantum dots from an ionic liquid and their applications in ultrasensitive detection of pesticides. Carbon 104: 33–39.

Li, K., W. Liu, Y. Ni, D. Li, D. Lin, Z. Su et al. 2017. Technical synthesis and biomedical applications of graphene quantum dots. J. Mater. Chem. B 5: 4811–4826.

Li, S.H., X.J. Wu, C.H. Liu, G.H. Yin, J.H. Luo and Z. Xu. 2016. Application of DNA aptamers as sensing layers for detection of carbofuran by electrogenerated chemiluminescence energy transfer. Anal. Chim. Acta 941: 94–100.

Li, X., Y. Zhou, Z. Zheng, X. Yue, Z. Dai, S. Liu et al. 2009. Glucose biosensor based on nanocomposite films of CdTe quantum dots and glucose oxidase. Langmuir 25(11): 6580–6586.

Lim, M.C. and Y.R. Kim. 2016. Analytical applications of nanomaterial in monitoring biological and chemical contamination in food. J. Microbiol. Biotechnol. 26(9): 1505–1516.

Lim, W.H., Y.K. Yap, W.Y. Chong and H. Ahmad. 2014. All-optical graphene oxide humidity sensors. Sensors 14: 24329–24337.

Lin, D. and B. Xing. 2007. Phytotoxicity of nanoparticles: inhibition of seed germination and root growth. Environ. Pollut. 150: 243–250.

Lin, Y.H., S.H. Chen, Y.C. Chuang, Y.C. Lu, T.Y. Shen, C.A. Chang et al. 2008. Disposable amperometric immunosensing strips fabricated by Au nanoparticles-modified screen-printed carbon electrodes for the detection of foodborne *Escherichia coli* O157:H7. Biosens. Bioelectron. 23(12): 1832–1837.

Lin, Y.X., Q. Zhou, D.P. Tang, R. Niessner and D. Knopp. 2017. Signal-on photoelectrochemical immunoassay for aflatoxin b-1 based on enzymatic product-etching MnO$_2$ nanosheets for dissociation of carbon dots. Anal. Chem. 89: 5637–5645.

Linting, Z., L. Ruiyi, L. Zaijun, X. Qianfang, F. Yinjun and L. Junkang. 2012. An immunosensor for ultrasensitive detection of aflatoxin B1 with an enhanced electrochemical performance based on graphene/conducting polymer/gold nanoparticles/the ionic liquid composite film on modified gold electrode with electrodeposition. Sens. Actuators B: Chem. 174: 359–365.

Liu, H., L. Zhang, M. Yan and J. Yu. 2017. Carbon nanostructures in biology and medicine. J. Mater. Chem. B 5: 6437–6450.

Liu, T., H.C. Su, X.J. Qu, P. Ju, L. Cui and S.Y. Ai. 2011. Acetylcholinesterase biosensor based on 3-carboxyphenylboronic acid/reduced graphene oxide-gold nanocomposites modified electrode for amperometric detection of organophosphorus and carbamate pesticides. Sens. Actuators B 160: 1255–1261.

Lv, S.Z., Y. Li, K.Y. Zhang, Z.Z. Lin and D.P. Tang. 2017. Carbon dots/g-C$_3$N$_4$ nanoheterostructures-based signal-generation tags for photoelectrochemical immunoassay of cancer biomarkers coupling with copper nanoclusters. ACS Appl. Mater. Interfaces 9: 38336–38343.

Maksimovic, J., J.A. Gagnon-Bartsch and T.P. Speed. 2015. Removing unwanted variation in differential methylation analysis of illumine human methylation array data. Nucleic Acid Res. 43(16): 106.

Marin, S. and A. Merkoci. 2012. Nanomaterials based electrochemical sensing applications for safety and security. Electroanalysis 24: 459–469.

Marslin, G., C.J. Sheeba and G. Franklin. 2017. Nanoparticles alter secondary metabolism in plants via ROS burst. Front. Plant. Sci. 8: 832.

Maurer, E.I., K.K. Comfort, S.M. Hussain, J.J. Schlager and S.M. Mukhopadhyay. 2012. Novel platform development using an assembly of carbon nanotube, nanogold and immobilized RNA capture element towards rapid, selective sensing of bacteria. Sensors 12: 8135–8144.

Mogha, N.K., V. Sahu, M. Sharma, R.K. Sharma and D.T. Masram. 2016. Biocompatible ZrO$_2$-reduced graphene oxide immobilized AChE biosensor for chlorpyrifos detection. Mater. Design 111: 312–320.

Mondal, A., R. Basu, S. Das and P. Nandy. 2011. Beneficial role of carbon nanotubes on mustard plant growth: an agricultural prospect. J. Nanopart. Res. 13: 4519–4528.

Mukherjee, A., S. Majumdar, A.D. Servin, L. Pagano, O.P. Dhankher and J.C. White. 2016. Carbon nanomaterials in agriculture: a critical review. Front. Plant. Sci. 7: 172.

Musameh, M.M., Y. Gao, M. Hickey and I.L. Kyratzis. 2012. Application of carbon nanotubes in the extraction and electrochemical detection of organophosphate pesticides: a review. Anal. Lett. 45: 783–803.

Nayak, M., A. Kotian, S. Marathe and D. Chakravortty. 2009. Detection of microorganisms using biosensors—A smarter way towards detection techniques. Biosens. Bioelectron. 25: 661–667.

Neethirajan, S., X. Weng, A. Tah, J.O. Cordero and K.V. Raghavan. 2018. Nanobiosensor platforms for detection of food allergens—New trends. Sensing and Biosens. Res. 18: 13–30.

Niwase, K., T. Tanaka, Y. Kakimoto, K.N. Ishihara and P.H. Shingu. 1995. Raman spectra of graphite and diamond mechanically milled with agate or stainless-steel ball-mill. Mater. Trans. JIM 36(2): 282–288.

Novoselov, K.S., A.K. Geim, S.V. Morozov, D. Jiang, Y. Zhang, S.V. Dubonos et al. 2004. Electric field effect in atomically thin films. Science 306(5696): 666–669.

Omanovic-Miklicanina, E. and M. Maksimovic. 2016. Nanosensors applications in agriculture and food industry. Bulletin Chemists Technol. Bosnia Herzegovina 47: 59–70.

Ozdemir, C., F. Yeni, D. Odaci and S. Timur. 2010. Electrochemical glucose biosensing by pyranose oxidase immobilized in gold nanoparticle-polyaniline/AgCl/gelatin nanocomposite matrix. Food Chemi. 119: 380–385.

Pan, M.F., L.J. Kong, B. Liu, K. Qian, G.Z. Fang and S. Wang. 2013. Production of multi-walled carbon nanotube/poly(aminoamide) dendrimer hybrid and its application to piezoelectric immunosensing for metolcarb. Sens. Actuators B 188: 949–956.

Pasinszki, T., M. Krebsz, T.T. Tung and D. Losic. 2017. Carbon nanomaterial's based biosensors for non-invasive detection of cancer and disease biomarkers for clinical diagnosis. Sensors 17: 1919.

Pelle, P.D. and D. Compagnone. 2017. Nanomaterial-based sensing and biosensing of phenolic compounds and related antioxidant capacity in food. Sensors 18: 462.

Perez-Lopez, B. and I.A. Merkoci. 2011. Nanomaterials based biosensors for food analysis applications. Trends Food Sci. Technol. 2: 625–639.

Petersen, E.J., R.A. Pinto, P.F. Landrum and J.W. Jr. Weber. 2009. Influence of carbon nanotubes on pyrene bioaccumulation from contaminated soils by earthworms. Environ. Sci. Technol. 43: 4181–4187.

Petersen, E.J., L. Zhang, N.T. Mattison, D.M. O'carroll, A.J. Whelton, N. Uddin et al. 2011. Potential release pathways, environmental fate, and ecological risks of carbon nanotubes. Environ. Sci. Technol. 45: 9837–9856.

Phillips, R.L., O.R. Miranda, C.C. You, V.M. Rotello and U.H. Bunz. 2008. Rapid and efficient identification of bacteria using gold nanoparticle-poly(para-phenyleneethynylene) constructs. Angew. Chem. Int. Ed. Engl. 47: 2590–2594.

Prakash-Deo, R., J. Wang, I. Block, A. Mulchandani, K.A. Joshi and M. Trojanowicz. 2005. Determination of organophosphate pesticides at a carbon nanotube/organophosphorus hydrolase electrochemical biosensor. Analytica Chimica Acta 530: 185–189.

Prameela, K.L. 2017. Nanomaterial's applications in agriculture. J. Chem. Pharma. Sci. 10: 593–596.

Rai, V., S. Acharya and N. Dey. 2012. Implications of nanobiosensors in agriculture. J. Biomat. Nanobiotech. 3: 315–324.

Ribeiro, F.W.P., M.F. Barroso, S. Morais, S. Viswanathan, P. de Lima-Neto, A.N. Correia et al. 2014. Simple laccase-based biosensor for formetanate hydrochloride quantification in fruits. Bioelectrochemi. 95: 7–14.

Rotariu, L., F. Lagarde, N. Jaffrezic-Renault and C. Bala. 2016. Electrochemical biosensors for fast detection of food contaminants-trends and perspective. TRAC Trends in Anal. Chem. 79: 80–87.

Saxena, M., S. Maity and S. Sarkar. 2014. Carbon nanoparticles in 'biochar' boost wheat (Triticum aestivum) plant growth. RSC Adv. 4: 39948.

Schnorr, J.M. and T.M. Swager. 2011. Emerging applications of carbon nanotubes. Chem. Mater. 23(3): 646–657.

Siddiqui, M.H., M.H. Al-Whaibi, M. Faisal and A.A. Al-Sahli. 2014. Nano-silicon dioxide mitigates the adverse effects of salt stress on Cucurbita pepo L. Environ. Toxicol. Chem. 33: 2429–2437.

Sinha, N., J. Ma and T.W. Yeow. 2006. Carbon nanotube-based sensors. J. Nanosci. Nanotechnol. 6: 573–590.

Sonawane, S.K., S.S. Arya, J.G. Le Blanc and N. Jha. 2014. Use of nanomaterials in the detection of food contaminants. Eur. J. Nutr. Food Saf. 4: 301–317.

Sonkar, S.K., M. Roy, D.G. Babar and S. Sarkar. 2012. Water soluble carbon nano-onions from wood wool as growth promoters for gram plants. Nanoscale 4: 7670.

Srivastava, S., V. Kumar, M.A. Ali, P.R. Solanki, A. Srivastava, G. Sumana et al. 2013. Electrophoretically deposited reduced graphene oxide platform for food toxin detection. Nanoscale 5(7): 3043–3051.

Srivastava, V., D. Gusain and Y.C. Sharma. 2015. Critical review on the toxicity of some widely used engineered nanoparticles. Ind. Eng. Chem. Res. 54: 6209–6233.

Stankovich, S., D.A. Dikin, G.H.B. Dommett, K.M. Kohlhaas, E.J. Zimney, E.A. Stach et al. 2006. Graphene-based composite materials. Nature 442: 282–286.

Stein, A., Z.Y. Wang and M.A. Fierke. 2009. Functionalization of porous carbon materials with designed pore architecture. Adv. Mater. 21: 265–293.

Sun, X., L. Qiao and X.Y. Wang. 2013. A novel immunosensor based on Au nanoparticles and polyaniline/multiwall carbon nanotubes/chitosan nanocomposite film functionalized interface. Nano-Micro Lett. 5: 191–201.

Talarico, D., F. Arduini, A. Amine, A. Cacciotti, D. Moscone and G. Palleschi. 2016. Screen-printed electrode modified with carbon black and chitosan: a novel platform for acetylcholinesterase biosensor development. Anal. Bioanal. Chem. 408: 7299–7309.

Tauhami, A. 2014. Biosensors and nanobiosensors: design and applications. Nanomedicine 374–403.

Tiwari, D.K., N. Dasgupta-Schubert, L.M. VillaseñorCendejas, J. Villegas, L. Carreto Montoya and S.E. BorjasGarcía. 2013. Interfacing carbon nanotubes (CNT) with plants: enhancement of growth, water and ionic nutrient uptake in maize (Zea mays) and implications for nanoagriculture. Appl. Nanosci. 4: 577–591.

Tong, Z., M. Bischoff, L. Nies, B. Applegate and R.F. Turco. 2007. Impact of fullerene (C60) on a soil microbial community. Environ. Sci. Technol. 41: 2985–2991.

Torabian, S., M. Zahedi and A.H. Khoshgoftar. 2016. Effects of foliar spray of nano-particles of FeSO$_4$ on the growth and ion content of sunflower under saline condition. J. Plant. Nutrition. 40: 615–623.

Tothill, I.E. 2009. Biosensors for cancer markers diagnosis. Sem. Cell Develop. Biol. 20: 55–62.

Tothill, I.E. 2011. Biosensors and nanomaterials and their application for mycotoxin determination. World Mycotoxin J. 4(4): 361–374.

Tripathi, S. and S. Sarkar. 2014. Influence of water soluble carbon dots on the growth of wheat plant. Appl. Nanosci. 5: 609–616.

Valdes, M.G., A.C.V. Gonzalez, J.A.G. Calzon and M.E. Dıaz-Garcıa. 2009. Analytical nanotechnology for food analysis. Microchim Acta 166: 1–19.

Verma, M.L., S.S. Kanwar and A.K. Jana. 2010. Bacterial biosensors for measuring availability of environmental pollutants. Int. Conf. Biomed. Eng. Assist. Technol. N. I. T. Jalandhar India December 17–19, 2010 pp. 1–5.

Verma, M.L. and S.S. Kanwar. 2010. Purification and characterization of a low molecular mass alkaliphilic lipase of Bacillus cereus MTCC 8372. Acta Microbiol. Immunol. Hung. 57: 191–207.

Verma, M.L. and S.S. Kanwar. 2012. Harnessing the potential of thermophiles: the variants of extremophiles. Dyn. Biochem. Process Biotechnol. Mol. Biol. 6: 28–39.

Verma, M.L., C.J. Barrow, J.F. Kennedy and M. Puri. 2012. Immobilization of β-d-galactosidase from Kluyveromyces lactis on functionalized silicon dioxide nanoparticles: characterization and lactose hydrolysis. Int. J. Biol. Macromol. 50: 432–437.

Verma, M.L., C.J. Barrow and M. Puri. 2013a. Nanobiotechnology as a novel paradigm for enzyme immobilisation and stabilisation with potential applications in biodiesel production. Appl. Microbiol. Biotechnol. 97: 23–39.

Verma, M.L., R. Chaudhary, T. Tsuzuki, C.J. Barrow and M. Puri. 2013b. Immobilization of β-glucosidase on a magnetic nanoparticle improves thermostability: application in cellobiose hydrolysis. Bioresour. Technol. 135: 2–6.

Verma, M.L., R. Rajkhowa, X. Wang, C.J. Barrow and M. Puri. 2013c. Exploring novel ultrafine Eri silk bioscaffold for enzyme stabilisation in cellobiose hydrolysis. Bioresour. Technol. 145: 302–306.

Verma, M.L., M. Naebe, C.J. Barrow and M. Puri. 2013d. Enzyme immobilisation on amino-functionalised multi-walled carbon nanotubes: structural and biocatalytic characterisation. PloS One 8: 73642.

Verma, M.L., M. Puri and C.J. Barrow. 2016. Recent trends in nanomaterials immobilised enzymes for biofuel production. Crit. Rev. Biotechnol. 36: 108–119.

Verma, M.L. 2017a. Enzymatic nanobiosensors in the agricultural and food industry. In Nanoscience in Food and Agriculture. Springer, Cham 4: 229–245.

Verma, M.L. 2017b. Nanobiotechnology advances in enzymatic biosensors for the agri-food industry. Environ. Chem. Lett. 15: 555–560.

Verma, M.L. 2017c. Fungus-mediated bioleaching of metallic nanoparticles from agro-industrial by-products. In Fungal Nanotechnology. Springer, Cham pp. 89–102.

Vinayaka, A.C., S. Basheer and M.S. Thakur. 2009. Bioconjugation of CdTe quantum dot for the detection of 2-4 dichlorophenoxyacetic acid by competitive fluoroimmunoassay based biosensor. Biosens. Bioelect. 24: 1615–1620.

Viswanathan, S., L.C. Wu, M.R. Huang and J.A. Ho. 2006. Electrochemical immunosensor for cholera toxin using liposomes and poly(3,4-thylenedioxythiophene)-coated carbon nanotubes. Anal. Chem. 78: 1115–1121.

Viswanathan, S., H. Radecka and J. Radecki. 2009. Electrochemical biosensors for food analysis. Monatsh Chem. 140: 891–899.

Wallace, P.R. 1947. The band theory of graphite. Phys. Rev. 71(91): 622.

Wang, G.C., X.C. Tan, Q. Zhou, Y.J. Liu, M. Wang and L. Yang. 2014. Synthesis of highly dispersed zinc oxide nanoparticles on carboxylic graphene for development a sensitive acetylcholinesterase biosensor. Sens. Actuators B 190: 730–736.

Wang, J., R.P. Deo and M. Musameh. 2003. Stable and sensitive electrochemical detection of phenolic compounds at carbon nanotube modified glassy electrode. Electroanal. 15: 1830–1834.

Wang, J. 2008. Electrochemical glucose biosensors. Chem. Rev. 108: 814–825.

Wang, J. and Y. Lin, 2008. Functionalized carbon nanotubes and nanofibers for biosensing applications. Trends Analyt. Chem. 27(7): 619–626.

Wang, X., H. Han, X. Liu, X. Gu, K. Chen and D. Lu. 2012. Multi-walled carbon nanotubes can enhance root elongation of wheat (Triticum aestivum) plants. J. Nanopart. Res. 14: 841.

Wang, Z. and Z. Dai. 2015. Carbon nanomaterial-based electrochemical biosensors: an overview. Nanoscale 7: 6420–6431.

Wei, T.X., Z.H. Dai, Y.H. Lin and D. Du. 2016. Electrochemical immunoassays based on graphene: a review. Electroanalysis 28: 4–12.

Williams, O.A. 2011. Nanocrystalline diamond. Diamond Relat. Mater. 20(5-6): 621–640.

Wilson, G.S. and R. Gifford. 2005. Biosensors for real-time *in vivo* measurements. Biosens. Bioelectron. 20(12): 2388–2403.

World Health Organization. 1998. Guidelines for Drinking-Water Quality, 2nd ed.; Addendum to Volume 1. Recommendations; W. H. O. Geneva, Switzerland, p. 13.

Wu, S., F.F. Huang, X.Q. Lan, X.Y. Wang, J.M. Wang and C.G. Meng. 2013. Electrochemically reduced graphene oxide and Nafion nanocomposite for ultralow potential detection of organophosphate pesticide. Sens. Actuators B 177: 724–729.

Wu, S., N. Duan, Z. Shi, C. Fang and Z. Wang. 2014. Simultaneous aptasensor for multiplex pathogenic bacteria detection based on multicolor up conversion nanoparticles labels. Analytical Chem. 86(6): 3100–3107.

Xia, N. and Y.P. Gao. 2015. Carbon nanostructures for development of acetylcholinesterase electrochemical biosensors for determination of pesticides. Int. J. Electrochem. Sci. 10: 713–724.

Xu, X.Y., R. Ray, Y.L. Gu, H.J. Ploehn, L. Gearheart, K. Raker et al. 2004. Electrophoretic analysis and purification of fluorescent single-walled carbon nanotube fragments. J. Am. Chem. Soc. 126(40): 12736–12737.

Yakes, B.J., R.J. Lipert, J.P. Bannantine and M.D. Porter. 2008. Detection of Mycobacterium avium subsp. paratuberculosis by a sonicate immunoassay based on surface-enhanced Raman scattering. Clin. Vaccine Immunol. 15: 227–34.

Yang, M., Y. Kostov and A. Rasooly. 2009. Carbon nanotubes based optical immunodetection of Staphylococcal enterotoxin B (SEB) in food. Int. J. Food Microbiol. 127: 78–83.

Yang, W., K.R. Ratinac, S.P. Ringer, P. Thordarson, J.J. Gooding and F. Braet. 2010. Carbon nanomaterials in biosensors: should you use nanotubes or graphene? Angew. Chem. Int. Ed. 49(12): 2114–2138.

Ye, W., J. Guo, X. Bao, T. Chen, W. Weng, S. Chen et al. 2017. Rapid and sensitive detection of bacteria response to antibiotics using nanoporous membrane and graphene quantum dot (gqds)-based electrochemical biosensors. Materials 10(6): 603.

Yu, G.X., W.X. Wu, Q. Zhao, X.Y. Wei and Q. Lu. 2015. Efficient immobilization of acetylcholinesterase onto amino functionalized carbon nanotubes for the fabrication of high sensitive organophosphorus pesticides biosensors. Biosens. Bioelectron. 68: 288–294.

Yudasaka, M., S. Iijima and V.H. Crespi. 2008. In carbon nanotubes. A. Jorio, G. Dresselhaus, M. S. Dresselhaus (eds.). Topics in Applied Physics. Springer-Verlag: Berlin, Heidelberg, Germany 111: 605–629.

Zhang, M., B. Gao, J. Chen and Y. Li. 2015a. Effects of grapheme on seed germination and seedling growth. J. Nanopart. Res. 17: 78.

Zhang, Q.Q., Q.C. Xu, Y.M. Guo, X. Sun and X.Y. Wang. 2016. Acetylcholinesterase biosensor based on the mesoporous carbon/ferroferric oxide modified electrode for detecting organophosphorus pesticides. RSC Adv. 6: 24698–24703.

Zhang, Q.R., G.F. Xu, L.S. Gong, H. Dai, S.P. Zhang, Y.L. Li et al. 2015b. An enzyme-assisted electrochemiluminescent biosensor developed on order mesoporous carbons substrate for ultrasensitive glyphosate sensing. Electrochim. Acta 186: 624–630.

Zhang, X., P. Geng, H. Liu, Y. Teng, Y. Liu, Q. Wang et al. 2009. Development of an electrochemical immunoassay for rapid detection of E. coli using anodic stripping voltammetry based on Cu@Au nanoparticles as antibody labels. Biosens. Bioelectron. 24: 2155–2159.

Zhang, Y., Z. Xia, Q.Z. Li, G.F. Gui, G.Y. Zhao and L.L. Lin. 2017. Surface controlled electrochemical sensing of chlorpyrifos in pinellia ternate based on a one step synthesis of palladium-reduced grapheme nanocomposites. J. Electrochem. Soc. 164: B48–B53.

Zhao, C.L., M. Qin and Q.A. Huang. 2011. Humidity sensing properties of the sensor based on graphene oxide films with different dispersion concentrations. Sensors. IEEE 129–132.

Zhao, X., L.R. Hilliard, S.J. Mechery, Y. Wang, R.P. Bagwe, S. Jin et al. 2004. A rapid bioassay for single bacterial cell quantitation using bioconjugated nanoparticles. Proc. Natl. Acad. Sci. USA 101: 15027–32.

Zheng, Y.Y., Z.M. Liu, H.J. Zhan, J. Li and C.C. Zhang. 2015. Development of a sensitive acetylcholinesterase biosensor based on a functionalized graphene-polyvinyl alcohol nanocomposite for organophosphorous pesticide detection. Anal. Methods 7: 9977–9983.

Zhou, Q., Y.X. Lin, J. Shu, K.Y. Zhang, Z.Z. Yu and D.P. Tang. 2017. Reduced grapheme oxide-functionalized FeOOH for signal-on photoelectrochemical sensing of prostate-specific antigen with bioresponsive controlled release system. Biosens. Bioelectron. 98: 15–21.

Zhou, Q., Y. Lin, K. Zhang, M. Li and D. Tang. 2018. Reduced graphene oxide/BiFeO$_3$ nanohybrids-based signal-on photoelectrochemical sensing system for prostate-specific antigen detection coupling with magnetic microfluidic device. Biosens. Bioelectron. 101: 146–152.

Zhu, Z. 2017. An overview of carbon nanotubes and graphene for biosensing applications. Nano-Micro Lett. 9: 25.

CHAPTER 9

Applications of Carbon Nanotubes-Based Electrochemical Sensing Strategies for Heavy Metals and Arsenic Quantification

Marcela C. Rodríguez,[1,]* *M. Dolores Rubianes,*[1]
Fabiana A. Gutierrez,[1] *Marcos Eguílaz,*[1] *Pablo R. Dalmasso,*[2]
M. Laura Ramírez,[1] *Cecilia S. Tettamanti,*[1] *Antonella Montemerlo,*[1]
Pablo Gallay[1] and *Gustavo A. Rivas*[1]

1. Introduction

One of the most important environmental concerns worldwide is the pollution of terrestrial and hydric resources with heavy metals (Liao et al. 2017, Abraham and Susan 2017, Islam et al. 2015). Heavy metals are defined as metals with an atomic weight between 63.5 and 200.6 g mol^{-1} and specific gravity greater than 5 g cm^{-3}. They are non-biodegradable and ubiquitously distributed, leading to a greater risk to human health and environment (da Costa Silva et al. 2011, Singh et al. 2010, Turdean 2011, Gao et al. 2013, Tag et al. 2007, Guascito et al. 2008, Li et al. 2013). They can be accumulated in the biosphere and enter living organisms through the alimentary chain (Rajaganapathy and Sreekumar 2011). In 2015, UNESCO created Agenda 2030, an action plan that proposes the improvement of the water quality by reducing or eliminating the dumping of chemical substances and toxic materials, decreasing the proportion of untreated wastewater, and increasing the possibility of recycling and reuse (ONU 2017). The main sources of heavy metals are waste products from different chemical industries, such as metal coating facilities, battery manufacturing, mining, fossil fuels, tannery,

[1] INFIQC, Departamento de Físico Química, Facultad de Ciencias Químicas, Universidad Nacional de Córdoba, Ciudad Universitaria, 5000 Córdoba, Argentina.
[2] CIQA, CONICET, Departamento de Ingeniería Química, Facultad Regional Córdoba, Universidad Tecnológica Nacional, 5016 Córdoba, Argentina.
* Corresponding author: marcela.rodriguez@fcq.unc.edu.ar

cosmetics, and fertilizers (Ihsanullah et al. 2016). Some of the metals, principally iron, zinc, copper, nickel, and manganese, are required in trace concentrations for the proper biological functioning of cells in living organisms, such as transportation and cell signaling (Valko et al. 2005). However, its presence at higher concentrations disrupts cellular functions, leading to toxicity. Moreover, there are a group of heavy metals ions (HMIs) without any beneficial effects, such as cadmium, mercury, lead, and chromium (Bagal-Kestwal et al. 2008, Patrick 2006). Consequently, they are considered environmental health hazard compounds even at trace or ultra-trace concentrations. In addition, there are some chemicals of great health concern, such as arsenic, which are considered as dangerous as HMIs (Wang and Yue 2017).

Based on toxicity of substance and potential exposure to contaminated air, water, and soil, arsenic, lead, mercury, and cadmium possess the greatest potential threat to human health, being ranked in the top 10, as the first, second, third, and seventh elements, respectively, while chromium (VI) holds the seventeenth position in the Priority List of Hazardous Substances (Agency for Toxic Substances and Disease Registry, ATSDR 2017). Many efforts are being done in order to evaluate the toxicity of these substances by several international agencies, such as World Health Organization (WHO) (Chen et al. 2013, Sharma et al. 2012, Jena and Raj 2008, Steinmaus et al. 2006), Centre for Disease Control (CDC) (Yantasee et al. 2007), Joint Food and Agricultural Organization (FAO)/WHO Expert Committee on Food Additives (JECFA), and International Agency for Research on Cancer (IARC) (Aragay et al. 2011). The primary sources of human exposure to these toxic ions are air, water, and food chain, while fish have become the most important cause of mercury toxicity (Gemma et al. 2006, da Costa Silva 2011, Prabhakar et al. 2012, Pujol et al. 2014, Wu et al. 2016). Major groups of human diseases suspected to result from exposure to lead, cadmium, mercury, arsenic, and chromium are several types of cancers, birth and immune system defects, mental retardation, behavioral abnormalities, immunotoxicity, low fertility, altered sex hormone balance, altered metabolism, and specific organ dysfunctions (Patrick 2006, Flora et al. 2008, Flora 2009, Wu et al. 2016, Del Rio et al. 2017, Pan et al. 2018). Furthermore, it has widely been accepted that environmental chemicals exert their toxicity as single entities and as mixtures (Wang and Fowler 2008, Kim et al. 2009, Wasserman et al. 2011). The toxicological mechanism of heavy metal and arsenic ions in humans, animals (Gemma et al. 2006, Sevcikova et al. 2011, Wu et al. 2016), and plants (Järup 2003, Singh et al. 2010, Dongwu et al. 2010, Tran et al. 2013) is in general through the inhibition of enzymes, oxidative stress, and damage of the antioxidant metabolism; that generates free radicals, which promote DNA damage, lipid peroxidation, and depletion of thiol-containing proteins (Flora 2009).

In this sense, many impressive efforts have been promoted in the last years in order to achieve improvements in the selective and sensitive quantification of these heavy metal and arsenic ions in natural and drinking water, tending to pull down the detection limits at sub-ppb levels. In consonance with this goal, WHO, US Environmental Protection Agency (USEPA), Joint Food and Agricultural Organization (FAO), Centre for Disease Control (CDC), and European Union have included HMIs and As as the hazardous priority substances to be monitored. All these organizations are in agreement and have established the permissible limits for the concentrations of these toxic elements in water following the environmental quality standards (EQS) (Gumpu et al. 2015, EC 2000, WHO 2017, USEPA 2018, Standard methods for the examination of water and wastewater 2017). For this reason, the development of extremely sensitive and selective methods for the quantification of trace or ultra-trace levels, ranging sub-parts per billion (sub-ppb), of these toxic ions in several complex matrices, such as natural and waste water, food, air and soil, as well as biological samples (blood, serum, saliva, etc.) is highly required.

Frequently, the highly sensitive quantification of these toxic substances in complex matrices is carried out by some spectroscopic techniques. Among them, it is possible to list atomic absorption spectroscopy (AAS) (Gong et al. 2016, Aragay and Merkoci 2012, Afkhami et al. 2013, Barbosa et al. 1999, Kenawy et al. 2000, Pohl 2009), flame atomic absorption spectrometry (FAAS) (Daşbaşı et al. 2016), electro thermal atomic absorption spectrometry (ETAAS), especially the graphite furnace atomic absorption spectrometry mode (GFAAS) (Ali et al. 2019, Chen et al. 2019, Cui et al. 2017, Fiorentini et al. 2019), and the high-resolution continuum source graphite furnace atomic absorption spectrometry mode (HR-CS GF AAS) (Paixão et al. 2019, Rovasi Adolfo et al. 2019), in addition to the hydride generation atomic absorption spectrometry (HG-AAS) (Ali et al. 2019, Büyükpınar et al. 2018, Maratta et al. 2018), and hydride generation atomic fluorescence spectrometry (HG-AFS) (Shishov et al. 2018, Wang et al. 2018, Zheng et al. 2019). Another group of relevant techniques for HMIs and As are inductively coupled plasma mass spectroscopy (ICP-MS) (Caroli et al. 1999, Silva et al. 2009, Koelmel and Amarasiriwardena 2012, Wang et al. 2015, Gong et al. 2016), inductively coupled plasma-optical emission spectrometry (ICP-OES) (Losev et al. 2015, Massadeh et al. 2016), inductively coupled plasma-atomic emission spectrometry (ICP-AES) (Sreenivasa Rao et al. 2002), X-ray Fluorescence Spectrometry (XRF) (Sitko et al. 2015), and Neutron activation analysis (NAA) (Losev et al. 2015). These techniques are useful in terms of simultaneous determination of the concentration of heavy metal ions for a large range of elements, achieving very low detection limits up to femtomolar levels (Pujol et al. 2014). However, these spectroscopic techniques are very complex, require highly-trained personnel, with long detection and sample preparation times, involving difficult analytical procedures, which lead to elevated procedural costs (Cui et al. 2015, Wang and Hu 2016, Ramnani et al. 2016, Bansod et al. 2017, Lu et al. 2018). In addition, these analytical methods are only suitable for quantitative analysis and have certain limitations for the metal ion speciation; therefore, they should be combined with other chromatographic techniques (Feldmann et al. 2009). Optical methods are used for the quantification of HMIs as well (Wang et al. 2007, Aragay and Merkoci 2012, Fang et al. 2018). As spectroscopic techniques, these methodologies require expensive and complex equipment (such as lasers and photo detectors, among others), high precision, and high power operations, thus becoming inconvenient analytical tools for in-field applications. Thus, the pursued goal is the development of fast, low cost, simple, and reliable techniques that could be suitable for *in-situ* and real-time measurements of heavy metal ions (Aragay and Merkoci 2012, Pujol et al. 2014, Cui et al. 2015). Excellent alternatives for overcoming the drawbacks of spectroscopic and optical methods are the electrochemical techniques, since they are affordable, easy-to-use, simple, fast, reliable, and efficient, making them convenient for in-field applications. At variance with spectroscopy, electrochemical techniques permit the use of miniaturization strategies for the construction of portable devices suitable for *in-situ* and real-time monitoring with the simplest and short-term analytical procedures of environmental water samples (Pujol et al. 2014, Ramnani et al. 2016, Lu et al. 2018) or biological fluids (Pei et al. 2014, Cai et al. 2017, Yao et al. 2019). On the other hand, the classic electrochemical techniques lack high sensitivities and exhibit higher detection limits compared to other spectroscopic or optical techniques. Voltammetric techniques could be classified as cyclic voltammetry (CV), linear sweep voltammetry (LSV), differential pulse voltammetry (DPV), and square wave voltammetry (SWV) (Bard and Faulkner 2001, Wang 2006, Pletcher et al. 2011, Osteryoung and Osteryoung 1985). The difference among these techniques lies in the respective perturbation functions applied, which produce a characteristic time waveform response (Osteryoung and Osteryoung 1985, Bard and Faulkner 2001, Pletcher et al. 2011). Moreover, among the mentioned techniques, DPV and SWV have the highest detection sensitivity and the possibility of reaching the lowest detection limits (Bard and Faulkner 2001, Wang 2006, Bansod et al. 2017). Several strategies have been proposed in order to improve sensitivity and detection limits in voltammetric methods (Bansod et al. 2017, Lu et al. 2018, Omanović et al. 2015). The most widely used strategy is stripping voltammetry techniques that include two steps: pre-concentration and re-dissolution (Wang 2006, Lu et al. 2018). Pre-concentration step involves the accumulation of the metal cations on the surface of the working electrodes using cathodic potential step or adsorption, and then obtaining a correlated signal by anodic stripping voltammetry (ASV) (Bansod et al. 2017, Lu et al. 2018, Omanović et al. 2015). During the faradaic reaction, the heavy metal cations are reduced to zero-valent metals under the constant negative potential, and then, they are deposited on the surface of the working electrode.

On the other hand, the adsorption is a reaction that could proceed between appropriate ligands and heavy metal cations on the surface of the working electrodes to produce complexes, and later reduce the heavy metal cations to zero-valent metals. After the pre-concentration step, the stripping process is carried out by sweeping the electrode potential to anodic potential values in order to re-oxidize the zero-valent metals to cations (Bansod et al. 2017, Lu et al. 2018). A high dissolution current peak is reached during the rapid oxidation process, and the stripping current peak potential changes according to the different kinds of HMIs and As. The surface of the working electrodes could be tuned using different materials, making them suitable for its application in the specific detection of HMIs and As (Ramnani et al. 2016). For this reason, the development of new strategies for the design of modified sensing platforms that allow to reach the lowest limits of detection with the highest sensitivity and selectivity in the HMIs and As detection is of paramount importance (Cui et al. 2015, Dali et al. 2018).

In the past, the traditional electrode materials used for the analysis of HMIs were based on mercury, mainly due to its reproducible surface, good negative potential window, and the ability to form amalgams with several HMIs (Bard and Faulkner 2001). However, owing to its toxicity and cost, mercury is not currently used for this purpose. Recently, exciting new strategies have been proposed, including the modification of different solid electrodes, which were applied to the analysis of different types of samples (Roy et al. 2014, Fang et al. 2018, Lin et al. 2018, Tu et al. 2018, Yao et al. 2019). In this respect, an example of alternative electrode material for the detection of HMIs is bismuth, an environment-friendly element, due to its attractive electrochemical features, including reproducible behavior, broad linear range, and good signal-to-background ratio (Pei et al. 2014, Borgo et al. 2015).

Nevertheless, the incorporation of nanomaterials-based metals, metal oxides, carbon, and their composites for the development of sensors and biosensors have opened a new era for Analytical Chemistry (Cui et al. 2015, Gumpu et al. 2015, Ramnani et al. 2016, Wang and Hu 2016, Bansod et al. 2017, Kempahanumakkagari et al. 2017, Lu et al. 2018). As it is well known, nanomaterials, especially carbon nanomaterials (CNMs), play a key role in the development of electrochemical sensors. In particular, single-walled (SWCNTs), multi-walled carbon nanotubes (MWCNTs), graphene (GR), carbon nanofibers (CNFs), carbon nanodots (CDs), nanodiamonds (NDs), and nanoporous carbon (NC) have been demonstrated to possess many interesting properties, such as high specific surface area and the ease to obtain a rich functionalized surface that promotes the enhancement of HMIs and As adsorption processes (Yang et al. 2019). CNMs hold good prospects for application in HMIs and As analysis due to their outstanding electrical, thermal, and mechanical properties, ultra-high electrical conductivity, chemical and physical stability, high heat and corrosion resistance, and wide electrochemical potential window (Li et al. 2019). Since their discovery by Iijima in 1991, as a novel type of synthetic nanomaterial with distinctive hollow and cylindrical structure (Iijima 1991), CNTs have been one of the most studied CNMs. Along these last years, CNTs have demonstrated a leading role in many sensing schemes, producing a real change in the paradigms of sensors and biosensors research fields (Rivas et al. 2017). Nowadays, the large scale production and low cost, in comparison to other CNMs, represent the main advantages of CNTs for technological applications in sensing devices and systems.

This chapter highlights the most interesting and relevant contributions of several research groups in the period 2014–2018 in the development of CNTs-based electrochemical sensing technology for the detection of Pb, Cd, Hg, Cr, and As. In the following sections, special attention is paid to the strategies of functionalization and immobilization of CNT, associated with pre-concentration and stripping methodologies that have provided improvements in the sensitive and selective electrochemical detection of HMIs and As. Table 9.1 summarizes, the most representative works published in this period have a special focus on the detection limit (DL) in view of the practical applications of the proposed sensors.

Table 9.1: Analytical performance of the most relevant HMIs and As CNTs-based electrochemical sensors reported in the period 2014–2018.

Analyte	Platform	Detection	LOD (ppb)	LR (ppb)	Real samples	Reference
Pb (II), Cd (II)	mPAD	SWASV	1 (Pb (II)), 1 (Cd (II))	5.0–150.0 (Pb(II)), 5.0–150.0 (Cd (II))	Digested particulate metals collected from dust	Rattanarat et al. 2014
Cd(II)	L-cMWCNTs-IL-CPE	DPASV	0.08	0.2–23	Tap water, industrial waste, tobacco, human hair, milk powder, edible fungi	Afkhami et al. 2014a
Pb (II), Cd (II)	GO–MWCNTs/Bi	DPASV	0.2 (Pb (II)), 0.1 (Cd (II))	0.5–30 (Pb (II)), 0.5–30 (Cd (II))	Electroplating effluent	Huang et al. 2014
Pb (II), Cd (II), Cu(II)	ooly(Arg)/cMWCNTs/PGE	DPASV	1.62 (Pb (II)), 1.03 (Cd (II)), 2.12 (Cu (II))	6.69–204.16 (Pb (II)), 4.16–205.92 (Cd (II)), 9.54–471.46 (Cu (II))	Soil, water, plant leaf, and blood	Roy et al. 2014
Pb (II), Cd (II)	Bi/MWCNTs-poly(PCV)/GCE	DPASV	0.40 (Pb (II)), 0.20 (Cd (II))	1.0–200.0 (Pb (II)), 1.0–300.0 (Cd (II))	Tap and well water	Chamjangali et al. 2015
Pb (II), Cd (II)	MWCNTs-poly(DAN)/Pt	SWASV	2.1 (Pb (II)), 3.2 (Cd (II))	4–150 (Pb (II)), 4–150 (Cd (II))	River water	Vu et al. 2015
Pb (II), Cd (II)	Bi/cMWCNTs-β-CD-Nafion/GCE	SWASV	0.21 (Pb(II)), 0.13 (Cd(II))	1–100 (Pb (II)), 1–100 (Cd (II))	Soil sample	Zhao et al. 2016
Pb (II), Cd (II)	engineered MWCNTs	SWASV	0.3 (Pb(II)), 0.4 (Cd(II))	2–50 (Pb (II)), 2–50 (Cd (II))	N/A	Li et al. 2016
Cd (II)	SWCNTs-Cys/GCE	LSASV	0.33	4–300	Groundwater	Gutierrez et al. 2017
Pb (II), Cd (II), Cu(II), Hg(II)	poly(furfural) film/MWCNTs/GCE	DPASV	0.01 (Pb(II)), 0.03 (Cd(II)), 0.06 (Cu(II)), 0.1 (Hg(II))	0.1–5.0 (Pb(II)), 0.5–5.5 (Cd(II)), 0.1–3.0 (Cu(II)), 1.5–6.0 (Hg(II))	Tap water	Huang et al. 2017
Pb (II), Cd (II)	MWCNTs-PARS/GCE	DPASV	0.47 (Pb(II)), 0.43 (Cd(II))	5.0–150.0 (Pb(II)), 5.0–160.0 (Cd(II))	Soil, waste and tap water	Chamjangali et al. 2017
Pb (II)	Chit-Mn(TPA)-SWCNTs/GCE	DPASV	7.87	20.7–289.9	Industrial waste water and human serum	Cai et al. 2017

Table 9.1 contd. ....

...Table 9.1 contd.

Analyte	Platform	Detection	LOD (ppb)	LR (ppb)	Real samples	Reference
Pb (II)	IIP-PAN-MWCNTs/GCE	DPASV	0.16	0.50–12.0	Tap and mineral water, physiological serum, and synthetic urine	Tarley et al. 2017
Pb (II)	Cys-SWCNTs/GCE	LSASV	0.69	5.0–125.0	Tap and rain water	Ramirez et al. 2018
Pb (II), Cd (II)	cSWCNTs-fungi/GCE	DPASV	2 (Pb(II)) 11 (Cd (II))	N/A	Ground water	Dali et al. 2018
Pb (II), Cd (II)	CNTs-PyTS/Nafion/PGE	DPASV	0.02 (Pb(II)) 0.8 (Cd(II))	1.0–110 (Pb(II)) 1.0–90 (Cd(II))	N/A	Jiang et al. 2018
Pb (II), Cd (II)	Bi/rGO-CNTs/Au	SWASV	0.2 (Pb(II)) 0.6 (Cd(II))	20–200 (Pb (II)) 20–200 (Cd (II))	Drinking water	Xuan and Park 2018
Cd (II)	XG-GSH-ZnSeQDs/MWCNTs/GCE	DPASV	2.2	11–560	Lake water	Ding et al. 2018
Hg(II) Tl(I)	MWCNTs-SnO$_2$NPs/GCE	SWASV	0.0012 (Hg(II)) 0.0009 (Tl(I))	0.004–400 (Hg(II)) 0.003–450 (Tl(I))	Surface water	Mnyipika and Nomngongo 2017
Hg(II)	SWCNTs-PhSH/AuE	SWASV	0.60	1–10	Tap, river, and waste water	Wei et al. 2014
Hg(II)	Zn/Al-LDH-MPP/SWCNTPE	CV	0.20	0.20–20	Fish, chili pepper, tomato, and gourd	Isa et al. 2015
Hg(II)	H$_4$tsdb-CNTs-PIGE	SWASV	0.16	0.48–44.13	Sea and lake water	Gayathri et al. 2018
Hg(II), Pb(II)	pmaba-CNTs-PIGE	SWASV	0.072 (Hg(II)) 0.23 (Pb(II))	0.66–13.24 (Hg(II)) 0.68–13.68 (Pb(II))	Sea, lake, and ground water	Selvan et al. 2018
Hg(II)	tbdocpdd-MWCNTs-IL-CPE	DPASV	0.010	0.040–52.16	Sea and industrial waste water, tuna fish, sausage, cream cheese, and raw milk	Afkhami et al. 2014
Hg(II)	BEPT-MWCNTs-CPE	Potentiometry	0.62	0.80–4.41×10^5	Laboratory, mineral, and distilled water	Mashhadizadeh et al. 2015
Hg(II)	v-SWCNTs(R)/AuE and a target recycling strategy using Exo III	SWV: indirect determination *via* oxidation current of methylene blue	6.02×10^{-7}	2.01×10^{-6}–201	Lake water and human serum	Shi et al. 2017

Cr(III)	MnOx/MWCNTs-Chit/GCE	Amperometry	15.60	156–4368	Water	Salimi et al. 2015
Cr(III)	Cr(III)-IIP/MWCNTs/CPE	Potentiometry	30.68	$52-52 \times 10^5$	Sea and river water, and soil	Alizadeh et al. 2017
Cr(VI)	clSWCNTs-BTMPPA/BD-SiE	Amperometry	0.01	0.01–10	Underground water	Deep et al. 2014
As(III)	MWCNTs/AuNPs/SPE	LSASV	0.5	10–550	N/A	Gamboa et al. 2014
As(III)	AuND/SWCNTs/MPL	ASV	0.4	0.5–80	Tap water	Duy et al. 2016

Abbreviations: mPAD: electrochemical microfluidic paper-based analytical device; L: new synthesized Schiff base; cMWCNTs: carboxylated multi-walled carbon nanotubes; IL: ionic liquid; CPE: carbon paste electrode; GO: graphene oxide; MWCNTs: multi-walled carbon nanotubes; Bi: bismuth film; poly(Arg): poly-arginine; PGE: pencil graphite electrode; poly(PCV): poly(pyrocatechol violet); GCE: glassy carbon electrode; poly(DAN): poly(1,5-diaminonaphthalene); Pt: sputtered planar platinum; β-CD: β-cyclodextrin; SWCNTs: single-walled carbon nanotubes; Cys: cysteine; PARS: poly-alizarin red S; Chit: chitosan; Mn(TPA): flake-shaped manganese-terephthalic acid metal-organic framework; IIP: ion-selective imprinted polymer; PAN: 1-2(-pyridylazo)-2-naphtol; cSWCNTs: carboxylated single-walled carbon nanotubes; fungi: fungal biomass of *Trichoderma asperellum*; CNTs: carbon nanotubes; PyTS: sodium pyrene-1,3,6,8-tetrasulfonate salt; rGO: reduced graphene oxide; Au: flexible gold substrate; XG: xanthan gum; GSH: glutathione; ZnSeQDs: zinc selenide quantum dots; PIGE: paraffin impregnated graphite electrode; AuE: gold electrode; SnO$_2$: tin oxide nanoparticles; PhSH: thiophenol; Zn/Al-LDH-MPP: Zn/Al layered double hydroxide-3(4-methoxyphenyl)propionate; SWCNTPE: SWCNT paste electrode; H$_4$tsdb: N,N′,N″,N‴-tetrasalicylidene-3,3′-diaminobenzidine; pmaba: N,N′-bis(pyrrole-2-ylmethylene)-2-aminobenzylamine; tbdocpdd: 1,3,5-triaza-2,4,7,8,11,12,15,16-tetrabenzo-9,14 dioxacycloheptadeca-1,5-diene; BEPT: 1,3-bis(2-ethoxyphenyl)triazene; v-SWCNTs(R): vertically aligned SWCNTs through self-assembled monolayers of 4-aminothiofenol; Exo III: exonuclease III; MnOx: manganese oxide nanoflakes; clSWCNTs: chlorinated SWCNTs; BTMPPA: bis(2,4,4-trimethylpentyl) phosphinic acid; BD-SiE: boron doped p-type silicon wafer; SPE: vibrating screen printed electrode; AuND: gold nanodendritic structure; SWASV: square wave anodic stripping voltammetry; AuNPs: gold nanoparticles; SWV: square wave voltammetry; CV: cyclic voltammetry; LSASV: Linear sweep anodic stripping voltammetry; ASV: anodic stripping voltammetry; DPASV: differential pulse anodic stripping voltammetry; N/A: Not available.

2. Carbon Nanotubes (CNTs)

CNTs are at the forefront as the most advantageous candidates for improving the electro-sensing performance and allowing the development of completely new nanoscale sensors and biosensors. Keeping in mind all these excellent properties, there is a growing interest for the use of CNTs in the development of new, simple, sensitive, reliable, low cost, and portable electrochemical sensors (Gupta et al. 2018). CNTs have attracted special interest and its use has been significantly increased in the environmental area compared to other materials (Sarkar et al. 2018, Wanekaya 2011). In particular, the application of CNTs for the detection/removal of HMIs is clearly reflected by the exponential growth in the number of publications in recent years (Figure 9.1).

CNTs are hollow cylinders of carbon atoms in an arrangement of periodic hexagon. They are like graphite sheets rolled up on themselves to form tubes, which have a diameter size in the range of nanoscale (Figure 9.2). According to the number of graphite sheets, CNTs are divided into main two types: single-walled carbon nanotubes (SWCNTs) and multi-walled carbon nanotubes (MWCNTs) (Ajayan 1999, Dresselhaus et al. 2004). SWCNTs show excellent chemical stability, good mechanical strength, and a wide range of electrical conductivity properties; MWCNTs show metallic electronic properties similar to metallic SWCNTs (Hu et al. 2010), which, in some aspects, makes them more suitable for electrochemical applications.

Recently, CNTs are almost becoming a benchmark for the performance of new nanomaterials in various applications (Farzin et al. 2018, Nasir et al. 2018, Sarkar et al. 2018, Gupta et al. 2018). However, CNTs have some limitations in their regular use as nanomaterials for applications as sensors and biosensors. CNTs have bond structure based on sp^2–sp^3 hybridizations (Blase et al. 1994). Thus, these properties have a close relation to the chirality, which are created by vacancies in the structure of CNTs. The existence of a small amount of defects on the surface of CNTs is responsible for their interaction with metal ions (Li et al. 2003). On the other hand, the bundle structure of CNTs is responsible for the diminution of surface area. In addition, the low specific capacitance of pristine CNTs restricts their use in many applications (Ghosh et al. 2012).

CNTs have some limitations toward ultra-trace detection of HMIs due to the high background currents accompanying their large surface area and the absence of functional groups on the CNTs surface. In order to improve the performance of CNTs-based electrodes, a variety of strategies have been explored. These are generally designed to provide better electrochemical properties, allowing superior selectivity and lower detection limits. CNTs modified with functional groups are a way to tackle these above-mentioned issues (Gao et al. 2012). In this sense, the ends and sidewalls of CNTs can be easily modified by attaching almost any desired chemical species.

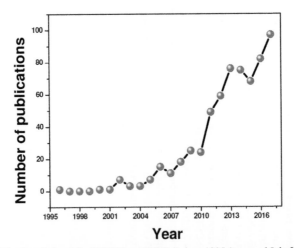

Figure 9.1: Graph of publications indexed for "CNTs" and "HMIs" since 1995 (accessed July 20, 2018). Source: Scopus.

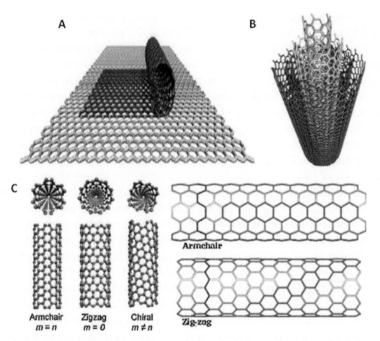

Figure 9.2: (A) Scheme of SWCNTs which is composed of a single layer of graphite. (B) Scheme of MWCNTs. (C) Different types of SWCNTs. Reprinted (adapted) from Yang, N., X. Chen, T. Ren, P. Zhang and D. Yang. 2015. Carbon nanotube based biosensors. Sens. Actuators B Chem. Part A 207: 690–715. Copyright with permission from Elsevier.

Pristine CNTs are insoluble in aqueous solutions, polymer resins, and most solvents due to their hydrophobic surfaces. Functionalization is a process in which certain molecules or functional groups are physically (non-covalent) or chemically (covalent) attached to the smooth sidewalls of CNTs without downgrading their desirable properties (Mallakpour and Soltanian 2016). Functionalized CNTs are more easily dispersible in liquids which have better biocompatibility and low toxicity (Marchesan et al. 2015). Covalent functionalization is done by forming bonds with the chemical groups generated at the oxidized nanotube sidewalls, whereas the non-covalent mode occurs through interaction between the hydrophobic domain of an amphiphilic molecule and the CNT surface. Therefore, it is important to take into account that the non-covalent functionalization of the sidewalls of CNTs is the best way to preserve the sp^2 nanotube structure and their electronic characteristics to enhance the sensors' performance towards diverse analytes (Ata et al. 2018).

3. Heavy Metals Ions (HMIs) and Arsenic (As)

3.1 Cadmiun (Cd) and Lead (Pb)

Cadmium (Cd(II)) and lead (Pb(II)) exhibit a characteristic slow toxicity process, and their degradation by microorganisms is difficult. Particularly, Cd is recognized as a metal posing threat to agricultural food quality due to its mobility in the water–soil–plant system. The main sources of soil contamination with Cd are phosphate fertilizers (as by-product), metal-working industries, waste incinerators, urban traffic, and cement factories. Cd enters into plant cells due to its similar chemical and physical characteristics to plant nutrients. It can cause many toxic symptoms, such as inhibition of growth and photosynthesis, deactivation or inhibition of enzymes, disturbances in plant water relationships, ion metabolism, and formation of free radicals (Kumar et al. 2008). As in the case of Cd, Pb is one of the HMIs that generates more concern in the field of health since it is highly toxic to living systems and may stay in the environment for a long time, due to its persistency (Wu et al. 2016). Pb is a widespread environmental and occupational xenobiotic, being hazardous to humans and environment (Jonasson and Afshari 2018, Sarwar et al. 2018, Wu et al. 2016). For humans, exposure to Pb from anthropogenic sources is mainly by ingestion through the mouth and inhalation from fumes and dust in the atmosphere (Jonasson and Afshari 2018, Kumar et al. 2018, Sarwar et al. 2018). Although Pb is used since ancient times, mainly for the construction of household utensils and plumbing systems (Jonasson and Afshari 2018), after the industrial revolution, the application of Pb was extended to painting, marine, automotive, nuclear, and fuel industries (Kumar et al. 2018, Sarwar et al. 2018). Pb accumulates in the organism, generating a multifactorial pathogenesis that includes mental disorders, central and periphery nervous system damage, several types of cancer, kidney diseases, and anemia (Jonasson and Afshari 2018, Kumar et al. 2018, Ramírez et al. 2018, Sarwar et al. 2018). As children exhibited high levels of hand-to-mouth activity associated with increased rates of lead absorption, they are highly susceptible to the effects of lead compared to adults (Jonasson and Afshari 2018, Pan et al. 2018, Ramírez et al. 2018). Its toxicological mechanism is based on an affinity for sulfhydryl groups, ability to bind to enzymes, such as delta-aminolevulinic acid dehydrase, glutathione reductase among others, and to induce DNA damage, which generates oxidative stress and alteration of gene expression (Gumpu et al. 2015, Wu et al. 2016). A decrease in trace mineral uptake, alteration of calcium homeostasis, and hypertension has also been reported (Patrick 2006).

As it was previously discussed, the application of CNT and its functionalization with chemical groups or molecules contributes to the enhancement of the selectivity and the lowering of the detection limits. Usually, the functionalization with carboxylic (–COOH), amino (–NH_2), and thiol groups (–SH) has been used to increase the pre-concentration ability towards metal ions on the electrode surface by coordination through the interaction between electron-rich ligands (Lewis bases) and electron-deficient (Lewis acids) HMIs (Wang and Yue 2017).

Most of the works reviewed in this section have a similar detection principle which includes three steps. The first one involves complex formation between the active electrode platform and the heavy metal ion (denoted as M^{2+}) at open circuit potential (Figure 9.3a). The second step is the reduction of the accumulated M^{2+} ions to M at closed circuit under negative potential (Figure 9.3b), and the last one is the stripping step where M is electrochemically stripped back into the solution at closed circuit by scanning towards positive potentials (Figure 9.3c). The resulting stripping peak is proportional to the amount of M^{2+} in the accumulation solution.

In Table 9.1, the contributions reviewed in this section are listed. As it can be seen, every electrochemical platform includes the functionalization of CNT with different molecules that allow the selective quantification of Cd(II) and Pb(II). These molecules have donor atoms which can interact with the HMIs by complexing process.

Rattanarat et al. 2014 proposed an interesting approach designing a simple and low-cost sensor for Cd(II) and Pb(II) detection in particulate matter collected on air sampling filters employing electrochemical microfluidic paper-based analytical devices (mPAD) through a three-dimensional configuration (Figure 9.4). The electrochemical mPAD was based on a three-electrodes system. The fabrication was based on MWCNTs-modified carbon inks that were screen-printed onto polyester in order to obtain a working electrode (WE). The WE was then modified with bismuth and ferricyanide to enhance the stripping signals of Cd and Pb, decreasing Cu interference. The top layer contains five wax-defined channels extending from an open central sample reservoir, where the electrochemical detection zone was placed towards the edges. Each channel works as both single sample pretreatment and detection zones, contributing to the increase of the selectivity and sensitivity of the detection strategy. Under optimum conditions, this mPAD is able to detect Cd(II) and Pb(II) in the level range of 5 to 150 ppb with a DL for both HMIs of 0.25 ng/L. Moreover, it was found that Mn(II), Mg(II), Zn(II), Fe(III), Fe(II), Al(III), Ba(II), and V(III) did not interfere with the assay. On the other hand, Ni(II), Co(II), Cu(II), and Cr(VI) interfere significantly. Authors proposed two ways for minimizing this problem: (i) dilution of the aerosol sample because it is a simple way to perform SWASV since the deposition process allows a preconcentration step for Ni(II), Co(II), and Cr(VI), and (ii) Cu (II) complexation with the appropriate concentration of ferricyanide in order to suppress the Cu(II) signal

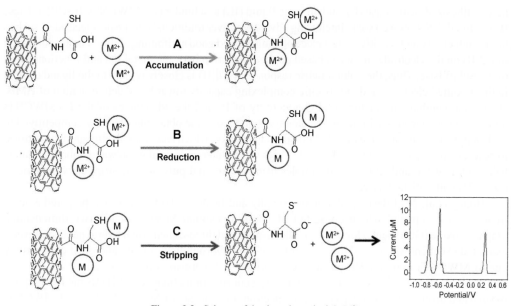

Figure 9.3: Scheme of the detection principle M^{2+}.

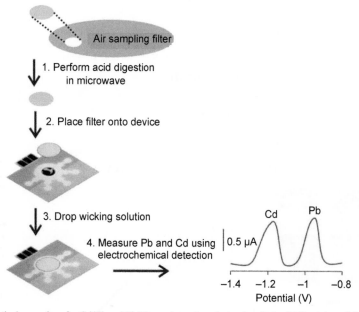

Figure 9.4: Analytical procedure for Cd(II) and Pb(II) sensing using electrochemical mPADs. Adapted from Rattanarat P., W. Dungchai, D. Cate, J. Volckens, O. Chailapakul and C.S. Henry. 2014. Multilayer paper-based device for colorimetric and electrochemical quantification of metals. Anal. Chem. 86: 3555–3562. Copyright American Chemical Society.

and recover Pb(II) and Cd(II) signals. The mPADs were used to measure particulate metal content in resuspended baghouse dust samples—the levels found being in good agreement with those obtained by traditional approaches.

Afkhami et al. 2014a reported the platform with the lowest DL for Cd(II). In this approach, the authors described the construction of a voltammetric sensor for the selective determination of Cd(II) based on a carbon paste electrode (CPE) modified with carboxylated MWCNTs (cMWCNTs), a new synthesized Schiff base (L), and an ionic liquid (IL) as a binder (L-cMWCNTs-IL-CPE). These IL-modified CPEs showed some interesting advantages over traditional composite electrodes, such as high conductivity and sensitivity, fast electron transfer, and good antifouling ability. The accumulation of Cd(II) on the electrode surface is based on the interaction between metal ion and functional sites of the Schiff base. Thus, the voltammetric response of Cd(II) is closely related to the ligand content in the modified electrode and the specific complexing capacity towards the determination of HMIs. Hence, the combination of the good conductivity of IL and the advantages of the L-cMWCNTs allowed the enhanced sensitivity of the electrode, and led to the obtaining of a very competitive DL of 0.08 ppb and a linear range of 0.2–23.0 ppb. This sensor, which can be easily prepared, has proved to be a selective sensor for the analysis of Cd(II) in tap water, industrial wastewater, tobacco, human hair, milk powder, and edible fungi samples using differential pulse anodic stripping voltammetry (DPASV) with satisfactory results.

Huang et al. 2014 developed a green-friendly and facile method. This group prepared a novel hybrid nanocomposite which consisted of one-dimensional MWCNTs and two-dimensional graphene oxide (GO) sheets (Figure 9.5). The as-prepared three-dimensional hybrid nanocomposites exhibited excellent water-solubility owing to the high hydrophilicity of GO and good conductivity due to MWCNTs immobilized on the GCE surface. A synergistic effect between MWCNTs and GO components was observed, enhancing its pre-concentration efficiency of HMIs mediated by the oxygenated groups of GO and accelerating electron transfer rate of MWCNTs. The GO-MWCNTs nanocomposite was used to simultaneously determine Pb(II) and Cd(II) with a linear range of 0.5 to 30.0 ppb for both ions. The DL was 0.2 ppb for Pb(II) while for Cd(II) was 0.1 ppb. It is worth

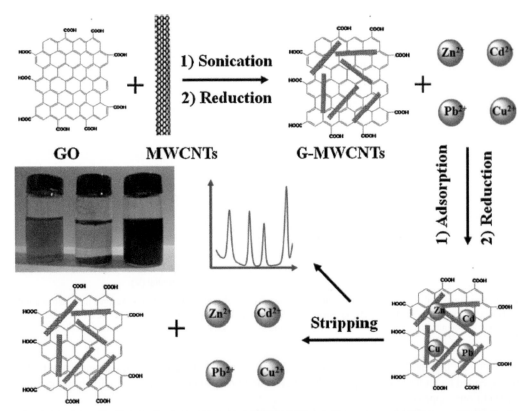

Figure 9.5: Schematic diagram for the synthesis of a GO-MWCNTs hybrid nanocomposite and its application for the electrochemical detection of Pb(II) and Cd(II). Reprinted (adapted) from Huang, H., T. Chen, X. Liu and H. Ma. 2014. Ultrasensitive and simultaneous detection of heavy metal ions based on three-dimensional graphene-carbon nanotubes hybrid electrode materials. Anal. Chim. Acta 852: 45–54. Copyright with permission from Elsevier.

mentioning that the GO-MWCNTs modified electrodes were successfully applied for the simultaneous detection of Cd(II) and Pb(II) in real electroplating effluent samples containing high amounts of surface active impurities.

Roy et al. 2014 proposed a sensor for the detection of Cd (II), Pb (II), and Cu (II), which relies on the modification of the surface of the pencil graphite electrode (PGE) with cMWCNTs. Then, a layer of conducting poly-arginine (poly-Arg) was used to coat the cMWCNTs with an electropolymerization technique. Poly-Arg functional groups promoted the binding of metal ions, contributing to the improvement of the sensing strategy. The simultaneous and selective electrochemical determination of Cd(II), Pb(II), and Cu(II) was performed under the same experimental conditions. It is interesting to remark that the oxidation signals for Cu(II), Cd(II), and Pb(II) can be clearly identified at potentials of 0.00, –0.50, and –0.20 V, respectively. The DPASV current response of the sensor for the individual analysis of Cd(II), Cu(II), and Pb(II) was linear over a concentration range from 4.16 to 205.92, 9.54 to 471.46, and 6.69 to 204.16 ppb, and the DLs were determined to be 1.03, 2.12, and 1.62 ppb, respectively. The proposed sensor was used to explore the total uptake of food-borne Cu(II), Cd(II), and Pb(II) in the human blood and the effect of this metal ion accumulation in soil, water, fruit, and vegetables, in the presence of interferents, such as ascorbic acid, glucose, iron, mercury, zinc, with excellent percentage of recovery.

Chamjangali et al. 2015 reported an interesting alternative for the development of a voltammetric electrochemical sensor based on glassy carbon electrodes (GCEs) coated with a MWCNTs-poly(pyrocatechol violet) (PCV) composite and a bismuth film. The Bi/MWCNTs-poly(PCV)/GCE sensor was used for the simultaneous determination of Cd(II) and Pb(II). PCV forms colored complexes with various metal ions, including Cd(II) and Pb(II). A synergistic effect caused by the enhanced surface area of MWCNTs, the conductivity, and chelation ability of the polymeric film, and the intrinsic advantages of bismuth coating, resulted in the improvement of sensitivity and performance of the sensor. Under optimal conditions, the electrode showed a good linear response for both metal ions with DLs of 0.20 and 0.40 ppb for Cd(II) and Pb(II), respectively. The sensor presented low interference for higher amounts (at least $\geq$ 10 times) of coexistent ions or compounds (Na(I), K(I), SO_4^{2-}, SO_3^{2-}, NO_3^-, $HCOO^-$, Ca(II), PO_4^{3-}, CO_3^{2-}, SCN^-, $S_2O_3^{2-}$, Cl^-, Cr(III), Ba(II), Mn(II), CN^-, Al(III), F^-, Zn(II), tartaric acid, and citric acid), since the change in the achieved peak current was lower than $\pm$ 3s. This behavior is due to the specific and selective recognition interaction of the polymer functionalities with Cd(II) and Pb(II). However, some metal ions, such as Cu(II), Fe (III), Ni(II), Ag(I), and Hg(II) had significant interferences. Thus, Chamjangali et al. reported the use of appropriate concentrations of SCN^-, $S_2O_3^{2-}$, tartaric acid, and citric acid in order to decrease the interference effect of Cu(II), Fe (III), Ni(II), and Ag(I). The proposed method was used for the simultaneous determination of the Cd(II) and Pb(II) content in tap and well water samples. Furthermore, the modified electrode showed good stability during one week for continuous measurements without refreshing the bismuth film.

Sputtered planar platinum electrodes modified with interpenetrated MWCNTs-poly(1,5-diaminonaphthalene) (MWCNTs-poly(DAN)) were employed by Vu et al. 2015 as a sensing interface for detection of Cd(II) and Pb(II). The electrochemical performance of this platform was evaluated by square wave anodic stripping voltammetry (SWASV). The sensor exhibited a linear response for Cd(II) and Pb(II) concentrations ranging from 4.0 to 150.0 ppb (Figure 9.6). Under the optimized conditions, the DLs were 3.2 and 2.1 ppb for Cd(II) and Pb(II), respectively. The interference studies showed that successive additions of 1.0 ppm of Na(I), Ca(II), Zn(II), Fe(II), Al(III), and Cu(II) each and 10.0 ppm of Cl^-, Br^-, and SO_4^{2-} each had no influence on the oxidation peak current of 100.0 ppb Pb(II). The performance of the MWCNTs-poly(DAN)/Pt sensor was comparable to those obtained by other reference methods, such as AAS and anodic stripping voltammetry using hanging drop mercury electrodes. This sensing platform was used for the quantification of Cd(II) and Pb(II) in river water samples, and the results were in good agreement with the values obtained by AAS.

A sensing platform based on GCEs modified with a bismuth film and a cMWCNT-β-cyclodextrin-Nafion nanocomposite (Bi/cMWCNTs-β-CD-Nafion/GCE) was fabricated and characterized by Zhao et al. 2016. The Bi/cMWCNTs-β-CD-Nafion/GCE sensor was used for the sensitive detection of

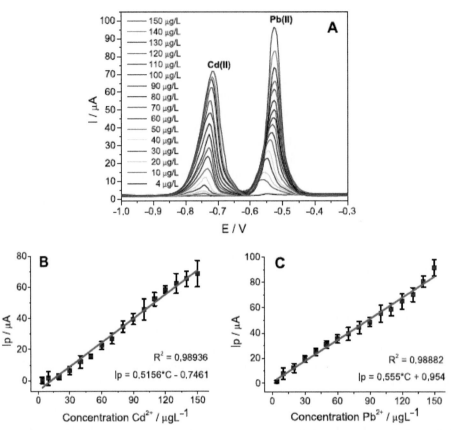

Figure 9.6: (A) SWASV response at MWCNTs-poly(DAN)/Pt for increasing concentrations of Cd(II) and Pb(II). Corresponding calibration plots for (B) Cd(II) and (C) Pb(II). Reprinted by permission Springer Nature Customer Service Centre GmbH: Springer Nature, Ionics, Vu, H.D., L.-H. Nguyen, T.D. Nguyen, H.B. Nguyen, T.L. Nguyen and D.L. Tran. 2015. Anodic stripping voltammetric determination of Cd^{2+} and Pb^{2+} using interpenetrated MWCNT/P1,5-DAN as an enhanced sensing interface. Copyright.

Cd(II) and Pb(II) using SWASV. Due to the synergistic effect of the sensing platform components, the obtained surface exhibited a propitious morphology for the improved efficiency of electrodeposition, accelerating the electron transfer rate for the determination of trace amounts of both HMIs. The DLs were 0.21 ppb and 0.13 ppb for Pb(II) and Cd(II), respectively. Besides, the sensor was further used for the detection of trace HMIs in soil samples with satisfactory results.

Li et al. 2016 synthesized and engineered MWCNTs via doping nitrogen atoms and grafting thiol groups into activated MWCNTs matrix. The engineered MWCNTs were immobilized at GCEs and used for the construction of an electrochemical sensor for the quantification of Cd(II) and Pb(II) by SWASV. Due to the introduction of nitrogen atoms and thiol groups, the engineered MWCNTs exhibited an improved dispersibility in distilled water by means of ultrasonic radiation. No significant signal changes were observed for the detection of Cd(II) and Pb(II) in the presence of Fe(III), Cu(II), Co(II), Ni(II), Zn(II), Mg(II), and Al(III), suggesting favorable and selective performance of the sensor for practical application. Furthermore, the developed electrode displayed good repeatability, stability, and reproducibility. This study demonstrated that the sensor based on engineered MWCNTs was very promising for potential application in the detection of heavy metal ions. However, the authors did not use it to study its analytical performance in real samples.

Gutierrez et al. 2017 focused on the development of an electrochemical sensor for the quantification of Cd(II) using GCEs modified with a dispersion of SWCNTs covalently functionalized with cysteine

(Cys). Cd (II) was preconcentrated at open circuit potential through the complex formation with Cys residues, minimizing the possibility of potential interference of other metallic cations. The combination of the SWCNTs electroactivity and the high affinity of Cys towards Cd(II) made possible the highly sensitive and selective quantification of Cd(II)-even in the presence of high concentrations of several heavy metal ions (Figure 9.7). The proposed sensor presented a DL of 0.3 ppb and a reproducibility of 1.7% using the same dispersion and 3.8% using 3 different dispersions. The SWCNTs-Cys/GCE sensor was successfully used for the quantification of Cd(II) in groundwater samples with excellent correlation with ICP-MS, demonstrating to be a very interesting alternative for further applications in environmental monitoring.

Huang et al. 2017 proposed a novel coordination matrix/signal amplifier strategy to construct an electrochemical sensor for the highly sensitive and simultaneous determination of Cd(II), Pb(II), Hg(II), and Cu(II) using a poly(furfural) film/MWCNTs-modified GCE. Poly(furfural) film possesses long conjugated π-electron backbones which can provide multiple π-electrons and can act as an excellent coordination matrix to capture HMIs. In this way, HMIs can be efficiently accumulated and deposited on the electrode surface where MWCNTs contribute, enhancing the anodic stripping current significantly. The poly(furfural) film/MWCNTs/GCE sensor provided low DLs for Cd(II) (0.03 ppb), Pb(II) (0.01 ppb), Cu(II) (0.06 ppb), and Hg(II) (0.1 ppb), which are much lower than the limits in drinking water suggested by WHO. Finally, the proposed sensor was successfully applied to simultaneously determine Cd(II), Pb(II), Cu(II), and Hg(II) in real tap water samples.

Another interesting contribution of Chamjangali et al. 2017 was to propose a continuous flow method for the simultaneous determination of the Cd(II) and Pb(II) using DPASV. The sensor was simply constructed by the immobilization of a homogeneous and uniform composite film made of MWCNTs and poly-alizarin red S (MWCNTs-PARS) on the surface of GCEs via *in situ* electro-polymerization (MWCNTs-PARS/GCE). The MWCNTs-PARS incorporation significantly enhanced the signals of Cd(II) and Pb(II) with respect to the bare GCE. This can be attributed to the high conductivity of the composite matrix and the chelating ability of the polymer film. The MWCNTs-PARS/GCE sensor showed a good sensitivity, acceptable reproducibility, and DL of 0.43 and 0.47 ppb for Cd(II) and Pb(II), respectively. Also, this sensor was successfully used for the simultaneous determinations of these HMIs in soil, wastewater, and tap water samples, which were in agreement with those obtained by the AAS method.

Cai et al. 2017 proposed a novel platform using metal-organic framework (MOF) which consists of a 2-D nanocomposite of flake-shaped manganese-terephthalic acid MOF/SWCNTs (Mn(TPA)-SWCNTs) deposited on GCEs, and it was used for the voltammetric determination of Pb(II). Through a theoretical modeling, the authors demonstrated that the Mn(TPA) component has stronger adsorption ability towards Pb(II) in comparison to other HMIs. This ability was in good agreement with the results displayed by the experimental electrochemical assays. The sensing platform was further modified with chitosan (Chit), which contributed to the efficient physical attachment of Mn(TPA)-SWCNT on GCE. Therefore, the Chit-Mn(TPA)-SWCNTs/GCE sensor evidenced the synergic effect of the structure and constitution of Mn(TPA), the electronic behavior of SWCNTs in terms of conductivity, and the additional efficient film-forming property of Chit that allowed a higher stability of the electrochemical signal. Under the optimal working conditions, this sensor presented a wide linear response from 0.1 to 14.0 mM and the DL was 38 nM. In addition, this platform possesses higher binding ability towards Pb(II), allowing its selective quantification even in the presence of several cations, such as Cu(II), Cd(II), Zn(II), Ni(II), Mg(II), Co(II), Ca(II), Mn(II), Al(III), and Fe(III). The proposed sensor was used for the recovery analysis of both diluted industrial wastewater and human serum samples, the percentages ranging between 99.5–103.0% and 98.1–106.8%, respectively, demonstrating the usefulness for practical application. The analytical response of the sensor was compared to ICP/MS measurements, giving a good agreement in terms of precision.

Another interesting approach for Pb(II) sensing was published by Tarley et al. 2017. The authors synthesized an ion-selective imprinted polymer (IIP) based on poly(methacrylic acid) loaded with

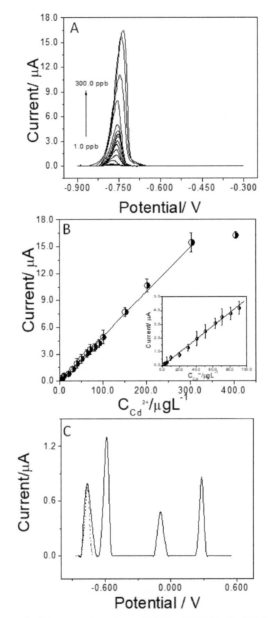

Figure 9.7: (A) LSASV response for different concentrations of Cd(II) at SWCNTs-Cys/GCE. (B) Corresponding calibration plot for Cd(II) obtained from the peak currents corresponding to the recordings. (C) LSV for Cd(II) in the absence (dashed line) and presence (solid line) of Pb(II), Hg(II), Co(II), Zn(II), Ni(II), Cr(III), As(III), Ir (IV), and Cu(II). Reprinted from Gutierrez, F.A., J.M. Gonzalez-Dominguez, A. Ansón-Casaos, J. Hernández-Ferrer, M.D. Rubianes, M.T. Martínez et al. 2017. Single-walled carbon nanotubes covalently functionalized with cysteine: A new alternative for the highly sensitive and selective Cd(II) quantification. Sens. Actuators B Chem. 249: 506–514. Copyright with permission from Elsevier.

chelating 1-2(-pyridylazo)-2-naphtol (PAN) in the presence of Pb(II). They proposed the construction of a nanocomposite based on IIP technology combining IIP-PAN with MWCNTs onto GCEs as a selective electrochemical sensing surface for Pb(II) in water samples. The electrochemical method started with a preconcentration of Pb(II) at open circuit potential for 20 minutes, followed by the reduction of Pb(II) at −1.2 V for 60 seconds, and the subsequent DPASV determination. The sensing surface was challenged with potential interfering metal ions with the same charge and similar radius and chemical properties as lead (Cu(II), Cd(II), Ni(II), Co(II), and Zn(II)). The imprinting effect for Pb(II) template was clearly observed since IIP had a specific size and shape, in addition to the availability of nitrogen and oxygen atoms present on the loaded PAN ligand, favoring the selective complex formation with Pb(II), which allowed its quantification even at trace levels. Under these working conditions, the DL was 0.16 ppb. The practical application of the developed sensor was investigated in recovery assays using water samples and synthetic urine with satisfactory recovery values ranging from 95 to 103%.

Recently, Ramírez et al. 2018 developed an electrochemical affinity sensor for Pb(II) quantification using a platform that combines GCEs and an aqueous dispersion of SWCNTs covalently modified with Cys. The biosensing protocol included the accumulation of Pb(II) at the electrode surface through the affinity interaction promoted by Cys residues at open circuit potential, followed by the reduction of the accumulated Pb(II) at –0.9 V, and the transduction step performed by linear sweep anodic stripping voltammetry (LSASV). Under the optimal working conditions, the SWCNTs-Cys/GCE sensor displayed a linear range from 5.0 to 125.0 ppb and a DL of 0.69 ppb. In addition, the selectivity of this sensor was evaluated in the presence of high concentrations of possible interferents, such as Cu(II), Cd(II), Ni(II), Hg(II), Rh(II), Ru(II), Zn(II), Ir(IV), Co(II), and As(III), demonstrating a high discrimination of Pb(II) in complex samples. The practical application of the biosensing platform was also evaluated with tap and rain water samples enriched with Pb(II), demonstrating its suitable application for environmental damage analysis and outstanding properties in terms of recovery percentages, showing an excellent agreement with ICP-MS analysis.

An original sensitive voltammetric method for detection of Cd(II) and Pb(II) was proposed by Dali et al. 2018 using a GCE modified with a mixture of carboxylated SWCNTs (cSWCNTs) and fungal biomass of *Trichoderma asperellum* used as a selective ligand for Cd(II) and Pb(II) (cSWCNTs-fungi/GCE). This fungus was selected from soils contaminated with metals. The cell walls of fungi have a large number of chemical groups, such as carboxyl, sulfonate, amine, hydroxyl, carbonyl, and imidazole, which mediate the capture of metal ions and therefore, allow their pre-concentration. It is interesting to remark that the sensitivity of the sensor was increased by the large surface area due to the presence of SWCNTs and consequently, a higher adsorption capacity of HMIs. The DLs were found to be 10^{-8} M and 10^{-7} M for Pb(II) and Cd(II), respectively, using DPASV. Real samples of ground water were analysed and quantified successfully, giving good recovery results (95.3–106.5%).

Jiang et al. 2018 used the novelty sodium pyrene-1,3,6,8-tetrasulfonate salt (PyTS) to functionalize CNTs (CNTs-PyTS) by a simple and easy sonochemical method using Nafion as a binder. CNTs-PyTS hosts active oxygenated groups, such as C=O, –OH, –COOH, and $-SO_3^{2-}$, which represents a three-dimensional network structure that can improve the electron transportation, and accumulates more HMIs at the electrode surface. The sensor was employed for the detection of Cd(II) and Pb(II), exhibiting good selectivity and sensitivity. Under the optimal condition, the stripping peak current of the CNTs-PyTS/Nafion/PGE reached DLs of 0.8 ppb for Cd(II) and 0.02 ppb for Pb(II). This sensor is a cost-effective platform for the facile and sensitive analysis of Cd(II) and Pb(II); however, high concentrations of Co(II), Fe(II), and Fe(III) had a negative influence on the analytical signal of Cd(II) and Pb(II), owing to their competing adsorption during the pre-accumulation step.

Xuan and Park 2018 fabricated a miniaturized, flexible, and fully integrated electrochemical micro-patterned sensor using a composite nanomaterial of reduced graphene oxide (rGO) and CNTs on a flexible gold substrate which was plated with Bi film. The Bi/rGO-CNTs/Au sensor exhibited well-defined and separate stripping peaks for Cd(II) and Pb(II) (Figure 9.8). rGO and CNTs contributed to the enlargement of the electrode surface area which, consequently, led to the enhancement of the analytical efficiency towards the target ions. Under optimal stripping conditions, high sensitivities and good DLs of 0.6 ppb and 0.2 ppb for Cd(II) and Pb(II) were obtained, respectively. This sensor was used for the detection of Cd(II) and Pb(II) in drinking water samples with recovery in the range of 92.0 to 104.4 percent. It is important to indicate that Cu(II) was found to reduce the response of

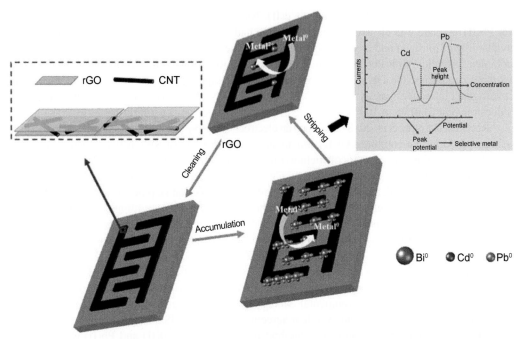

Figure 9.8: Schematic illustration of the sensing principle for detection of Cd(II) and Pb(II) at Bi/rGO-CNTs/Au. Reprinted from Xuan, X. and J.Y. Park. 2018. A miniaturized and flexible cadmium and lead ion detection sensor based on micro-patterned reduced graphene oxide/carbon nanotube/bismuth composite electrodes. Sens. Actuators B Chem. 255: 1220–1227. Copyright with permission from Elsevier.

target HMIs due to the competition between electroplating Bi and Cu on the electrode surface as a result of the close reduction potentials of Cu and Bi.

In the design of a sensing strategy for Cd(II), Ding et al. 2018 took advantage of water soluble glutathione capped zinc selenide quantum dots (GSH-ZnSeQDs) to develop composite films with xanthan gum (XG), which is a negatively charged polysaccharide. The surfaces of GCEs were modified through casting a thin layer of MWCNTs, followed by a complex layer of GSH-ZnSeQDs stabilized by XG. The attachment of the XG-GSH-ZnSeQDs complex to the MWCNTs greatly improved the electrocatalytic activity, stability, and binding capability resulting from larger surface area, higher porosity, and higher affinity via interaction with the carboxylic acid groups of the GSH, which led to a more sensitive detection of Cd(II). The sensor showed a DL of 2.2 ppb and a high selectivity for Cd(II) with negligible interference from other co-existing metal ions (Ag(I), Pb(II), Ni(II), Co(II), Fe(II), Zn(II), and Bi(III)). The applicability of the proposed method for Cd(II) detection was evaluated in lake water samples, which clearly demonstrated its reliability for environmental monitoring.

3.2 Mercury (Hg)

Mercury (Hg(II)) is considered one of the highly toxic HMIs which can be found in different chemical forms in the environment, travels long distances through the atmosphere and the oceans, bioaccumulates in microorganisms, and biomagnifies along the food chain (Beckers and Rinklebe 2017). Thus, Hg and its derivatives can cause a variety of toxic effects to plants, animals, and human health, including nephrotoxicity, teratogenicity, and damage to the cardiovascular system (Beckers and Rinklebe 2017, Kim et al. 2016, Mahbub et al. 2017). In this sense, the European Water Framework Directive classifies Hg as one of the 33 "priority pollutants" (EC 2000) and the World Health Organization (WHO) has established a limit of 1 ppb Hg in water for human consumption (WHO 2017). Due to the extensive use of this heavy metal in gold mining within the last decades, the environmental perturbations from anthropogenic Hg emissions have remained a major public health and safety concern (Esdaile and Chalker 2018). As aqueous Hg(II) cations are environmentally more persistent and can be transformed to methylated forms that are more bioaccumulative (ATSDR 2017, Mason et al. 2012), their rapid detection at low levels is critical for fast response and protection of biota and humans. Therefore, it is mandatory to develop sensing strategies for the recognition and detection at ultra-traces of Hg(II) in biological and environmental systems. In this sense, we present the most significant information about CNTs-based electrochemical sensors for Hg(II) and their analytical characteristics, summarized, in Table 9.1.

Mnyipika and Nomngongo 2017 developed an electrochemical sensor based on a GCE modified with MWCNTs-tin oxide nanoparticles (MWCNTs-SnO_2NPs) nanocomposite for the simultaneous determination of Hg(II) and Tl(I) using SWASV. The combination of both nanomaterials improved the electroactivity of the modified electrode compared to individual materials due to the high electrical conductivity and surface area of the resulting nanocomposite. The electrochemical sensor MWCNTs-SnO_2NPs/GCE showed a wide linear range of five orders of magnitude for both ions and very low DLs of 0.0012 ppb and 0.0009 ppb for Hg(II) and Tl(I), respectively. Nevertheless, Cd(II) is a great interference in the determination of Hg(II) and Tl(I).

Different chelating agents have also been used as modifiers combined with CNTs for the electrochemical determination of Hg(II). For example, a novel electrochemical sensor for the detection of Hg(II) was fabricated by Wei et al. 2014 via the immobilization of thiophenol-functionalized SWCNTs (SWCNTs-PhSH) onto AuE surfaces. The combination of the outstanding physicochemical properties of SWCNTs and the strong chelating ability of thiol groups to mercury ions allowed the selective and sensitive determination of Hg(II) by SWASV. After the mercury deposition at –0.1 V for 120 seconds, this electrochemical sensor exhibited a wide linearity range from 5.0 to 50 nM Hg(II), with a DL of 3.0 nM. Moreover, the SWCNTs-PhSH/AuE sensor showed a very good reproducibility (RSD 3.8%) and the interference from other heavy metal ions, such as Cr(II), Mn(II), Fe(II), Co(II), Ni(II), Cu(II), and Zn(II) ions on the stripping signal of Hg(II) was effectively inhibited. Regarding its practical application, the proposed sensor was successfully applied to determine Hg(II) in tap water, river water, and waste water samples with recovery ranging from 97.8% to 103.6 percent.

The novel modification of a SWCNT paste electrode (SWCNTPE) with a Zn/Al layered double hydroxide-3(4-methoxyphenyl)propionate (Zn/Al-LDH-MPP) nanocomposite was used by Isa et al. 2015 to prepare an electrochemical sensor for the determination of Hg(II) in real samples by CV. Although SWCNTs enhanced the conductivity of this sensor, the CVs for mercury ions at chemically modified SWCNTPE indicated that the presence of the nanocomposite in the paste was essential, since the redox process took place after the accumulation of Hg(II) at the electrode surface via complexation with Zn/Al-LDH-MPP. In this sense, the Hg(II) electrooxidation was found to be a surface adsorption-controlled process. Concerning the analytical performance, this sensor presented: (i) a linear range from 1.0 nM to 0.1 mM Hg(II) with a DL of 1 nM, (ii) a very good reproducibility (RSD 1.2%), (iii) a high selectivity, since Mg(II), Ca(II), Sr(II), Ba(II), Mn(II), Co(II), and Cd(II) ions showed no influence on the voltammetric response of Hg(II), and (iv) a good reliability to determine Hg(II) in food samples, such as fish, chili pepper, tomato, and gourd, with recoveries in the range of 98.2% to 99.4 percent.

The Schiff base compounds have proven to be very reactive towards Hg(II) ions and are considered excellent chelating agents due to the relative ease of preparation, synthetic flexibility, and the special property of C=N groups. CNTs have been combined with different Schiff base ligands to develop electrochemical sensors for Hg(II) quantification. For example, Gayathri et al. 2018 investigated a paraffin-impregnated graphite electrode (PIGE) modified with CNTs and the symmetric novel octadentate ligand N, N, N″, N‴-tetrasalicylidene-3,3′-diaminobenzidine (H4tsdb) for determination of Hg(II) in water samples by SWASV. Due to the combination of the conductivity of CNTs and the ability of hydroxyl and imine groups of H4tsdb ligand to coordinate Hg(II) in solution during the pre-concentration step, the modified electrode exhibited high sensitivity and selectivity for Hg(II) determination. Under optimized conditions, the electrochemical sensor showed a linear range from

2.4 to 220.0 nM, and a DL of 0.8 nM for Hg(II), with a very good reproducibility (RSD 3.0%) using four different electrodes, and a long-time stability of 3 weeks (RSD 3.5%). In addition, the proposed electrode showed an excellent selectivity with no influence on the signal of Hg(II) in the presence of Pb(II), Cd(II), Zn(II), Ni(II), Mn(II), and Cu(II). The same group (Selvan et al. 2018) proposed another electrochemical platform based on a PIGE modified with CNTs and the asymmetrical tetradentate Schiff base N,N'-bis(pyrrole-2-ylmethylene)-2-aminobenzylamine (pmaba) as ligand for the simultaneous ASV determination of Hg(II) and Pb(II). The modified electrode was used for the simultaneous determination of Hg(II) and Pb(II), and displayed higher peak current responses for lead and mercury compared to bare electrode. The anodic stripping peak currents obtained after 180 seconds of pre-concentration under open circuit condition and reduction by applying –0.8 V, showed DLs of 0.36 nM and 1.1 nM for Hg(II) and Pb(II), respectively, and linear ranges of 3.3–66.0 nM for both HMIs. Moreover, the asymmetric modified electrode showed good electrochemical stability and reusability, and it was successfully employed for the determination of Hg(II) and Pb(II) in spiked water samples, with average recovery between 101 and 105 percent.

Afkhami et al. 2014b also used a CNTs-modified CPE for the voltammetric determination of Hg(II) in both water and food samples. MWCNTs and the newly synthesized Schiff base "1,3,5-triaza-2,4,7,8,11,12,15,16-tetrabenzo-9,14 dioxacycloheptadeca-1,5-diene" were used as modifiers of CPEs, and the ionic liquid "1-butyl-1-methylpyrrolidinium bis(trifluoromethylsulfonyl)imide ([BMPyr] [NTf2])" was used as the binder. The new Schiff base is a macrocyclic chelating agent that allowed the selective detection of Hg(II) by DPASV. In addition, the synergistic effect of MWCNTs and ionic liquid present in the composite film further accelerated the electron transfer rate of mercury. The anodic peak current for different concentrations of Hg(II) after 120 seconds of accumulation at –0.8 V showed a wide linear dynamic range, between 0.2–260.0 nM, the limits of detection and quantification being 0.05 nM and 0.18 nM, respectively. In addition, the proposed voltammetric sensor offered attractive properties, such as high selectivity, sensitivity, and reproducibility, and it was used for the determination of Hg(II) in sea water and industrial waste water samples, as well as in food matrices, such as tuna fish, sausage, cream cheese, and raw milk.

Another CNTs-based potentiometric sensor for Hg(II) sensing was reported by Mashhadizadeh et al. 2015. Hereby, a chemically modified CPE was developed using MWCNTs to enhance the electrical conductivity and sensitivity of the composite electrode, and 1,3-bis(2-ethoxyphenyl) triazene (BEPT) as a selective ionophore to form stable complexes with Hg(II). A linear relationship potential versus logarithm of Hg(II) concentration was obtained in the range from 4.0 nM to 2.2 mM for the proposed sensor, with a DL of 3.1 nM. In addition, the authors also reported (i) no change in the potentiometric response after storage for about 30 days, (ii) a high performance for the selective detection of Hg(II) even in the presence of several interfering metal ions, such as Na(I), K(I), Mg(II), Ca(II), Ag(I), Cd(II), and Pb(II), and (iii) an acceptable recovery (99–102%) for the determination of mercury ions in some aqueous samples.

Shi et al. 2017 proposed an interesting DNA-biosensor based on AuEs modified with vertically aligned SWCNTs through self-assembled monolayers of 4-aminothiofenol (v-SWCNTs(R)), and a novel Hg(II)-recycling strategy using exonuclease III (Exo III) (Figure 9.9). In this strategy, report-probes with methylene blue tags (R-probes) were hybridized with amine-probes (A-probes) covalently immobilized to magnetic bead (MBs) through partial base pairing, leaving free thymine-thymine pairs (T-T) for Hg(II) recognition through the formation of stable thymine-Hg(II)-thymine (T-Hg(II)-T). The addition of Exo III initiated a nicking reaction, resulting in the dissociation of R-probes and the release of Hg(II), which was available for the formation of more T-Hg(II)-T structures and the beginning of a new nicking cycle. The R-probes were then adsorbed to the v-SWCNTs(R) through p-p interactions, and the resulting signal of methylene blue was used to measure Hg(II) concentration. This strategy allowed an extremely wide linear range of eight orders of magnitude (10 fM–1 mM) and an ultra low DL of 3 fM, with a good reproducibility obtained with independent experiments (RSD 3.4%) and a good long-time stability of the biosensors stored at 4°C with a signal change of 3.6%

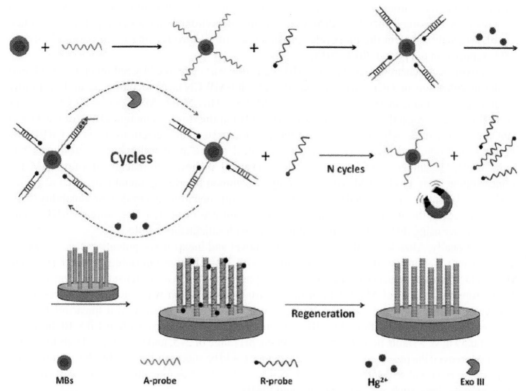

Figure 9.9: Design of an Exo III-assisted sensing scheme for Hg(II) detection based on v-SWCNTs(R). Republished with permission of Royal Society of Chemistry from Shi, L., Y. Wang, Z. Chu, Y. Yin, D. Jiang, J. Luo et al. 2017. A highly sensitive and reusable electrochemical mercury biosensor based on tunable vertical single-walled carbon nanotubes and a target recycling strategy. J. Mater. Chem. B 5: 1073–1080. Copyright permission conveyed through Copyright Clearance Center, Inc.

after 12 weeks. The excellent performance of the biosensor was attributed to the high binding capacity of thymine for Hg(II) and the greater amount of R-probes effectively bound to the v-SWCNTs(R). In addition, negligible interference was observed for the determination of Hg(II) in the presence of Co(II), Zn(II), Mg(II), Ca(II), Pb(II), Mn(II), Cu(II), Ni(II), and Cd(II). Satisfactory recoveries were obtained for lake water (from 98.0% to 102.9%) and human serum.

3.3 Chromium (Cr)

Chromium is another highly toxic heavy metal which is now considered a major inorganic pollutant due to its toxicity for ecological, nutritional, and environmental reasons (Mishra and Bharagava 2016). Chromium compounds are naturally occurring and widely distributed on earth. Due to extensive anthropogenic uses, they are found in significant amounts in industrial effluents related to electroplating, metal refining, leather tanning, and textile dyes (Silva et al. 2016). Cr is commonly found as Cr(III) and Cr(VI), which have significantly different physicochemical properties and toxicities. While Cr(III) at trace levels is essential for humans and animals since it is involved in the metabolism of sugars and fats, high concentration can trigger allergic skin reactions (Cohen et al. 1993, Eastmond et al. 2008). Cr(VI) is non-essential and highly toxic, and recognized as a human lung carcinogen by the United States Environmental Protection Agency (USEPA), which is also toxic to many plants, aquatic animals, and microorganisms (Cohen et al. 1993, Rakhunde et al. 2012). WHO 2017 and USEPA 2018 have recommended a drinking water standard of 50 ppb and 100 ppb for Cr(VI) and total chromium, respectively. Therefore, new analytical innovations are necessary to meet the ever-growing demand for chromium sensing in environmental water at ppb levels. In that regard, we analyze herein the most relevant strategies for electro-sensing of chromium based on CNTs reported in the period 2014–2018.

A novel electrochemical sensor for Cr(III) detection was developed by Salimi et al. 2015 based on the modification of GCEs with a thin film of Chit-MWCNTs nanocomposite and uniformly electrodeposited manganese oxide nanoflakes (MnOx). This MnOx/MWCNTs-Chit/GCE sensor showed high electrocatalytic ability for Cr(III) oxidation due to the synergic effect of CNTs and MnOx nanostructures, which acted as a suitable mediator to shuttle electrons between Cr(III) and working electrode, and facilitated the electrochemical regeneration following electron exchange with Cr(III). The amperometric detection of Cr(III) at 0.9 V allowed to reach a DL of 0.3 mM and a linear range from 3 to 84 mM with selectivity to common interfering metal ions, such as Fe(II), Co(II), Ni(II), Cu(II), Zn(II), Cd(II), Hg(II), and Pb(II), and good intra-electrode reproducibility (RSD 4.7%). Moreover, the applicability of the proposed sensor was evaluated determining Cr(III) in city drinking water using differential pulse voltammetry with satisfactory results.

Additionally, Alizadeh et al. 2017 reported a novel and inexpensive potentiometric sensor for Cr(III) determination based on a CPE modified with ion-selective imprinted polymer (IIP) and MWCNTs. Thus, the presence of the IIP as an ionophore resulted in a Cr(III)-selective sensor, while the appropriate amount of MWCNTs in the electrode composition was found to be necessary to observe Nernstian response. This Cr(III)-IIP/MWCNTs/CPE sensor showed a linear relationship between potential and logarithm of Cr(III) concentration in the range from 1.0 to 1.0×10^5 mM with a DL of 0.59 mM, a stable potential response in the pH range of 2–5, and good reproducibility (RSD 4.5%). Moreover, the proposed potentiometric device could be successfully used for the quantification of Cr(III) in river and sea water samples, as well as soil without significant interference of common divalent metal ions, such as Cu(II), Hg(II), and Pb(II).

Deep et al. 2014 developed a sensitive and selective electrochemical sensor for the determination of Cr(VI) via a simple covalent functionalization route of chlorinated SWCNTs (clSWCNTs) with bis(2,4,4-trimethylpentyl) phosphinic acid (BTMPPA), which is a widely used metal ion extractant. This sensor was constructed by spin-casting a clSWCNTs-BTMPPA dispersion on the boron doped p-type silicon wafer surface (clSWCNTs-BTMPPA/BD-SiE). The analytical performance of the proposed sensor was evaluated by amperometry at 0.2 V, obtaining a wide linear range of calibration for Cr(VI) from 0.01 to 10.00 ppb and a limit of quantification of 0.01 ppb. The availability of functional groups P=O and the retention of electronic properties of the BTMPPA-functionalized clSWCNTs allowed to obtain an electrochemical device for Cr(VI) sensing with high sensitivity (up to sub-ppb levels), a remarkable selective response (in the co-presence of Cr(III), Al(III), Fe(III), Mn(II), Ni(II), Cu(II), Zn(II), and Cd(II)), very good intra and inter-electrode reproducibility, and practical utility for the analysis of hexavalent chromium in both electroplating effluent and groundwater samples.

3.4 Arsenic (As)

Arsenic is a widespread element found in trace amounts, which has deeply influenced human history and has received considerable attention during the past 50 years due to its high harmfulness (Luong et al. 2014, Kaur et al. 2015, Kempahanumakkagari et al. 2017). Its high concentrations can be attributed to both geochemical processes and anthropogenic activities (Luong et al. 2014, Wu et al. 2016). In nature, As is present in both organic and inorganic forms, having different chemical species. The inorganic form is most toxic and could be accumulated in the exposed organisms. Hence, the main ions in terms of higher toxicity and greater mobility are arsenate (As(V)) and arsenite (As(III)). As(III) is significantly more toxic than As(V) due to its higher affinity for thiol groups of proteins, such as the enzymes of the human respiratory system (Liu and Huang 2014, Wu

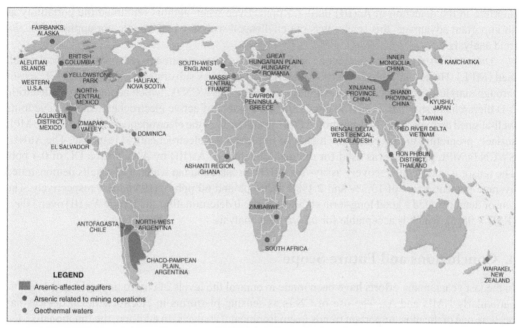

Figure 9.10: Distribution of documented world problems with As in groundwater in major aquifers, as well as water and environmental problems related to mining and geothermal sources. Reprinted from Smedley, P.L. and D.G. Kinniburgh. 2002. A review of the source, behavior, and distribution of arsenic in natural waters. Appl. Geochem. 17: 517–569. Copyright with permission from Elsevier.

Color version at the end of the book

et al. 2016, Kempahanumakkagari et al. 2017). Arsenic affects about 140–200 million people in 70 different countries on all six inhabited continents (Figure 9.10), mainly due to the contamination of drinking and ground waters (Luong et al. 2014, Antonova and Zakharova 2016). Intake of drinking water with high As content over a short period of time can cause diarrhea, nausea, muscle pain, and hyperkeratosis on the palm or feet. Long periods of ingesting drinking water with high concentrations of As have been associated with chronic arsenicosis, such as "blackfoot" disease, atherosclerosis, hypertension, and different cancer forms of bladder, kidney, liver, lung, and skin (Das Sarma et al. 2017). Additionally, As alters the immune system and it is linked to birth defects, serious reproductive problems, and cardiovascular diseases (Tchounwou et al. 2012, Gumpu et al. 2015, Chen et al. 2013). In this sense, WHO 2017 and USEPA 2018 have specified that the maximum As concentration in drinking water is 10 ppb. Therefore, due to its high poisonous power and harmful effects, accurate, sensitive, and selective analysis of As in ground and drinking waters is of importance to identify the contaminated source, the efficiency of water treatment, as well as to provide public health regulatory agencies with pertinent important information.

In the last 5 years, only two As sensing electrochemical strategies using CNTs by ASV have been published. The first one was reported by Gamboa et al. 2014 designing an original low cost strategy based on a MWCNTs/gold nanoparticles (AuNPs)-modified vibrating screen printed electrode for the determination of As(III). MWCNTs provided the surface for the attachment of AuNPs, while the AuNPs contributed to the electrochemical response of As. The innovation was based on a vibrating motor extracted from an iPhone, which was attached to a commercial screen printed in order to create a portable and autonomous system with enhanced mass transfer during the preconcentration step. Therefore, the main advantages of this system are reflected in its portability, low energy consumption (6 V), and internal control of convection, which led to more accurate results. After 120 seconds of deposition time, the DL was 0.5 ppb, while the limit of quantification was 1.5 ppb. Cu(II) does not interfere in the detection of As(III) until 1.5 ppm. Even when authors remarked the portability as an important advantage, this sensor was not challenged with any real sample nor applied to *in situ* field analysis.

Duy et al. 2016 proposed the second approach for As determination based on a mechanical pencil lead (MPL). Here, constant size carbon material was used as substrate to construct a MPL sensor through simple electrodeposition of Au in the presence of SWCNTs, which adopted a nanodendritic 3-D hierarchical structure (AuND), increasing the number of active electrochemical sites due to its well-aligned terraces. SWCNTs promoted the firm adhesion of the electrodeposited AuND to the MPL surface, promoting the ruggedness which also improved the electrochemical response. The AuND/SWCNTs/MPL platform was used for the quantification of As(III) by ASV with a DL of 0.4 ppb. The sensor was tested in recovery assays for As(III) prepared in tap water. The results demonstrated average percent errors of 10.3% and 2.1% for 20 ppb and 60 ppb As(III) added, respectively. The sensor demonstrated a good long-term stability up to 50 determinations of 10 ppb As(III) over 3 days (RSD 8.98%), which is acceptable for *in situ* field analysis.

4. Conclusions and Future Scope

In the last years, many efforts have been made to control the levels of environmental hazard agents, particularly HMIs and As. The use of CNTs as sensing platforms in electrochemical devices has become one of the most important trends in environmental analysis. In addition, the functionalization of CNTs, to furnish them with chemical groups, is a crucial topic in the improvement of specific interaction and efficient preconcentration of HMIs and As.

However, despite the high number of publications regarding the outstanding performance of CNTs-based electrochemical sensors and biosensors, the construction of advanced, highly selective, accurate, portables and integrated detection devices for in-field applications are scarce. Unfortunately, these sensing devices still remain laboratory prototypes in preliminary stages quite far away from commercial applications. In this sense, the main critical drawbacks to be solved for in-field applications of sensors are robustness, long-term stability, ease of construction, and surface reutilization. Probably, these challenges could be overcome with an intense collaborative effort between science and engineering disciplines to design and fabricate competitive and easy to use sensors for remote real-time monitoring of environmental hazard agents. Therefore, CNTs-based electrochemical sensing systems hold a formidable and promising future towards the sensitive detection of environmental analytes.

Acknowledgments

This work was partly supported by SECyT-UNC, CONICET and ANPCyT.

References

Abraham, M.R. and T.B. Susan. 2017. Water contamination with heavy metals and trace elements from Kilembe copper mine and tailing sites in Western Uganda; implications for domestic water quality. Chemosphere 169: 281–287.

Afkhami, A., F. Soltani-Felehgari, T. Madrakian, H. Ghaedi and M. Rezaeivala. 2013. Fabrication and application of a new modified electrochemical sensor using nanosilica and a newly synthesized Schiff base for simultaneous determination of Cd^{2+}, Cu^{2+} and Hg^{2+} ions in water and some foodstuff samples. Anal. Chim. Acta 771: 21–30.

Afkhami, A., H. Khoshsafar, H. Bagheri and T. Madrakian. 2014a. Construction of a carbon ionic liquid paste electrode based on multi-walled carbon nanotubes-synthesized Schiff base composite for trace electrochemical detection of cadmium. Mater. Sci. Eng. C 35: 8–14.

Afkhami, A., H. Khoshsafar, H. Keypour, H. Zeynali and T. Madrakian. 2014b. Novel sensor fabrication for the determination of nanomolar concentrations of Hg^{2+} in some foods and water samples based on multi-walled carbon nanotubes/ionic liquid and a new Schiff base. Food Anal. Methods 7: 1204–1212.

Ajayan, P.M. 1999. Nanotubes from carbon. Chem. Rev. 99: 1787–1800.

Ali, A.S.M., H.A.M. Ahmed, H.A.E. Emara, M.N. Janjua and N. Alhafez. 2019. Estimation and bio-availability of toxic metals between soils and plants. Polish J. Environ. Stud. 28: 15–24.

Alizadeh, T., S. Mirzaee and F. Rafiei. 2017. All-solid-state Cr(III)-selective potentiometric sensor based on Cr(III)-imprinted polymer nanomaterial/MWCNTs/carbon nanocomposite electrode. Int. J. Environ. Anal. Chem. 97: 1283–1297.

Antonova, S. and E. Zakharova. 2016. Inorganic arsenic speciation by electroanalysis. From laboratory to field conditions: A mini-review. Electrochem. Comm. 70: 33–38.

Aragay, G., J. Pons and A. Merkoci. 2011. Recent trends in macro-, micro-, and nanomaterial-based tools and strategies for heavy-metal detection. Chem. Rev. 111: 3433–3458.

Aragay, G. and A. Merkoci. 2012. Nanomaterials application in electrochemical detection of heavy metals. Electrochim. Acta 84: 49–61.

Ata, M.S., R. Poon, A.M. Syed, J. Milne and I. Zhitomirsky. 2018. New developments in non-covalent surface modification dispersion and electrophoretic deposition of carbon nanotubes. Carbon 130: 584–598.

ATSDR (Agency for Toxic Substances and Disease Registry). 2017. Summary data for 2017 priority list of hazardous substances. Available at: http://www.atsdr.cdc.gov/spl. (Website accessed July 2018).

Bagal-Kestwal, D., M.S. Karve, B. Kakade and V.K. Pillai. 2008. Invertase inhibition based electrochemical sensor for the detection of heavy metal ions in aqueous system: application of ultra-microelectrode to enhance sucrose biosensor's sensitivity. Biosens. Bioelectron. 24: 657–664.

Bansod, B.K., T. Kumar, R. Thakur, S. Rana and I. Singh. 2017. A review on various electrochemical techniques for heavy metal ions detection with different sensing platforms. Biosens. Bioelectron. 94: 443–455.

Barbosa, F., F.J. Krug and É.C. Lima. 1999. On-line coupling of electrochemical preconcentration in tungsten coil electrothermal atomic absorption spectrometry for determination of lead in natural waters. Spectrochim Acta B 54: 1155–1166.

Bard, A.J. and L.R. Faulkner. 2001. Electrochemical Methods: Fundamentals and Applications. 2nd edition. John Wiley & Sons, Inc. New York, USA.

Beckers, F. and J. Rinklebe. 2017. Cycling of mercury in the environment: Sources, fate, and human health implications: A review. Crit. Rev. Environ. Sci. Technol. 47: 693–794.

Blase, X., L.X. Benedict, E.L. Shirley and S.G. Louie. 1994. Hybridization effects and metallicity in small radius carbon nanotubes. Phys. Rev. Lett. 72: 1878–1881.

Borgo, S.D., V. Jovanovski, B. Pihlar and S.B. Hocevar. 2015. Operation of bismuth film electrode in more acidic medium. Electrochim. Acta 155: 196–200.

Büyükpınar, Ç., B. Bekar, E. Maltepe, D.S. Chormey, F. Turak, N. San et al. 2018. Development of a sensitive closed batch vessel hydride generation atomic absorption spectrometry method for the determination of cadmium in aqueous samples. Instrum. Sci. Technol. 46: 645–655.

Cai, F., Q. Wang, X. Chen, W. Qiu, F. Zhan, F. Gao et al. 2017. Selective binding of Pb^{2+} with manganese-terephthalic acid MOF/SWCNTs: Theoretical modeling, experimental study and electroanalytical application. Biosens. Bioelectron. 98: 310–316.

Caroli, S., G. Forte, A.L. Iamiceli and B. Galoppi, 1999. Determination of essential and potentially toxic trace elements in honey by inductively coupled plasma-based techniques. Talanta 50: 327–336.

Chamjangali, M.A., H. Kouhestani, F. Masdarolomoor and H. Daneshinejad. 2015. A voltammetric sensor based on the glassy carbon electrode modified with multi-walled carbon nanotube/poly(pyrocatechol violet)/bismuth film for determination of cadmium and lead as environmental pollutants. Sens. Actuators B Chem. 216: 384–393.

Chamjangali, M.A., S. Boroumand, G. Bagherian and N. Goudarzi. 2017. Construction and characterization a non-amalgamation voltammetric flow sensor for online simultaneous determination of lead and cadmium ions. Sens. Actuators B Chem. 253: 124–136.

Chen, Y., F. Wu, M. Liu, F. Parvez, V. Slavkovich and M. Eunus. 2013. A prospective study of arsenic exposure, arsenic methylation capacity, and risk of cardiovascular disease in Bangladesh. Environ. Health Perspect. 121: 832–838.

Chen, Y., X. Xu, Z. Zeng, X. Lin, Q. Qin and X. Huo. 2019. Blood lead and cadmium levels associated with hematological and hepatic functions in patients from an e-waste-polluted area. Chemosphere 220: 531–538.

Cohen, M.D., B. Kargacin, C.B. Klein and M. Costa. 1993. Mechanisms of chromium carcinogenicity and toxicity. Crit. Rev. Toxicol. 23: 255–281.

Cui, L., J. Wu and H. Ju. 2015. Electrochemical sensing of heavy metal ions with inorganic, organic and bio-materials. Biosens. Bioelectron. 63: 276–286.

Cui, H., W. Guo, L. Jin, Q. Guo and S. Hu. 2017. Direct speciation of Cr in drinking water by *in-situ* thermal separation ETAAS. Anal. Methods 9: 1307–1312.

da Costa Silva, L.M., A. Melo and A.M. Salgado. 2011. Biosensors for environmental applications. pp. 1–16. *In*: V. Somerset (ed.). Environmental Biosensors. IntechOpen, Cape Peninsula, South Africa. http://www. intechopen.com/books/environmental-biosensors/biosensor-for-environmental-applications.

Dali, M., K. Zinoubi, A. Chrouda, S. Abderrahmane, S. Cherrad and N. Jaffrezic-Renault. 2018. A biosensor based on fungal soil biomass for electrochemical detection of lead (II) and cadmium (II) by differential pulse anodic stripping voltammetry. J. Electroanal. Chem. 813: 9–19.

Das Sarma, S., A. Hussain and J. Das Sarma. 2017. Advances made in understanding the effects of arsenic exposure on humans. Curr. Sci. 112: 2008–2015.

Daşbaşı, T., Ş. Saçmacı, N. Çankaya and C. Soykan. 2016. A new synthesis, characterization and application chelating resin for determination of some trace metals in honey samples by FAAS. Food Chem. 203: 283–291.

Deep, A., A.L. Sharma, S.K. Tuteja and A.K. Paul. 2014. Phosphinic acid functionalized carbon nanotubes for sensitive and selective sensing of chromium (VI). J. Hazard Mater. 278: 559–565.

Del Rio, M., J. Alvarez, T. Mayorga, S. Dominguez and C. Sobin. 2017. A comparison of arsenic exposure in young children and home water arsenic in two rural West Texas communities. BMC Public Health 17: 850. doi 10.1186/s12889-017-4808-4.

Ding, Y., X. Hao, H. Yin, I.L. Kyratzis, S. Shen, K. Sun et al. 2018. Ultrasensitive and selective detection of Cd(II) using ZnSe-xanthan gum complex/CNT modified electrodes. Electroanalysis 30: 877–885.

Dongwu, L., C. Zhiwei, X. Hongzhi and D. Xifeng. 2010. Bioaccumulation of lead and the effects of lead on catalase activity, glutathione levels and chlorophyll content in the leaves of wheat. Commun. Soil Sci. Plant Anal. 41: 935–944.

Dresselhaus, M.S., G. Dresselhaus and A. Jorio. 2004. Unusual properties and structure of carbon nanotubes. Annu. Rev. Mater. Res. 34: 247–278.

Duy, P.K., J.-R. Sohn and H. Chung. 2016. A mechanical pencil lead-supported carbon nanotube/Au nanodendrite structure as a electrochemical sensor for As(III) detection. Analyst 141: 5879–5885.

Eastmond, D.A., J.T. MacGregor and R.S. Slesinski. 2008. Trivalent chromium: assessing the genotoxic risk of an essential trace element and widely used human and animal nutritional supplement. Crit. Rev. Toxicol. 38: 173–190.

EC. 2000. Directive 2000/60/EC of the European Parliament and of the Council. 23 October 2000 EU Water Framework Directive. Waternotes 1–10.

Esdaile, L.J. and J.M. Chalker. 2018. The mercury problem in artisanal and small-scale gold mining. Chem. Eur. J. 24: 6905–6916.

Fang, X., B. Zong and S. Mao. 2018. Metal-organic framework-based Sensors for environmental contaminant sensing. Nano-Micro Lett. 10: 64. doi: doi.org/10.1007/s40820-018-0218-0.

Farzin, L., M. Shamsipur, L. Samandari and S. Sheibani. 2018. Advances in the design of nanomaterial-based electrochemical affinity and enzymatic biosensors for metabolic biomarkers: A review. Microchim. Acta 185: 276. doi: 10.1007/s00604-018-2820-8.

Feldmann, J., P. Salaun and E. Lombi. 2009. Critical review perspective: Elemental speciation analysis methods in environmental chemistry—moving towards methodological integration. Environ. Chem. 6: 275–289.

Fiorentini, E.F., B.V. Canizo and R.G. Wuilloud. 2019. Determination of As in honey samples by magnetic ionic liquid-based dispersive liquid-liquid microextraction and electrothermal atomic absorption spectrometry. Talanta 198: 146–153.

Flora, S.J.S., M. Mittal and A. Mehta. 2008. Heavy metal induced oxidative stress & its possible reversal by chelation therapy. Indian J. Med. Res. 128: 501–523.

Flora, S.J.S. 2009. Structural, chemical and biological aspects of antioxidants for strategies against metal and metalloid exposure. Oxid. Med. Cell Longev. 2: 191–206.

Gamboa, J.C.M., L. Cornejo and J.A. Squella. 2014. Vibrating screen printed electrode of gold nanoparticle-modified carbon nanotubes for the determination of arsenic (III). J. Appl. Electrochem. 44: 1255–1260.

Gao, C., Z. Guo, J.H. Liu and X.J. Huang. 2012. The new age of carbon nanotubes: An updated review of functionalized carbon nanotubes in electrochemical sensors. Nanoscale 4: 1948–1963.

Gao, C., X.Y. Yu, S.Q. Xiong, J.-H. Liu and X.J. Huang. 2013. Electrochemical detection of arsenic(III) completely free from noble metal: Fe_3O_4 microspheres-room temperature ionic liquid composite showing better performance than gold. Anal. Chem. 85: 2673–2680.

Gayathri, J., K.S. Selvan and S.S. Narayanan. 2018. Fabrication of carbon nanotube and synthesized Octadentate ligand modified electrode for determination of Hg (II) in Sea water and Lake water using square wave anodic stripping voltammetry. Sens. Biosensing Res. 19: 1–6.

Gemma, F., M.L. Juan, B. Ana and D.L. Jose. 2006. Daily intake of arsenic, cadmium, mercury, and lead by consumption of edible marine species. J. Agric Food Chem. 54: 6106–6112.

Ghosh, D., S. Giri, S. Kalra and C.K. Das. 2012. Synthesis and characterisations of TiO coated multiwalled carbon nanotubes/graphene/polyaniline nanocomposite for supercapacitor applications. Open J. Appl. Sci. 2: 70–77.

Gong, T., J. Liu, X. Liu, J. Liu, J. Xiang and Y. Wu. 2016. A sensitive and selective platform based on CdTe QDs in the presence of L-cysteine for detection of silver, mercury and copper ions in water and various drinks. Food Chem. 213: 306–312.

Guascito, M.R., C. Malitesta, E. Mazzotta and A. Turco. 2008. Inhibitive determination of metal ions by an amperometric glucose oxidase biosensor. Sens. Actuators B Chem. 131: 394–402.

Gumpu, M.B., S. Sethuramanb, U.M. Krishnanb and J.B.B. Rayappana. 2015. A review on detection of heavy metal ions in water—An electrochemical approach. Sens. Actuators B Chem. 213: 515–533.

Gupta, S., C.N. Murthy and C.R. Prabha. 2018. Recent advances in carbon nanotube based electrochemical biosensors. Int. J. Biol. Macromol. 108: 687–703.

Gutierrez, F.A., J.M. Gonzalez-Dominguez, A. Ansón-Casaos, J. Hernández-Ferrer, M.D. Rubianes, M.T. Martínez et al. 2017. Single-walled carbon nanotubes covalently functionalized with cysteine: A new alternative for the highly sensitive and selective Cd(II) quantification. Sens. Actuators B Chem. 249: 506–514.

Hu, P.A., J. Zhang, L. Li, Z. Wang, W. O'Neill and P. Estrela. 2010. Carbon nanostructure-based field-effect transistors for label-free chemical/biological sensors. Sensors 10: 5133–5159.

Huang, H., T. Chen, X. Liu and H. Ma. 2014. Ultrasensitive and simultaneous detection of heavy metal ions based on three-dimensional graphene-carbon nanotubes hybrid electrode materials. Anal. Chim. Acta 852: 45–54.

Huang, J., S. Bai, G. Yue, W. Cheng and L. Wang. 2017. Coordination matrix/signal amplifier strategy for simultaneous electrochemical determination of cadmium(II), lead(II), copper(II), and mercury(II) ions based on polyfurfural film/multi-walled carbon nanotube modified electrode. RSC Adv. 7: 28556–28563.

Ihsanullah, A., Abbas, A.M. Al-Amer, T. Laoui, M.J. Al-Marri, M.S. Nasser et al. 2016. Heavy metal removal from aqueous solution by advanced carbon nanotubes: Critical review of adsorption applications. Sep. Purif. Technol. 157: 141–161.

Iijima, S. 1991. Helical microtubules of graphitic carbón. Nature 354: 56–58.

Isa, I.M., S.N.M. Sharif, N. Hashim and S.A. Ghani. 2015. Amperometric determination of nanomolar mercury(II) by layered double nanocomposite of zinc/aluminium hydroxide-3(4-methoxyphenyl)propionate modified single-walled carbon nanotube paste electrode. Ionics 21: 2949–2958.

Islam, M.S., M.K. Ahmed, M. Raknuzzaman, M.H.-Al-Mamun and M.K. Islam. 2015. Heavy metal pollution in surface water and sediment: A preliminary assessment of an urban river in a developing country. Ecol. Indic. 48: 282–291.

Järup, L. 2003. Hazards of heavy metal contamination. Br. Med. Bull. 68: 167–182.

Jena, B.K. and C.R. Raj. 2008. Gold nanoelectrode ensembles for the simultaneous electrochemical detection of ultratrace arsenic, mercury, and copper. Anal. Chem. 80: 4836–4844.

Jiang, R., N. Liu, S. Gao, X. Mamat, Y. Su, T. Wagberg et al. 2018. A facile electrochemical sensor based on PyTS–CNTs for simultaneous determination of cadmium and lead ions. Sensors 18: 1567. doi:10.3390/s18051567.

Jonasson, M.E. and R. Afshari. 2018. Historical documentation of lead toxicity prior to the 20th century in English literature. Hum. Exp. Toxicol. 37: 775–788.

Kaur, H., R. Kumar, J.N. Babu and S. Mittal. 2015. Advances in arsenic biosensor development—A comprehensive review. Biosens. Bioelectron. 63: 533–545.

Kempahanumakkagari, S., A. Deep, K.-H. Kim, S.K. Kailasa and H.-O. Yoon. 2017. Nanomaterial-based electrochemical sensors for arsenic—A review. Biosens. Bioelectron. 95: 106–116.

Kenawy, I.M.M., M.A.H. Hafez, M.A. Akl and R.R. Lashein. 2000. Determination by AAS of some trace heavy metal ions in some natural and biological samples after their preconcentration using newly chemically modified chloromethylated polystyrene-PAN ion-exchanger. Anal. Sci. 16: 493–500.

Kim, K.-H., E. Kabir and S.A. Jahan. 2016. A review on the distribution of Hg in the environment and its human health impacts. J. Hazard Mater. 306: 376–385.

Kim, Y., B.N. Kim, Y.C. Hong, M.S. Shin, H.J. Yoo, J.W. Kim et al. 2009. Co-exposure to environmental lead and manganese affects the intelligence of school-aged children. Neurotoxicology 30: 564–571.

Koelmel, J. and D. Amarasiriwardena. 2012. Imaging of metal bioaccumulation in Hayscented fern (Dennstaedtia punctilobula) rhizomes growing on contaminated soils by laser ablation ICP-MS. Environ. Pollut. 168: 62–70.

Kumar, S., U.J. Mehta and S. Hazra. 2008. Accumulation of cadmium in growing peanut (Arachis hypogaea L.) seedlings—Its effect on lipid peroxidation and on the antioxidative enzymes catalase and guaiacol peroxidase. J. Plant Nutr. Soil Sci. 171: 440–447.

Kumar, S.S., P. Muthuselvam, V. Pugalenthi, N. Subramanian, K.M. Ramkumar, T. Suresh et al. 2018. Toxicoproteomic analysis of human lung epithelial cells exposed to steel industry ambient particulate matter (PM) reveals possible mechanism of PM related carcinogenesis. Environ. Pollut. 239: 483–492.

Li, M., H. Gou, I. Al-ogaidi and N. Wu. 2013. Nanostructured sensors for detection of heavy metals: A review. ACS Sustain Chem. Eng. 1: 713–723.

Li, X., H. Zhou, C. Fu, F. Wang, Y. Ding and Y. Kuang. 2016. A novel design of engineered multi-walled carbon nanotubes material and its improved performance in simultaneous detection of Cd(II) and Pb(II) by square wave anodic stripping voltammetry. Sens. Actuators B Chem. 236: 144–152.

Li, X., J. Ping and Y. Ying. 2019. Recent developments in carbon nanomaterial-enabled electrochemical sensors for nitrite detection. TrAC—Trends Anal. Chem. 113: 1–12.

Li, Y.-H., J. Ding, Z. Luan, Z. Di, Y. Zhu, C. Xu et al. 2003. Competitive adsorption of Pb^{2+}, Cu^{2+} and Cd^{2+} ions from aqueous solutions by multiwalled carbon nanotubes. Carbon 41: 2787–2792.

Liao, J., J. Chen, X. Ru, J. Chen, H. Wu and C. Wei. 2017. Heavy metals in river surface sediments affected with multiple pollution sources, South China: Distribution, enrichment and source apportionment. J. Geochemical Explor. 176: 9–19.

Lin, X., Z. Lu, W. Dai, B. Liu, Y. Zhang, J. Li et al. 2018. Laser engraved nitrogen-doped graphene sensor for the simultaneous determination of Cd(II) and Pb(II). J. Electroanal. Chem. 28: 41–49.

Liu, Z.-G. and X.-J. Huang. 2014. Voltammetric determination of inorganic arsenic. Trends Anal. Chem. 60: 25–35.

Losev, V.N., O.V. Buyko, A.K. Trofimchuk and O.N. Zuy. 2015. Silica sequentially modified with polyhexamethylene guanidine and arsenazoI for preconcentration and ICP-OES determination of metals in natural waters. Microchem. J. 123: 84–89.

Lu, Y., X. Liang, C. Niyungeko, J. Zhou, J. Xua and G. Tiana. 2018. A review of the identification and detection of heavy metal ions in the environment by voltammetry. Talanta 178: 324–338.

Luong, J.H.T., E. Lam and K.B. Male. 2014. Recent advances in electrochemical detection of arsenic in drinking and ground waters. Anal. Methods 6: 6157–6169.

Mahbub, K.R., K. Krishnan, R. Naidu, S. Andrews and M. Megharaj. 2017. Mercury toxicity to terrestrial biota. Ecol. Ind. 74: 451–462.

Mallkpour, S. and S. Soltanian. 2016. Surface functionalization of carbon nanotubes: Fabrication and applications. RSC Adv. 6: 109916–109935.

Maratta, A., B. Carrizo, V. Bazán, G. Villafañe, L.D. Martínez and P. Pacheco. 2018. Antimony speciation analysis by hydride trapping on hybrid nanoparticles packed in a needle trap device with electro-thermal atomic absorption. J. Anal. At. Spectrom. 33: 2195–2202.

Marchesan, S., K. Kostarelos, A. Bianco and M. Prato. 2015. The winding road for carbon nanotubes in nanomedicine. Mater. Today 18: 12–19.

Mashhadizadeh, M.H., S. Ramezani and M.K. Rofouei. 2015. Development of a novel MWCNTs–triazene-modified carbon paste electrode for potentiometric assessment of Hg(II) in the aquatic environments. Mater. Sci. Eng. C 47: 273–280.

Mason, R.P., A.L. Choi, W.F. Fitzgerald, C.R. Hammerschmidt, C.H. Lamborg, A.L. Soerensen et al. 2012. Mercury biogeochemical cycling in the ocean and policy implications. Environ. Res. 119: 101–117.

Massadeh, A.M., A.A. Alomary, S. Mir, F.A. Momani, H.I. Haddad and Y.A. Hadad. 2016. Analysis of Zn, Cd, As, Cu, Pb, and Fe in snails as bioindicators and soil samples near traffic road by ICP-OES. Environ. Sci. Pollut. Res. 23: 13424–13431.

Mishra, S. and R.N. Bharagava. 2016. Toxic and genotoxic effects of hexavalent chromium in environment and its bioremediation strategies. J. Environ. Sci. Health C Environ. Carcinog. Ecotoxicol. Rev. 34: 1–32.

Mnyipika, S.H. and M.P. Nomngongo. 2017. Square wave anodic stripping voltammetry for simultaneous determination of trace Hg(II) and Tl(I) in surface water samples using SnO_2@MWCNTs modified glassy carbon electrode. Int. J. Electrochem. Sci. 12: 4811–4827.

Nasir, S., M.Z. Hussein, Z. Zainal and N.A. Yusof. 2018. Carbon-based nanomaterials/allotropes: A glimpse of their synthesis, properties and some applications. Materials 11: 1–24.

Omanović, D., C. Garnier, K. Gibbon-Walsh and I. Pizeta. 2015. Electroanalysis in environmental monitoring: Tracking trace metals—A mini review. Electrochem. Commun. 61: 78–83.

ONU. 2017. The Sustainable Development Goals Report, United Nations. 1–56. doi:10.18356/3405d09f-en.

Osteryoung, J.G. and R.A. Osteryoung. 1985. Square wave voltammetry. Anal. Chem. 57(1): 101A–110A.

Paixão, L.B., G.C. Brandão, R. Geovanny, O. Araujo, M. Graças and A. Korn. 2019. Assessment of cadmium and lead in commercial coconut water and industrialized coconut milk employing HR-CS GF AAS. Food Chem. 284: 259–263.

Pan, S., L. Lin, F. Zeng, J. Zhang, G. Dong, B. Yang et al. 2018. Effects of lead, cadmium, arsenic, and mercury co-exposure on children's intelligence quotient in an industrialized area of southern China. Environ. Pollut. 235: 47–54.

Patrick, L. 2006. Lead toxicity. Part II: The role of free radical damage and the use of antioxidants in the pathology and treatment of lead toxicity. Altern. Med. Rev. 11: 114–127.

Pei, X., W. Kang, W. Yue, A. Bange, W.R. Heineman and I. Papautsky. 2014. Improving reproducibility of lab-on-a-chip sensor with bismuth working electrode for determining Zn in serum by anodic stripping voltammetry. J. Electrochem. Soc. 161: B3160–B3166.

Pletcher, D., R. Greff, R. Peat, L.M. Peter, J. Robinson, Southampton Electrochemistry Group. 2011. Instrumental Methods in Electrochemistry. Woodhead Publishing Limited, New Delhi, India.

Pohl, P. 2009. Determination of metal content in honey by atomic absorption spectroscopy. Trends Anal. Chem. 28: 117–128.

Prabhakar, S., A.K. Singh and D.S. Pooni. 2012. Effect of environmental pollution on animal and human health: A review. Ind. J. Anim. Sci. 82: 244–255.

Pujol, L., D. Evrard, K.G. Serrano, M. Freyssinier, A. Ruffien-Cizsak and P. Gros. 2014. Electrochemical sensors and devices for electrochemical assay in water: The French groups' contribution. Front Chem. 2: 19. doi: 10.3389/fchem.2014.00019.

Rajaganapathy, X. and M.P. Sreekumar. 2011. Heavy metal contamination in soil, water and fodder and their presence in livestock and products: A review. J. Environ. Sci. Technol. 4: 234–249.

Rakhunde, R., L. Deshpande and H.D. Juneja. 2012. Chemical speciation of chromium in water: A review. Crit. Rev. Environ. Sci. Technol. 42: 776–810.

Ramírez, M., C. Tettamanti, F. Gutierrez, J. Gonzalez-Domínguez, A. Ansón-Casaos, J. Hernández-Ferrer et al. 2018. Cysteine functionalized bio-nanomaterial for the affinity sensing of Pb(II) as an indicator of environmental damage. Microchem. J. 141: 271–278.

Ramnani, P., N.M. Saucedo and A. Mulchandani. 2016. Carbon nanomaterial-based electrochemical biosensors for label-free sensing of environmental pollutants. Chemosphere 143: 85–98.

Rattanarat, P., W. Dungchai, D. Cate, J. Volckens, O. Chailapakul and C.S. Henry. 2014. Multilayer paper-based device for colorimetric and electrochemical quantification of metals. Anal. Chem. 86: 3555–3562.

Rivas, G.A., M.D. Rubianes, M.C, Rodríguez, N.F. Ferreyra, G.L. Luque, M.L. Pedano et al. 2007. Carbon nanotubes for electrochemical biosensing. Talanta 74: 291–307.

Rivas, G.A., M.C. Rodríguez, M.D. Rubianes, F.A. Gutierrez, M. Eguílaz, P.R. Dalmasso et al. 2017. Carbon nanotubes-based electrochemical (bio)sensors for biomarkers. Appl. Mater. Today 9: 566–588.

Rovasi Adolfo, F., P. Cícero do Nacimento, G. Camera Leal, D. Bohrer, C. Viana, L. Machado De Carvalho et al. 2019. Simultaneous determination of iron and nickel as contaminants in multimineral and multivitamin supplements by solid sampling HR-CS GF AAS. Talanta 195: 745–751.

Roy, E., S. Patra, R. Madhuri and P.K. Sharma. 2014. Simultaneous determination of heavy metals in biological samples by a multiple-template imprinting technique: An electrochemical study. RSC Adv. 4: 56690–56700.

Salimi, A., B. Pourbahram, S. Mansouri-Majd and R. Hallaj. 2015. Manganese oxide nanoflakes/multi-walled carbon nanotubes/chitosan nanocomposite modified glassy carbon electrode as a novel electrochemical sensor for chromium (III) detection. Electrochim. Acta 156: 207–215.

Sarkar, B., S. Mandal, Y.F. Tsang, P. Kumar, K.H. Kim and Y.S. Ok. 2018. Designer carbon nanotubes for contaminant removal in water and wastewater: A critical review. Sci. Total Environ. 612: 561–581.

Sarwar, F., R.N. Malik, C.W. Chow and K. Alam. 2018. Occupational exposure and consequent health impairments due to potential incidental nanoparticles in leather tanneries: An evidential appraisal of south Asian developing countries. Environ. Int. 117: 164–174.

Selvan, K.S. and S.S. Narayanan. 2018. Synthesis and characterization of carbon nanotubes/asymmetric novel tetradentate ligand forming complexes on PIGE modified electrode for simultaneous determination of Pb(II) and Hg(II) in sea water, Lake water and well water using anodic stripping voltammetry. J. Electroanal. Chem. 810: 176–184.

Sevcikova, M., H. Modra, A. Slaninova and Z. Svobodova. 2011. Metals as a cause of oxidative stress in fish: A review. Vet. Med. 56: 537–546.

Sharma, P., V. Bihari, S.K. Agarwal, V. Verma, C.N. Kesavachandran and B.S. Pangtey. 2012. Groundwater contaminated with hexavalent chromium[Cr(VI)]: A health survey and clinical examination of community inhabitants (Kanpur, India). PLoS ONE 7: e47877. doi: 10.1371/journal.pone.0047877.

Shi, L., Y. Wang, Z. Chu, Y. Yin, D. Jiang, J. Luo et al. 2017. A highly sensitive and reusable electrochemical mercury biosensor based on tunable vertical single-walled carbon nanotubes and a target recycling strategy. J. Mater. Chem. B 5: 1073–1080.

Shishov, A., M. Wieczorek, D. Dudek-adamska, A. Telk, L. Moskvin and A. Bulatov. 2018. An automated continuous homogeneous microextraction for the determination of selenium and arsenic by hydride generation atomic fluorescence spectrometry. Talanta 181: 359–365.

Silva, B., I.C. Neves and T. Tavares. 2016. A sustained approach to environmental catalysis: Reutilization of chromium from wastewater. Crit. Rev. Environ. Sci. Technol. 46: 1622–1657.

Silva, E.L., P.S. Roldan and M.F. Gine. 2009. Simultaneous preconcentration of copper, zinc, cadmium, and nickel in water samples by cloud point extraction using 4-(2-pyridylazo)-resorcinol and their determination by inductively coupled plasma optic emission spectrometry. J. Hazard Mater. 171: 1133–1138.

Singh, A., R.K. Sharma, M. Agrawal and F.M. Marshall. 2010. Health risk assessment of heavy metals via dietary intake of food stuffs from the wastewater irrigated site of a dry tropical area of India. Food Chem. Toxicol. 48: 611–619.

Sitko, R., P. Janik, B. Zawisza, E. Talik, E. Margui and I. Queralt. 2015. Green approach for ultra-trace determination of divalent metal ions and arsenic species using total reflection X-ray fluorescence spectrometry and mercapto-modified graphene oxide nanosheets as a novel adsorbent. Anal. Chem. 87: 3535–3542.

Sreenivasa Rao, K., T. Balaji, T. Prasada Rao, Y. Babu and G.R.K. Naidu. 2002. Determination of iron, cobalt, nickel, manganese, zinc, copper, cadmium and lead in human hair by inductively coupled plasma-atomic emission spectrometry. Spectrochim. Acta B At Spectrosc. 57: 1333–1338.

Standard Methods for the Examination of Water and Wastewater, 23rd Edition. 2017. American Public Health Association, American Water Works Association, Water Environment Federation. Editors: E.W. Rice, R.B. Baird, A.D. Eaton. Denver, USA.

Steinmaus, C.M., C.M. George, D.A. Kalman and A.H. Smith. 2006. Evaluation of two new arsenic field test kits capable of detecting arsenic water concentrations close to 10 microg/L. Environ. Sci. Technol. 40: 3362–3366.

Tag, K., K. Riedel, H.-J. Bauer, G. Hanke, K.H.R. Baronian and G. Kunze. 2007. Amperometric detection of Cu^{2+} by yeast biosensors using flow injection analysis (FIA). Sens. Actuators B Chem. 122: 403–409.

Tarley, C.R.T., A.M. Basaglia, M.G. Segatelli, M.C. Prete, F.A.C. Suquila and L.L.G. de Oliveira. 2017. Preparation and application of nanocomposite based on imprinted poly(methacrylic acid)-PAN/MWCNT as a new electrochemical selective sensing platform of Pb^{2+} in water samples. Journal of Electroanalytical Chemistry 801: 114–121.

Tchounwou, P.B., C.G. Yedjou, A.K. Patlolla and D.J. Sutton. 2012. Heavy metals toxicity and the environment. EXS 101: 133–164.

Tran, T.A., L.P. Popova, F. Stromeyer and K. Hermann. 2013. Functions and toxicity of cadmium in plants: recent advances and future prospects. Turk. J. Bot. 37: 1–13.

Tu, J., Y. Gao, T. Liang, H. Wan and P. Wang. 2018. A miniaturized electrochemical system for high sensitive determination of chromium(VI) by screen-printed carbon electrode with gold nanoparticles modification. Sens. Actuators B: Chem. 272: 582–588.

Turdean, G.L. 2011. Design and development of biosensors for the detection of heavy metal toxicity. Int. J. Electrochem. 2011: 3–5.

USEPA (United States Environmental Protection Agency). 2018 Edition. Drinking Water Standards and Health Advisories Tables March 2018. Website accessed June 2018.

Valko, M., H. Morris and M.T.D. Cronin. 2005. Metals, toxicity and oxidative stress. Curr. Med. Chem. 12: 1161–1208.

Vu, H.D., L.-H. Nguyen, T.D. Nguyen, H.B. Nguyen, T.L. Nguyen and D.L. Tran. 2015. Anodic stripping voltammetric determination of Cd^{2+} and Pb^{2+} using interpenetrated MWCNT/P1,5-DAN as an enhanced sensing interface. Ionics 21: 571–578.

Wanekaya, A.K. 2011. Applications of nanoscale carbon-based materials in heavy metal sensing and detection. Analyst 136: 4383–4391.

Wang, G. and B.A. Fowler. 2008. Roles of biomarkers in evaluating interactions among mixtures of lead, cadmium and arsenic. Toxicol. Appl. Pharmacol. 233: 92–99.

Wang, J. 2006. Analytical Electrochemistry 3rd Edition. John Wiley & Sons, Inc. Hoboken, New Jersey, USA.

Wang, Y. and S. Hu. 2016. Applications of carbon nanotubes and graphene for electrochemical sensing of environmental pollutants. J. Nanosci. Nanotechnol. 6: 7852–7872.

Wang, H., Z.K. Wu, B.B. Chen, M. He and B. Hu. 2015. Chip-based array magnetic solid phase micro extraction on-line coupled with inductively coupled plasma mass spectrometry for the determination of trace heavy metals in cells. Analyst 140: 5619–5626.

Wang, S., E.S. Forzani and N.J. Tao. 2007. Detection of heavy metal ions in water by high-resolution surface plasmon resonance spectroscopy combined with anodic stripping voltammetry. Anal. Chem. 79(12): 4427–4432.

Wang, T. and W. Yue. 2017. Carbon nanotubes heavy metal detection with stripping voltammetry: a review paper. Electroanalysis 29: 2178–2189.

Wang, Y., Y. Li, K. Lv, X. Chen and X. Yu. 2018. A simple and sensitive non-chromatographic method for quantification of four arsenic species in rice by hydride generation–atomic fluorescence spectrometry. Spectrochim. Acta Part B 149: 197–202.

Wasserman, G.A., X.H. Liu, F. Parvez, P. Factor-Litvak, H. Ahsan, D. Levy et al. 2011. Arsenic and manganese exposure and children's intellectual function. Neurotoxicology 2: 450–457.

Wei, J., D. Yang, H. Chen, Y. Gao and H. Li. 2014. Stripping voltammetric determination of mercury(II) based on SWCNT-PhSH modified gold electrode. Sens. Actuators B Chem. 190: 968–974.

WHO. 2017. Guidelines for drinking-water quality: fourth edition incorporating the first addendum. Geneva: World Health Organization; 2017. Licence: CC BY-NC-SA 3.0 IGO.

Wu, X., S.J. Cobbina, G. Mao, H. Xu, Z. Zhang and L. Yang. 2016. A review of toxicity and mechanisms of individual and mixtures of heavy metals in the environment. Environ. Sci. Pollut. Res. 23: 8244–8259.

Xuan, X. and J.Y. Park. 2018. A miniaturized and flexible cadmium and lead ion detection sensor based on micro-patterned reduced graphene oxide/carbon nanotube/bismuth composite electrodes. Sens. Actuators B Chem. 255: 1220–1227.

Yang, N., X. Chen, T. Ren, P. Zhang and D. Yang. 2015. Carbon nanotube based biosensors. Sens. Actuators B Chem. 207, Part A: 690–715.

Yang, X., Y. Wan, Y. Zheng, F. He, Z. Yu, J. Huang et al. 2019. Surface functional groups of carbon-based adsorbents and their roles in the removal of heavy metals from aqueous solutions: A critical review. Chem. Eng. J. 366: 608–621.

Yantasee, W., Y. Lin, K. Hongsirikarn, G.E. Fryxell, R. Addleman and C. Timchalk. 2007. Electrochemical sensors for the detection of lead and other toxic heavy metals: the next generation of personal exposure biomonitors. Environ. Health Perspect. 115: 1683–1690.

Yao, Y., H. Wub and J. Pinga. 2019. Simultaneous determination of Cd(II) and Pb(II) ions in honey and milk samples using a single-walled carbon nanohorns modified screen-printed electrochemical sensor. Food Chem. 274: 8–15.

Zhao, G., H. Wang, G. Liu and Z. Wang. 2016. Simultaneous determination of Cd(II) and Pb(II) based on bismuth film/carboxylic acid functionalized multi-walled carbon nanotubes-β-cyclodextrin-nafion nanocomposite modified electrode. Int. J. Electrochem. Sci. 11: 8109–8122.

Zheng, H., J. Hong, X. Luo, S. Li, M. Wang, B. Yang et al. 2019. Combination of sequential cloud point extraction and hydride generation atomic fluorescence spectrometry for preconcentration and determination of inorganic and methyl mercury in water samples. Microchem. J. 145: 806–812.

CHAPTER 10

Carbon Encapsulated Functional Magnetic Nanoparticles for Life Sciences

Clara Marquina[1,2,]* and *M. Ricardo Ibarra*[1,2,3,4,]*

1. Introduction

Nanoscience and Nanotechnology are wide and interdisciplinary fields that have experienced a strong development during the last years. Nanoscience, that is, the study and manipulation of matter at nanometer scale, has revealed how the size reduction leads to new physical phenomena (not observable at macroscopic length scales) and therefore, to important changes in the electrical, magnetic, optical, thermal, and mechanical properties of the materials. Nanotechnology allows the design and fabrication of new materials and devices, making use of these new properties. Therefore, Nanoscience and Nanotechnology are at the origin of a new revolution in fields, such as electronics, information and telecommunication technologies, energy storage, environment, chemical and pharmaceutical industries, among others. One of the most exciting areas of research is how to bring this new scientific revolution to Life Sciences, making use of these new phenomena, materials, and devices in Biochemistry, Biomedicine, Bioengineering, Biotechnology, Cell biology, Botany, and other branches of Biology. In this regard, since more than a decade, there is a great deal of activity on the use of nanoparticles, and especially of magnetic nanoparticles, in Biomedicine and Biotechnology. This chapter focuses on a particular type of magnetic nanoparticles, the carbon-coated iron nanoparticles, and gives an overview of the interdisciplinary research that we have carried out in the aforementioned fields.

Nanoparticles are moieties made of inorganic or organic materials with dimensions between 1 and 100 nm (i.e., between 1×10^{-9} and 1×10^{-7} m) (Arruebo et al. 2007a). In the particular case of magnetic nanoparticles, the starting materials for their synthesis are ferromagnetic metals, such as

[1] Instituto de Ciencia de Materiales de Aragón (ICMA), Consejo Superior de Investigaciones Científicas (CSIC) - Universidad de Zaragoza, Zaragoza, Spain.

[2] Departamento de Física de la Materia Condensada, Facultad de Ciencias, Universidad de Zaragoza; Pedro Cerbuna 12, 50009-Zaragoza, Spain.

[3] Instituto de Nanociencia de Aragón (INA), Universidad de Zaragoza; Pedro Cerbuna 12, 50009-Zaragoza, Spain.

[4] Laboratorio de Microscopias Avanzadas (LMA), Universidad de Zaragoza; Mariano Esquillor s/n, 50018-Zaragoza, Spain.

* Corresponding authors: clara@unizar-es; ibarra@unizar.es

iron cobalt and nickel, and some of their magnetic oxides and alloys (Cullity 1974, De Teresa et al. 2005). An important characteristic of all of them is their high magnetization at room temperature and the ability they have to interact with magnetic fields, both static and alternating. This is what makes the magnetic nanoparticles suitable for their application in Biomedicine. They can act as:

- Contrast agents in Magnetic Resonance Imaging (MRI) (Hahn et al. 1990, Weissleder et al. 1990)
- Heating agents for magnetic hyperthermia (i.e., for the thermal ablation of tumors) (Pankhurst et al. 2003, Mehdaoui et al. 2011, Lima et al. 2014, Sanz et al. 2017)
- Magnetic vectors for magnetically targeted drug delivery (Arruebo et al. 2006a, b, Arruebo et al. 2007a), or for gene therapy (Plank et al. 2011)
- Labels for quantitative immune-chromatographic sensors (for example, to detect infectious diseases, drugs, bacterial contamination, etc.) (De Teresa et al. 2005, Marquina et al. 2012, Serrate et al. 2012).

One of the important issues when working with nanoparticles in all those *in vitro* and *in vivo* applications that involve the interaction of the nanoparticles with cells and tissues is the necessary biocompatibility of the material. Besides, all these applications require avoiding the oxidation of the nanoparticles in order to preserve their magnetic properties (De Teresa et al. 2005). The application of the nanoparticles in the organism requires their use in suspension, as a magnetic fluid that has to also be biocompatible (Ledezma et al. 2009), and in which the magnetic nanoparticles have to be stable. The nanoparticles should be adequately functionalized in order to avoid the formation of large aggregates, and consequently their subsequent precipitation. In addition, when dealing with magnetic nanoparticles, it is also necessary to avoid the aggregation due to the dipolar magnetic interactions between their respective magnetic moments (De Teresa et al. 2005, Fernández-Pacheco 2008). This functionalization is optimized when the magnetic nanoparticles are encapsulated either in organic or inorganic materials. The resulting nanoparticles are called core-shell nanoparticles (Klabunde et al. 1994). The shell allows the functionalization of the surface with bioactive components: antibodies (Arruebo et al. 2007b, Arenal et al. 2013), biomolecules that can act, for example, as cell surface markers for tumor detection (Berry 2005), therapeutic drugs to be delivered at a particular organ (Alexiou et al. 2000), etc. These moieties are attached to the particle shell by chemical bonding or by adsorption. There are a large number of molecules of biological interest and a vast list of functionalization protocols to attach them to the desired nanoparticle with a particular shell (Puertas et al. 2010). Therefore, the proper election of the coating broadens the field of application of these nanomaterials.

Organic materials are the most commonly used as coatings: polysaccharides (such as dextran (Grüttner et al. 2001)), and polymers (as polyvinyl alcohol (PVA) (Pardoe et al. 2001) and poly(ethylene glycol) (PEG) (Zahr et al. 2006)). Nanoparticles with an inorganic coating are also being considered promising for biomedical applications; in particular, encapsulation *in silica* has been widely studied (Aliev et al. 1999, Fernández-Pacheco et al. 2006). Another approach is the encapsulation in carbon. This material provides biocompatibility, a chemically stable surface, easy functionalization, and prevents inter-particle magnetic interactions, among other advantages (Kuznetsov et al. 1999, Blazewicz 2001, Fernández-Pacheco 2008, Iturrioz-Rodriguez et al. 2017). The carbon-coated nanoparticles constitute the basis of this chapter.

In this contribution, we offer a comprehensive survey of the research carried out on a specific type of carbon-coated magnetic nanoparticles, whose use is intended for applications in Biomedicine and Biotechnology. The first section is devoted to their synthesis and physicochemical characterization, which includes the study of their magnetic properties. The following sections report the characterization of the nanoparticles performed in order to test that they are suitable for application, in particular in life sciences. A detailed study of the interaction of the nanoparticles with animal cells and in animal models is presented. The last part of this chapter is devoted to the research in the plant kingdom, which envisages a great impact on agriculture, ecology, and food industry.

2. Synthesis and Characterization of Fe@C

2.1 *Synthesis of nanoparticles*

Since the last decade of the last century, following the discovery of fullerenes (Kroto et al. 1985) and carbon nanotubes (Iijima 1991), a vast variety of methods has been developed for the synthesis of carbon-coated nanoparticles. These include the radio frequency plasma torch route (Turgut et al. 1998), IR laser co-pyrolysis (Dumitrache et al. 2004), modified flame spray synthesis (Athanassiou et al. 2006), chemical vapor deposition (CVD) (Seo et al. 2006, Castrillón et al. 2012, 2013), ball-milling (Fernández-Pacheco 2008), and hydrothermal synthesis (Zhao et al. 2016), among others.

The following sections of the chapter will focus on the research carried out with carbon-coated magnetic nanoparticles obtained by a modification of the arc discharge method designed by Krätschmer-Huffman in 1990 for the production of fullerenes and nanotubes (Krätschmer et al. 1990, McHenry et al. 1994, Cadek et al. 2002). Besides, for the production of carbon-coated magnetic nanoparticles, this method has also been successfully used for the production of silica-coated magnetic nanoparticles (Fernández-Pacheco et al. 2006). This method makes use of a plasma furnace. A scheme of the arc-furnace is shown in Figure 10.1.

It is a water refrigerated cylindrical chamber, whose sidewall is a quartz tube, and the upper section is traversed by a mobile electrode ending in a graphite rod. The lower section (the stationary electrode) is a graphite crucible containing a drilled graphite cylinder filled with several micrograms of 10-micron iron powders. We apply a DC electrical current between the graphite electrodes in a helium atmosphere. This current originates an arc, ionizing the gas and generates high temperature plasma. The moveable electrode is sublimed and builds up a carbonaceous deposit or *collaret* around it at the same time that soot is deposited on the inner surface of the chamber walls (Fernández-Pacheco et al. 2007, 2009, Fernández-Pacheco 2008).

The X-ray Diffraction pattern of the obtained powder revealed that besides carbon, most of the sample contains α-Fe. We also observed a very small fraction of maghemite/magnetite due to the oxidation of some Fe^0 taking place during the synthesis in the discharge chamber, as further confirmed by Transmission Electron Microscopy.

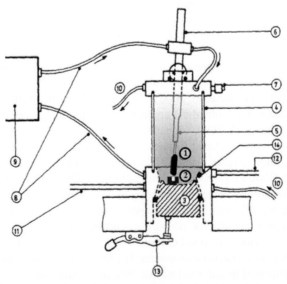

Figure 10.1: Scheme of the arc-furnace for Fe@C nanoparticle synthesis. (1) Movable graphite rod; (2) drilled graphite cylinder; (3) graphite crucible; (4) cylindrical chamber; (5) movable electrode; (6) insulating handle for movable electrode handling; (7) safety valve; (8) water refrigerated power cables; (9) D.C. power supply; (10) refrigeration inlet/outlet; (11) vacuum pump connection; (12) gas inlet; (13) clamp; (14) titanium getter for chamber degassing.

2.2 Transmission electron microscopy characterization

A detailed morphological and structural characterization was performed by Transmission Electron Microscopy (TEM). The TEM images were obtained on samples produced by arc discharge, collected from the *collaret* deposited on the cathode and soot collected from the chamber walls. They showed a mixture of carbon-coated magnetic nanoparticles and different carbon nanostructures. In addition, we found non-encapsulated or partially-encapsulated metallic particles. The magnetic nanoparticles are encapsulated either in carbon nanotubes (this are mainly found in the cathode), in concentric graphitic shells ("onions"), or in amorphous carbon. In the case of samples obtained by high energy milling, TEM shows that the final powder is composed of a mixture of carbon-coated nanoparticles together with the free carbon and non-encapsulated magnetic metallic particles.

In Figure 10.2, we show images of samples produced by the Krätschmer-Huffman method. The magnetic nanoparticles are encapsulated in several layers of graphitic carbon and surrounded by amorphous carbon. High Resolution HR-TEM images show that in this case, the coating of the magnetic particles is complete, thus rendering them biocompatible. HR-TEM also makes it possible to see the atomic planes of the nanoparticle metallic core.

The diameter of the particles is obtained from the analysis of several images. The obtained diameter probability distribution function is plotted as a size distribution histogram in Figure 10.3.

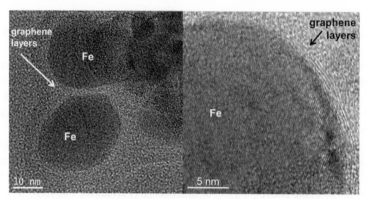

Figure 10.2: (a) HR-TEM images of Fe@C nanoparticles, and (b) detail of a particle. The graphene layers of the coating as well as the atomic planes of the Fe metallic core are clearly visible.

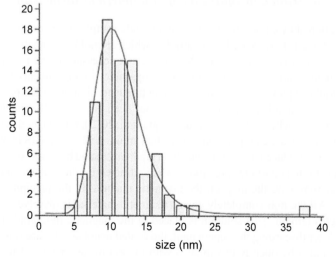

Figure 10.3: Particle size histogram obtained from the analysis of TEM images (De Teresa et al. 2005).

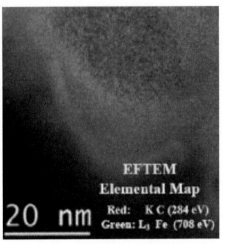

Figure 10.4: Elementary map of a carbon-coated nanoparticle, showing the spatial distribution of the carbon, coating a α-Fe$_2$O$_3$ core.

Color version at the end of the book

As it can be seen, a sample produced by arc discharge contains particles of diameters ranging from 5 nm up to 50 nm, the center of the size distribution being at 10 nm.

We performed Electron Filter (EF)-TEM in order to get a better insight into the chemical nature of the synthesized material. EF-TEM images have been taken selecting energy windows of the Electron Energy Loss Spectroscopy (EELS) of the characteristic peaks: for iron the 708 eV-L3 peak, and for carbon the 284 eV-K peak (Fernández-Pacheco 2008). Once both images are obtained, an elementary map can be drawn, showing a colored distribution of each element (see Figure 10.4). The EF-TEM confirms the presence of iron content in the core of the nanoparticles, which is completely surrounded by a carbon coating. The crystallographic structure of the samples was analyzed by electron diffraction. According to the patterns, most of the nanoparticle cores consisted of α-Fe, except a small fraction that crystallized in the cubic phase γ-Fe$_2$O$_3$, or maghemite (Fernández-Pacheco 2008) appearing as a passivation layer at the surface.

2.3 *Sample purification and synthesis of nanoparticle suspensions*

TEM characterization showed that the Fe@C nanoparticle samples prepared by arc discharge also contain non-coated or partially-coated nanoparticles, which are not biocompatible and therefore not suitable to be used for *in vivo* applications. Moreover, it is necessary to get rid of all those carbon magnetic nanostructures not containing magnetic nanoparticles in order to enrich the concentration of magnetic material in a powder sample or in a magnetic suspension. With this purpose, a combined magnetic and chemical sample purification protocol has been developed (Fernández-Pacheco et al. 2007, 2009, Fernández-Pacheco 2008, Pérez-de-Luque et al. 2012). For magnetic purification, stable suspensions of the particles are prepared in a surfactant solution (2.5 g of SDS in 500 ml of distilled water). The suspension is placed in the magnetic field gradient created by means of a 3 KOe permanent magnet. The magnetic nanoparticles precipitate when attracted by the magnet, and the supernatant is discarded. After removing the magnet, the magnetic material in the precipitate is washed with HCl 3 M at 80ºC. All the non-completely coated magnetic material is dissolved, whereas the fully coated nanoparticles remain. This solution is placed in the magnetic field gradient to separate the coated particles. After discarding the supernatant, the coated nanoparticles are washed several times in distilled water and subsequently precipitated with the aid of the magnet. The resulting powder sample is finally heated at 350°C in order to evaporate the remaining free amorphous carbon structures (Fernández-Pacheco 2008).

As a result of the chemical etching of the carbon, the nanoparticles get some hydroxyl groups (–OH) on their surface that could be substituted with more reactive chemical chains, able to bind covalently to amine (NH_2) or carboxyl (COOH) terminal groups. In comparison to what has been found in silica nanoparticles, the number of hydroxyl groups on the carbon surface is quite low, and much less reactive (Fernández-Pacheco 2008). This fact makes the carbon shell of the nanoparticles obtained after the purification stage not very attractive for chemical functionalization. However, this carbon surface is highly hydrophobic, and therefore it will easily adsorb any organic molecule in the solution. As it will be discussed in a following section, this fact makes these Fe@C nanoparticles potentially suitable as carriers for magnetically targeted drug delivery.

Most applications of nanoparticles require their use in suspension which, for *in vivo* applications, has to be biocompatible. In the case of Fe@C nanoparticles, three different biocompatible magnetic fluids were prepared, dispersing the nanoparticles in water, gelafundine, and mannitol (De Teresa et al. 2005, Fernández-Pacheco et al. 2007, Fernández-Pacheco 2008, González-Melendi et al. 2008, Cifuentes et al. 2010). The aggregation state of the nanoparticles in the suspension is an important parameter to control. In particular, for their use *in vivo*, the size of the aggregates must be small enough so that they can travel in the bloodstream, or in the circulatory system of a plant, etc. Dynamic Light Scattering measurements showed that when the purified powder of Fe@C nanoparticles was dispersed in water, the aggregates had an average size about 200 nm, which is in principle lower than that of the smallest capillaries of the human vascular system (Arruebo et al. 2007a). Likewise, the aggregation state of the nanoparticles in gelafundine and mannitol suspension was also adequate for their use *in vivo* (Fernández-Pacheco et al. 2007, Escribano et al. 2012) and *in planta* (González-Melendi et al. 2008, Cifuentes et al. 2010) experiments.

2.4 Magnetic characterization

As mentioned in the introduction, there are a large number of biomedical techniques and applications that are based on the use of magnetic materials (in many cases, nanoparticles) and on their response to a magnetic field. This response is what determines whether the material is suitable for a given application or not. The so called magnetization isotherm, a characteristic of each material, shows how its magnetization varies when a dc magnetic field is applied. There are materials in which at zero applied magnetic field, the macroscopic magnetization is zero. When turning on the field they will magnetize, and will eventually reach their saturation magnetization value above a certain saturation field. Another important parameter is the remanence of the material, i.e., the remaining magnetization after the field has been turned off. The coercive field is the field necessary to apply again to make the macroscopic magnetization vanish. Such an isotherm is the so-called hysteresis loop, and is the typical behavior of the ferromagnetic materials, one of the largest groups in which bulk magnetic materials are classified, together with the diamagnetics and paramagnetics. Reducing the dimensions of a magnetic material to the nanometer scale and below certain critical size makes these nanoparticles behave as superparamagnets. The characteristic isotherm does not show either remanence or coercivity, as their magnetization completely vanishes at zero magnetic field (Cullity 1974). If the dimensions of the nanoparticles are above the critical size for a given material, they will behave as a ferromagnet. The use of ferromagnetic nanoparticles is inadvisable in those biomedical applications; for instance, when used as nanocarriers for drug delivery. Furthermore, we should avoid the dipolar magnetic interaction between the individual nanoparticles. This is particularly important in *in vivo* and *in planta* applications, since the magnetic interactions could give rise to aggregates whose size is not adequate for circulating in the bloodstream, or through the vascular system of a plant. The use of superparamagnetic nanoparticles is, in principal, a way to avoid these problems.

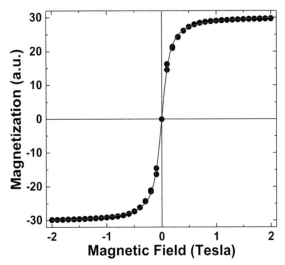

Figure 10.5: Magnetization isotherm of Fe@C nanoparticles at room temperature. The solid line represents the fit to the Langevin law (De Teresa et al. 2005).

The magnetic characterization of the purified Fe@C nanoparticles was performed by magnetization measurements carried out in a high sensitivity magnetometer equipped with Superconducting Quantum Interference Device (SQUID) detection. As an example, the magnetization isotherm at room temperature is displayed in Figure 10.5.

This field dependence of the magnetization shows no coercivity and no remanent magnetization, which indicates that at room temperature, the synthesized Fe@C nanoparticles were in the superparamagnetic state. Therefore, given their size and magnetic behavior, the obtained carbon coated nanoparticles were the adequate for *in vivo* applications. The experimental magnetization results can be analyzed in the framework of Langevin's theory for superparamagnetism. According to it, the dependence of the magnetization of a nanoparticle, m, on temperature, T, and applied magnetic field, H, is given by

$$m(T, H) = m_s \left(\coth\left(\frac{\mu H}{k_B T} \right) - \frac{k_B T}{\mu H} \right)$$

where m_s is the saturation magnetization of the particle, μ is the magnetic moment of the particle, and k_B is the Boltzman constant (Cullity 1974). Introducing the value of m_s (in our case, iron, 1740 emu/cm^3), and fitting the experimental magnetization data to the expression above, the particle size can be derived from the calculated $\mu = m_s V$, with V being the particle volume and assuming nanoparticles of spherical shape. The comparison of the theoretical value and the one derived from TEM images was satisfactory (De Teresa et al. 2005).

The magnetic characterization was completed by Mössbauer spectroscopy (Coaquira et al. 2001, Rechenberg et al. 2001). This technique allows the identification and quantification of the different magnetic and non-magnetic phases in a sample. In the present case, the Mössbauer spectra, besides confirming the α-Fe phase, revealed the presence of non-magnetic γ-Fe and Fe$_3$C (Coaquira et al. 2001, Rechenberg et al. 2001).

3. Evading the Reticulo-endothelial System: Protein and Polymeric Coating of Fe@C Nanoparticles

Phagocytosis is a major limitation in any *in vivo* application based on the intravenous delivery of nanoparticles. Phagocytosis describes the process in which phagocytes, which are specialized cells in the Reticuloendothelial system (RES), destroy pathogens and cell debris in the organism (such as bacteria, dead tissue cells, small mineral particles, etc.). In an *in vivo* application, once the nanoparticles are injected in the bloodstream, they become recognizable to the immune system, and therefore they are subsequently cleared from circulation by phagocytosis after opsonization. Opsonines are plasma proteins and other components of the blood that immediately coat the nanoparticles. This process increases the size of the nanoparticles and allows their capture by the RES (Berry 2005). Therefore, strategies have to be developed in order to increase the lifetime of the nanoparticles in the blood circulation. One possibility is taking advantage of electrostatic and steric repulsion. For example, coating the nanoparticle with Bovine Serum Albumin (BSA) will charge the nanoparticle surface negatively at the physiological pH, impeding the specific recognition of receptors on the phagocyte membrane, which is also negatively charged. An alternative is coating the nanoparticle with poly(ethylene glycol) (PEG). PEG molecules on the surface of a nanoparticle will form a dynamic molecular cloud over the particle surface. This cloud of mobile and flexible PEG molecules causes the steric repulsion of opsonines, making their adsorption onto PEG molecules energetically unfavorable (Zahr et al. 2006, Fernández-Pacheco 2008).

3.1 Protein coating with serum albumin: BSA and RSA coating

The details of the protocol followed for coating the Fe@C nanoparticles with BSA can be found in Fernández-Pacheco (2008). The nanoparticles were incubated for one hour with different BSA concentrations to estimate the maximum capacity of adsorption. After incubation, the nanoparticles coated with BSA were separated by centrifugation, and the remaining protein was detected by VIS-UV spectroscopy using the bicinchoninic acid (BCA) Protein Assay for protein quantification. The maximum adsorption was achieved for a BSA concentration of 2 mg/ml. The spontaneous desorption of the protein was studied by increasing the ionic force of the medium, and also by direct adsorption competition by adding detergents, such as tween® or triton®. The percentage of BSA that desorbs from the surface of the particles was minimum even at the strongest conditions of desorption.

In order to avoid rejection reactions in *in vivo* experiments, the serum albumin for the coating should be extracted from the same animal in which the nanoparticles are going to be injected. Due to the intrinsic difficulties for isolating the protein from/for each individual, we used commercial serum albumin. For example, Fe@C nanoparticles injected in New Zealand rabbits were successfully coated with commercial Rabbit Serum Albumin (RSA), following the same procedure as for the BSA coating (Fernández-Pacheco 2008).

3.2 Polymeric coating with PAMAO-PEG

As already mentioned, an outer polymeric shell would prevent the adsorption of opsonins as well, by yielding an electrically neutral and hydrophilic nanoparticle surface, and hence their recognition by the immune system would be delayed (Yu et al. 2006). In order to successfully avoid opsonization by steric repulsion, the choice of the PEG molecular weight, and therefore the polymer chain length, are crucial. It was reported that the hydration of coiled, flexible PEG chains from 1500 to 5000 Da helped to avoid protein adsorption (Stolnik et al. 1995). Also, Zahr and co-workers studied the dependence of protein adsorption and *in-vitro* uptake by macrophages of PEG-coated nanoparticles, for polymers between to 2 and 20 kDa (Zahr et al. 2006). Hunter and co-workers suggested that the stability of the polymer coating could determine the distribution of a colloidal carrier in a biological environment (Hunter et al. 1983). Besides, methylated PEG (mPEG) was previously used to decrease immunogenicity and proteolysis of a number of various macromolecules, such as albumin, interferon-γ, interleukin-1, colony stimulating factor, and superoxide dismutase, as well as to increase their circulation half-lives (Monfardini and Veronese 1998). There are three common ways to attach PEG molecules to the surface of a particulate carrier: incorporation of PEG during the synthesis of the particles, covalent attachment of activated PEG molecules to the reactive surface of previously produced particles, and physical adsorption of appropriate PEG derivatives, such as poly-(lactyde-co-glycolide)-PEG to nanospheres. Particles with a poorly reactive surface and a rapid desorption of physically adsorbed PEG are severe limitations to these protocols. Desorption of PEG leads to the formation of steric barriers when particles are in contact with blood, and this process provokes a quick capture of the particles by macrophage (Moghimi and Hunter 2001).

The procedure followed for coating the Fe@C nanoparticles with PEG is explained in detail in Fernández-Pacheco (2008). This protocol makes use of methoxy-modified PEG and an amphiphilic molecule, poly(maleic anhydride-alt-1-octadecene) (PMAO). The long hydrophobic chain of the PMAO adsorbs on the carbon hydrophobic surface, whereas the hydrophilic carboxyl groups at the other side of the molecule covalently bind to the amino groups of the methoxy-modified polymer. As coating with PEG should produce a neutralization of the particle surface charge, the measurement of zeta potential before and after the coating reaction is a useful tool to determine the effectiveness of the polymer attachment to the particle. The formation of the copolymer after reaction of PMAO and PEG was also followed by Infra-Red spectrometry. We followed the evolution of the characteristic peaks due to the decomposition of anhydride (1775 cm^{-1}) and the release of carboxyl groups (1715 cm^{-1}), respectively (Kang et al. 2000, Atici et al. 2001). This analysis was completed by X-ray Photoelectron Spectroscopy (XPS), comparing the spectrum of a Fe@C nanoparticle sample to that of the PEG-PMAO functionalized particles. In this one, the appearance of a peak in the N 1s threshold region (at 397 eV) is characteristic of the nitrogen associated to the amino groups of the polymer (Popat et al. 2004). A peak in the O 1s region (at 534 eV) indicated the presence of PEG on the particle surface. In addition, besides the characteristic hydrocarbon C-C peak at 285 eV in the C 1s region, the C-O peak appeared (at 289 eV), which is characteristic of PEG coupling (Popat et al. 2004). Further evidence of a proper coating of the particles was obtained studying their ability to adsorb the protein BSA. Fe@C nanoparticles, which are highly hydrophobic, adsorb BSA rapidly; however, once the carbon surface was coated with PMAO-PEG, no BSA was detected on the particle surface (Fernández-Pacheco 2008).

4. Fe@C Nanoparticle uptake by Dendritic Cells–DCs as Trojan Horse

We have developed a new strategy in order to avoid the action of the RES for *in vivo* applications. The nanoparticles are *in vitro* internalized in dendritic cells that will be used as selective biological vectors. The success of this strategy depends on the selectivity of the cells to reach the target, and on the amount of particles with which the cells can be loaded without compromising their viability.

The diagnosis and treatment of cancerous tumors is one of the most important applications of magnetic nanoparticles, either by using them as MRI contrast agents for their detection, or as drug carrier of therapeutic agents, or for magnetic hyperthermia. Dendritic cells (DCs) are very promising as carriers to deliver the magnetic nanoparticles in the target organ. DCs are the main antigen presenting cells (APCs) that activate the immune system. They are present in small quantities in tissues that are in contact with the external environment, mainly in the peripheral tissues. They can also be found in an immature state in the blood. DCs are normally in immature state, having a poorly developed antigen presentation activity. Under the influence of certain stimuli, DCs are activated, and evolve to a mature state. At this stage they grow the dendrites (which give the cell its name), and the ability to process and present antigens significantly increases. Maturation of DCs involves changes that allow them to migrate to secondary lymphoid organs, where they interact with T cells and B cells to initiate and shape the adaptive immune response (Steinman 1991, Steinman and Cohn 1973, Fearnley et al. 1997, Hart 1997, Banchereau and Steinman 1998, Foti et al. 1999, Hopken and Lipp 2004). The presence of dendritic cells has been identified in various types of tumors, such as head and neck tumors, breast cancer, cervical or endometrial cancer (Coppola et al. 1998), gastric cancer (Kakeji et al. 1993) or renal cancer (Thurnher et al. 1996). Clinical pathologists have also associated the presence of a great number of DCs with a better prognosis (Ambe et al. 1987, Furihata et al. 1992). Dendritic cells can also have an important role in angiogenesis mechanisms irrigating tumor cells. Angiogenesis is a physiological process involving the growth of new blood vessels from pre-existing vessels. It is a normal process in growth and development, as well as in wound healing (Prior et al. 2004). However, dendritic cell precursors are also involved in tumor vasculogenesis through the cooperation of beta-defensins and vascular endothelial growth factor-A (Vegf-A) (Conejo-García et al. 2004). Therefore, the inclusion of particles inside dendritic cells would play the role of a Trojan horse, to arrive unnoticed at the tumor site. Once there, the carriers will perform their function, whether it is the release of a chemotherapy agent, acting as a contrast agent, or selectively heating a specific area by applying an alternate magnetic field. Tumor targeting with nanoparticle-loaded DCs is a potentially interesting strategy because DCs extracted from a patient would pass unnoticed to the patient's immune system.

With this purpose as a final goal, the uptake of Fe@C particles inside DCs was studied, testing the viability of DCs charged with magnetic nanoparticles and the efficiency of the nanoparticle internalization (Fernández-Pacheco 2008, Goya et al. 2008). DCs were obtained from blood monocytes of healthy volunteers and induced to maturation. They were subsequently cultured with 15 µl/ml of a 5% suspension of Fe@C nanoparticles for 7 days. The detailed protocols can be found in Fernández-Pacheco (2008). Finally, cells were collected, washed twice with PBS, and re-suspended in the same buffer and centrifuged for 25 minutes at 400 G with Ficoll histopaque. DCs were isolated from the PBS/Ficoll interface, whereas the nanoparticles that were not incorporated into the cells remained at the bottom of the centrifuge tube. Incorporation of magnetic nanoparticles was assessed by confocal microscopy and TEM. For the latter, DCs were processed in pellets, and 0.1 µm sections were cut using a microtome, to be placed on the TEM sample holders. TEM images showed that the Fe@C nanoparticles formed intra-lysosomal aggregates within DCs. Given the magnetic character of the nanoparticles, measurement of magnetization is an alternative to detect their incorporation into DCs. This simple method also offers the possibility of quantifying the amount of magnetic material inside the cells, which will be proportional to the magnetization of the sample. For these measurements, samples of DCs isolated from the PBS/Ficoll interface (those assumed to contain the Fe@C nanoparticles) were conditioned in closed containers to cool them down to ~ 265 K. The pellets (i.e., all the magnetic particles not included in the DCs) and were also prepared in the same way. A commercial SQUID magnetometer was used to perform static magnetization measurements as a function of the field and temperature. Magnetization was measured in dc applied magnetic fields up to 1 T, between 5 K and 250 K to avoid the melting of the solid matrix (solvent). These results were compared to those obtained in samples of DCs as cultured, i.e., performed before adding the Fe@C nanoparticles to the cell culture, and with the magnetization curves of DCs inside Fe@C nanoparticles.

Figure 10.6 shows the magnetic response of the DCs before and after the internalization of magnetic nanoparticles. It is worthy to observe the linear response with negative magnetic susceptibility that corresponds to the diamagnetic behavior of the DCs without nanoparticles. The main magnetic contribution observed is due to the magnetic Fe@C nanoparticles. As they form aggregates in the cytoplasm, hysteresis appears in the magnetization cycle. After normalizing both magnetization measurements (those of the DCs with and without nanoparticles) to the total culture volume, the diamagnetic signal of the cells can be subtracted from the total sample signal to obtain a net saturated magnetic contribution (constant magnetization). Both, the as-prepared Fe@C nanoparticles and the nanoparticle loaded DCs, display the same magnetic parameters, indicating that the particle magnetization and size distribution remains essentially unaltered after the incorporation into DCs. Knowing the amount of DCs in the culture and the concentration of Fe@C nanoparticles added, the final amount of nanoparticles incorporated to the cells can be calculated. Moreover, the amount of magnetic material incorporated by the DCs yielded a magnetic signal (10^{-3} emu) larger than the diamagnetic signal from biological tissues (10^{-4}–10^{-5} emu for DCs at low fields), and quite above the detection limit of SQUID (10^{-7} emu). This situation makes our separation process potentially powerful for the detection of small amounts of all types of targeted cells as is, for example, the case

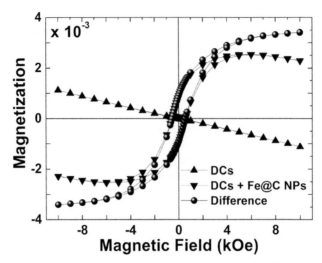

Figure 10.6: Magnetization of as-cultured dendritic cells (DCs) and of DCs incubated with Fe@C nanoparticles (DCs+Fe@C NPs) at 265 K. The difference between both magnetizations (Difference) corresponds to the signal of the Fe@C nanoparticles internalized by the cells.

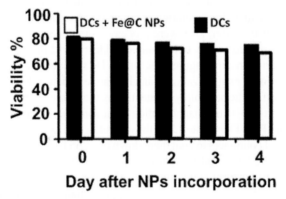

Figure 10.7: Percentage of viable dendritic cells before (DCs), and after the incorporation of Fe@C nanoparticles (DCs+Fe@C NPs) (Goya et al. 2008).

of metastatic cells within the blood stream or lymph nodes (Goya et al. 2008). Once the capacity of the DCs to internalize Fe@C nanoparticles was confirmed, the cell viability was studied. For this purpose, DCs were cultured in a 12-well plate, and the nanoparticles were added to the culture as described above. Cells were collected from different plates 0, 1, 2, 3, and 4 days after adding the nanoparticles (i.e., on days 7, 8, 9, 10, and 11). Cell viability was assessed by the exclusion of trypan blue (Fernández-Pacheco 2008). Cell viability was 80% on day 0 after incorporation of nanoparticles, and decreased slowly with the time, remaining above 75% for four days (see Figure 10.7). Therefore, the fraction of viable cells after nanoparticle incorporation was not significantly affected.

5. Biocompatibility Tests-Haematology Results

Before any *in vivo* application, the biocompatibility of circulating particles in blood has to always be tested. Therefore, the interaction of Fe@C particles with blood cells and platelets and the main hematological and rheological parameters were studied (Fernández-Pacheco et al. 2007, Fernández-Pacheco 2008). For the study of hemato-rheological parameters, the viscosity tests recommended for

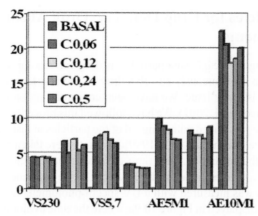

Figure 10.8: Mean values of human blood viscosity (V, in centipoises) at different shear rates (S = 5.7 and 230 s^{-1}); Erythrocyte aggregation index (AE) measured for M1 minutes under low shear rate (M1 = 5, 10 minutes); C = Particle concentration (mg/ml); BASAL = human blood without nanoparticles.

conventional radiological contrast media and for paramagnetic nanoparticles used as MRI contrast agents were adapted for nanoparticles in circulation (Nageswari et al. 1998, Lind et al. 2002, Reinhart et al. 2005). Nanoparticles are similar in shape and length to large circulating plasma proteins (such as fibrinogen or immunoglobulins) that have a decisive influence on plasma and blood viscosity. Therefore, the effect of nanoparticles on these two parameters was studied, applying different shear stress to simulate circulating blood in veins, capillaries, and arteries. In addition, their effect on blood erythrocyte aggregation, which could have a dramatic effect on the micro-vessel circulation, was also investigated. For these purposes, a suspension of 100 mg of Fe@C nanoparticles in 100 ml gelafundine was prepared and mixed at several dilutions with human blood from 5 healthy volunteer donors (see the protocol in Fernández-Pacheco et al. (2007) and in Fernández-Pacheco (2008)). The results in Figure 10.8 correspond to the experiment carried out mixing 1 ml of blood with the particle suspension in different nanoparticle dilutions (0.06, 0.12, 0.24, and 0.5 mg/ml).

Other tests include the complete blood count, including differential leukocyte count, and the usual coagulation tests (Prothrombin time, thromboplastin time, and fibrinogen and D-Dimer immuno-turbidimetry). All the tests before and after mixing human blood with different concentrations of nanoparticles were in the normal range and there were neither statistical, nor clinical significant differences between control blood and blood with different concentrations of particles (Fernández-Pacheco 2008).

Similar protocols were followed in *ex vivo* tests with rabbit blood samples: blood from 10 New Zealand rabbits (test rabbits) was analyzed 10 minutes, 30 minutes, and 24 hours after intravenous injection of 1 ml of the Fe@C nanoparticles suspension. The results were compared to tests performed on 10 non-injected New Zealand rabbits (as control). As in the case of the tests performed with human blood, no important differences were found between tests and control samples. Some of the results showed a very mild hypo-coagulation state (probably due in part to the dispersant), although that did not exceed the recommended value for prevention of thrombosis during surgery, and was within the tolerable limits. Fibrinogen was moderately elevated as well. However, it was in the upper-normal range and could be attributed to the stress produced for invasive procedures (Fernández-Pacheco et al. 2007).

As a first approximation, all the test results showed the good biological compatibility of the Fe@C nanoparticles in gelafundine suspension, although further experiments have to be carried out to look for very mild adverse effects, in particular looking for the administration of nanoparticles to human patients.

6. Fe@C Nanoparticles for Drug Delivery: Drug Absorption and Release Experiments

With the objective of using the Fe@C nanoparticles as drug carriers for targeted drug delivery, the ability of these particles to spontaneously adsorb a drug on their surface, as well as the subsequent spontaneous drug release is a key issue. We have been concentrating on the use of doxorubicin as test drug (Fernández-Pacheco et al. 2007). We have followed this procedure: a 100 mg/ml solution of doxorubicin hydrochloride was mixed with 1 mg/ml of the particles in distilled water and placed on a shaker at room temperature (Kuznetsov et al. 1999). Aliquots were taken 5, 15, 30, 60, 90, 120, and 180 minutes after starting the incubation. Each time, the adsorbent (i.e., the Fe@C nanoparticles) was magnetically separated in a gradient field produced with a 3 kOe permanent magnet, and the optical density of the supernatant was measured with a UV spectrophotometer at 498 nm wavelength. The corresponding spectra are displayed in Figure 10.9a, where the time evolution of the absorption peaks of the doxorubicin (at 296 and 498 nm) is clearly seen. The concentration of the adsorbed drug was then calculated from the intensity of the absorption peaks, and represented as a function of the time. According to Figure 10.9b, 39% of the initial drug concentration was adsorbed in 3 hours, and after that the saturation was reached. To study the doxorubicin desorption kinetics in *in vivo* conditions, 5 milligrams of particles were incubated in a solution of doxorubicin. Once the drug adsorption had taken place, the nanoparticles were separated magnetically, and subsequently mixed with 5 ml of human plasma at 36°C under mild stirring. The magnetic phase was magnetically separated at 2 hours intervals during the first day, and every 8 hours thereafter. At each interval, the optical density of the supernatant was measured with a UV spectrophotometer at 296 nm to calculate the amount of released drug (see Figure 10.9c). According to Figure 10.9d, doxorubicin was released from the

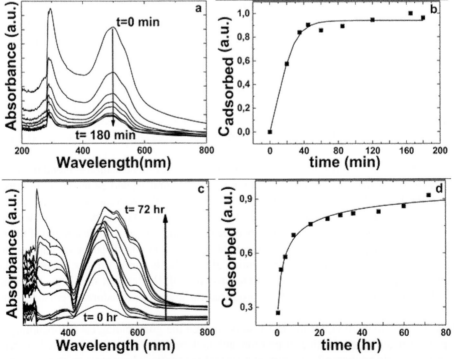

Figure 10.9: (a) Absorption spectra corresponding to the doxorubicin adsorption process on Fe@C nanoparticles and (b) kinetics of the adsorption process. (c) Absorption spectra corresponding to the doxorubicin desorption process and (d) kinetics of the desorption process. Arrows in (a) and (b) indicate the time sequence in which the experiments were done (after Fernandez-Pacheco et al. 2007).

surface of nanoparticles at a rate slow enough to reach the target organ. After 3 days, 84% of the initially adsorbed drug was released. Therefore, from the point of view of drug adsorption and release, the Fe@C nanoparticles are suitable carriers for drug delivery.

7. Response to Magnetic Field Gradients Created by Magnetic Implants

In order to use the Fe@C nanoparticles as drug carriers for targeted drug delivery, besides their capability to absorb and desorb a drug, it is also necessary to evaluate whether it is possible to target a certain place in the body. Once this objective is reached, the nanoparticles should remain there and release the drug in a controlled way (Widder et al. 1978a, b, Lubbe et al. 1996, 2001, Alexiou et al. 2000, Pankhurst et al. 2003, Gupta and Gupta 2005). In our case we use magnetic forces, and therefore, it is necessary to study the carrier response to the magnetic field created at the place of interest. Another key issue, as already mentioned in previous sections, is the response of the immune system once the nanoparticles have been administrated systemically to the patient. This could be one of the major drawbacks of the method and therefore, we should be sure that a significant fraction of the administered carriers reach the tissue to perform the chemotherapeutic effect there.

Guiding the magnetic carriers towards the target is done by applying magnetic fields. Here we can consider two approaches followed in different research groups. One is exposing the patient to an external focused applied magnetic field (Alexiou et al. 2000). This methodology has some drawbacks. First of all, given the decrease of the magnetic field with the distance, the therapy would only be advisable for the treatment of superficial organs. Second, the geometry of the magnetic field has to be carefully designed to achieve the perfect focusing of the magnetic field in the region of interest, to concentrate the nanoparticles there, and to promote an effective drug release (Lubbe et al. 2001, Shinkai 2002, Alexiou et al. 2005). Finally, the drug loaded magnetic particles have to be injected in a blood vessel close to the target organ.

An alternative to the use of external magnets is producing the magnetic field by a small permanent magnet implanted in the organ of interest, and therefore creating an internal magnetic field gradient. This alternative was tested in *in vivo* models as New Zealand rabbits (Fernández-Pacheco et al. 2007) and CD-1 mice (Escribano et al. 2012). In Fernández-Pacheco et al. (2007), the authors used a small cylindrical gold-coated neodymium–iron–boron permanent magnet (4 mm diameter and 2 mm height). Such a magnet can create a large field gradient of the order of 5 kOe/cm. For the experiments, a biocompatible magnetic fluid was prepared, suspending 100 mg particles in 100 ml gelafundine. Two milliliters of this nanoparticle suspension was injected intravenously in the marginal ear vein of 36 New Zealand rabbits, in whose left kidney (target organ) a magnet had previously been implanted by laparoscopic surgery. The right kidney (control organ) was also punctured (although no magnet was inserted) to produce the same tissue damage as in the other kidney. Groups of animals were sacrificed by terminal anesthesia at half an hour, 2, 7, 14, 21, and 28 days after the injection of the magnetic suspension. Their lungs, liver, spleen, kidneys, and spine were removed, and tissue samples were processed for histological study. All the procedures are described in detail in Fernández-Pacheco et al. (2007) and Fernández-Pacheco (2008). No iron was detected in any of the studied tissues after Perls' Prussian blue staining, which confirmed the good encapsulation of the magnetic particle core inside the graphitic carbon shell, in good agreement with the physicochemical characterization of the particles. As an example, Figure 10.10 shows an optical microscopy image of tissue (after haematoxylin-eosin staining) from a left kidney of a rabbit sacrificed 2 days after nanoparticle injection. The black spots pointed by arrows in Figure 10.10 are nanoparticle agglomerates. These features were common to all the optical microscopy images corresponding to tissue samples taken from the left kidneys.

Despite the difficulties of this method for an accurate quantification of the nanoparticle concentration in each tissue, the amount of nanoparticles found in the left kidney was always higher than that found in the control kidney. Also, in the samples corresponding to the left kidney, the nanoparticles inside each aggregate were clearly oriented in the direction of the magnetic field

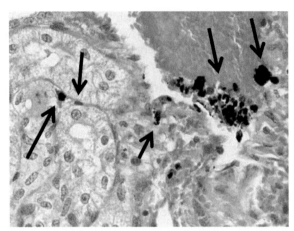

Figure 10.10: Optical microscopy image of tissue from the left kidney (with magnet implant) of a rabbit sacrificed 2 days after nanoparticle injection (haematoxylin-eosin stain; X 400 magnification). Arrows point to Fe@C nanoparticle aggregates.

gradient. This clearly evidences a good response of the Fe@C nanoparticles to the magnetic field, and reciprocally, the capability of the implanted magnets to create a magnetic field gradient large enough to target the particles to the desired organ. Implanted magnets are therefore a promising and effective alternative to the use of external magnetic fields in targeted drug delivery therapies. However, a relatively large amount of the particles was also observed in tissue samples corresponding to liver and spleen. A very small amount was detected in samples taken from the lungs of some animals. These observations suggest that once the Fe@C nanoparticles are in the bloodstream, they are recognized and captured by the macrophages of the respective organs, despite the excellent biocompatibility of the nanoparticles shown by the *in vitro* hematology experiments. This reaction of the reticuloendothelial system should be avoided, although the capture of the nanocarriers by the liver or spleen macrophages could also be helpful for other applications, such as treatment of liver or spleen tumors, or for the treatment of leishmaniasis (Chellat et al. 2005).

There are several alternatives to avoid the capture of nanoparticles by the immune system. As already mentioned, some strategies aim at improving the nanoparticle coating and decreasing the size of the aggregates in suspension. *In vivo* experiments injecting PEG-coated Fe@C nanoparticles in New Zealand rabbits under conditions identical to those described above showed a significant decrease of nanoparticles in liver and spleen tissues. This indicated the effectiveness of PEG coating in improving the nanoparticle suspension, and increased the residence time of the magnetic carriers in the bloodstream (Fernández-Pacheco 2008). Another strategy tested with the Fe@C nanoparticles was to inject them directly into the carotid and abdominal aorta and in the renal arteries of the rabbits (Fernández-Pacheco 2008). The analysis of the tissues extracted after having administered the nanoparticles by these routes showed a significant increase in the amount of nanoparticles in the test kidney, compared to that found in the other organs.

An alternative method that prevents the capture by the macrophages is the direct supply of the nanocarriers via direct puncture and delivery into the target organ. These experiments were performed using 1 ml of a 1% Fe@C nanoparticle suspension in gelafundine that was injected directly into the test kidney of a group of rabbits. In these cases, the magnet was implanted by laparoscopic techniques, as already explained, and the particles were injected with the aid of an ultrasound scan equipment (eco-guided injection) (Fernández-Pacheco 2008). This facilitated the exact visualization of the kidney and the position of the magnet. This ultrasound scan-assisted injection technique also helps to position the particles as close to the magnet as possible, so that they remain there for a long time. The concentration of nanoparticles in the vicinity of the magnet implies an increase of metallic nuclei around it, which leads to a widening of the ultrasound echo. The images obtained from the

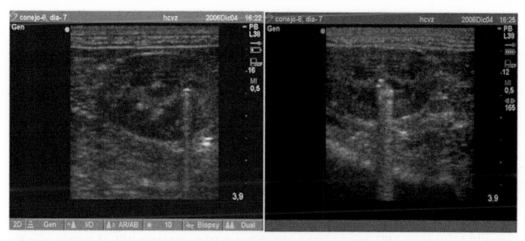

Figure 10.11: The ultrasound scan displays the image of a goldcoated magnet inserted inside a kidney (left image). After the injection, magnetic nanoparticles remain attracted to the magnet, the amount of metallic material increases, and the echo of the ultrasound is broadened (right image).

abdomen of the different animals showed that the largest amount of particles was located in the renal parenchyma (see Figure 10.11). Some amount of nanoparticles was still detected in liver and spleen (Fernández-Pacheco 2008).

8. *In Vivo* Study of the Bio-distribution of Fe@C Nanoparticles: A Magnetic Resonance Imaging Study

Besides the *in vitro* biocompatibility and cell viability tests, it is necessary to study the bio-distribution of Fe@C nanoparticles and the time evolution after *in vivo* systemic administration. With such a purpose, we performed a Magnetic Resonance Imaging (MRI) study in a preclinical animal model. MRI is one of the most powerful diagnostic tools in medicine, due to its non-invasive nature and spatial resolution (Livingston 1996, Browne and Semelka 1999, Elster and Burdette 2001). Due to their magnetic properties, Fe@C core-shell nanoparticles are very promising materials to synthesize biocompatible magnetic fluids, able to modify the longitudinal T1 and transversal T2 proton relaxation of water in body tissues. Consequently, these nanoparticles are able to modify the contrast in MRI images at the place where the nanoparticles accumulate (Weissleder et al. 1990, Lawaczeck et al. 1997). For these bio-distribution studies, a suspension of Fe@C nanoparticles in gelafundine at 1% wt concentration was prepared, and a group of New Zealand rabbits were injected by the marginal ear vein with 2 ml of this suspension (which amounts to approximately 0.088 mMol Fe/kg bw). MRI scans were performed in a 1.0 T whole-body scanner (NT Intera 1.0 T Power R.11, Philips Medical Systems International B.V.), equipped with a quadrature knee/foot coil. The experimental procedure was previously evaluated by the Ethics Committee for Animal Research of the University of Zaragoza (Spain). T1- and T2-weighted images of the rabbit abdomen were generated by Turbo Spin-Echo (TSE) sequences. Scans were performed in sagittal and coronal orientation, with the animal positioned in lateral decubitus. The parameters for the sagittal TSE-T1 sequences are: spin echo, turbo factor: 3, TR = 700–870 ms, TE = 18 ms, voxel size for acquisition = 0.59 × 0.74 × 3 mm, pixel size for final image (reconstruction) = 0.3125 × 0.3125 mm. For the coronal TSE-T1 sequences, we used: spin echo, turbo factor: 3, TR = 350 ms, TE = 18 ms, voxel size for acquisition = 0.59 × 0.74 mm, pixel size for final image (reconstruction) = 0.3125 × 0.3125 mm. For the sagittal and coronal TSE-T2, the parameters were the following: spin echo, turbo factor: 11, TR = 4036 ms, TE = 100 ms, voxel size for acquisition = 0.59 × 0.74 mm × 3 mm; pixel size for final image (reconstruction) = 0.37 × 0.37 mm. We obtained images 15 minutes after injection of the nanoparticle suspension, and periodically repeated over eighteen months. The amount of nanoparticles in liver, kidney, bone marrow, and para-vertebral muscle was qualitatively estimated from each image. We determined the variation in the signal intensity caused by the presence of nanoparticles in the respective tissues. This allowed following the distribution of the Fe@C nanoparticles in the organism, and its evolution over the time.

The images obtained using T1-weighted TSE sequences turned out to be the most suitable for this study, given the value of the magnetization of the particles and their concentration in the fluid. In T1-weighted TSE images, the presence of metallic objects in an organ causes a decrease in the MRI signal. Figures 10.12a and b show the T1-weighted TSE image of the rabbit abdomen obtained before injecting the Fe@C suspension, and the image obtained 15 minutes post-administration, respectively. The comparison of both images clearly shows a darkening of the rabbit liver (outlined by a dotted line) 15 minutes after the injection. This intensity decrease is due to the accumulation of nanoparticles in this organ, and suggests that they are cleared in a few minutes by the Reticulo-Endothelial System (RES) cells, being captured by the liver macrophages (Lawaczeck et al. 1997, Pankhurst et al. 2003). Figure 10.12c shows a MRI image of the rabbit abdomen taken approximately 18 months after the administration of the nanoparticle suspension. In comparison to Figure 10.12b, the intensity of the liver in the image has now increased, indicating that the amount of nanoparticles in this organ has decreased. This indicates that nanoparticles have been either metabolized in the liver, so that iron has become a part of compounds that can be assimilated by the organism (Escribano et al. 2012), or have left the liver, being excreted in the urine and feces (Jain et al. 2008, Yu and Zheng 2015).

From the inspection of the different MRI images of the abdomen, it was found that, as a general trend, liver and bone marrow were the organs in which the accumulation of nanoparticles 15 minutes after injection was higher. The accumulation was lower in kidneys (although the intensity in the images was slightly affected by the respiratory movements of the anesthetized animal) and in the para-vertebral muscle. To study the time evolution of the bio-distribution of the nanoparticles in more detail, measurements of the signal intensity were carried out on the images. The intensity corresponding to each organ resulted from averaging three values obtained when measuring in three different places of the respective organ, taking the Regions of Interest always with area equal to 0.2 cm². Each intensity value was then divided by the intensity of the para-vertebral muscle, taken as reference for the calculation of the intensity index. The intensity indices corresponding to the liver, kidney, and bone marrow were plotted as a function of the time, showing the time evolution of the accumulation of the Fe@C nanoparticles in each organ. The results are displayed in Figure 10.13.

In all cases, after slight variations during the first 25 days after nanoparticle injection, the MRI signal increases with time, which indicates that the nanoparticles leave the respective organs. According to this graph, the organ where most nanoparticles accumulate is the bone marrow, followed by kidney and liver, whose macrophages seem to capture a very similar amount. In addition, the variations

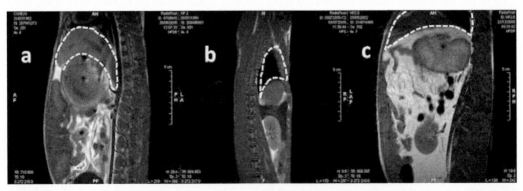

Figure 10.12: T1-weighted TSE images of the rabbit abdomen obtained (a) before injecting the Fe@C suspension, (b) 15 minutes, and (c) 530 days post-administration, respectively. The discontinuous white line surrounds the rabbit's liver.

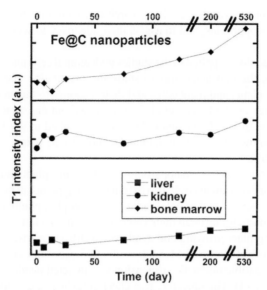

Figure 10.13: Time dependence of the intensity index (in arbitrary units) calculated from T1-weighted TSE images, corresponding to the time variation of the amount of Fe@C nanoparticles accumulated in the rabbit's bone marrow (upper panel), kidney (middle panel), and liver (lower panel).

observed during the 25 days post-administration appear to be correlated with each other, especially in the case of liver and kidney.

A more accurate calculation of the nanoparticles accumulated in each organ could be done by taking images of phantoms in which a known quantity of nanoparticles had been injected and using the intensity of the signal obtained for further calibration. However, this was beyond the scope of the study presented here. The results obtained showed that the MRI technique is an excellent tool to study the distribution of magnetic nanoparticles within an organism and the evolution of this distribution along the time. In addition, these experiments proved that Fe@C nanoparticles create a high enough contrast in MRI images, and therefore Fe@C biocompatible fluids could be used as a contrast agent. First of all, it will be desirable to improve the quality of the suspensions in order to avoid the nanoparticles being cleared from the circulatory system by the macrophages, and therefore increase the residence time of the nanoparticles in blood. Secondly, a detailed study of the suspension relaxivity by Time Domain NMR *in vitro* experiments will be also necessary to assess the effectiveness of the magnetic biocompatible fluid as contrast agent. The nanoparticles could be conjugated with a biomolecule of interest (an antibody, tumor marker receptor, etc.) linked to their coating. This would allow the specificity of the MRI contrast agent for a particular organ or tumor marker. Moreover, making use of the capability of storing a chemotherapy drug adsorbed onto the carbon coating, as seen in a previous section, confers the nanoparticles a multifunctional character, making them useful for both diagnosis and therapy.

9. Fe@C Nanoparticles for Plant Cell Research and Agronomy

The ongoing research and results obtained till now regarding the use of magnetic nanoparticles for biomedical applications open a wide range of possibilities for their use in other disciplines of Biology, for example in general, plant research, agronomy, and biotechnology. The nanoparticles could be used to detect fungi, bacteria, or parasites in a crop, to fight pests by releasing pesticides or fungicides inside a plant, for the improvement of species through genetic manipulation, etc. The promising results obtained with Fe@C nanoparticles in *in vitro* and *in vivo* experiments with animal cells and animal models suggest that these nanoparticles can work as well in the applications involving plant cell and crops. We present an overview of the research carried out to study the interaction between plant cells and Fe@C nanoparticles, and to assess their adequacy to work in the above mentioned fields in this section.

Following the methodology applied in the studies with animal cells and animal models presented in the previous sections, the first stage was to study the penetration and transport of the Fe@C nanoparticles into living plants and plant cells, and their response to magnetic field gradients. After the earlier work by Torney et al. (Torney et al. 2007), carried out on isolated plant cells and intact leaves, the experiments with Fe@C nanoparticles being described in the following sections were, for the first time to our knowledge, carried out inside whole living plants (González-Melendi et al. 2008, Corredor et al. 2009).

To study the transport of Fe@C nanoparticles inside living plants and their interaction with magnetic fields, a series of experiments were carried out on living pumpkin plants, which were injected with Fe@C nanoparticles suspended in gelafundine. The pumpkin plants were chosen to facilitate the transport of the nanoparticles through their vascular system, given the large size of their vessels. The procedures regarding plant growing and synthesis of the nanoparticle gelafundine suspension can be found in some works (González-Melendi et al. 2008, Corredor et al. 2009, Pérez-de-Luque et al. 2012). This biocompatible magnetic suspension was injected inside the internal hollow of the leaf petiole (see Figure 10.14). The objective was checking whether the suspension penetrated into the plant and translocated to different parts, travelling through the plant vascular system. In addition, small magnets (discs about 5 mm diameter) were placed in different regions of the plant to create magnetic field gradients, making Fe@C nanoparticles move and concentrate at the position of the magnets. These magnets were placed on the leaf petiole opposite to the injection point and on the roots (see Figure 10.14).

Samples of several plant tissues were collected 24, 48, 72, and 168 hours after the injection of the magnetic gelafundine suspension, and processed for microscopy analysis. Tissue samples were collected from the stem and leaf petiole at the injection point. Roots and petiole samples were collected at the point where magnets were placed, but also before and after the magnet position (i.e., facing the expected movement of the nanoparticles from the injection point towards the magnet). The samples were subsequently observed by light microscopy, by a confocal laser scanning microscope, and by TEM. Sample processing and observing protocols are thoroughly described in González-Melendi et al. (2008) and Corredor et al. (2009).

Figure 10.15 shows the images corresponding to sections of petioles (upper row) and of roots (lower row), taken at the injection point (A), before the position of the magnet (D), at the magnet point (B and E), and after the magnet position (C and F). Figure 10.15a shows a detail of the vascular

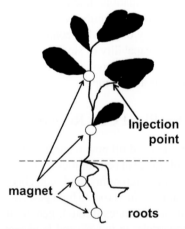

Figure 10.14: Sketch of a pumpkin plant for the experiments. Arrows point to the magnet positions. The leave petiole for the injection of the Fe@C nanoparticle suspension is also marked.

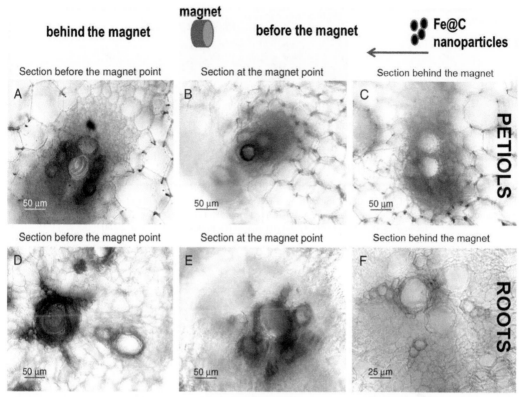

Figure 10.15: Light microscopy images of vascular tissue of the petiole cut (A) at the application point; (B) adjacent to a magnet; (C) opposite to a magnet. Light microscopy images of vascular tissue of the root tissue cut (D) before the position of the magnet; (E) at the position of the magnet; (F) behind the magnet (after González-Melendi et al. 2008).

tissue at the application point. The dark areas in the images were identified as nanoparticle aggregates. The fact that these dark accumulations also appear in Figures 10.15b, d, and e evidenced that the Fe@C nanoparticles moved from the injection point towards the other parts of the plant where the magnets were placed, and that the magnets concentrated them in the vascular tissues there. Images in Figures 10.15d and e demonstrated that the nanoparticles are able to travel long distances (i.e., even down to the roots) due to the attraction of the magnetic field gradient. No aggregates were identified in Figures 10.15c and f, which means that the magnets trap the nanoparticles in the suspension, preventing them from travelling beyond.

Tissue samples were also analyzed by confocal and fluorescence microscopies. Nanoparticle aggregates were clearly visualized by the differential interference contrast (DIC) or Nomarski technique on a projection of a 3-D confocal stack, and also illuminating the samples with the Ar laser (488 nm) and collecting the reflected signal in the range of 485–500 nm. Figure 10.16 shows the images taken in the confocal microscope of a cell of the stem cortex next to the internal hollow of the petiole just before the place where the magnet is located, collected 72 hours after the injection. A nanoparticles aggregate was clearly visualized by both Nomarski (Figure 10.16a) and reflection (Figure 10.16b) techniques. Figure 10.16c shows the overlay of Figures 10.16a and b with an almost complete co-localization. These structures were not observed in control samples (i.e., from non-injected plants or plants injected only with gelafundine), discarding that the structures visualized in the images were artefacts (due for example to sample processing, etc.). In addition, nanoparticles were also detected in the cell wall of the xylem vessels.

Tissues from different parts of the plants taken at different periods after the administration of the nanoparticle suspension were analyzed by correlative light and TEM studies. Nanoparticle

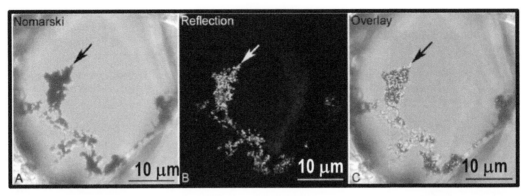

Figure 10.16: Projection of 3-D confocal stacks from a cell of the stem at the position of the magnet, 72 hours after nanoparticle administration. Aggregates of nanoparticles (marked by arrows) are detected by (A) Nomarski and (B) reflection techniques. (C) Overlay of (A) and (B) images (González-Melendi et al. 2008).

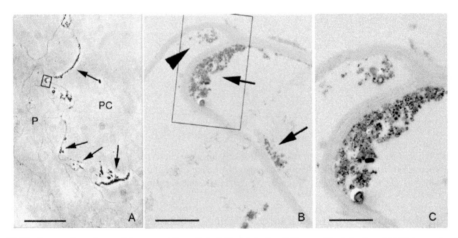

Figure 10.17: Images from tissue of the stem at the point of administration 24 hours after injection. (A) Phase contrast image of the parenchymatic cells (P) closer to the pith cavity (PC). The arrows point to nanoparticle aggregates. (B) TEM image of the region selected in (A). The arrows point to nanoparticle aggregates in the cell wall surrounding the pith cavity and into the cytoplasm of the first cell layer. (C) High magnification of the region squared in (B). Scale bars are 40 μm (in A), 2 μm (in B), 1 μm (in C) (Corredor et al. 2009).

aggregates were detected in the internal wall of the pith cavity of the stem at the point of administration 24 hours after injection. They were found in the cell wall as well as in the cytoplasm (see Figure 10.17).

Inspection of the images taken 48 hours after administration around the injection point showed that these nanoparticles had migrated into the stem parenchyma. Images taken from the vascular core (see Figure 10.18) showed that nanoparticles accumulated inside the xylem vessels 48 hours after injection. These images suggested that xylem vessels were probably the path followed by the nanoparticles in their long distance movement from the point of injection towards the roots.

TEM analysis of samples collected 48 hours after injection, close to the magnets placed far from the point of injection, revealed the presence of individual nanoparticles in the cytoplasm of cells close to the vascular system.

The careful inspection of TEM images allowed measuring the size of individual nanoparticles. Most of the Fe@C nanoparticles were equal or below 10 nm. The largest particles had a diameter around 50 nm, although they were very few in number. This fact suggests that in plants there is a size selection mechanism (probably involving cell walls and waxes) by which larger particles do not penetrate in plant tissues. However, more studies are necessary to clarify this issue. TEM images

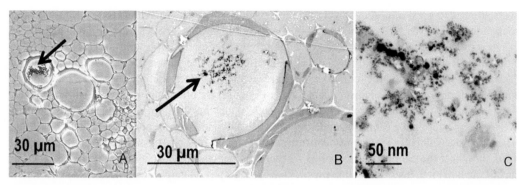

Figure 10.18: (A) Phase contrast image of a vascular core cut 48 hours after the administration of the nanoparticles; arrow points to a nanoparticle aggregate inside of a xylem vessel. (B) TEM image of the xylem vessel. (C) Detail of a region in (B) (Corredor et al. 2009).

from tissues from trichomes in the outer surface of the plant showed the presence of nanoparticles both inside and outside of the trichomes, already 24 hours after the application of the nanoparticle suspension. This fact points to a sort of detoxification mechanism by which the plant would get rid of the excess nanoparticles. The images taken 168 hours after administration showed almost no nanoparticles either in the pit cavity or in the stem tissues.

Looking for other ways of administration of the nanoparticles, which were more practical for real applications (e.g., for plague and parasite control), the same gelafundine suspension was sprayed on the leaves of a pumpkin plant (Corredor et al. 2009). Tissue samples (close and far from the application point) were also collected 24, 48, and 168 hours after spraying. However, nanoparticles were only observed in samples from the petioles close to the sprayed leaves where a magnet was placed, and collected 168 hours after administration. The amount of nanoparticles was very small in all the cases, so nanoparticles were only detected when analyzing the samples by TEM. The TEM images revealed that the nanoparticles were accumulated in cells from the first epidermal layer of the petioles, although no large aggregates were observed. The most likely nanoparticle penetration route appeared to be through the stomata and sub-estomatic chambers. The fact that pathogens of different species choose this path to enter some plants supports this assumption, although more tests are needed to confirm it. Moreover, the nanoparticle suspension can still be optimized to facilitate the penetration of the nanoparticles. An alternative would be suspending the nanoparticles in other biocompatible liquid carrier, as for example mannitol (Cifuentes et al. 2010). Besides, the functionalization of the carbon shell of the nanoparticles with the appropriate biomolecules could also improve the penetration of the suspension trough the leaves, making spray administration an easy method to use in practice by plant researchers, and also for agronomists in field applications.

Finally, the absorption and translocation of Fe@C nanoparticles was also studied when the nanoparticle suspension was applied by the plant roots (Cifuentes et al. 2010). In addition, in this case, no magnets were placed to study whether the Fe@C nanoparticles were able to move freely inside the plant, and to study the distribution in the different organs, in a way similar to the study carried out with the same particles in the *in vivo* animal model discussed in the previous sections. Four living crop plants of different families were selected for these experiments, with the objective of making a comparative study of the paths followed by the nanoparticles, and of the organs and tissues where nanoparticles tended to concentrate. The plants selected were sunflower (*Helianthus annuus*) from the family *Compositae*, tomato (*Lycopersicum sculentum*) from the *Solanaceae*, pea (*Pisum sativum*) from the *Fabaceae*, and wheat (*Triticum aestivum*) from the *Triticeae*. In this case, to have a proper visualization of the roots, the plants were grown *in vitro* in a rhizotron, a petri dish system. The carbon-coated nanoparticles were suspended in 1% manitol solution (Etxeberria et al. 2007). To ensure that the experiments were made from plants at the same stage of development, they were allowed to grow until they developed their second pair of leaves. At this time the nanoparticle mannitol solution was applied, immersing only some of their roots. Tissues from different parts of the plants were taken after 24 and 48 hours, and fixed for further microscopic analysis by a conventional light microscope. The protocol for sample preparation and observation conditions can be found in Cifuentes et al. (2010). For the four crops, large areas stained in black staining corresponding to Fe@C nanoparticle aggregates were observed in root tissues collected after 24 hours. Tissues from roots not exposed directly to the suspension were also inspected under the microscope. Nanoparticle aggregates were detected within the central cylinder of roots located diametrically opposite to the treated roots. Therefore, the nanoparticles had probably moved through the phloem (Oparka and Cruz 2000). These facts show that the nanoparticles applied by the roots penetrate much faster than when applied by any of the two methods discussed in the previous paragraphs. Nanoparticles were easily detected in the xylem vessels of all the specimens studied, although there were some differences from crop to crop. Pea roots accumulated higher contents of nanoparticles than sunflower or wheat, suggesting that pea roots could be more permeable to the nanoparticles. To study whether the nanoparticles could travel from the root up to the aerial part of the plant, tissues were taken from the plant crown 24 and 48 hours after the immersion of the roots into the nanoparticle suspension. In all the cases, 24 hours after the immersion, the images of the xylem vessels in the upper areas showed the presence of black deposits, indicating that the vascular system of the plants was the path followed by the nanoparticles heading for the aerial part. The concentration of nanoparticles was the highest in the vascular tissues of the crown of pea and wheat, which suggests a higher nanoparticle transportation rate in these cases. In the case of sunflower, it seems that the nanoparticles uptake through the roots is much slower than in the other species. To study whether the analyzed crops developed any kind of nanoparticle excretion system as observed in the experiments with pumpkin plants (González-Melendi et al. 2008, Corredor et al. 2009), the tissues from leaf trichomes were processed and observed under the light microscope as well. Very large accumulations of nanoparticles (larger than those observed in pumpkin plants) were detected only in the leaf trichomes of the wheat plants. This fact suggests differences in the physiology of the monocots (group to which the wheat belongs) and of the dicots (as it is the case of pea, sunflower, and tomato). The samples taken 48 hours after nanoparticle application showed no significant differences from crop to crop, and an intense accumulation of nanoparticles was detected in all the cases. According to the images, the nanoparticles also moved towards the leaves and leaf petioles. Therefore, these results and those obtained in pumpkin plants showed that the Fe@C nanoparticles are adequate to prepare biocompatible suspensions to be administered to whole living plants, by injection into the leave petioles, by spraying on the leaves, and by immersion of the roots in the particle suspension. This last way of administration seems to be the most reliable and efficient in terms of the amount of nanoparticles delivered to the different plant tissues. The Fe@C nanoparticles can travel through the vascular system of the plant from the root to the aerial parts, with or without the aid of magnetic field gradients created by small magnets. All these results mean there is already a great advance towards the definitive application of magnetic nanoparticles, and in particular of the core-shell Fe@C nanoparticles in plant research, agriculture, and agronomy. However, there is still a wide field open for research. In particular, it is necessary to clarify the mechanisms by which nanoparticles can penetrate into the plant cells, to explore the potential of nanoparticles as smart treatment-delivery carriers in plants. Another important issue is the potential toxicity induced by the nanoparticles to the plants. No damage was observed in any of the crops in between the administration of the nanoparticles and the time at which the sampling took place. In addition, some of the treated plants not used in the experiments were transplanted in pots, and grew until maturity like normal plants, without apparent signs of damage. Nevertheless, cell damage at local level without external evidences was reported by other authors (Pavel et al. 1999, Cotae and Creanga 2005, Pavel and Creanga 2005), which highlights the need to perform cytotoxicity studies. In addition, the effect of nanoparticles on soils and entering the animal and human food chain must be considered and thoroughly studied.

10. Summary

This review reports on the relevance of the carbon-coated functional magnetic nanostructures and, in particular, presents a broad description of the role that the carbon encapsulation plays in the functionalization of magnetic nanoparticles. This constitutes a scenario for the incorporation of the carbon-coated magnetic nanoparticles in life sciences. Of particular interest is the biocompatibility of such nanostructures for being used in new focalized therapies and diagnosis. The work presented here clearly demonstrates their capability as drug carriers to adsorb and desorb some specific therapeutic agents. Their systemic administration needs an additional agent to retain these specific nanocarriers at the target organ. With this in mind, we have resorted to the strategy of generating an internal magnetic field gradient to guide and retain the magnetic carriers at the site of interest, taking advantage of their magnetic properties. This has been tested in animal models, generating the gradient by magnets implanted using laparoscopic techniques. The carbon-coated magnetic nanoparticles are also promising as contrast agents for MRI tomography, and might also be tracked by ultrasounds, broadening the potential applications of these nanosystems. These applications have been considered on the basis of the fundamental role played by the graphitic and amorphous encapsulation in protecting the magnetic core and in providing it with the necessary biocompatibility. We have demonstrated the cell uptake of these carbon-coated nanoparticles and the subsequent cell viability, thus achieving magnetic cells. A dendritic cell model has led to a new method to carry the magnetic nanoparticles towards a metastatic region with high angiogenesis activity. The morphology and magnetic properties of these nanoparticles also constitute the basis for a signaling method to track and accumulate specific phytosanitary agents in plants. This is a scarcely investigated field of research, which may have a great impact on the field of environmental sciences.

Acknowledgments

The authors wish to thank the following authors and colleagues: R. Fernández-Pacheco, M. Arruebo, G.F. Goya, D. Serrate, A. Tres, J.G. Valdivia, A. Viloria, M. Gutiérrez, T. Higuera, L. Asín, B. Saez, R. Cornudella, A. García, J.A. García de Jalón, A. Laborda, M.S. Romero, Z. Cifuentes, M.J. Coronado, E. Corredor, J.M. de la Fuente, P. González-Melendi, A. Pérez de Luque, P.S. Testillano, M.C. Risueño, and D. Rubiales. The authors also acknowledge financial support from the Spanish Ministerio de Economía y Competitividad (MINECO) through project MAT2016-78201-P, and from the Department of Innovation, Research and University of the Government of Aragon through the Research Groups grants program co-financed by the FEDER Operational Program Aragón 2014–2020 "Building Europe from Aragon". Support from the Advanced Microscopy Laboratory (LMA) and Servicio General de Apoyo a la Investigación of the University of Zaragoza is also acknowledged. The authors also thank the following publishers and journals: Inderscience Publishers and International Journal of Nanotechnology (https://www.inderscience.com/jhome.php?jcode=ijnt), as the original source of Figure 10.5 published in De Teresa et al. 2005, doi: 10.1504/IJNT.2005.006971; Elsevier Ltd. and Cell Biology International (www.elsevier.com/locate/cellbi), as the original source of Figure 10.7, published in Goya et al. 2008, doi:10.1016/j.cellbi.2008.04.001; Elsevier B.V. and Journal of Magnetism and Magnetic Materials (www.elsevier.com/locate/jmmm), as original source of Figure 10.9, published in Fernández-Pacheco et al. 2007, doi: 10.1016/j.jmmm.2006.11.192; Oxford University Press and Annals of Botany (www.aob.oxfordjournals.org), as original source of Figures 10.15 and 10.16, published in González-Melendi et al. 2008, doi:10.1093/aob/mcm283; and BioMed Central and BMC Plant Biology (http://www.biomedcentral.com/1471-2229/9/45), as the original source of Figures 10.17 and 10.18, published in Corredor et al. 2009, doi:10.1186/1471-2229-9-45.

References

Alexiou, C., W. Arnold, R.J. Klein, F.G. Parak, P. Hulin, C. Bergemann et al. 2000. Locoregional cancer treatment with magnetic drug targeting. Cancer Res. 60: 6641–6648.

Alexiou, C., R. Jurgons, R. Schmid, A. Hilpert, C. Bergemann, F. Parak et al. 2005. *In vitro* and *in vivo* investigations of targeted chemotherapy with magnetic nanoparticles. J. Magn. Magn. Mater. 293: 389–393.

Aliev, F.G., M.A. Correa-Duarte, A. Mamedov, J.W. Ostrander, M. Giersig, L.M. Liz-Marzan et al. 1999. Layer-by-layer assembly of core-shell magnetite nanoparticles: Effect of silica coating on interparticle interactions and magnetic properties. Adv. Mater. 11: 1006–1010.

Ambe, K., M. Mori and M. Enjoji. 1987. Early gastric-carcinoma with multiple endocrine cell micronests. Am. J. Surg. Pathol. 11: 310–315.

Arenal, R., L. De Matteis, L. Custardoy, A. Mayoral, M. Tence, V. Grazu et al. 2013. Spatially-resolved EELS analysis of antibody distribution on biofunctionalized magnetic nanoparticles. Acs Nano 7: 4006–4013.

Arruebo, M., M. Galan, N. Navascues, C. Tellez, C. Marquina, M.R. Ibarra et al. 2006a. Development of magnetic nanostructured silica-based materials as potential vectors for drug-delivery applications. Chem. Mat. 18: 1911–1919.

Arruebo, M., R. Fernández-Pacheco, S. Irusta, J. Arbiol, M.R. Ibarra and J. Santamaria. 2006b. Sustained release of doxorubicin from zeolite-magnetite nanocomposites prepared by mechanical activation. Nanotechnology 17: 4057–4064.

Arruebo, M., R. Fernández-Pacheco, M.R. Ibarra and J. Santamaria. 2007a. Magnetic nanoparticles for drug delivery. Nano Today 2: 22–32.

Arruebo, M., R. Fernández-Pacheco, B. Velasco, C. Marquina, J. Arbiol, S. Irusta et al. 2007b. Antibody-functionalized hybrid superparamagnetic nanoparticles. Adv. Funct. Mater. 17: 1473–1479.

Athanassiou, E.K., R.N. Grass and W.J. Stark. 2006. Large-scale production of carbon-coated copper nanoparticles for sensor applications. Nanotechnology 17: 1668–1673.

Atici, O.G., A. Akar and R. Rahimian. 2001. Modification of poly(maleic anhydride-co-styrene) with hydroxyl containing compounds. Turk. J. Chem. 25: 259–266.

Banchereau, J. and R.M. Steinman. 1998. Dendritic cells and the control of immunity. Nature 392: 245–252.

Berry, C.C. 2005. Possible exploitation of magnetic nanoparticle-cell interaction for biomedical applications. J. Mater. Chem. 15: 543–547.

Blazewicz, M. 2001. Carbon materials in the treatment of soft and hard tissue injuries. Eur. Cells Mater.: 21–9.

Browne, M. and R.C. Semelka. 1999. MRI: Basic Principles and Applications. Wiley, New York.

Cadek, M., R. Murphy, B. McCarthy, A. Drury, B. Lahr, R.C. Barklie et al. 2002. Optimisation of the arc-discharge production of multi-walled carbon nanotubes. Carbon 40: 923–928.

Castrillón, M., A. Mayoral, C. Magen, J.G. Meier, C. Marquina, S. Irusta et al. 2012. Synthesis and characterization of ultra-small magnetic FeNi/G and NiCo/G nanoparticles. Nanotechnology 23: 085601 (10 pp).

Castrillón, M., A. Mayoral, A. Urtizberea, C. Marquina, S. Irusta, J.G. Meier et al. 2013. Synthesis and magnetic behavior of ultra-small bimetallic FeCo/graphite nanoparticles. Nanotechnology 24: 505702 (11 pp).

Chellat, F., Y. Merhi, A. Moreau and L. Yahia. 2005. Therapeutic potential of nanoparticulate systems for macrophage targeting. Biomaterials 26: 7260–7275.

Cifuentes, Z., L. Custardoy, J.M. de la Fuente, C. Marquina, M.R. Ibarra, D. Rubiales et al. 2010. Absorption and translocation to the aerial part of magnetic carbon-coated nanoparticles through the root of different crop plants. J. Nanobiotechnol. 8: 26.

Coaquira, J.A.H., H.R. Rechenberg, C. Marquina, M.R. Ibarra, A.M. Benito, W. Maser et al. 2001. Hyperfine and magnetic characterization of Fe particles hosted in carbon nanocapsules. Hyperfine Interact. 134: 103–108.

Conejo-García, J.R., F. Benencia, M.C. Coureges, E. Kang, A. Mohamed-Hadley, R.J. Buckanovich et al. 2004. Tumor-infiltrating dendritic cell precursors recruited by a beta-defensin contribute to vasculogenesis under the influence of Vegf-A. Nat. Med. 10: 950–958.

Coppola, D., L. Fu, S.V. Nicosia, S. Kounelis and M. Jones. 1998. Prognostic significance of p53, bcl-2, vimentin, and S100 protein-positive Langerhans cells in endometrial carcinoma. Hum. Pathol. 29: 455–462.

Corredor, E., P.S. Testillano, M.J. Coronado, P. Gonzalez-Melendi, R. Fernandez-Pacheco, C. Marquina et al. 2009. Nanoparticle penetration and transport in living pumpkin plants: *in situ* subcellular identification. BMC Plant Biol. 9: 45.

Cotae, V. and L. Creanga. 2005. LHC II system sensitivity to magnetic fluids. J. Magn. Magn. Mater. 289: 459–462.

Cullity, B.D. 1974. Introduction to Magnetic Materials. Addison-Wesley, Reading MA, USA.

De Teresa, J.M., C. Marquina, D. Serrate, R. Fernandez-Pacheco, L. Morellon, P.A. Algarabel et al. 2005. From magnetoelectronic to biomedical applications based on the nanoscale properties of advanced magnetic materials. Int. J. Nanotechnol. 2: 3–22.

Dumitrache, F., I. Morjan, R. Alexandrescu, R.E. Morjan, I. Voicu, I. Sandu et al. 2004. Nearly monodispersed carbon coated iron nanoparticles for the catalytic growth of nanotubes/nanofibres. Diam. Relat. Mat. 13: 362–370.

Elster, A.D. and J.H. Burdette. 2001. Questions and Answers in Magnetic Resonance Imaging. Elsevier–Health Sciences Division, St. Louis, USA.

Escribano, E., R. Fernández-Pacheco, J.G. Valdivia, M.R. Ibarra, C. Marquina and J. Queralt. 2012. Effect of magnet implant on iron biodistribution of Fe@C nanoparticles in the mouse. Arch. Pharma. Res. 35: 93–100.

Etxeberria, E., P. González and J. Pozueta-Romer. 2007. Mannitol-enhanced, fluid-phase endocytosis in storage parenchyma cells of celery (Apium graveolens; Apiaceae) petioles. Am. J. Bot. 94: 1041–1045.

Fearnley, D.B., A.D. McLellan, S.I. Mannering, B.D. Hock and D.N.J. Hart. 1997. Isolation of human blood dendritic cells using the CMRF-44 monoclonal antibody: Implications for studies on antigen-presenting cell function and immunotherapy. Blood 89: 3708–3716.

Fernández-Pacheco, R., M. Arruebo, C. Marquina, M.R. Ibarra, J. Arbiol and J. Santamaria. 2006. Highly magnetic silica-coated iron nanoparticles prepared by the arc-discharge method. Nanotechnology 17: 1188–1192.

Fernández-Pacheco, R., C. Marquina, J.G. Valdivia, M. Gutierrez, M.S. Romero, R. Cornudella et al. 2007. Magnetic nanoparticles for local drug delivery using magnetic implants. J. Magn. Magn. Mater. 311: 318–322.

Fernández-Pacheco, R. 2008. Synthesis, Characterisation and Functionalisation of Magnetic Nanoparticles for Biomedical Applications. Ph:D. Thesis, Universidad de Zaragoza, Zaragoza, Spain.

Fernández-Pacheco, R., J.G. Valdivia and M.R. Ibarra. 2009. Magnetic nanoparticles for local drug delivery using magnetic implants. pp. 559–570. In: J.W. Lee and R.S. Foote (eds.). Micro and Nano Technologies in Bioanalysis: Methods and Protocols. Humana Press-Springer Science+Business Media, New York, USA.

Foti, M., F. Granucci, D. Aggujaro, E. Liboi, W. Luini, S. Minardi et al. 1999. Upon dendritic cell (DC) activation chemokines and chemokine receptor expression are rapidly regulated for recruitment and maintenance of DC at the inflammatory site. Int. Immunol. 11: 979–986.

Furihata, M., Y. Ohtsuki, E. Ido, J. Iwata, H. Sonobe, K. Araki et al. 1992. HLA-DR antigen-positive and S-100 protein-positive dendritic cells in esophageal squamous-cell carcinoma—their distribution in relation to prognosis. Virchows Arch. 61: 409–414.

González-Melendi, P., R. Fernández-Pacheco, M.J. Coronado, E. Corredor, P.S. Testillano, M.C. Risueño et al. 2008. Nanoparticles as smart treatment-delivery systems in plants: Assessment of different techniques of microscopy for their visualization in plant tissues. Ann. Bot. 101: 187–195.

Goya, G.F., I. Marcos-Campos, R. Fernández-Pacheco, B. Saez, J. Godino, L. Asin et al. 2008. Dendritic cell uptake of iron-based magnetic nanoparticles. Cell Biol. Int. 32: 1001–1005.

Grüttner, C., S. Rudershausen and J. Teller. 2001. Improved properties of magnetic particles by combination of different polymer materials as particle matrix. J. Magn. Magn. Mater. 225: 1–7.

Gupta, A.K. and M. Gupta. 2005. Synthesis and surface engineering of iron oxide nanoparticles for biomedical applications. Biomaterials 26: 3995–4021.

Hahn, P.F., D.D. Stark, R. Weissleder, G. Elizondo, S. Saini and J.T. Ferrucci. 1990. Clinical-application of superparamagnetic iron-oxide to MR imaging of tissue perfusion in vascular liver-tumors. Radiology 174: 361–366.

Hart, D.N.J. 1997. Dendritic cells: Unique leukocyte populations which control the primary immune response. Blood 90: 3245–3287.

Hopken, U.E. and M. Lipp. 2004. All roads lead to Rome: Triggering dendritic cell migration. Immunity 20: 244–246.

Hunter, R.J., R. Matarese and D.H. Napper. 1983. Rheological behavior of polymer flocculated latex suspensions. Colloids Surf. 7: 1–13.

Iijima, S. 1991. Helical microtubules of graphitic carbon. Nature 354: 56–58.

Iturrioz-Rodriguez, N., E. Gonzalez-Dominguez, E. Gonzalez-Lavado, L. Marin-Caba, B. Vaz, M. Perez-Lorenzo et al. 2017. A biomimetic escape strategy for cytoplasm invasion by synthetic particles. Angew. Chem. Int. Edit. 56: 13736–13740.

Jain, T.K., M.K. Reddy, M.A. Morales, D.L. Leslie-Pelecky and V. Labhasetwar. 2008. Biodistribution, clearance, and biocompatibility of iron oxide magnetic nanoparticles in rats. Mol. Pharm. 5: 316–327.

Kakeji, Y., Y. Maehara, D. Korenaga, S. Tsujitani, M. Haraguchi, A. Watanabe et al. 1993. Prognostic-significance of tumor-host interaction in clinical gastric-cancer-relationship between DNA ploidy and dendritic cell infiltration. J. Surg. Oncol. 52: 207–212.

Kang, Y., Y.H. Seo and C.J. Lee. 2000. Synthesis and conductivity of PEGME branched poly(ethylene-alt-maleimide) based solid polymer electrolyte. Bull. Korean Chem. Soc. 21: 241–244.

Klabunde, K.J., D. Zhang, G.N. Glavee, C.M. Sorensen and G.C. Hadjipanayis. 1994. Encapsulated nanoparticles of iron metal. Chem. Mat. 6: 784–787.

Krätschmer, W., L.D. Lamb, K. Fostiropoulos and D.R. Huffman. 1990. Solid C-60—A new form of carbon. Nature 347: 354–358.

Kroto, H.W., J.R. Heath, S.C. Obrien, R.F. Curl and R.E. Smalley. 1985. C-60—Buckminsterfullerene. Nature 318: 162–163.

Kuznetsov, A.A., V.I. Filippov, O.A. Kuznetsov, V.G. Gerlivanov, E.K. Dobrinsky and S.I. Malashin. 1999. New ferro-carbon adsorbents for magnetically guided transport of anti-cancer drugs. J. Magn. Magn. Mater. 194: 22–30.

Lawaczeck, R., H. Bauer, T. Frenzel, M. Hasegawa, Y. Ito, K. Kito et al. 1997. Magnetic iron oxide particles coated with carboxydextran for parenteral administration and liver contrasting–Pre-clinical profile of SH U555A. Acta Radiol. 38: 584–597.

Ledezma, R., D. Bueno and R.F. Ziolo. 2009. Preparation and characterization of stable aqueous ferrofluids using low molecular weight sulfonated polystyrene. Macromol. Symp. 283-84: 307–310.

Lima, E., E. De Biasi, R.D. Zysler, M.V. Mansilla, M.L. Mojica-Pisciotti, T.E. Torres et al. 2014. Relaxation time diagram for identifying heat generation mechanisms in magnetic fluid hyperthermia. J. Nanopart. Res. 16: 2791 (11 pp).

Lind, K., M. Kresse, N.P. Debus and R.H. Muller. 2002. A novel formulation for superparamagnetic iron oxide (SPIO) particles enhancing MR lymphography: Comparison of physicochemical properties and the *in vivo* behaviour. J. Drug Target. 10: 221–230.

Livingston, J.D. 1996. Driving Force: The Natural Magic of Magnets. Harvard University Press, Cambridge, USA.

Lubbe, A.S., C. Bergemann, W. Huhnt, T. Fricke, H. Riess, J.W. Brock et al. 1996. Preclinical experiences with magnetic drug targeting: Tolerance and efficacy. Cancer Res. 56: 4694–4701.

Lubbe, A.S., C. Alexiou and C. Bergemann. 2001. Clinical applications of magnetic drug targeting. J. Surg. Res. 95: 200–206.

Marquina, C., J.M. de Teresa, D. Serrate, J. Marzo, F.A. Cardoso, D. Saurel et al. 2012. GMR sensors and magnetic nanoparticles for immuno-chromatographic assays. J. Magn. Magn. Mater. 324: 3495–3498.

McHenry, M.E., S.A. Majetich, J.O. Artman, M. Degraef and S.W. Staley. 1994. Superparamagnetism in carbon-coated Co particles produced by the Krätschmer carbon-arc process. Phys. Rev. B 49: 11358–11363.

Mehdaoui, B., A. Meffre, J. Carrey, S. Lachaize, L.M. Lacroix, M. Gougeon et al. 2011. Optimal size of nanoparticles for magnetic hyperthermia: a combined theoretical and experimental study. Adv. Funct. Mater. 21: 4573–4581.

Moghimi, S.M. and A.C. Hunter. 2001. Recognition by macrophages and liver cells of opsonized phospholipid vesicles and phospholipid head groups. Pharm. Res. 18: 1–8.

Monfardini, C. and F.M. Veronese. 1998. Stabilization of substances in circulation. Bioconjugate Chem. 9: 418–450.

Nageswari, K., R. Banerjee, R.V. Gupte and R.R. Puniyani. 1998. Hemorheological parameters for biocompatibility evaluation. J. Biomater. Appl. 13: 74–80.

Oparka, K.J. and S.S. Cruz. 2000. The great escape: Phloem transport and unloading of macromolecules. Annu. Rev. Plant Physiol. Plant Molec. Biol. 51: 323–347.

Pankhurst, Q.A., J. Connolly, S.K. Jones and J. Dobson. 2003. Applications of magnetic nanoparticles in biomedicine. J. Phys. D-Appl. Phys. 36: R167–R181.

Pardoe, H., W. Chua-anusorn, T.G. St Pierre and J. Dobson. 2001. Structural and magnetic properties of nanoscale iron oxide particles synthesized in the presence of dextran or polyvinyl alcohol. J. Magn. Magn. Mater. 225: 41–46.

Pavel, A., M. Trifan, I.I. Bara, D.E. Creanga and C. Cotae. 1999. Accumulation dynamics and some cytogenetical tests at Chelidonium majus and Papaver somniferum callus under the magnetic liquid effect. J. Magn. Magn. Mater. 201: 443–445.

Pavel, A. and D.E. Creanga. 2005. Chromosomal aberrations in plants under magnetic fluid influence. J. Magn. Magn. Mater. 289: 469–472.

Pérez de Luque, A., Z. Cifuentes, C. Marquina, J.M. de la Fuente and M.R. Ibarra. 2012. Synthesis, application, and tracking of magnetic carbon-coated nanoparticles in plants. pp. 263–272. *In*: M. Soloviev (ed.). Nanoparticles in Biology and Medicine: Methods and Protocols Humana Press-Springer Science+Business Media, New York, USA.

Plank, C., O. Zelphati and O. Mykhaylyk. 2011. Magnetically enhanced nucleic acid delivery. Ten years of magnetofection—Progress and prospects. Adv. Drug. Deliver. Rev. 63: 1300–1331.

Popat, K.C., S. Sharma and T.A. Desai. 2004. Quantitative XPS analysis of PEG-modified silicon surfaces. J. Phys. Chem. B 108: 5185–5188.

Prior, B.M., H.T. Yang and R.L. Terjung. 2004. What makes vessels grow with exercise training? Journal of Applied Physiology 97: 1119–1128.

Puertas, S., M. Moros, R. Fernández-Pacheco, M.R. Ibarra, V. Grazu and J.M. de la Fuente. 2010. Designing novel nano-immunoassays: antibody orientation versus sensitivity. J. Phys. D-Appl. Phys. 43: 474012 (8 pp).

Rechenberg, H.R., J.A.H. Coaquira, C. Marquina, B. García-Landa, M.R. Ibarra, A.M. Benito et al. 2001. Mössbauer and magnetic characterisation of carbon-coated small iron particles. J. Magn. Magn. Mater. 226: 1930–1932.

Reinhart, W.H., B. Pleisch, L.G. Harris and M. Lutolf. 2005. Influence of contrast media (iopromide, ioxaglate, gadolinium-DOTA) on blood viscosity, erythrocyte morphology and platelet function. Clin. Hemorheol. Microcirc. 32: 227–239.

Sanz, B., M.P. Calatayud, T.E. Torres, M.L. Fanarraga, M.R. Ibarra and G.F. Goya. 2017. Magnetic hyperthermia enhances cell toxicity with respect to exogenous heating. Biomaterials 114: 62–70.

Seo, W.S., J.H. Lee, X.M. Sun, Y. Suzuki, D. Mann, Z. Liu et al. 2006. FeCo/graphitic-shell nanocrystals as advanced magnetic-resonance-imaging and near-infrared agents. Nat. Mater. 5: 971–976.

Serrate, D., J.M. De Teresa, C. Marquina, J. Marzo, D. Saurel, F.A. Cardoso et al. 2012. Quantitative biomolecular sensing station based on magnetoresistive patterned arrays. Biosens. Bioelectron. 35: 206–212.

Shinkai, M. 2002. Functional magnetic particles for medical application. Journal of J. Biosci. Bioeng. 94: 606–613.

Steinman, R.M. and Z.A. Cohn. 1973. Identification of a novel cell type in peripheral lymphoid organs of mice. J. Exp. Med. 137: 1142–1162.

Steinman, R.M. 1991. The dendritic cell system and its role in immunogenicity. Annu. Rev. Immunol. 9: 271–296.

Stolnik, S., L. Illum and S.S. Davis. 1995. Long circulating microparticulate drug carriers. Adv. Drug Deliv. Rev. 16: 195–214.

Thurnher, M., C. Radmayr, R. Ramoner, S. Ebner, G. Bock, H. Klocker et al. 1996. Human renal-cell carcinoma tissue contains dendritic cells. Int. J. Cancer 68: 1–7.

Torney, F., B.G. Trewyn, V.S.Y. Lin and K. Wang. 2007. Mesoporous silica nanoparticles deliver DNA and chemicals into plants. Nat. Nanotechnol. 2: 295–300.

Turgut, Z., J.H. Scott, M.Q. Huang, S.A. Majetich and M.E. McHenry. 1998. Magnetic properties and ordering in C-coated FexCo1-x, alloy nanocrystals. J. Appl. Phys. 83: 6468–6470.

Weissleder, R., G. Elizondo, J. Wittenberg, C.A. Rabito, H.H. Bengele and L. Josephson. 1990. Ultrasmall superparamagnetic iron-oxide—characterization of a new class of contrast agents for MR Imaging. Radiology 175: 489–493.

Widder, K.J., A.E. Senyei and D.G. Scarpelli. 1978a. Magnetic microspheres—model system for site specific drug delivery *in vivo*. Proc. Soc. Exp. Biol. Med. 158: 141–146.

Widder, K.J., A.E. Senyei, S.D. Reich and D.F. Ranney. 1978b. Magnetically responsive microspheres as a carrier for site-specific delivery of adriamycin. Proceedings of the American Association for Cancer Research 19: 17–17.

Yu, M.X. and J. Zheng. 2015. Clearance pathways and tumor targeting of imaging nanoparticles. Acs Nano 9: 6655–6674.

Yu, W.W., E. Chang, C.M. Sayes, R. Drezek and V.L. Colvin. 2006. Aqueous dispersion of monodisperse magnetic iron oxide nanocrystals through phase transfer. Nanotechnology 17: 4483–4487.

Zahr, A.S., C.A. Davis and M.V. Pishko. 2006. Macrophage uptake of core-shell nanoparticles surface modified with poly(ethylene glycol). Langmuir 22: 8178–8185.

Zhao, H., H.J. Cui and M.L. Fu. 2016. A general and facile method for improving carbon coat on magnetic nanoparticles with a thickness control. J. Colloid Interface Sci. 461: 20–24.

CHAPTER 11

Chromatographic Techniques for Characterization of Carbons and Carbon Composites

Adam Voelkel* and Beata Strzemiecka

1. Introduction

Surface properties of carbon materials are crucial in almost all their applications. Nitrogen adsorption is most often used to assess surface activity of carbon. However, in some cases, it is necessary to possess more detailed knowledge about carbon surface, e.g., its ability for dispersive interactions or specific interactions. The magnitude of these interactions may influence, for instance, the distribution of the filler particles in the polymer matrix (Strzemiecka et al. 2013). Chromatographic techniques seem to be useful in this case. Activity of active carbon, carbon blacks, carbon fibers, and carbon nanotubes was examined by means of inverse gas chromatography both at infinite dilution and finite concentration. Surface activity of carbon filler in the polymeric composite will influence the magnitude of adhesion between the filler particles and polymeric matrix. It will be also discussed in this chapter. Physicochemical examination of carbon or carbon-containing materials by chromatographic techniques is usually supported by other techniques, e.g., TGA-MS, XPS, microscopic techniques, contact angle measurements, and others (Strzemiecka et al. 2014). The use of Hansen Solubility Parameters leads to interesting results in the examination of carbon nanotubes and carbon fibers/ epoxy resin composites.

2. Carbon and Carbon Composites Examined by Chromatographic Techniques

The variety of carbons which were characterized by chromatographic techniques inspired us to prepare a short collation presenting this field of activity. It must be stressed that authors do not claim that

Poznań University of Technology, Institute of Chemical Technology and Engineering, Berdychowo 4, 60-965 Poznań, Poland.
Emails: Beata.Strzemiecka@put.poznan.pl
* Corresponding author: Adam.Voelkel@put.poznan.pl

inverse gas chromatography is the only (unique) technique which might be used in characterization of this kind of materials. However, we would like to show that information carried out by results from IGC broadens our knowledge of the properties of carbon materials.

Inverse gas chromatography (IGC) is one of the techniques often applied in physicochemical characterization of various types of carbon materials as well as composites containing carbon fillers. One can find papers describing the use of chromatographic techniques for characterization of such materials as: activated carbon GF 40 was chemically activated using phosphoric acid (Diaz et al. 2004), many forms of carbon being manufactured in laboratories and industry—here heterogeneous forms as coals, cokes, and chars (Grajek 2007). These types of carbon are described as graphitic and non-graphitic, depending on the degree of crystallographic ordering, and represent an intermediate between the organic precursor and single crystal graphite.

Pérez-Mendoza et al. (Perez-Mendoza et al. 2008) reported on the dispersive as well as adsorptive surface properties of the series of active carbons from A- and HK-series obtained from poly(ethylene terephthalate) (PET). P-series was obtained by carbonization of polyfurfuryl alcohol, C- and H-series from carbonization (in nitrogen) and the activation (in carbon dioxide) of almond shells and olive stones, respectively. GAe was a commercial active carbon, and GAe-ox1 and GAe-ox2 are the oxidation products of the former with ammonium peroxidisulfate and hydrogen peroxide, respectively. A commercial graphite manufactured by Sigma–Aldrich was also used. This sample was demineralized by consecutive treatment with HF and HCl aqueous solutions. Another group of activated carbons was used by Bagreev et al. (Bagreev et al. 1999). These were wood-based activated carbons BAX-1500, WVA-900, WVA 1100, and UMC, as well as based on RB3 carbon (coal-based) and S208c carbon (coconut shell). The second group of these carbons was oxidized using various oxidizing agents. Inverse Gas Chromatography as well as nitrogen adsorption were used to study influence of temperature, pH, and surface chemistry of the studied carbons on the adsorption of H_2S.

Park and Donnet [Park and Donnet 1998] examined polyacrylonitrile-based high strength carbons which were used to manufacture fibers. These virgin fibers (Type I, surface nontreated and nonsized) had been submitted to an anodic treatment in various conditions (Types II, III and IV). Donnet et al. (Donnet et al. 2002) also investigated the series of carbon blacks of different grades and structures, as indicated by the specific surface areas and dibutylphthalate (DBP) adsorption values. The samples were extracted in a Soxhlet device with water and toluene to eliminate contaminants of different origins. The carbon blacks, in pellet form, with the granule size between 250 and 400 mm, were selected.

Commercial activated carbons labeled P, S, N5, N6, N8, and N10 were selected by Cossarutto et al. (Cossarutto et al. 2001) to examine adsorption ability in the gas cleaning processes. Nx carbons were obtained from coconut shell and were further modified by steam activation after the carbonization process. P and S carbon were wood-based and chemically activated with H_3PO_4. In the case of N10 sample, the N-enrichment process was performed by reaction with formamide and following heat treatment. Active carbon spheres (ACSs) with different porous structures prepared in the laboratory were characterized by static adsorption studies and inverse gas chromatographic (IGC) technique (Singh et al. 2004). Active carbon spheres (ACSs) were prepared from a polymeric precursor, polystyrene sulfonate resin, having 8 wt% divinyl benzene copolymers. Samples were prepared under different sets of flow rates of activating agent CO_2 and duration of activation. The particle size of the final product obtained was in the range of 0.3–0.6 mm. Two activated carbons were examined by Diaz et al. (Diaz et al. 2005). GF-40 was an olive stone-based carbon, chemically-activated carbon using the phosphoric acid process. RB1 was a peat-based steam-activated carbon. Authors used several purification (washing) procedures to remove the impurities from the activated carbons. In all cases, the purified activated carbon was dried overnight at 100°C.

Vagner et al. (Vagner et al. 2003) examined cokes which were produced by the co-pyrolysis of a coal-tar pitch (CTP) with different amounts of polyacrylonitrile (PAN) as a possible method to synthesize carbonaceous materials enriched in nitrogen.

The carbon fibers used by Thielmann and Burnett (Thielmann and Burnett 2004) were PAN-based high tensile strength carbon fibers C320.00A. These authors also examined composites containing carbons fibers. These were PAN-based carbon fibers-untreated (CA), an oxidized (AS4), and an oxidized and epoxy-sized (AS4-GP) material (Thielmann et al. 2005). The interaction between carbon fibers with different surface treatments and thermoplastic polymer matrices were investigated. Papirer et al. (Papirer et al. 1999) examined three types of carbon: graphite, carbon black, and fullerene (C60) samples. Those samples differed in their structures and also in their surface properties.

Four commercial carbon blacks, Corax N774, Corax N326, CoraxN134, and Corax N220 (named C32, C71, C159, and C178, respectively) of similar structure but with different specific surface area and surface chemistry were examined by Strzemiecka et al. (Strzemiecka et al. 2014). Composites containing these carbon fillers were also investigated in their earlier work (Strzemiecka et al. 2013). The influence of the carbon filler content and its character on mechanical, viscoelastic, thermal, and adhesion properties of composites were also examined (Garcia and Martin-Martinez 2015). The physical affinity between an epoxy matrix and oxidized, unsized carbon fibers has been evaluated using Hansen Solubility Parameters (HSP) (Launay et al. 2007). The liquid epoxy resin (bisphenol F-epichlorhydrin epoxy/bisphenol A-diglycidylether) and the hardener (polyoxyalkyleneamine/3-aminomethyl-3,5,5-trimethylcyclohexylamine) were mixed at a ratio of 10/3 by weight at room temperature. The curing time was 48 hours at room temperature. Planar glassy carbon (SIGRADUR G) was purchased from HTW Hochtemperatur.

Carbon nanotubes were also examined by the discussed techniques. Menzel et al. (Menzel et al. 2009) used CNT materials produced via CVD growth. Commercial CNT materials were obtained from Arkema SA (Lacq-Mourenx, France) and Nanocyl SA (Sambreville, Belgium) and labeled Arkema and Nanocyl CNTs, respectively. Authors also characterized carbon nanotubes synthesized in their laboratory employing typical CVD-growth conditions. The as-received Arkema CNTs were modified by high temperature annealing. CNTs were modified by grafting with methyl methacrylate, employing a thermochemical treatment. The behavior of single-walled carbon nanotubes (SWNTs) in various solvents was examined by Bergin et al. (Bergin et al. 2009). They applied the concept of Hansen Solubility Parameters to predict the surface activity of examined nanomaterials. Various techniques were applied to examine the influence of several surfactants on the stability of water dispersions of single-walled carbon nanotubes (Sun et al. 2008).

Garnier et al. (Garnier et al. 2007) assessed the energetic surface heterogeneity of four different activated carbons by the parallel probing at the solid/liquid and solid/gas interfaces. They applied Filtrasorb F400 carbon Chemviron, Maxsorb (non-commercial sample) from The Kansai Coke and Chemicals Co. Ltd., while Tanac 7500 and 4500 were provided by Tanac S.A.

Zhao et al. (Zhao et al. 2004) investigated diamond-like carbon (DLC) and tetrahedral amorphous carbon (ta-C) coatings on stainless steel sheets, which were prepared using an unbalanced magnetron sputtering system and a filtered cathodic vacuum arc system, respectively.

Moreover, it is worthy to note that carbon-based composites are widely applied in various chromatographic techniques, such as gas–liquid chromatography, high-performance liquid chromatography, and electrically driven separation techniques for the separation, quantitative determination, and identification of a wide variety of compounds in complex matrices (Cerháti 2009, Gierak et al. 2006).

The above short "report" discloses the variety of carbon materials which were characterized mainly by inverse gas chromatography. It is worthy to note that the mentioned papers described the carbon materials of various origins, obtained by different way of synthesis, and of different applications. The authors of these papers decided to use inverse gas chromatography as a complementary technique, whose results enable the tracking of the changes of the properties of the materials, being results of various technological processes.

3. Inverse Gas Chromatography

Inverse gas chromatography (IGC)—a well-known and accepted technique of gas chromatography—is extensively used in the physicochemical characterization of a large variety of materials (Danner et al. 1998, Thielmann 2004, Santos and Guthrie 2005). The short packed chromatographic column is filled with a non-volatile material of interest playing the role of a stationary phase in such a gas–solid or gas–liquid system. This short chromatographic column is mounted in a gas chromatograph oven. Carefully selected test solutes are injected and transported over the material (stationary phase) by mobile phase (gas). Retention data (retention time, net retention volume, specific retention volume) is further converted into the parameters describing the required property of the examined substance.

The examined material may be solid, and in such a case its particles are placed in a chromatographic column by the usual procedure applied in gas chromatography. Liquid materials should be coated on to solid carrier–standard commercial chromatographic supports. However, in the examination of carbon materials, only the first methodology of column filling was applied.

Depending on the properties of the examined substance, a different retention mechanism must be considered. The adsorption mechanism is valid in the case of solid materials. In this case, the obtained retention data might depend on the experiment conditions, for example, when porous species are examined. Inverse gas chromatography may be realized in two modes: (i) at finite concentration of test solutes, FC-IGC, or (ii) in the region of infinitive dilution ID-IGC. In the first case, the examination of heterogeneity of the examined surface is also possible.

Recently, the manufacturing of a sophisticated combination of inverse gas chromatograph and "sorptomate" facilitated the examination of various materials at a different surface coverage (Ylä-Mäihäniemi et al. 2008, Ho et al. 2010, Burnett et al. 2012, Das and Stewart 2012, Gamble et al. 2012, Ali et al. 2013, Pérez-Mendoza et al. 2008). The examination of materials at infinite concentration of the test solutes leads to the information of their interactions with highly energetic active sites on the surface. The assumption that the test solute molecules interact only with the surface must be valid. In Finite Concentration mode, i.e., at higher surface coverage of the stationary phase by test solute molecules, they interact with a larger population of active sites in the surface layer. It will be discussed later.

The properties of solid materials in IGC are mainly expressed by the surface energy parameters and its components corresponding to the ability of the dispersive and specific interactions (Thielmann 2004, Santos and Guthrie 2005, Voelkel et al. 2009a, b, Voelkel 2012). IGC experiment might also be carried out using pulse or frontal technique. In a pulse mode the given amount of the test solute is injected into the stream of carrier gas (mobile phase) and transported through a column filled with the stationary phase (examined material). In frontal technique, test solute is continuously added to the mobile phase, which leads to the formation of the breakthrough curve on the chromatogram. The use of pulse technique is suggested for systems where the equilibrium is quickly established. It happens most often when interactions between the test solute and examined material are relatively weak. The alternative, for a system of "slow" equilibrium, is the frontal mode (Thielmann 2004).

The dispersive component of surface free energy, γ_S^D, is most often used for the characterization of the surface layer of examined material by means of IGC. It expresses the ability of the examined materials to non-specific interactions. There are several methods used for determination of γ_S^D. Two of them are commonly used in IGC at infinite dilution: (i) the Schultz and Lavielle (Sun and Berg 2003, Lavielle and Schultz 1991) method, (ii) the Dorris and Gray (Dorris and Gray 1980, Voelkel et al. 2009b) method.

In the method of Schultz and Lavielle, γ_S^D is calculated from the equation (1)

$$R \cdot T \cdot \ln V_N = 2 \cdot N \cdot a_p \cdot \sqrt{\gamma_S^D \cdot \gamma_L^D} + C \tag{1}$$

where symbol γ_S^D denotes the dispersive of component of surface free energy of the solid, symbol γ_L^D denotes the dispersive of component of surface free energy of the test solute, symbol a_p denotes the area occupied by adsorbing molecule, symbol V_N denotes the net retention volume of the probe, symbol R is the ideal gas constant, symbol T denotes the measurement temperature, and symbol N is the Avogadro number. Equation (1) represents the straight line where slope equals to $2 \cdot N \cdot \sqrt{\gamma_S^D}$. It allows the determination of the dispersive component of surface free energy of the examined material, which in our case is the carbon material.

Dorris and Gray calculated γ_S^D according to the equation (2):

$$\gamma_S^D = \frac{-R^2 \cdot T^2 \cdot \left[\ln \left[\frac{V_N^{(C_{n+1}H_{2n+4})}}{V_N^{(C_nH_{2n+2})}} \right] \right]^2}{4 \cdot N^2 \cdot (a_{CH_2})^2 \cdot \gamma_{(CH_2)}} \tag{2}$$

where a_{CH_2}—the surface area of a methylene group, and a variation of the value of this parameter leads to significant differences of estimation of γ_S^D; N—Avogadro's number ($6.023 \cdot 10^{23}$ [1/mol]), $V_N^{(C_{n+1}H_{2n+4})}$—the net retention volume of alkane, $C_{n+1}H_{2n+4}$; $V_N^{(C_nH_{2n+2})}$—the net retention volume of alkane, C_nH_{2n+2}; $\gamma_{(CH_2)}$—the surface energy of the polyethylene-type polymers with a finite molecular weight [mJ/m²]. The value $\gamma_{(CH_2)}$ is calculated according to the following equations (3) or (4):

$$\gamma_{(CH_2)} = 34.0 - 0.058 \cdot t \tag{3}$$

or

$$\gamma_{(CH_2)} = 35.6 + 0.058(293 - T) \tag{4}$$

where t is the temperature [°C] and T is the temperature [K].

The ability of the carbon material surface for specific interactions by means of inverse gas chromatography is most often determined by the injection of the polar probes of known characteristics and collection of their retention data. The IGC experimental data is further elaborated using LSER (linear solvation energy relationship) technique (Vagner et al. 2003) or the linear free energy relationship (LFER) (Park and Donnet 1998). The term used by Park and Donnet seems to be appropriate, considering the way of calculation of the parameters characterizing the examined material.

The ability of the examined surface for specific interactions is expressed by the specific component of adsorption energy ΔG^S, and further, the specific component of the enthalpy of adsorption. ΔG^S is determined as the difference between the adsorption energy of polar compound, ΔG_{polar}, and adsorption energy of hypothetical alkane, ΔG_{ref} (equations 5 and 6), having the same selected property as the polar test solute (e.g., vapor pressure in Papirer method or $a_p \sqrt{\gamma_L^D}$ value in Schultz and Lavielle method) as polar compound:

$$\Delta G^S = \Delta G_{polar} - \Delta G_{ref} \tag{5}$$

where

$$\Delta G_{polar} = -RT \ln V_{N_{polar}} + C \quad \text{and} \quad \Delta G_{ref} = -RT \ln V_{N_{refr}} + C \tag{6}$$

The problem of the selection of the reference state in the determination of the specific component of energy of adsorption was discussed elsewhere (Voelkel et al. 2009b). The graphical interpretation of ΔG^S estimation is given in Figure 11.1. The data presented in Figure 11.1 is for carbon black C71 described in Strzemiecka et al. (2014).

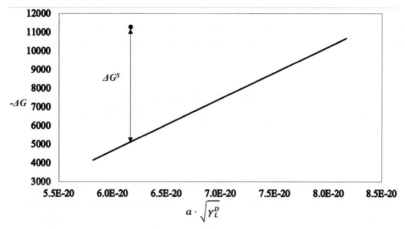

Figure 11.1: The graphical interpretation of ΔG^S estimation for butan-1-ol on the basis of data for carbon black C71 described in Strzemiecka et al. (2014).

ΔH^S is calculated from the dependence of ΔG^S on temperature (equation 7):

$$\Delta G^S = \Delta H^S - T \cdot \Delta S^S \tag{7}$$

where ΔG^S—specific component of the free adsorption energy, ΔS^S—specific component of the free adsorption entropy of polar compound onto the surface of investigated solid.

Plotting $\Delta G^S/T$ against $1/T$ gives a straight line with the slope ΔH^S. ΔH^S should be determined for, at least, four test compounds, and ΔG^S should be determined at, at least, three temperatures.

Specific component of enthalpy of adsorption of polar compound, ΔH^S related to acceptor and donor numbers describing the electron acceptor (AN^*) and electron donor (DN) properties of the test compound (equation 8):

$$\Delta H^S = DN \cdot K_A + AN^* \cdot K_D \tag{8}$$

K_A and K_D parameters express the ability of the examined material to act as electron acceptor and electron donor, respectively.
Plotting $\Delta H^S/AN^*$ against DN/AN^*:

$$\frac{\Delta H^S}{AN^*} = \frac{DN}{AN^*} \cdot K_A + K_D \tag{9}$$

one obtains the straight line with the slope of K_A (equation 9). As the estimation of K_D from the intercept of equation (9) may lead to a significant error, one should determine this value as the slope of the following relationship (equation 10):

$$\frac{\Delta H^S}{DN} = \frac{AN^*}{DN} \cdot K_D + K_A \tag{10}$$

K_A and K_D parameters are dimensionless. On the basis of their values, one can assess whether the surface is acidic or basic, but it is not possible to compare these values to the dispersive component of the surface energy and to estimate total surface energy and its components. Good-van Oss equation (equation 11) enables to express the acidic and basic parameters of a solid surface, γ_s^+ and γ_s^-, respectively, in the same units as for γ_s^d (in mJ/m²):

$$\Delta G^S = 2 \cdot N_A \cdot a_p \cdot (\gamma_l^+ \cdot \gamma_s^-)^{1/2} + (\gamma_l^- \cdot \gamma_s^+)^{1/2}) \tag{11}$$

In equation (11), γ_l^+, γ_l^- are the electron acceptor and donor parameters of the probe molecules, respectively. ΔG^s is the specific component of the free energy of adsorption of polar compound. The way of determination of ΔG^s value is described in many papers, for instance, Voelkel (2009a, b). For calculation of γ_s^+, γ_s^- dichloromethane (DM) and ethyl acetate (EA) were used as test compounds. DM is monopolar acid and γ_{DM}^- is 0.0 mJ/m². Equation (11) is condensed to equation (12):

$$\gamma_s^- = \Delta G_{DM}^s / (4 \cdot N_A^2 \cdot a_{DM}^2 \cdot \gamma_{DM}^+) \tag{12}$$

and γ_s^- can be easily calculated. The value of γ_{DM}^+ was established as 5.2 mJ/m² on the basis of van Oss et al. (1988). Similarly, EA is a monopolar base and γ_{EA}^+ is 0.0 mJ/m², and from equation (13), γ_s^+ parameter for examined solid can be calculated:

$$\gamma_s^+ = \Delta G_{EA}^s / (4 \cdot N_A^2 \cdot a_{EA}^2 \cdot \gamma_{EA}^-) \tag{13}$$

The value of γ_{EA}^- was established as 19.2 mJ/m² (Das et al. 2011). However, in the literature, there are different values of the components and parameters of the surface free energy for the test compounds (Jańczuk et al. 1999). van Oss gave values of only γ_I^- for EA equal to 6.2 mJ/m² different from this in Das et al. (2011). Moreover, one cannot find the value of γ_I^- for dichloromethane (DM), but only for chloroform (CH) 1.5 mJ/m² therein (van Oss 2006). The influence of this physicochemical data on the resulting values of γ_s^+, γ_s^- parameters and further on the values of γ_s^{sp}, γ_s and then W_A^{ab} was discussed (Strzemiecka and Voelkel 2012). In that paper, the values of γ_s^+, γ_s^- for studied materials were calculated for the data of γ_I^+, γ_I^- from both these sources (Della Volpe and Siboni 1997, Siboni and Della Volpe 2015).

The specific component of the surface free energy was also used to calculate the parameter of specific interaction of polar solutes (I_{sp}). This parameter involves the surface properties in terms of potential and acid–base interactions, and may be determined from the difference of free energy of adsorption, $\Delta(\Delta G)$, between a polar solute and the real or hypothetical n-alkane with the same surface area (Diaz et al. 2005).

$$I_{sp} = \frac{\Delta(\Delta G)}{N a_p} \tag{14}$$

where a_p is the probe surface area, N—Avogadro number. In the present work, a_p, is calculated from the liquid density, ρ, and the molar weight of the molecule, M, assuming a spherical shape of the molecule in a hexagonal close-packed configuration (Diaz et al. 2005) equation (15):

$$a_p = 1.09 \cdot 10^{14} \left(\frac{M}{\rho N} \right)^{2/3} \tag{15}$$

3.1 Hansen solubility parameters

Cohesive energy (E_{coh}) related to a molar volume is called *cohesive energy density c* (equation 16):

$$c = \frac{-E_{coh}}{V} \tag{16}$$

and the square root of cohesive energy density is called *solubility parameter δ* (equation 17). This term proposed by Hildebrand for non-polar systems, used as a measure of intermolecular forces of different solvents, is related to the enthalpy of an evaporation ΔH_w (equation 17):

$$\delta = \sqrt{c} = \sqrt{\frac{E_{coh}}{V}} = \sqrt{\left(\frac{\Delta H_w - RT}{V} \right)} \tag{17}$$

where δ is the solubility parameter, E_{coh} is the cohesive energy, V is the molar volume of a pure liquid, R is the gas constant, T is the temperature. The solubility parameter expressed by equation (17) is called the Hildebrand solubility parameter.

The most widely accepted concept of solubility parameter (total solubility parameter) related to more complex systems has been proposed by Hansen (equation 18) (Hansen 1967a, b, Hansen and Skaarup 1967):

$$\delta_T^2 = \delta_d^2 + \delta_p^2 + \delta_h^2 \tag{18}$$

where δ_d, δ_p, and δ_h denote dispersive, polar, and hydrogen bonding contribution, respectively.

The determination of solubility parameter and Hansen solubility parameters (HSPs) by means of inverse gas chromatography was discussed earlier (Voelkel et al. 2009b).

Ito and Guillet (Ito and Guillet 1979) proposed the procedure of the estimation of the solubility parameter for polymeric materials above their T_g (equation 19),

$$\frac{\delta_{1i}^2}{RT} - \frac{\chi_{(12)i}^{\infty}}{V_{1i}} = \frac{2\delta_2}{RT}\delta_{1i} - \left(\frac{\delta_2^2}{RT} + \frac{\chi_s^{\infty}}{V_{1i}}\right) \tag{19}$$

For polymers with high molecular weight, equation (19) represents the straight line and its slope is proportional to δ_{1i}.

$\chi_{(12)i}^{\infty}$ is Flory-Huggins interaction parameter calculated from chromatographic retention data using equation (20) for the series of "i" test solutes having molar volume V_{1i} and solubility parameter δ_{1i}. Flory-Huggins interaction parameter is related to IGC retention parameters by equation (20):

$$\chi_{12}^{\infty} = \ln\left(\frac{273.15 \cdot R}{p_1^o \cdot V_g \cdot M_1}\right) - \frac{p_1^o}{R \cdot T} \cdot \left(B_{11} - V_1\right) + \ln\left(\frac{\rho_1}{\rho_2}\right) - \left(1 - \frac{V_1}{V_2}\right) \tag{20}$$

where *1* denotes the solute and *2* denotes examined material, M_1 is the molecular weight of the solute, p_1^o is the saturated vapor pressure of the solute, B_{11} is the second virial coefficient of the solute, V_i is the molar volume, ρ_i is the density, and R is the gas constant.

Price (Price 1989, Price and Shillcock 2002) applied this procedure for the estimation of solubility parameter values for the low-molecular weight compounds (Price 1989) and liquid crystalline systems (Price and Shillcock 2002). Price reported that the experimental relation between the left hand-side of equation (19) and solubility parameters of test solutes is different from the linear relationship. The significant curvature of the equation (19) was observed. The downward curvature for alkanes and upward curvature for other compounds was found. Therefore, the tendency for alkanes leads to underestimation of δ_2, while tendency for polar compounds may cause the overestimation of the value of δ_2.

Adamska et al. (Adamska et al. 2008) proposed to calculate the HSP data using a model proposed by Lindvig et al. (Lindvig et al. 2002), combining experimental data of Flory-Huggins interaction parameter χ_{12}^{∞} with components of solubility parameter for the examined material (equation 21):

$$\chi_{12}^{\infty} = \alpha \frac{V_1}{RT}\left(\left(\delta_{1,d} - \delta_{2,d}\right)^2 + 0,25\left(\delta_{1,p} - \delta_{2,p}\right)^2 + 0,25\left(\delta_{1,h} - \delta_{2,h}\right)^2\right) \tag{21}$$

where α, V_1, R, T are a corrective coefficient, molar volume of the test solute, gas constant, and temperature of measurement, respectively. This equation describes how the differences between Hansen Solubility Parameters for different test solutes ($\delta_{1,d}$, $\delta_{1,p}$, $\delta_{1,h}$) and HSPs values for the examined material ($\delta_{2,d}$, $\delta_{2,p}$, $\delta_{2,h}$) influence the value of Flory-Huggins interaction parameter. For the purpose of IGC, experimentally obtained χ_{12}^{∞} values can be used for the determination of the HSP for the examined material by applying the above relation (Adamska et al. 2008).

The determination of solubility parameter for solid materials by means of inverse gas chromatography is based on the model of adsorption described by Snyder and Karger, and requires the knowledge of value of adsorption energy for the respective test solutes (Karger and Snyder 1976, Karger et al. 1978, Nguyen et al. 1986). According to Karger and Snyder's (Karger and Snyder 1976) model, the molecule of test solute "*i*" is adsorbed onto the surface of solid adsorbent "2". From the energy balance, the following expression is derived for the energy of adsorption ΔE^A (equation 22):

$$-\Delta E_i^A = V_i(\delta_i \delta_2) \tag{22}$$

Introducing Hansen's concept of solubility parameter, one obtains (equation 23):

$$-\Delta E_i^A = V_i(\delta_{id}\delta_{2d} + \delta_{ip}\delta_{2p} + \delta_{ih}\delta_{2h}) \tag{23}$$

The energy of adsorption is related to the specific retention volume by the following equation (equation 24):

$$lnV_{ig} = -(\Delta E_i^A/RT) + const \tag{24}$$

For *N* test solutes, a system of *N* equations is obtained (equations 25 and 26):

$$
\begin{pmatrix} -\Delta E_1^A \\ ... \\ -\Delta E_n^A \\ ... \\ -\Delta E_N^A \end{pmatrix} =
\begin{pmatrix} V_1\delta_{1d} & V_1\delta_{1p} & V_1\delta_{1h} \\ ... & ... & ... \\ V_n\delta_{nd} & V_n\delta_{np} & V_n\delta_{nh} \\ ... & ... & ... \\ V_N\delta_{Nd} & V_N\delta_{Np} & V_N\delta_{Nh} \end{pmatrix}
*
\begin{pmatrix} \beta_1 \\ ... \\ \beta_n \\ ... \\ \beta_N \end{pmatrix} +
\begin{pmatrix} \varepsilon_1 \\ ... \\ \varepsilon_n \\ ... \\ \varepsilon_N \end{pmatrix}
\tag{25}
$$

$$Y = X\beta + \varepsilon \tag{26}$$

- *Y* is the column vector containing the *N* values of measured values of the energy of adsorption $(-\Delta E_n^A)$ of *N* solutes.
- *X* is the experimental matrix, formed of elements (X_{nk}), where $X_{nk} = V_n\delta_{nk}$.
- V_n is the molar volume of the *nth* solute, and δ_{nk} is one of the Hansen solubility parameters of type *k* (*k* = *d*, *p*, or *h*) of the respective solute.
- The β vector contains the real values of HSPs of the adsorbent, i.e., $\delta_{2d}, \delta_{2p}, \delta_{2h}$.
- The ε vector corresponds to the experimental errors, ε_n.

Values of the components of solubility parameter might be found using equation (24), and δ_T is calculated from equation (18).

4. Properties of Carbon Materials Estimated by Means of IGC

The activity of carbon materials examined by means of IGC is most often expressed by using the dispersive component of surface free energy γ_S^D. This data is obtained by different procedures. Dorris-Gray and Schultz-Lavielle procedures seem to be the most popular. It is probably connected to the relatively simple way of collection of retention data for test solutes here, n-alkanes. However, also during the examination of carbon materials, researchers met several problems. These will be shortly described in this chapter.

Pérez-Mendoza et al. (Perez-Mendoza et al. 2008) reported results for the series of ACs, where γ_S^D was calculated using Dorris-Gray procedure. They found very high values of γ_S^D for examined materials. There was a very good linear fit of these values versus temperature, which enabled the extrapolation of γ_S^D to 298 K. Authors realized that these values were higher than those found by other techniques (e.g., contact angle measurements). This observation is well-known, as usually γ_S^D from IGC measurements are also higher for other types of materials. This difference is due to two factors: (i) the different experimental conditions during IGC experiment in comparison to other techniques, (ii) the microporosity of active carbons. A much higher temperature of IGC experiment may influence, for instance, the value of the area occupied by adsorbing methylene group a_{CH_2} (see equation 27). This value is usually taken as constant, and equal to the value at 298 K, but at elevated temperatures (e.g., $\sim$ 400–500 K) and under infinite conditions of IGC experiment, the probe (test solute) can be treated as perfect gas. Therefore, it seems to be justified to calculate a_{CH_2} from the relationship, taking into account the influence of temperature (equation 27):

$$a_{CH_2} = a_{CH_{2(298)}} \left(\frac{T}{298} \right)^{2/3} \tag{27}$$

However, Pérez-Mendoza et al. (Perez-Mendoza et al. 2008) showed that even after such corrections, γ_S^D values from IGC were still significantly larger than those determined for ACs by contact angle measurements. They attributed this difference to the different experimental conditions between IGC and other techniques. Contact angle measurements and capillary wetting measurements supply data describing the mean energy of various active sites. In the case of infinite dilution, IGC test solute molecules interact (adsorb) mainly on high energy sites. The increase of concentration of test solute molecules in the gas phase allows the interactions with the active sites of lower energy, which leads to the decrease of γ_S^D (Rojewska et al. 2017). Microporosity of the active carbons might influence the ability of the adsorbed molecules to interact in pores of the dimension close to their own. The closeness of the opposite walls in micropores may produce a cooperative effect (Pérez-Mendoza et al. 2008) leading to the more exothermic adsorption process.

Diaz et al. (Diaz et al. 2004) presented γ_S^D values for zeolites and activated carbon higher than those obtained for the adsorption of the same compound on polymer surfaces, $\sim$ 60 mJ/m² or compounds such as theophylline and caffeine, $\sim$ 50 mJ/m². They also attributed these high values of γ_S^D to a high interaction potential in the micropores. They show that values of γ_S^D decrease with the increase of the average pore diameter. This relation indicates that the interactions are stronger in the micropores. Several authors have reported the high values of the dispersive component for activated carbons (Herry et al. 2001, Jagiełło et al. 1992).

Menzel et al. (Menzel et al. 2009) examined the series of carbon nanotubes of different origins. They found that the values for various CNTs are similar and lie in the range 110–115 mJ/m², which the authors discussed as being close to those found for other carbon nanotubes (114 mJ/m²) or graphite powder (105 mJ/m²). Authors indicated some limitations in the examination of these materials caused by a strong retention of test solutes at low temperatures (long retention times, non-symmetrical peaks, or even irreversible adsorption of higher alkanes). Despite similar values describing their ability for dispersive interactions, examined material were characterized by significantly different values of the specific component of surface energy (ΔG^S). The differences in ΔG^S values were attributed to varying compositions of the surface layer. Authors also examined the influence of surface modification on IGC-derived properties due to the annealing or thermal oxidation (Menzel et al. 2009).

It is also worthy to note the paper discussing the influence of inverse gas chromatography measurement conditions on elution peaks on activated carbon (Wu et al. 2004).

Cossarutto et al. (Cossarutto et al. 2001) examined series of ACs from various origins. They showed that it was possible to discriminate AC produced from wood and those from coconut-shell. The values of the dispersive component of surface free energy γ_S^D reported for the examined materials were in the range from 310 to 500 mJ/m². The authors explained that such high values are result of multiple interactions characteristic for microporous solids. The magnitude of these interactions depends on the structure and shape of pores. However, one should remember that carbon activity is not limited to non-specific interactions. It was shown just above (Menzel et al. 2009).

Vagner et al. (Vagner et al. 2003) characterized cokes obtained by the co-pyrolysis of a coal-tar pitch (CTP) with different amounts of polyacrylonitrile (PAN) by means of IGC. They applied the dispersive component of surface free energy to express the ability of these materials for dispersive interaction and parameters derived from LSER (linear solvation energy relationship) to show their ability for specific interaction. Authors found that both methods are equivalent when used to characterize dispersive interactions. However, the sensitivity of the LSER method does not allow discriminating between these three samples in terms of specific interactions.

Twenty years ago, Park and Donnet (Park and Donnet 1998) presented the application of inverse gas chromatography to the evaluation of acid-base interaction parameter, I_{a-b}, between two constitutive elements in a composite system. Specific interaction parameter, I_{a-b}, which describes the acid-base interaction between the fiber (f) and the matrix (m), knowing the K_A and K_D values of the fiber and matrix (from equations 9 and 10), was defined by equation (28)

$$I_{a-b} = K_{Af} * K_{Dm} + K_{Am} * K_{Dm} \tag{28}$$

in arbitrary units. In this work, the acid-base interaction parameters between fibers and matrix are greatly correlated with the results of the interfacial shear strength carried out on single fibers and the interlaminar shear strength of the composites.

Thielmann and Burnett measured surface energy and acid-base characteristics for the series of fluorinated carbon fibers and PVDF polymer (Thielmann and Burnett 2004). The results were used to calculate the work of adhesion between fibers and polymers. The work of adhesion was calculated from equation (29)

$$W_A^{total} = 2 \cdot (\gamma_1^d + \gamma_2^d)^{1/2} + 2 \cdot (\gamma_l^+ \cdot \gamma_s^-)^{1/2} + (\gamma_l^- \cdot \gamma_s^+)^{1/2}) \tag{29}$$

A comparison with apparent interfacial shear strength numbers shows that the practical adhesion due to fluorination increases less significantly than expected from energy measurements. This is most likely due to the existence of an optimum fluorine/carbon ratio, which is lower than the one for the samples investigated. Interfacial shear strength was estimated by the "pull-out" test. This test was described in Bismarck and Schulz (2003). It was performed using a home-made apparatus. The "single fiber composites" for the pull-out tests were prepared in a special embedding machine, which allows a fiber orientation exactly perpendicular to the matrix surface at a defined embedded fiber length. The shear strength τ_{IFSS} was calculated from the maximum pull-out force F_{max} and fiber embedded area in the matrix using the following equation:

$$\tau_{IFSS} = F_{max}/\pi \cdot d_f \cdot L \tag{30}$$

where L is the embedded length and d_f is the fiber diameter. More details on the fiber pull-out technique are described in Drzal et al. (2001). Such comparison of the thermodynamic work of adhesion determined by IGC technique with a pull-out test gives the strength of practical adhesion.

Similar systems were examined by Strzemiecka et al. (Strzemiecka et al. 2013). IGC-derived data indicated that the work of adhesion increased by increasing the surface area of the carbon black, but the opposite trend was found in W_A/W_{coh} and work of cohesion. According to the W_A/W_{coh} values, the filler particles should be well-dispersed into the polyurethane matrix, giving homogenous composites. The carbon black-thermoplastic polyurethane interactions determined by plate–plate rheology showed the same trend as that for the W_A/W_{coh} values. In this case, the thermodynamic work of adhesion values derived from IGC were not in agreement with the carbon black-polyurethane interfacial interactions, likely due to the dominant effect of the carbon black in reducing the crystallinity and increasing the degree of phase separation of the system consisting of carbon filler and thermoplastic polyurethane. Surface properties of carbon black were also estimated using various techniques. The acidity of the carbon blacks (expressed by K_A, K_A/K_D, γ_s^+) is related to the presence of COO-groups (expressed by the amount of CO_2). On the other hand, the activity expressed as γ_s^D and total surface energy (γ_s^{total}) value depends on the H_2O amount, i.e., the oxygen atom in H_2O molecule can form weak hydrogen bond with the hydrogen atoms of alkanes, acting as electron donor for acid test probes. On the other hand, the hydrogen atoms of H_2O molecule can interact with the lone pair of electrons of the basic test probes, leading to an increase in surface acidity. The work of adhesion between carbon black and thermoplastic polyurethane is higher than 200 mJ/m^2. The work of adhesion increases by increasing the surface energy and the specific surface area of the carbon black. Furthermore, the work of adhesion between carbon black and thermoplastic polyurethane results from specific interactions W_A^{ab} (donor–acceptor interactions), rather than from dispersive interactions W_A^D, as the contribution of the specific interactions to the work of adhesion W_A^{ab} is about two-fold higher than the dispersive contribution W_A^D.

The properties of thermoplastic polyurethane adhesives (TPU) can be altered by adding fillers. Garcia and Martin-Martinez (Garcia and Martin-Martinez 2015) examined the composites of TPU with different amounts (0.5–10 wt%) of carbon black and the rheological, mechanical, thermal, and adhesion properties of the filled polyurethane adhesives were compared. Authors found that the addition of 0.5–2 wt% carbon black filler to the polyurethane adhesive caused an increase in rheological, viscoelastic, thermal, mechanical, and adhesion properties.

To express the ability of examined chemically activated carbons and steam activated carbons as well as carbons after acidic and basic washing, Diaz et al. (Diaz et al. 2005) used the parameter of specific interaction of polar solutes (I_{sp}) related with the specific component of surface free energy according to equation (14). Authors were able to indicate the differences between the examined materials and to follow the changes after the respective material treatment. Regretfully, I_{sp} values depend on the nature of used test solute. Therefore, one obtains various values of I_{sp} parameter (using different test solutes) for the same material. It makes the discussion somewhat problematic. They also indicated the parameter expressing the potential of a solid to undergo London or dispersive types of interactions, γ_S^D, is slightly more sensitive to surface changes than the heat of adsorption.

Zhou et al. (Zhou et al. 2010) indicated that the surface energy of pyrolytic carbon black (TWPC) modified by titanate coupling agent, especially the specific interaction γ_S^{sp} determined by the specific probe molecule benzene, shows a strong interaction between the TWPC and the polyisoprene chains from the rubber matrix. Mechanical testing results confirmed the IGC prediction concerning the stronger interactions between TMPC modified by titanate coupling agent and rubber matrix than that found for the system using standard natural gas semireinforcing furnace black (SRF) filler. They estimated γ_S^{sp} values for only one test solute (benzene). It was a result of the definition of this parameter:

$$\gamma_s^{sp} = \frac{\Delta G_A^{sp}}{N a_p} \tag{31}$$

The potential usefulness of this parameter is limited due to the use of single test solute (benzene). Benzene is able to interact with the examined material by only a restricted range of intermolecular interactions (excluding for instance, acid-base or hydrogen bonding interactions).

Zhao et al. (Zhao et al. 2004) examined diamond-like carbon (DLC) and tetrahedral amorphous carbon (ta-C). The surface free energy components were calculated using the Owens–Wendt geometric mean approach, the Wu harmonic mean approach, the van Oss acid–base approach, and the equation of state. The experimental results showed that the total surface free energy and dispersive surface free energy of the ta-C coatings, DLC coatings, stainless steel 304, and titanium decreased with increasing surface temperature, while the acid–base SFE component increased with increasing temperature. The total surface free energy values of the coatings measured by using the different methods are statistically and significantly different (P-value < 0.05). The Owens–Wendt approach was recommended to determine the surface energy and its components for this specific case.

The balance between filler-polymer adhesion and filler-filler cohesion is crucial for good dispersion of fillers in the polymer matrix. In Strzemiecka et al. (2013), the thermodynamic cohesion between CB-CB particles, W_{coh}, as well as the thermodynamic adhesion between carbon-polymer, W_A, were estimated. The components of the W_{coh} were calculated similarly as the components of W_A using following equations (equations 32–34):

$$W_{coh}^D = 2 \cdot \sqrt{\gamma_{s,CB}^D \cdot \gamma_{s,CB}^D} \tag{32}$$

$$W_{coh}^{ab} = 2 \cdot ((\gamma_{s,CB}^{+} \cdot \gamma_{s,CB}^{-})^{1/2} + (\gamma_{s,CB}^{-} \cdot \gamma_{s,CB}^{+})^{1/2}) \tag{33}$$

$$\text{thus, } W_{coh}^{total} = 2 \cdot (\gamma_{s,CB}^{D} \cdot \gamma_{s,CB}^{D})^{1/2} + 2 \cdot ((\gamma_{s,CB}^{+} \cdot \gamma_{s,CB}^{-})^{1/2} + (\gamma_{s,CB}^{-} \cdot \gamma_{s,CB}^{+})^{1/2}). \tag{34}$$

$W_{A}^{total}/W_{coh}^{total}$ ratio values are more appropriate when assessing thermodynamic interfacial adhesion. In such a case, the thermodynamic adhesion/cohesion values follow the rheological data. However, results presented in Strzemiecka et al. (2013) indicate that additional factors to the thermodynamics also play a role in the interactions between the carbon black and the polyurethane chains.

Zabihi et al. (Zabihi et al. 2017) also used γ_{s}^{D}, γ_{s}^{+}, γ_{s}^{-}, and γ_{s}^{total} for the assessment of interactions of carbon fiber and modified carbon fibers with epoxy resin. Authors were able to show that nanoclay-based modification does not change the tensile strength and Weibull modulus of CF significantly, even though both specific and dispersive components of surface free surface energy, obtained by inverse gas chromatography technique, were increased.

Donnet et al. (Donnet et al. 2002) used inverse gas chromatographic technique at finite concentration to determine the distribution of surface energy sites. Authors applied three molecular probes (hexane, 1-pentanol, 1-pentylamine) to examine the series of carbon blacks. The carbon black surface has a good affinity with hydrocarbon chains, such as hexane. In the case of pentanol, weaker interactions were detected. The possibility of hydrogen bond formation between pentanol and carbon black surface is small. Stronger interactions were detected for pentylamine, revealing acidic character of examined carbon blacks. Authors were able to present good correlations between BET nitrogen surface areas and the number of the sites on carbon black surfaces.

An interesting review on adsorbate–adsorbent and adsorbate–adsorbate interactions having a decisive influence on the distribution of adsorbate between gas and solid phases in inverse gas chromatography (IGC) was demonstrated by Grajek (Grajek 2007). He gave a deep thermodynamic analysis of processes occurring in chromatographic column during IGC experiments. In his conclusions, Grajek stressed that chromatographic measurements are usually performed at temperatures higher than those which are employed in static tests. It results in the decrease of the enthalpy of physical adsorption. The coverage of adsorbent surface with adsorbate is usually low during chromatographic tests. At higher coverage of the surface of examined material, adsorbate–adsorbate interactions may be ignored for lower hydrocarbons, but not for higher hydrocarbons. The chromatographic determination of the isosteric enthalpy allows a quick, simple, and reasonably precise estimation of the enthalpy of adsorption, provided the chromatographic tests are performed at ideal linear, and ideal non-linear chromatographic conditions.

Papirer et al. (Papirer et al. 1999) examined graphite, carbon black, and fullerene (C60) at finite concentration conditions. They applied IGC data not only for the determination of adsorption isotherms, but also to determine the value of the first derivative of adsorption isotherm. This value is necessary for the calculation of the adsorption energy distribution function. They concluded that the results given by IGC at finite concentrations are in fair agreement with those obtained by applying IGC at infinite dilution conditions. Authors reported also that major differences between the fullerene and other carbons were found when the population of adsorption sites leading to the highest values of adsorption energies were considered.

It is well known that there are many structural defects on the carbon black surface and the surface energy is heterogeneously distributed. A group from Dublin extensively examined the influence of defects and basal plane contribution on the surface energy of graphite (Ferguson et al. 2016). This procedure was also applied to other types of materials, i.e., non-carbon samples (molybdenum disulphide–MoS_2 and boron nitride BN) (Caffrey 2017). Caffrey stated that the surface energy at full coverage is equal to the average surface energy value of the whole distribution, the decay constant is proportional to the fraction of high energy sites, and that the difference between values of dispersive component of surface energy at zero and full coverage depends on the standard deviation of the Gaussian curve. It was demonstrated that it is possible to infer the surface energy distribution from plots of energy values versus the fraction of high energy sites.

Wu et al. (Wu et al. 2002) tried to clarify the influence of surface roughness on the thermodynamic interactions between carbon particles and polymer melts. They proved that the carbon surfaces possess energetic heterogeneity, with the most active sites at the graphite crystalline edges, and the interactions in play are van der Waals in nature.

Programmed IGC was applied in two papers on the properties of carbon blacks (McMahon et al. 2002, Liauw et al. 2005). Authors suggested that it is an extension of traditional inverse gas chromatography! However, it is difficult to agree with such a statement. The proposed procedures do not lead to significant characteristics of the examined material. The authors neglected the solute-solute interactions when injecting multiple probes (mixtures of test solutes)! Their values of adsorption enthalpy and energy are questionable. Moreover, they (McMahon et al. 2002) presented, in fact, a doubtful modification of OLD Rohrschneider-McReynolds method. They said that it can easily distinguish between two examined materials using described figures based on their proposal. It is difficult to call it IGC. Moreover, they "diluted" the examined material with diatomaceous earth Chromosorb W-AW–DMCS calling the "solvent" inert. They should remember that their results come from the interactions between solute molecules of different types, interaction of solute molecules with two components of the examined stationary phase. It means that in such a case, mixed retention mechanism should be considered. One should mean here that interactions between the molecules of solute 1 and solute 2 should be taken into account. Therefore, the use of "standard" set of equations for the characterization of these carbon blacks is not justified. Liauw et al. (Liauw et al. 2005) examined mixed carbon black/silica materials also using "multiple probe temperature programmed inverse gas chromatography". Their probes were selected from non-polar (linear, branched, and cyclic alkanes) and polarizable (linear and cyclic alkenes and benzene) hydrocarbons. In practise, no polar probes were used to predict the activity of the material. They again used Chromosorb W-AW–DMCS as a "diluent"! Presented values of enthalpy of adsorption of test solute are burdened by solute-solute interactions.

5. Conclusions

The aim of this chapter was to focus the attention of readers on the advantages of using inverse gas chromatography in characterization of the important set of materials—various carbon materials. We were able to present these possibilities and prove that this technique is an effective source of physicochemical data, enabling extensive characterization of various types of carbon materials and composites containing carbon fillers. Thanks to the IGC technique, it is possible to describe in detail adsorption properties and surface activity of carbon materials. It is crucial that IGC enables studying carbon materials in real system, which means no special conditions or sample preparation is needed (e.g., preheating, performance of measurements in vacuum). Moreover, the carbon materials can be studied by IGC in any form as particles or fibers. IGC is the complementary method for other techniques giving detailed information about chemistry of carbon surface, such as XPS, TG-MS. Combining information from IGC, nitrogen adsorption, contact angle measurements, XPS analysis or TG-MS, and other techniques, it is possible to describe very detail and explain many phenomena which occur on carbon surface and predict its behavior in different industry applications.

It is worthy to note that up to now there is no report discussing the application of IGC in characterization of carbon materials. These are often characterized by nitrogen adsorption BET, and their properties are deduced from estimated surface area. However, it is known that surface area is only one of the parameters deciding the activity of carbons of various types. The information of their (carbons) activity through non-specific and specific interaction is crucial for prediction of further applicability. Results obtained by means of inverse gas chromatography might be used in tracking changes occurring during the technological processes connected, e.g., with modification (activation) of various carbon materials.

Acknowledgments

This work was supported by 03/32/DSPB/0800 grant.

References

Adamska, K., R. Bellinghausen and A. Voelkel. 2008. New procedure for the estimation of Hansen solubility parameters by means of inverse gas chromatography. J. Chromatogr. A 1195: 146–149.

Ali, S.S.M., J.Y.Y. Heng, A.A. Nikolaev and K.E. Waters. 2013. Introducing inverse gas chromatography as a method of determining the surface heterogeneity of minerals for flotation. Powder Technol. 249: 373–377.

Bagreev, A., F. Adib and T. Bandosz. 1999. Initial heats of H_2S adsorption on activated carbons: effect of surface features. J. Colloid Interface Sci. 219: 327–332.

Bergin, S., Z. Sun, D. Rickard, P.V. Streich, J.P. Hamilton and J.N. Coleman. 2009. Multicomponent solubility parameters for single-walled carbon nanotube solvent mixtures. ACS Nano 3: 2340–2350.

Bismarck, A. and E. Schulz. 2003. Adhesion and friction behaviour between carbon fibres and poly(vinylidene fluoride. J. Mater. Sci. 38: 4965–4972.

Burnett, D.J., J. Khoo, M. Naderi, J.Y.Y. Heng, G.D. Wang and F. Thielmann. 2012. Effect of processing route on the surface properties of amorphous indomethacin measured by inverse gas chromatography. AAPS Pharm. Sci. Technol. 13: 1511–1517.

Caffrey, I. 2017. Inferring the Surface Energy Distribution of Low Dimensional Materials. Ph.D. thesis, University of Dublin, Trinity College, Ireland.

Cerháti, T. 2009. Carbon-based sorbents in chromatography. New achievements. Biomed. Chromatogr. 23: 111–118.

Cossarutto, L., C. Vagner, G. Finqueneisel, J.V. Weber and T. Zimny. 2001. Surface free energy (γ_S^D) of active carbons determined by inverse gas chromatography: influences of the origin of precursors, the burn off level and the chemical modification. Appl. Surf. Sci. 177: 207–211.

Danner, R.P., F. Tihminlioglu, F.K. Surana and J.L. Duda. 1998. Inverse gas chromatography applications in polymer-solvent systems. Fluid Phase Equilibria 148: 171–188.

Das, S.C., I. Larson, D.A.V. Morton and P.J. Stewart. 2011. Determination of the polar and total surface energy distributions of particulates by inverse gas chromatography. Langmuir 27: 521–523.

Das, S.C. and P.J. Stewart. 2012. Characterising surface energy of pharmaceutical powders by inverse gas chromatography at finite dilution. J. Pharm. 64: 1337–1348.

Della Volpe, C. and S. Siboni. 1997. Some reflections on acid-base solid surface free energy theories. J. Colloid Interface Sci. 195: 121–136.

Diaz, E., S. Ordonez, A. Vega and J. Coca. 2004. Adsorption characterisation of different volatile organic compounds over alumina, zeolites and activated carbon using inverse gas chromatography. J. Chromatogr. A 1049: 139–146.

Diaz, E., S. Ordóñez, A. Vega and J. Coca. 2005. Comparison of adsorption properties of a chemically activated and a steam-activated carbon, using inverse gas chromatography. Microporous Mesoporous Mat. 82: 173–181.

Donnet, J.B., E. Custodéro, T.K. Wang and G. Hennebert. 2002. Energy site distribution of carbon black surfaces by inverse gas chromatography at finite concentration conditions. Carbon 40: 163–167.

Dorris, G.M. and P. Gray. 1980. Adsorption of n-alkanes at zero surface coverage on cellulose paper and wood fibers. J. Colloid. Interf. Sci. 77: 353–358.

Drzal, L.T., P.J. Herrera-Franco and H. HO. 2001. In: Test Methods, Nondestructive Evaluation and Smart Composites. Vol. 5, edited by L.A. Carlsson; Comprehensive Composite Materials. edited by A. Kelly and C. Zweben (Pergamon, Amsterdam).

Ferguson, A., I. Caffrey, C. Backes, J.N. Coleman and S.D. Bergin. 2016. Differentiating defect and basal plane contributions to the surface energy of graphite using inverse gas chromatography. Chemistry of Materials 28: 6355–6366.

Gamble, J.F., M. Leane, D. Olusanmi, M. Tobyn, E. Šupuk, J. Khoo et al. 2012. Surface energy analysis as a tool to probe the surface energy characteristics of micronized materials: a comparison with inverse gas chromatography. Int. J. Pharm. 422: 238–244.

Garcia, S.A. and J.M. Martin-Martinez. 2015. Effect of the carbon black content on the thermal, rheological and mechanical properties of thermoplastic polyurethanes. J. Adhesion Sci. Technol. 29: 1–19.

Garnier, C., T. Görner, F. Villiéras, Ph. De Donato, M. Polakovič, J.-L. Bersillon et al. 2007. Activated carbon surface heterogeneity seen by parallel probing by inverse liquid chromatography at the solid/liquid interface and by gas adsorption analysis at the solid/gas interface. Carbon 45: 240–247.

Gierak, A., M. Seredych and A. Bartnicki. 2006. The preparation, properties and application of carbon fibers for SPME. Talanta 69: 1079–1087.

Grajek, H. 2007. Rediscovering the problem of interpretation of chromatographically determined enthalpy and entropy of adsorption of different adsorbates on carbon materials. Critical appraisal of literature data. J. Chromatogr. A 1145: 1–50.

Hansen, C.M. 1967a. The three dimensional solubility parameter—key to paint component affinities II. J. Paint Technol. 39: 505–510.

Hansen, C.M. 1967b. The three dimensional solubility parameter—key to paint component affinities I. J. Paint Technol. 39: 104–117.

Hansen, C.M. and K. Skaarup. 1967. The three dimensional solubility parameter—key to paint component affinities II. J. Paint Technol. 39: 511–514.

Herry, C., M. Baudu and D. Raveau. 2001. Estimation of the influence of structural elements of activated carbons on the energetic components of adsorption. Carbon 39: 1879–1889.

Ho, R., A.S. Muresan, G.A. Hebdink and J.Y.Y. Heng. 2010. Influence of fines on the surface energy heterogeneity of lactose for pulmonary drug delivery. Int. J. Pharm. 388: 88–94.

Ito, K. and J.E. Guillet. 1979. Estimation of solubility parameters for some olefin polymers and copolymers by inverse gas chromatography. Macromolecules 12: 1163–1167.

Jagiello, J., T.J. Bandosz and J.A. Schwarz. 1992. Inverse gas chromatographic study of activated carbons: The effect of controlled oxidation on microstructure and chemical functionality. J. Colloid Interface Sci. 151: 433–445.

Jańczuk, B., T. Białopiotrowicz and A. Zdziennicka. 1999. Some remarks on the components of the liquid surface free energy. J. Colloid. Interface Sci. 211: 96–103.

Karger, B.L. and L.R. Snyder. 1976. An expanded solubility parameter treatment for classification and use of chromatographic solvents and adsorbents. Parameters for dispersion, dipole and hydrogen bonding interactions. J. Chromatogr. 125: 71–88.

Karger, B.L. L.R. Snyder and C. Eon. 1978. Expanded solubility parameter treatment for classification and use of chromatographic solvents and adsorbents. Anal. Chem. 50: 2126–2136.

Launay, H., C.M. Hansen and K. Almdal. 2007. Hansen solubility parameters for a carbon fiber/epoxy composite. Carbon 45: 2859–2865.

Lavielle, L. and J. Schultz. 1991. Surface properties of carbon fibers determined by inverse gas chromatography: Role of pretreatment. Langmuir 7: 978–981.

Liauw, C.M., G.C. Lees, A.W. McMahon, R.N. Rothon, C.A. Rego and P.J. McLaughlin. 2005. Surface activity studies on carbon-silica dual phase fillers using flow microcalorimetry and multiple probe temperature programmed inverse gas chromatography. Composite Interfaces 12: 201–220.

Lindvig, T., M.L. Michelsen and G.M. Kontogeorgis. 2002. A Flory-Huggins model based on Hansen solubility parameters. Fluid Phase Equilib. 203: 247–260.

McMahon, A.W., D.G. Kelly and P.J. McLaughlin. 2002. Characterisation of heterogeneous solid surfaces by multiple probe, temperature-programmed inverse gas chromatography (IGC). A feasibility study. The Analyst 127: 17–21.

Menzel, R., A. Lee, A. Bismarck and M.S.P. Shaffer. 2009. Inverse gas chromatography of as-received and modified carbon nanotube. Langmuir 25: 8340–8348.

Nguyen, H.P., R.P.T. Luu, A. Munafo, P. Ruelle, N.T. Ho, M. Buchmann et al. 1986. Determination of partial solubility parameters of lactose by gas-solid chromatography. J. Pharm. Sci. 75: 68–72.

Papirer, E., E. Brendle, F. Ozil and H. Balard. 1999. Comparison of the surface properties of graphite, carbon black and fullerene samples, measured by inverse gas chromatography. Carbon 37: 1265–1274.

Park, S.J. and J.B. Donnet. 1998. Anodic surface treatment on carbon fibers: determination of acid-base interaction parameter between two unidentical solid surfaces in a composite system. J. Colloid Interface Sci. 206: 29–32.

Pérez-Mendoza, M., M.C. Almazán-Almazán, L. Méndez-Liñán, M. Domingo-Garcia and F.J. López-Garzón. 2008. Evaluation of the dispersive component of the surface energy of active carbons as determined by inverse gas chromatography at zero surface coverage. J. Chromatogr. A 1214: 121–127.

Price, G.J. 1989. Calculation of solubility parameters by inverse gas chromatography. pp. 48–58. *In*: D.R. Lloyd, T.C. Ward and H.P. Schreiber (eds.). Inverse Gas Chromatography. Characterization of Polymers and Other Materials. ACS Symposium Series. 391. Washington, USA.

Price, G.J. and I.M. Shillcock. 2002. Inverse gas chromatographic measurements of solubility parameters in liquid crystalline systems. J. Chromatogr. A 964: 199–204.

Rojewska, M., A. Bartkowiak, B. Strzemiecka, A. Jamrozik, A. Voelkel and K. Prochaska. 2017. Surface properties and surface free energy for cellulosic, etc. mucoadhesive polymers. Carbohydrate Polymers 171: 152–162.

Santos, J.M.R.C.A. and J.T. Guthrie. 2005. Analysis of interactions in multicomponent polymeric systems: the key-role of inverse gas chromatography. Mater. Sci. Eng. R 50 50: 79–107.

Siboni, S. and C. Della Volpe. 2015. Acid-base theory of adhesion: a critical review. Rev. Adhesion Adhesives 3(3): 253–310.

Singh, G.S., D. Lal and V.S. Tripathi. 2004. Study of microporosity of active carbon spheres using inverse gas chromatographic and static adsorption techniques. J. Chromatogr. A 1036: 189–195.

Strzemiecka, B. and A. Voelkel. 2012. Estimation of the work of adhesion by means of inverse gas chromatography for polymer complex systems. Int. J. Adh. Adh. 38: 84–88.

Strzemiecka, B., A. Voelkel, J. Donate-Robles and J.M. Martin-Martinez. 2013. Estimation of polyurethane-carbon black interactions by means of inverse gas chromatography. J. Chromatogr. A 1314: 249–254.

Strzemiecka, B., A. Voelkel, J. Donate-Robles and J.M. Martin Martinez. 2014. Assessment of the surface chemistry of carbon blacks by TGA-MS, XPS and inverse gas chromatography using statistical chemometric analysis. Appl. Surf. Sci. 316: 315–323.

Sun, C. and J.C. Berg. 2003. A review of the different techniques for solid surface acid-base characterization. Adv. Coll. Int. Sci. 105: 151–175.

Sun, Z., V. Nicolosi, D. Rickard, S.D. Bergin, D. Aherne and J.N. Coleman. 2008. Quantitative evaluation of surfactant-stabilized single-walled carbon nanotubes: dispersion quality and its correlation with zeta potential. J. Phys. Chem. 112: 10692–10699.

Thielmann, F. 2004. Introduction into the characterization of porous materials by inverse gas chromatography. J. Chromatogr. A 1037: 115–123.

Thielmann, F. and D. Burnett. 2004. Determination of Carbon Fibre-Polymer Interactions by Inverse Gas Chromatography. Unpublished material https://www.researchgate.net/publication/265561107.

Thielmann, F., D. Burnett and A. Bismarck. 2005. Measuring adhesion phenomena between carbon fibers and model polymer matrices. International SAMPE Symposium and Exhibition (Proceedings). 50: 1551–1558. 50th International SAMPE Symposium and Exhibition; Long Beach, CA, USA.

Vagner, C., G. Finqueneisel, T. Zimny, P. Burg, B. Grzyb, J. Machnikowski et al. 2003. Characterization of the surface properties of nitrogen-enriched carbons by inverse gas chromatography methods. Carbon 41: 2847–2853.

Van Oss, C.J. 2006. Interfacial forces in aqueous media. Taylor & Francis Group, LLC. Boca Raton, London, New York.

Van Oss, C., R. Good and M. Chaudhury. 1988. Additive and nonadditive surface tension components and the interpretation of contact angles. Langmuir 4: 884–891.

Voelkel, A., B. Strzemiecka, K. Adamska, K. Milczewska and K. Batko. 2009a. Surface and bulk characteristics of polymers by means of inverse gas chromatography. pp. 71–102. In: A.B. Nastasović and S.M. Jovanović (eds.). Polymeric Materials. Transworld Research Network. Kerala, India.

Voelkel, A., B. Strzemiecka, K. Adamska and K. Milczewska. 2009b. Inverse gas chromatography as a source of physicochemical data. J. Chromatogr. A 1216: 1551–1566.

Voelkel, A. 2012. Physicochemical measurements (Inverse gas chromatography). pp. 477–494. In: C.F. Poole (ed.). Gas Chromatography. Elsevier, Amsterdam, Holland.

Wu, G., S. Asai, M. Sumita and H. Yui. 2002. Entropy penalty-induced self-assembly in carbon black or carbon fiber filled polymer blends. Macromolecules 35: 945–951.

Wu, Y., Z. Li and H. Xi. 2004. Effects of inverse gas chromatography measurement conditions on elution peaks on activated carbon. Carbon 42: 3003–3042.

Ylä-Mäihäniemi, P.P., J.Y.Y. Heng, F. Thielmann and D.R. Williams. 2008. Inverse gas chromatographic method for measuring the dispersive surface energy distribution for particulates. Langmuir. 24: 9551–9557.

Zabihi, O., M. Ahmadi, Q. Li, S. Shafei, M.G. Huson and M. Naebe. 2017. Carbon fibre surface modification using functionalized nanoclay: A hierarchical interphase for fibre-reinforcd polymer composites. Composites Sci. Technol. 148: 49–58.

Zhao, Q., Y. Liu and E.W. Abel. 2004. Effect of temperature on the surface free energy of amorphous carbon films. J. Colloid Interface Sci. 280: 174–183.

Zhou, J., T. Yu, S. Wu, Z. Xie and Y. Yang. 2010. Inverse gas chromatography investigation of rubber reinforcement by modified pyrolytic carbon black from scrap tires. Ind. Eng. Chem. Res. 49: 1691–1696.

CHAPTER 12

Fullerenes and Polycyclic Aromatic Hydrocarbons in Separation Science

Yoshihiro Saito,[1,*] *Koki Nakagami,*[1] *Ohjiro Sumiya*[1] *and Ikuo Ueta*[2]

1. Introduction

Separation science is a research field over a wide range of science and technology. The scope of the separation science includes not only the research and development in analytical chemistry, but also other related research areas, such as pharmaceutical, biomedical, environmental analyses, along with the a huge number of academic and industrial applications. As one of the main interests in the separation science, the development of effective separation techniques has been intensively studied on the basis of the development of new materials and state-of-the-art analytical instruments. Among various modern separation methods, liquid chromatography (LC) is the most powerful method of choice for the separation and isolation of various chemical compounds in complex mixtures (Gehrke et al. 2001). No one doubts the fact that the resulting industrial applications are necessary for our modern life. In contrast to the extensive applications, the entire separation mechanism in the chromatographic process has not been clear yet. This is mainly because the complicated separation mechanism contains many parameters for the molecular recognition of the analyte in the chromatographic process.

For the majority of the LC applications, a reversed-phase (RP) condition has been employed, where a chemically-modified silica-based stationary phase is normally used. As the representative of the RP stationary phases, octadecylsilica phases (ODSs) are quite dominant for the general separation of the typical mixtures due to a wide availability and an affordable price. The ODSs can be divided into two types, depending on the bonding chemistry during the phase synthesis. One is polymeric, which is typically synthesized with trifunctional (or difunctional) silanes as the starting material in aqueous conditions, and another is monomeric synthesized from monofunctional silanes in non-aqueous conditions. In contrast to the number of reports for the selectivity and the ordering differences between polymeric and monomeric ODS phases (Sander and Wise 1984, Jinno et al. 1987), however, the systematic understanding and the resulting theory for the intermolecular interaction between solutes and bonded phase ligands has not been established well.

[1] Department of Applied Chemistry and Life Science, Toyohashi University of Technology, 1-1 Hibarigaoka, Tempaku-cho, Toyohashi 441-8580, Japan.
[2] Department of Applied Chemistry, University of Yamanashi, 4-3-11 Takeda, Kofu 400-8511, Japan.
Email: iueta@yamanashi.ac.jp
* Corresponding author: saito@tut.jp

During the past several decades, various types of newly synthesized chemically-bonded stationary phases have been developed in LC. Introducing the recent innovations in the synthetic and bonding chemistries (Pesek and Cash 1989, Kirkland et al. 1989, Welch and Pirkle 1992, Kimata et al. 1993), many novel stationary phases could be realized on the basis of the design of the bonded phase ligands. Therefore, a more systematic characterization of the stationary phases has been increasingly important to develop a more powerful stationary phase with a desirable separation performance for a particular separation problem. To precisely characterize the chemically-bonded stationary phases on the surface of the support, such as a porous silica material, many approaches have been investigated, for example, systematic analysis of the retention behaviors for a certain group of solutes, various spectroscopic measurements of the stationary phases, and theoretical calculations for the bonded phase structure. As one of the approaches to the systematic retention behavior analysis, polycyclic aromatic hydrocarbons (PAHs) and fullerenes (Figure 12.1) have been introduced as a group of the

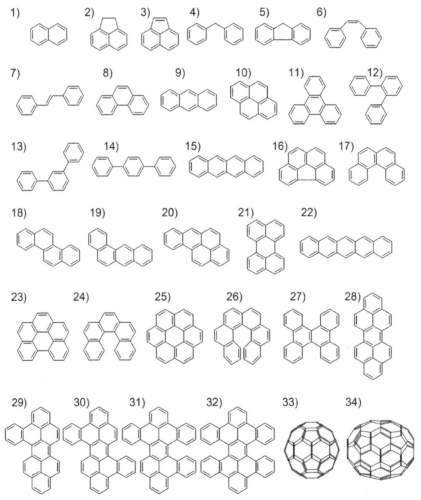

Figure 12.1: Chemical structures of various PAHs, including C_{60} and C_{70} fullerenes. (1) naphthalene, (2) acenaphthene, (3) acenaphthylene, (4) diphenylmethane, (5) fluorene, (6) *cis*-stilbene, (7) *trans*-stilbene, (8) phenanthrene, (9) anthracene, (10) pyrene, (11) triphenylene, (12) *o*-terphenyl, (13) *m*-terphenyl, (14) *p*-terphenyl, (15) naphthacene, (16) benzo[*ghi*] fluoranthene, (17) benzo[*c*]phenanthrene, (18) chrysene, (19) benz[*a*]anthracene, (20) benzo[*a*]pyrene (BaP), (21) perylene, (22) pentacene, (23) benzo[*ghi*]perylene, (24) dibenzo[*c,g*]phenanthrene, (25) coronene, (26) phenanthro[*3,4-c*]phenanthrene (PhPh), (27) dibenzo[*g,p*]chrysene (tetrabenzonaphthalene; TBN), (28) dibenzo[*cd,lm*]perylene, (29) tribenzo[*a,cd,lm*]perylene, (30) tetrabenzo[*a,cd,j,lm*]perylene, (31) pentabenzo[*a,cd,f,j,lm*]perylene, (32) hexabenzo[*a,cd,f,j,lm,o*]perylene, (33) C_{60}, and (34) C_{70}.

sample probes in LC (Jinno 1996, 1999). Since there are many structural isomers in PAHs having aromatic rings of more than four, the total numbers of the homologous compounds including the respective isomers are quite large when compared to other class of compounds. However, at the same time, these relatively simple molecular structures are quite suitable for the systematic analysis of the retention behavior on the basis of the molecular shape recognition process than real samples to be separated in typical LC applications.

On the other hand, fullerene molecules, as typically known as C_{60} and C_{70}, can be regarded as very large PAHs with specific features, molecular dimensions, and molecular weight. That means that the combination of the results obtained from the retention behavior study for PAHs with that for fullerenes could allow a systematic interpretation about molecular shape recognition process on the stationary phase. The characteristic information for the stationary phases from the consideration of retention behavior can be further applied to the next synthetic strategies of the chemically bonded stationary phase in LC.

In this chapter, the use of various PAHs, including fullerenes as the sample probes in the retention behavior analysis in LC, has been reviewed. Taking advantage of a variety of homologues and the corresponding many structural isomers, PAHs have played an important role for the surface characterization of the stationary phase and the analysis of intermolecular interaction during the actual separation system, and also offered important information to make promising strategies for the development of novel stationary phases based on the molecular shape recognition concept.

2. Fullerenes and PAHs as the Sample Probes for the Systematic Analysis of Retention Behavior in LC

2.1 PAHs as the sample analyte on ODS stationary phases in LC

PAHs is a group of aromatic compounds having a unique molecular shape and size, as illustrated in Figure 12.1, and there are many homologues as typically recognized as naphthalene ($C_{10}H_8$), anthracene ($C_{14}H_{10}$), and naphthacene ($C_{18}H_{12}$), with increasing the number of aromatic rings. At the same time, all the PAHs consisting of more than three aromatic rings have corresponding isomer(s). For example, the PAH with the molecular formula of $C_{18}H_{12}$ consisted of five isomers, triphenylene, benzo[*c*] phenanthrene, benz[*a*]anthracene, chrysene, and naphthacene. An increased number of isomers can be expected with increasing the number of aromatic rings in PAHs. However, it suggests that the difference in the entire molecular structure of isomers will be relatively smaller when increasing the number of aromatic rings. To systematically describe the molecular structure of PAHs, especially in the analysis of retention behavior, several size and shape parameters have been proposed.

One of the molecular size descriptors, F, is introduced by Schbron et al. (Schbron et al. 1977), defined as follows: F = (number of double bonds) + (number of primary and secondary carbons) − 0.5 × (number of non-aromatic rings). As can be calculated from this equation, the F of a certain PAH molecule will be the same as the total number of double bonds in the PAH molecule if there is no primary and secondary carbons without any non-aromatic ring, for example naphthalene (F = 5), anthracene (F = 7), naphthacene (F = 9), and so on. With the F number, the trend in the retention of various PAHs was studied in conventional RPLC conditions using a monomeric ODS stationary phase, and a good linear correlation between the logarithmic retention factor (log k) and F number has been reported (Jinno and Kawasaki 1983).

In contrast to a clear correlation normally found between F and the logarithmic retention factor for a set of homologous PAHs on the typical monomeric ODS stationary phases, a different trend has been reported for polymeric ODS phases (Sander and Wise 1984, 1986). There is a negative deviation for the PAHs having a non-planar molecular shape on the polymeric ODS phases, that is, less interaction between non-planar PAHs and the bonded stationary phase synthesized with multifunctional silanes. On the other hand, a larger retentivity for the PAHs with a shape similar to a "rod" was observed on the polymeric ODS stationary phases, indicating an enhanced interaction between the PAHs having a "rod-like" molecular shape, as can be recognized as a larger aspect ratio of the analyte molecule. As described earlier, the monomeric phase is synthesized from monofunctional silanes in non-aqueous conditions, while the polymeric one is typically synthesized with trifunctional silanes as the starting material in aqueous conditions, as shown in Figure 12.2. Then, the partial polymerization or origomerization of silanization reagent (trichlorooctadecylsilane) is initiated for polymeric synthesis by the existence of water in the system. Since the silane molecule is trifunctional, one or more silanols could be formed during the synthetic process in the presence of water. In addition, a partial polymerization and/or cross-linking of C_{18} ligands could form a rigid phase structure on the silica support surface. On the contrary, the C_{18} ligands could be attached to the support surface by single siloxane bond in the monomeric phase (Sander and Wise 1984). From the above trend, one can assume that a slight difference in the molecular shape of PAHs could be recognized well on polymeric ODS phases, although the fundamental trend in the correlation between the retention and the molecular size is somewhat similar in both monomeric and polymeric ODS phases. Therefore, the fundamental evaluation of the shape selectivity, such as planar/non-planar selectivity for PAHs, can be made by these plots using F number as a molecular size descriptor.

The selectivity for the two-dimensional shape has been further studied with another molecular shape descriptor—length-to-breadth (L/B) ratio proposed by Wise et al. (Wise et al. 1981) and Kaliszan et al. (Kaliszan and Lamparczyk 1978), as illustrated in Figure 12.3. The L/B is defined as the maximized two-dimensional length-to-breadth ratio of the analyte molecule when it is projected on the flat surface. Thus L/B value of a PAH molecule expresses the corresponding two-dimensional

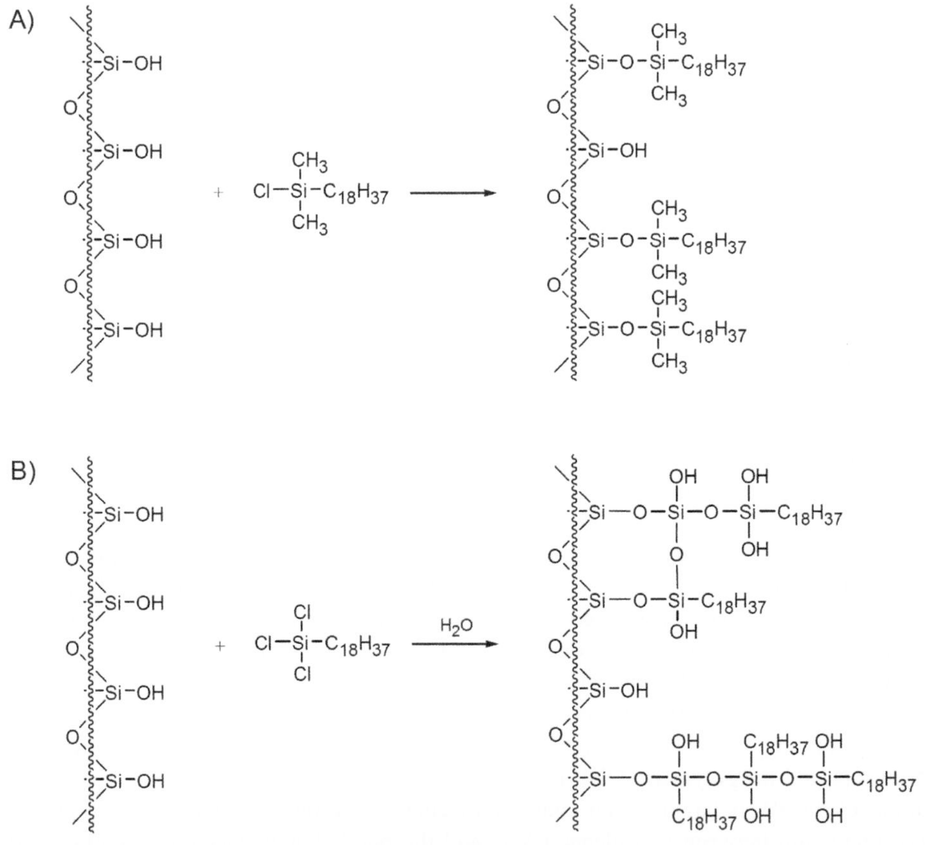

Figure 12.2: Typical synthetic scheme of (A) monomeric and (B) polymeric ODS phases commonly employed as the stationary phase in LC.

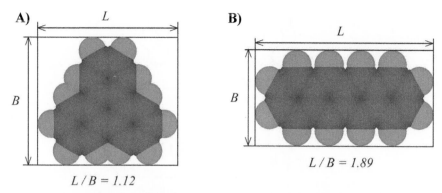

Figure 12.3: Illustrations for the calculation of *L/B* values for (A) triphenylene and (B) naphthacene.

Color version at the end of the book

shape of the molecule, and it can be used to quantitatively classify "rod-like" PAH molecules from that of "square-like" shape. The comprehensive employment of these two size and shape descriptors, *F* and *L/B*, have been successfully done for the analysis of the characterization of the chemically bonded stationary phase ligands on the support surface. The classification of monomeric and polymeric ODS phases was reported on the basis of this evaluation approach (Jinno 1996, Sander and Wise 1986).

For the evaluation of the molecular shape selectivity between planar and non-planar solutes, Tanaka et al. (Tanaka et al. 1982) and Jinno et al. (Jinno and Mae 1990) introduced a set of two aromatic compounds, *o*-terphenyl and triphenylene. Strictly speaking, *o*-terphenyl might be classified as a somewhat different aromatic compound from *ortho*-fused PAHs such as triphenylene, because two phenyl groups are linked to an aromatic ring with a single "C-C" bond in the case of *o*-terphenyl. However, these two compounds have quite a similar molecular size, as confirmed with the same *F* number (*F* = 9) along with almost the same *L/B*: 1.11 and 1.12 for *o*-terphenyl and triphenylene, respectively. These two solutes have a different planarity; *o*-terphenyl has a non-planar shape and triphenylene has a planar shape, although the molecular size and two-dimensional shape are almost the same as described above.

Similar shape selectivity analysis was proposed by the research group of Sander et al. (Sander and Wise 1986). They introduced several pair of solutes having different planarity but similar *F* and *L/B* values: benzo[*ghi*]fluoranthene and benzo[*c*]phenanthrene, benzo[*ghi*]perylene and dibenzo[*c,g*] phenanthrene, and coronene and phenanthro[*3,4-c*]phenanthrene (PhPh), where the molecular distortion for generating the non-planar molecular shape was increased from benzo[*c*]phenanthrene to PhPh due to the increasing steric hindrance, as can be expected from these PAHs molecular structures in Figure 12.1. Due to a steric hindrance of two hydrogen atoms bonded inside of the "bay-like" structure, a distorted molecular conformation for benzo[*c*]phenanthrene is expected, although the molecular size and the two-dimensional aspect ratio of benzo[*ghi*]fluoranthene and benzo[*c*] phenanthrene are almost the same. In the case of the planar/non-planar pair of benzo[*ghi*]perylene and dibenzo[*c,g*]phenanthrene, an increased molecular distortion of non-planar analyte, dibenzo[*c,g*] phenanthrene, could be estimated because of the increased steric hindrance when compared to the molecular shape of benzo[*c*]phenanthrene. With this set of test probes, a more precise evaluation of the planarity recognition capability of the stationary phase in LC was possible.

Another systematic survey for the molecular shape selectivity of various ODS phases was also carried out by the above authors with a test mixture containing benzo[*a*]pyrene (BaP), tetrabenzonaphthalene (TBN), and PhPh (Sander and Wise 1987). In this test mixture, TBN and PhPh have the same molecular formula, $C_{26}H_{16}$, and therefore, the same *F* number along with similar *L/B* values: 1.09 and 1.07 for TBN and PhPh, respectively (Oña-Ruales et al. 2018). These two

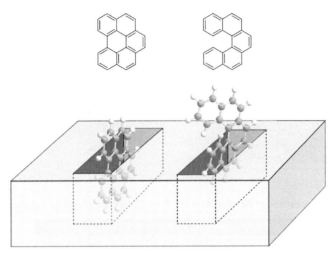

Figure 12.4: Slot model for the explanation of the shape selectivity on polymeric ODS phases.

isomers, however, have a different molecular planarity. As the reference compound, a planar solute, BaP was also included in the test mixture. From the careful investigations on the retention behaviors of these aromatic sample probes, it has been proposed that the polymeric ODS phases form a rigid ligand structure, so-called "slot-like" structure on the silica support. In Figure 12.4, the proposed retention model was on a typical polymeric ODS phase with the "slot-like" phase structure on the surface of the silica gel support. A planar solute can interact with more than a non-planar one, and at the same time, for planar solutes, a longer (typically recognized as a larger L/B value) solute molecule can also interact more than that having a smaller L/B value with a "square-like" molecular shape. In the ODS phase, the good shape selectivity of polymeric phases could be well explained by the retention model.

The results from above chromatographic studies have a good agreement with the interpretation of the phase structure differences depending on the synthetic processes-monomeric or polymeric. The results also demonstrate the validity of the systematic analysis of the retention behavior studies of PAHs for the characterization of the actual structure of the ODS phases during the separation, although many other parameters for describing the nature of the solutes and bonded phases (Radecki et al. 1979, Jinno and Kawasaki 1984), such as the electronic and hydrophobic properties, should be further considered to apply the results for other class of compounds and types of bonded phases. The estimation of the rigid bonded phase structure of the polymeric ODS phases has been further supported by the chromatographic retention measurements at different column temperatures (Sander and Wise 1986), and also by several spectroscopic techniques, including infrared spectroscopy (IR) (Jinno et al. 1993a) and nuclear magnetic resonance spectroscopy (NMR) (Sander et al. 1983, Jinno at al. 1989, Jinno 1989).

2.2 *Retention behavior analysis using fullerenes as the sample analytes*

The unique chemical structures of fullerenes and their chemical reactivities have been studied since the observation of a stable carbon cluster ion containing 60 carbon atoms was first reported in 1985 (Kroto et al. 1985). In addition to many publications dealing with the separation/isolation and analysis of the spherical carbon molecules called "fullerenes" using various separation and spectrometric techniques, the retention behaviors of fullerenes in LC have been studied in the last several decades (Pirkle and Welch 1991, Jinno et al. 1992a). This is because, especially at the first stage of the fullerenes research, the separation and purification of fullerenes are inevitable procedures to characterize those interesting and unique compounds before the applications in various fields of science and technology. At the same time, fullerene molecules, such as C_{60} and C_{70}, could be regarded as very large PAHs with specific shape and size, and therefore, they seemed to be quite an attractive group of compounds as the sample probes for the analysis of retention behavior in LC. The systematic analysis of the separation mechanism of fullerenes on various chemically bonded stationary phases in LC may make it possible to design the new stationary phases to effectively separate fullerenes based on the molecular recognition (Saito et al. 2004a).

Since fullerenes can be regarded as a group of very large PAHs, the preparative separation on conventional stationary phases was studied, where many types of ODS phases have been introduced along with bare silica stationary phases in an open column LC configuration. From the results for the separation of two most popular fullerene molecules, C_{60} and C_{70}, and also those of larger than C_{70}, typically recognized as higher fullerenes (Jinno et al. 1993b, 1995a), monomeric ODS phases with a high carbon content often showed a good separation for C_{60} and C_{70}. That means, if the silica support of the ODS phase is the same, a better separation of C_{60} and C_{70} is expected on the ODS phase having a higher surface coverage. More C_{18} ligands per unit surface area of the silica support allowed an increased selectivity between C_{60} and C_{70}, as long as the base silica is the same. Typical polymeric phases, however, exhibited only quite limited retention values for these fullerene solutes, although some types of polymeric phase showed a better shape selectivity for the separation of higher fullerenes, such as C_{76}, C_{78}, and C_{84}. These trends are basically consistent with the retention behavior for PAHs: a monomeric phase can separate the analytes with molecular size, while polymeric phase can additionally distinguish the shape differences, but sometimes offers only a limited retention for very bulky analytes (Jinno et al. 1996, Wise and Sander 1996). In contrast to a typical molecular size of PAHs, fullerenes have an extremely large molecular size with a bulky shape. The interval of the C_{18} ligands on the silica support surface might be nearly comparable to the diameter of spherical fullerene molecules. It can be assumed that the surface coverage of bonded C_{18} ligands could play an important role in adjusting the interval among the C_{18} ligands on the silica gel.

2.3 Retention of fullerenes and PAHs at subambient temperatures

The bonded phase structure on the silica support has been studied with a group of well-designed analytes, including PAHs and fullerenes. On the other hand, the temperature dependence of the shape selectivity for these test probes was also studied, because in general, a less shape selectivity to a set of critical solute pair can be expected at an elevated temperature. Thus, if the column temperature is reduced, an increased selectivity could be obtained on the basis of the rigid bonded ligands of the stationary phase. The temperature dependence of the LC retention has been investigated by a number of authors, and it has been found that the retention of analytes increases with decreasing the column temperature (Colin et al. 1978, Jinno and Hirata 1982, Zarzycki et al. 2006, Ohta et al. 2017), and the resulting selectivity between two analytes can be enhanced. On the contrary, some exceptions have been reported, where an abnormal retention change with decreasing the column temperature was reported for the separation of C_{60} and C_{70} (Pirkle and Welch 1991, Jinno et al. 1993c).

Ohta et al. studied the retention of PAHs and fullerenes on ODS phases over the temperature range between –70°C and 80°C (Ohta et al. 1994, 1995, 2000), where a unique temperature dependence for the retention of fullerenes with monomeric ODSs was confirmed. In these investigations, the existence of a critical temperature was reported, at which the retention of each fullerene molecule attained the maximum values, and at the same time, the temperature for the maximum retention was somewhat different on the basis of the surface coverage, number of C_{18} ligands per unit surface of the ODS phases. In the case of conventional C_{18} phase, it is apparent that the elution order of the compounds in a mixture (containing C_{60} and C_{70} in toluene as the solvent) changes from toluene, C_{60}, C_{70} to C_{60}, C_{70}, toluene at the temperature of about –50°C.

The unique temperature dependence of the retentions of C_{60} and C_{70} fullerenes was further confirmed with a comparison of the respective van't Hoff plots of the C_{18} phase for these fullerene molecules and PAHs. From the above results, it was considered that a change in the configuration of the bonded C_{18} ligands as a function of temperature is the most dominant factors for determining the retention of fullerenes in LC, and that the elution order at low temperatures is caused by an exclusion effect of the rigid alkyl chains which do not have enough space between adjacent C_{18} ligands for an effective interaction with bulky fullerenes molecules. The C_{18} functional groups on the silica support are easily relaxed and movable at ambient or higher temperature, and then, they can be interacted with fullerene molecules without any significant restriction. As the distance between C_{18} ligands is about 0.76 nm for typical ODS phase having a high carbon content, the ligand interval can be regarded as quite similar to the diameter of C_{60}. With decreasing the carbon content, that is, the decrease in the surface coverage, there was a bigger interval between the adjacent C_{18} ligands, resulting in lesser interaction between the stationary phase and the fullerenes.

When the column temperature is elevated, the bonded C_{18} ligands have a satisfactory freedom to move, and as the result, the molecular recognition capability for fullerenes on different ODS phases, having different carbon content, will be similar, although the possibility of interaction is still dominant. The C_{18} ligands of ODSs become more rigid and ordered at lower temperatures. The ligand distance becomes very crucial, which allows the ligands to interact with the fullerene molecule. In these conditions, any mismatch of ligand distance and the diameter of the fullerene molecule may result in a weak interaction. The above unique temperature effect and the interpretation of the retention mechanism were also supported by the solid state NMR measurements with cross-polarization and magic angle spinning techniques (Ohta et al. 1996). The influence of the column temperature on the shape selectivities for PAHs and fullerenes was also reviewed (Sander and Wise 2001).

3. Development of Novel Stationary Phases on the Basis of the Retention Behavior Studies with PAHs

In the history of chromatography and the innovations as a separation technique, various types of stationary phases have been developed and commercialized to satisfy the demands from the users. The stationary phases in chromatography have been traditionally developed mainly on the basis of a wide variety of experience and numerous numbers of experimental findings by the chromatographers. However, various approaches to the development of the stationary phase have been studied on the basis of a systematic interpretation of the retention behavior with a group of well-designed sample probes as the analytes. For the development of novel stationary phase that can effectively separate the target analytes typically contained in a complex mixture, a number of stationary phases should be experimentally synthesized, and the resulting separation performance should also be evaluated to get a valuable feedback for planning the next synthetic strategy of a more effective and selective stationary phase. In order to effectively proceed to the above synthetic approach, down-sizing the separation system is one of the advantageous challenges. Microcolumn LC has been regarded as a powerful technique for the evaluation of many new stationary phases, where a small amount of experimentally prepared stationary phases can be evaluated with valuable sample analytes, as they typically have quite a limited availability (Ishii 1988, Yang 1989, Saito et al. 2004c). In this section, the development of various novel stationary phases for LC separations on the basis of the retention behavior of PAHs and fullerenes is described.

3.1 Alkylphenyl and multi-phenyl bonded stationary phases

The effect of the interval between C_{18} ligands on the silica support surface was investigated with a set of stationary phases having phenyl functional group(s) at the root of the ligand, as shown in Figure 12.5, where the reference stationary phases without phenyl functional group were also synthesized with the same silica gel support. Changing the molecular structure of the silane coupling reagents, two phenyl functionalities were introduced at the bottom part of the bonded ligand (Saito et al. 1995a, Ohta et al. 1996), where octadecyldiphenychlorosilane, $C_{18}H_{37}Si(C_6H_6)_2Cl$, was used as the silanization reagent

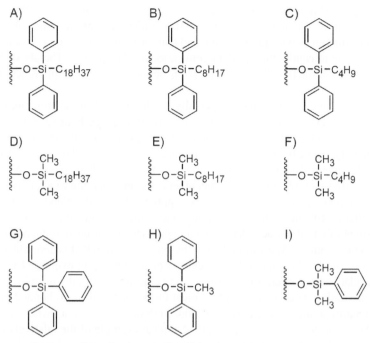

Figure 12.5: A group of stationary phases for the evaluation of the effect of the interval on the silica surface. (A) C18 Diph, (B) C8 Diph, (C) C4 Diph, (D) C18, (E) C8, (F) C4, (G) Triph, (H) Diph, and (I) Monoph.

of the silica gel support. The resulting bonded phase have a minimal interval between the ligands on the silica due to the steric hindrance of the bulky two phenyl groups at the bottom of the long C_{18} chain. Using this approach, Saito et al. prepared octadecyldiphenyl- (C_{18} Diph), octyldiphenyl- (C_8 Diph) and butyldiphenyl- (C_4 Diph) silica phases, and the retention behaviors were compared to that obtained by octadecyldimethyl- (C_{18}), octyldimethyl- (C_8) and butyldimethyl- (C_4) silica phases synthesized with the same silica gel support. The results indicated that two phenyl rings bonded to the silicon atom at the bottom of the bonded ligands induced a kind of uniformity of a relatively wide interval among C_{18} alkyl chains, allowing a good retention power for fullerenes on the phase when compared to a conventional phase synthesized by dimethyloctadecylchlorosilane, $C_{18}H_{37}Si(CH_3)_2Cl$ (Saito et al. 1995a). The results also suggest a certain contribution of the alkyl ligand length to the retention of fullerenes. A similar trend is confirmed with other ODS phases having different surface coverage values (Ohta et al. 1996).

The retention behavior of fullerenes at subambient column temperature was also studied with above alkyldiphenyl bonded phases with different length of the alkyl chains—C_{18} Diph, C_8 Diph, and C_4 Diph silica phases (Ohta et al. 1996). In contrast to the expected general trend from the data on various ODS phases at low temperatures, the change of the elution order was not observed with the C_{18} Diph phase, where two bulky phenyl functional groups attached at the bottom of the bonded C_{18} ligand on the silica surface could contribute to the uniform ligand interval between the C_{18} chains. It can be interpreted from the results that the relatively uniform ligand interval allowed an effective interaction of these bulky fullerenes between the ligands if the interval was similar to the size of the fullerenes, while the interaction with alkyl bonded ligands attached on the silica support without a uniform interval might be reduced, especially at an elevated temperature.

On the other hand, a systematic comparison of the retention behavior of C_{60} and C_{70} on a group of stationary phases, Triph, Diph, and Monoph shown in Figure 12.5, was also carried out (Saito et al. 1994a). These three bonded phases were synthesized with the same silica support. The results demonstrated a large retentivity towards C_{60} and C_{70} on the Triph phase, although there was no significant difference in the corresponding selectivities to these fullerenes on these three phases. From the temperature study on the Triph phase, it was also confirmed that the temperature dependence of the retention of fullerenes was quite limited, and that the change in the corresponding selectivity (C_{70}/C_{60}) was also significantly smaller than conventional ODS phases. The above trends clearly suggest an effective interaction between the Triph bonded phase and fullerenes including π-π interaction.

On the basis of the extensive study, especially on the bonded phase structure of monomeric and polymeric ODS phases as described above, the ordering of the bonded phase ligands on the support surface has been regarded as one of the key parameters for controlling the molecular shape recognition capability. As the flexible chemical structure of the C_{18} bonded phase, the ordering of these alkyl chains on the support could significantly affect the molecular shape recognition for the analyte having a slightly different molecular shape than others. The introduction of a relatively rigid ligand structure on the support is, therefore, a prospective approach for the development of new stationary phases, which offer an enhanced shape selectivity. One of the ultimate goals of this approach might be the development of chiral bonded phase with a highly specific molecular interaction only to one of the corresponding isomers. Apart from the one-by-one interaction that is normally required in the typical separations of chiral compounds on a chiral bonded phase, a novel bonded phase with an improved molecular shape recognition capability can be designed, and the resulting bonded phases allow a better separation when compared to typical ODS phases. Several bonded phases with an aromatic functionality in the structure were introduced (Saito et al. 2003a), where a rigid chemical structure could be expected in the aromatic ring. A typical example of the relatively rigid structure in the bonded phase ligands was observed for the phases illustrated in Figure 12.6. Introducing C_{60} and C_{70} fullerenes as the sample probes, the formation of a cavity-like three-dimensional structure

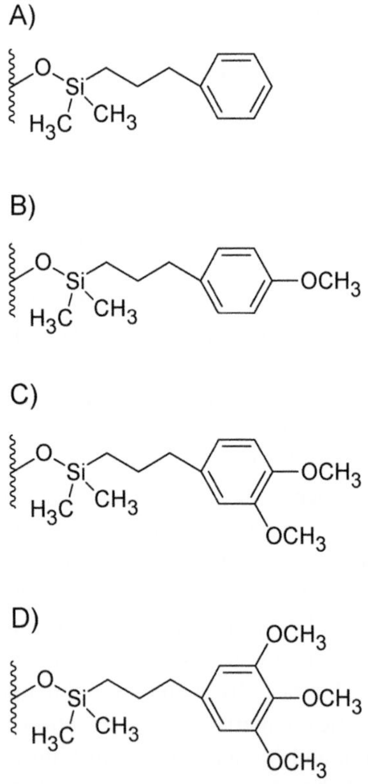

Figure 12.6: Chemical structures of (A) phenylpropyl (PP), (B) methoxyphenylpropyl (MPP), (C) dimethoxyphenylpropyl (DMPP), and (D) trimethoxyphenylpropyl (TMPP) phases.

of the bonded ligand on the silica support has been suggested. The observed good molecular shape recognition power of 3,4-dimethoxylated phenyl bonded phase (DMPP) could be explained by the energy minimized conformation of the phase and the resulting effective interaction with the fullerenes having a curved molecular surface (Jinno et al. 1993d). Jinno et al. observed exceptionally small changes in the retention of these carbon clusters on the DMPP phase during the operations at different column temperatures (Jinno et al. 1993c). Due to a good separation performance, the DMPP-bonded silica phase was also studied as a stationary phase for the separation of higher fullerenes at elevated column temperature (Jinno et al. 1993c).

3.2 *Liquid-crystal bonded silica phases*

Among a wide variety of compounds exhibiting a specific phenomenon of liquid-crystal nature, a rod-like molecule is often found, with a rigid structure and also with polar functional group(s) therein. Since the ordering in the molecular arrangement under a certain environment is an important factor to determine the liquid-crystal property, one can expect a specific molecular shape recognition capability of liquid-crystal compounds when they are used as a bonded stationary phase ligand, although the liquid-crystal nature may not be observed after the bonding reaction due to a limited movement of the bonded ligand. Pesek et al. synthesized liquid-crystal bonded phases containing rigid and rod-like bonded phase ligands in the chemical structures, and the contribution of the bonded phase rigidity and ordering on the shape selectivity have been studied (Jinno et al. 1991, Saito et al. 1994b, 1995b). The results suggest that these phases having a rigid structure on the basis of aromatic functionalities clearly demonstrate a better molecular shape recognition capability for PAHs when comparing other phases typically employed as a general stationary phase.

The increased molecular shape recognition power of the liquid-crystal bonded phases can be explained by an ordered bonded ligand arrangement on the silica support (Saito et al. 1994b). In contrast to conventional C_{18} bonded phase, these phases include two phenyl rings in the structures, offering a relatively rigid ligand structure, along with a site of π-π interaction with aromatic analytes. As normally expected for a phenyl bonded phase, the π-π interaction between the bonded phenyl ligand and PAHs would help an increased retentivity on the phase along with a hydrophobic interaction under a reversed-phase condition. In addition, the rigid phenyl rings at the top of the bonded phase could contribute to the increased shape selectivity toward PAHs. From the systematic analysis of the retention data on the liquid-crystal bonded phases, it can be assumed that the "slot-like" structure is formed by the ordered arrangement of the ligands, providing good molecular shape selectivity to isomeric PAHs with quite small differences in their molecular shape. The improved molecular shape recognition power of the liquid-crystal bonded phase was compared to that observed on monomeric and polymeric ODS phases (Saito et al. 1994b, 1995b).

Thermodynamic behaviors during the retention of analytes have been well investigated in LC, and in general, a correlation between the logarithmic retention factor and the reciprocal column temperature, which is widely recognized as the van't Hoff plot, has been obtained in the temperature range normally employed in LC separations. In the general thermodynamic trend, the retention factor will decrease by increasing the column temperature. If a linear van't Hoff plot is obtained in a range of column temperature, it is normally suggested that the retention mechanism remains the same in the temperature range, where the enthalpy of solute transfer from the mobile phase to the stationary phase can be estimated from the slope of the linear van't Hoff plot. From the retention behavior of PAHs on the liquid-crystal phase at different column temperatures, the phase has an excellent molecular shape recognition capability even at a high column temperature, although the molecular shape recognition power is gradually decreased with increasing the column temperature. The obtained van't Hoff plots for typical PAHs could be regarded as a linear relationship, and therefore, the retention mechanism in the temperature range was kept constant. The above trend has quite a good agreement with the selectivity study for ODS phases at various column temperatures: a good molecular shape selectivity can be obtained at a low column temperature, while the recognition ability will be decreased by increasing the temperature. That means that the fundamental trend on the shape selectivity of the liquid-crystal bonded phase is regarded as similar to that on typical ODS phases, although a better shape recognition power to a certain pair of PAHs can be observed on the liquid-crystal bonded phase due to the rigid ligand structure.

For the retention behavior of fullerenes, however, a unique temperature dependence of the retention was reported on the liquid-crystal bonded phase (Saito et al. 1995b), where an increased retentivity to fullerenes was observed on one of the liquid-crystal bonded phases when the column temperature was elevated from room temperature to 80°C. In addition, only a slight decrease in the selectivity between C_{70} and C_{60} was observed on the liquid-crystal bonded phase. Taking into account the retention behavior of the liquid-crystal bonded phase for PAHs, and the above unusual temperature dependence for fullerenes, the authors interpreted the unique retention behavior of the liquid-crystal bonded phase with a retention model consistent with the interaction of a bulky fullerene molecule and a pair of bonded stationary phase ligands on the silica support (Saito et al. 1995b).

3.3 Cholesteryl bonded phase

From the systematic retention behavior analysis, especially on the polymeric ODS and other phases having a rigid ligand structure, it is expected that a novel stationary phase with a rigid ligand structure and conformation must offer a good molecular shape recognition capability to isomeric compounds. One of the examples could be a cholesterol-bonded stationary phase, where cholesteryl-10-undecenolate bonded silica phase was employed as a stationary phase in LC. As expected, the rigid ligand structure could generate a unique shape recognition power for various compounds, such as benzodiazepines, barbiturates, and steroidal derivatives (Catabay et al. 1998a, 1999). A similar unique selectivity for PAHs with various shapes and planarity was also reported on the phase in LC (Catabay et al. 1997, 1998b). In addition to the remarkable results in LC, the cholesterol-bonded stationary phase demonstrated an excellent separation power for a certain class of drugs in capillary electrochromatography (CEC) (Catabay et al. 1998c).

3.4 Multi-legged phenyl bonded phases

In order to enhance the molecular shape recognition capability of the stationary phase in LC, various new firmly-bonded phenyl phases with a significantly restricted movement of the phenyl functional group(s) were designed, as typically illustrated in Figure 12.8. These phases were bonded with more than two bonds to the silica support. Only quite limited conformational change can be allowed to the aromatic functionality in the resulting multi-point-bonded phases (Jinno et al. 1990a). The successful synthesis of these bonded phase structures on the silica support material was spectroscopically confirmed by ^{13}C and ^{29}Si NMR measurements (Jinno et al. 1990a, 1992b, c, 1995b), although the number of bonded ligands on the silica support was somewhat smaller than that of conventional stationary phases having aromatic functionality, such as a phenylpropyl-bonded phase. Compared to other conventional bonded stationary phases, the multi-legged phenyl phases have an extraordinary structure, allowing a horizontal arrangement of phenyl group(s) on the silica support surface.

Among these multi-legged phenyl-bonded phases, an extraordinary shape recognition capability was observed in the TP-bonded phase in Figure 12.8. The selectivity for triphenylene/o-terphenyl under a reversed-phase LC condition, with a mobile phase consisting of methanol and water, was calculated as 0.65, clearly indicating a unique favorable interaction with the non-planar analyte, o-terphenyl (Jinno et al. 1992c). In the numerous number of stationary phases developed, the above selectivity can be regarded as exceptionally unique, because it has been reported that the selectivities obtained with conventional phases, including ODSs, are usually larger than unity (i.e., more than 1.0) under similar experimental conditions in LC.

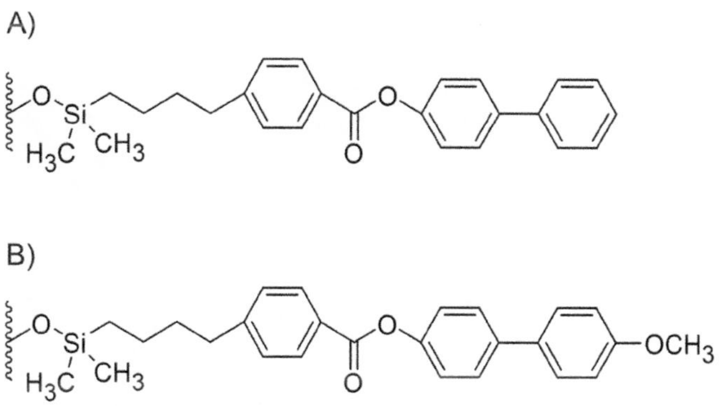

Figure 12.7: Chemical structures of two liquid-crystal bonded phases. (A) LC-1 and (B) LC-2, synthesized by hydrosilylation reactions.

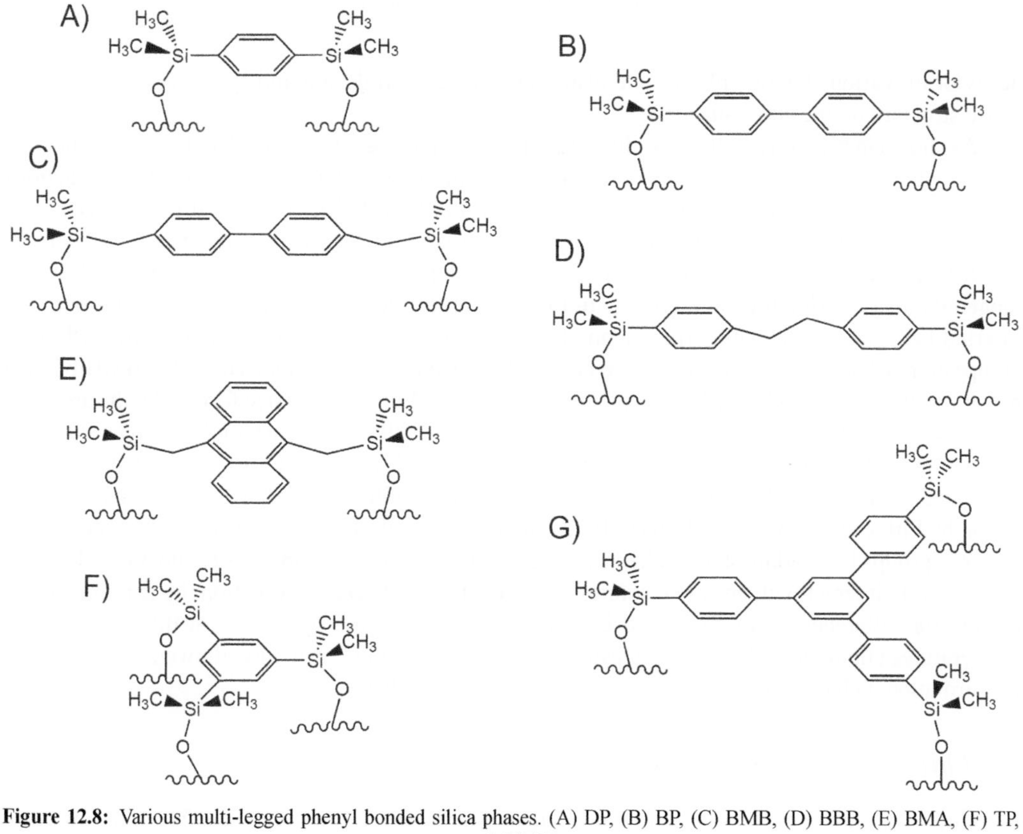

Figure 12.8: Various multi-legged phenyl bonded silica phases. (A) DP, (B) BP, (C) BMB, (D) BBB, (E) BMA, (F) TP, and (G) QP.

From the results, the retention models of TP phase are illustrated in Figure 12.9a, where the non-planar solute (*o*-terphenyl) can interact more effectively than the planar counterpart (triphenylene) on the basis of the rigid bonded ligand having a "cavity-like" structure with aromatic ring at the bottom. All the methyl functional groups at the edge of the "cavity-like" structure could generate a kind of steric hindrance to the planar triphenylene during the interaction with the phenyl functionality at the bottom of the cavity. It can be regarded that the above exceptionally unique "non-planarity" recognition power of the TP phase has been realized on the systematic interpretation of the retention

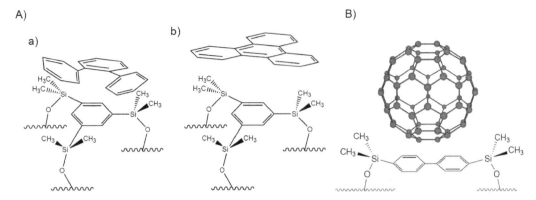

Figure 12.9: Interaction models for multi-legged bonded phases. (A): (a) An effective interaction of non-planar analyte (*o*-terphenyl) and (b) a limited interaction of planar analyte (triphenylene) on TP bonded phase. (B): An estimated interaction between bulky C_{60} molecule and BP bonded phase having a cavity-like structure which is suitable for the interaction a bulky analyte having the close to C_{60} molecule.

behavior on various bonded phases and also on the successful design and synthesis of the bonded phase structure on the silica surface.

As shown in Figure 12.8, these multi-legged phases have been synthesized to develop the bonded phases with various cavity structures of different size and shape, allowing characteristic interactions with bulky solute molecules, as fullerenes (Jinno et al. 1992b, 1995b). The specially-designed cavity structure on the silica support demonstrated a good separation power for fullerenes. For example, the BMB phase was designed from the results on the BP phase. In order to enhance the mobility of biphenyl moiety during the interaction with fullerene molecules, a methylene functional group was inserted between the biphenyl and silicon atom, generating a larger cavity having two methylene groups more when compared to BP-bonded phase. In fact, an excellent retentivity of the BMB-bonded phase was demonstrated for fullerenes (Jinno et al. 1995b). As the cavity size formed by the aromatic ligand was well-designed to be comparable to the size of bulky fullerenes molecules, an effective interaction of the bonded phase ligand and the fullerene molecules could be expected, as typically illustrated in Figure 12.9b (Jinno et al. 1995b). The concept of the bonded phase synthesis is similar to that by Pirkle et al. (Welch and Pirkle 1992) and Kimata et al. (Kimata et al. 1993), although these research groups have additionally taken into account the electronic properties of the bonded phases to effectively interact with aromatic solutes. Jinno et al. also observed exceptionally small retention change for fullerenes on multi-legged biphenyl phase during the operations at different column temperatures (Jinno et al. 1992b), suggesting that the "cavity-like" structure allowed only a limited conformational change and/or moving of the ligand on the silica.

3.5 *Phthalocyanine bonded phases*

Introducing a phthalocyanine as the bonded ligand structure, a horizontal configuration of large aromatic moiety was realized as the stationary phase in LC (Jinno et al. 1992d, 1996). A remarkable planarity recognition power of one of the phthalocyanine-bonded phases could be interpreted on the basis of an outstanding π-π interaction between the bonded phthalocyanine ligand and planar PAHs, such as BaP. Similar large retentivity was also observed for fullerenes on these phthalocyanine-bonded phases. In addition to the good molecular shape recognition power for planar PAHs, a good retentivity for fullerenes was obtained, where the selectivity for C_{60} and C_{70} was further increased with the additional introduction of the alkyl chains to form a cavity-like structure. Introducing some alkyl chains surrounding the phthalocyanine moiety, the molecular shape and size selectivity was enhanced due to the effective interaction of fullerenes with these alkyl ligands with an appropriate chain length to interact with these bulky molecules during the interaction with phthalocyanine moiety at the bottom of the cavity structure (Jinno et al. 1996). The validity of the proposed retention model was supported by the retention studies at high column temperatures, where the selectivity was maintained on the phase even at high column temperature conditions. The results have a good agreement with that obtained on the multi-legged phenyl bonded phases described earlier.

3.6 Fine powdered dicoronylene stationary phase

The contribution of a large aromatic environment in the bonded stationary phase structure to the retentivity and also to the specific selectivity has also been investigated. One of the examples is dicoronylene molecules as the stationary phase in LC (Jinno et al. 1988, 1990b). As can be expected from the general trend in the physicochemical properties of various organic compounds, the solubility of PAHs will be significantly decreased by increasing the molecular weight of the PAHs. Only negligible solubility could be expected for very large PAHs in typical organic solvents commonly employed as the mobile phase in LC, allowing a possibility of extremely large PAHs as the stationary phase material in LC. The chemical structure of dicoronylene, $C_{48}H_{20}$, is illustrated in Figure 12.10. The molecular weight of dicoronylene is 596. It can be slightly soluble in chlorobenzene, while insoluble in conventional organic solvents as the mobile phase in LC, such as dichloromethane and methanol.

Taking advantage of the non-soluble nature of dicoronylene, it could be directly packed into a blank fused-silica capillary tube. Without using a support material, such as silica gel particles, these dicoronylene molecules themselves were employed as the stationary phase having a large aromatic moiety. Since the stationary phase particle was assumed to be a kind of very large aggregated cluster of dicoronylene molecules, the structure of dicoronylene could be found on the surface of the particles. A strong selectivity for the molecular planarity is easily expected form the planar molecular shape of dicoronylene, as similar to other stationary phases having a large aromatic functionality. From the careful considerations on the molecular shape recognition capability of the phase, it has been demonstrated that the dicoronylene phase possessed a strong molecular planarity recognition capability over typical polymeric ODS phases, clearly suggesting the possibility of other novel materials as the stationary phase in chromatography. The retentivity to planar PAHs was quite remarkable to allow the use of dichloromethane as the main component of the mobile phase solvent, suggesting a strong interaction of these planar PAHs with dicoronylene molecules, probably exposed on the surface of the dicoronylene cluster. The results also indicate that other types of fine powdered solid materials having a non-soluble character to the mobile phase solvent could be adopted as a novel stationary phase in LC.

3.7 Particulate C_{60} and chemically bonded C_{60} phases

Due to the unique molecular shape and dimensions, fullerenes have been employed as the sample probes to interpret the retention behavior of various types of stationary phases in LC. These bulky

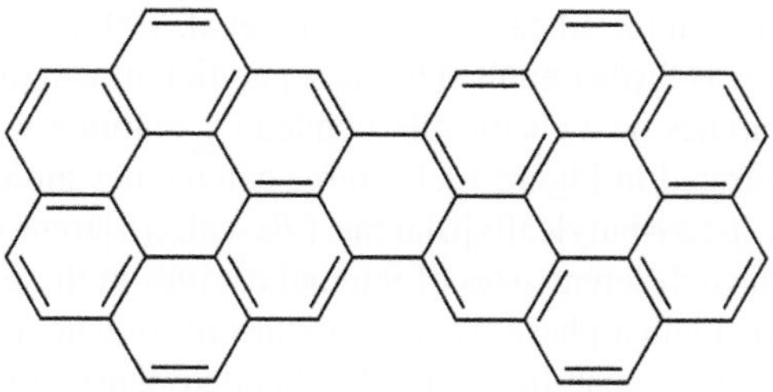

Figure 12.10: Chemical structure of dicoronylene having molecular formula of $C_{48}H_{20}$ (M.W. 596).

carbon cluster molecules can compensate for the lack of large PAHs having spherical shape when using them as the sample probes for the investigation of molecular size and shape recognition process in LC. On the other hand, these fullerene molecules could be introduced as a stationary phase. This approach is similar to a traditional development technique of chiral stationary phases, where a certain solute molecule that is preferably retained on a bonded stationary phase ligand can be applied as the new stationary phase ligand having a good retentivity to the molecule with a similar molecular structure to the above original bonded phase ligand. Therefore, if the C_{60} fullerene is employed as a stationary phase in LC, a screening process can be carried out to find novel stationary phase ligand structure with a specific selectivity to C_{60}.

On the basis of the above concept, and also taking advantage of a microcolumn technique (Saito and Jinno 2003, Saito et al. 2004c), solid C_{60} was introduced as a novel stationary phase material in LC (Jinno et al. 1992e, 1993e), where fine powdery solid C_{60} was packed into a short fused-silica capillary by a slurry method, with cyclohexane as the dispersion solvent. As similar to the phenomenon on the dicoronylene phase mentioned earlier, it was assumed that the C_{60} molecules were embedded on the particle surface of the cluster of C_{60}. A unique molecular shape selectivity was confirmed for the retention of various PAHs having different planarity and dimensions. The results indicated a larger retentivity to the PAHs with a "square-like" molecular shape in the case of planar and small PAHs, while a remarkable large retention was observed for certain non-planar PAHs, such as tetrabenzo[a,cd,j,lm]perylene (TEBP) (Jinno et al. 1992e, 1993e). In case of relatively small PAHs, such as triphenylene having 4 aromatic rings, the entire molecule could be effectively interacted with the C_{60} molecule exposed on the surface of the C_{60} cluster. However, at the same time, there can be a larger retention on the basis of an effective π-π interaction between the C_{60} molecule and large non-planar TEBP molecule due to a good fitting of the curvature of TEBP to that of C_{60}. The above trend has a good agreement with the retention tendency on multi-legged phenyl bonded phase and dicoronylene phase mentioned earlier.

Upon successful introduction of solid C_{60} particle as the stationary phase in microcolumn LC, further investigations have been carried out on a chemically-bonded C_{60} phase, as illustrated in Figure 12.11. In order to synthesize the chemically-bonded C_{60} phase, several silane coupling reagents were developed anew (Nagashima et al. 1994, 1995, Saito et al. 1995c). As can be expected from the difference between a well-designed chemically bonded phase structure and particulate solid clusters of C_{60} itself, the function of C_{60} molecule as the stationary phase material was significantly improved. For the retention of small PAHs having less than five aromatic rings therein, a high linear correlation was confirmed between C_{60} bonded phase and fine powdered C_{60} solid cluster phase. On the C_{60} bonded silica phase, characteristic molecular shape selectivities were observed for several solute pairs, such as phenanthrene/anthracene ($F = 7$) and triphenylene/naphthacene ($F = 9$). In general, the elution order of phenanthrene and anthracene on a commercially-available ODS phase is: phenanthrene first and then anthracene. With the C_{60} bonded phase, the opposite elution order of these isomeric solute pair was observed, i.e., anthracene first, followed by phenanthrene. The same trend was found for triphenylene/naphthacene, where naphthacene eluted first and then triphenylene on the C_{60} phase. The above results suggest a preferable interaction of the bonded C_{60} ligand with PAHs having a partial molecular structure of C_{60} molecule. In addition to the spectroscopic analysis of the C_{60} bonded stationary phase, the chromatographic results also demonstrated the successful modification with C_{60} ligands on the silica support (Saito et al. 1995c).

The C_{60} bonded phase was further applied to the separation of calixarenes. Saito et al. reported a separation of three calixarenes on a chemically bonded C_{60} stationary phase in microcolumn LC (Saito et al. 1996). As illustrated in Figure 12.12, three calixarenes molecules, 4-*tert*-butylcalix[4]arene (*'Bu*-calix[4]arene), 4-*tert*-butylcalix[6]arene (*'Bu*-calix[6]arene), and 4-*tert*-butylcalix[8]arene (*'Bu*-calix[8]arene) have different sizes of internal cavities in these structures. From a typical separation on the C_{60} bonded silica phase using a mixture of toluene and methanol as the mobile phase, a satisfactory separation capability of the C_{60} bonded phase could be confirmed for these three calixarenes, where the elusion order was *'Bu*-calix[4]arene, *'Bu*-calix[6]arene, and then

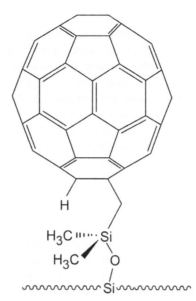

H

H₃C ⋯⋯ Si

H₃C ◢ O

Si

Figure 12.11: Chemical structure of the C_{60} bonded silica phase (Nagashima et al. 1995, Saito et al. 1995c).

A)   B)

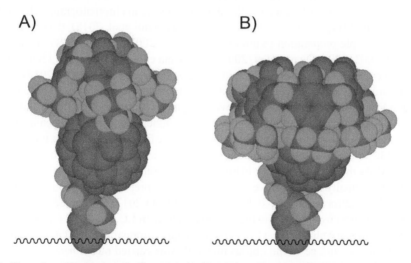

Figure 12.12: The estimated interaction of calixarenes with the C_{60} phase shown in Figure 12.11. (A) *'Bu*-calix[4]arene and (B) *'Bu*-calix[8]arene on the C_{60} bonded silica phase (Saito et al. 1996).

Color version at the end of the book

'Bu-calix[8]arene. Taking into account the cavity size of these calixarenes and also the size of the C_{60} bonded ligand, a set of retention models have been proposed, as depicted in Figure 12.12 (Saito et al. 1996). In the case of *'Bu*-calix[4]arene, the cavity size is small enough to interact with the spherical C_{60} bonded ligand, while *'Bu*-calix[6]arene has a larger cavity size than *'Bu*-calix[4]arene for an increased interaction with the C_{60} moiety. However, in the case of *'Bu*-calix[8]arene, the size of the cavity is well comparable to the molecular dimension of the C_{60} molecule attached on the surface of the silica gel support, allowing even larger interaction with the bonded C_{60} ligand. The above interpretation of the retention behavior was further supported by the results for other types of calixarenes without a cavity-like opening to effectively interact with the C_{60} ligand.

The unique molecular shape and size selectivities for isomeric PAHs were also confirmed on other chemically-bonded C_{60} silica phases (Jinno et al. 1997a, b). From the systematic comparison of the retention and selectivity for various PAHs on several C_{60} bonded silica phases synthesized with different bonding chemistry and carbon content, the unique molecular shape selectivity of the C_{60} bonded ligand can be attributed as the shape and size of the C_{60}, although the density of the bonded C_{60} ligands on the silica support can also contribute to the selectivity to a certain class of bulky solutes. Kubo et al. recently synthesized several types of C_{60} and C_{70} bonded phase (Kubo et al. 2014, 2015a, b, Kanao et al. 2017). They also reported the synthesis of corannulene-bonded silica monolith, where a partial molecular structure of C_{60} is attached to the silica monolith backbone (Kanao et al. 2019). Significantly, unique molecular shape selectivities on the corannulene-bonded stationary phase have been reported.

3.8 Fibrous stationary phases

Majority of the stationary phases developed for the use in typical LC separations could be classified as silica-based materials, where well-designed ligands are attached to the surface of silica gel particles. In the preparation of those silica-based stationary phases, porous silica particles are commonly employed to enhance the carbon content of the stationary phase, and in general, an increased retentivity is obtained on the bonded phase with higher carbon content, if the same bonded ligand structure is introduced. As an alternative approach to increase the surface area of the stationary phase, fine fibrous materials have been introduced as the stationary phase in chromatography (Kiso et al. 1986, 1989, Nakane et al. 2011, Saito et al. 2003b, 2004b, Nakagami et al. 2018), the extraction medium in miniaturized sample preparation (Saito et al. 2000a, b, 2002a, c, 2004d, Saito and Jinno 2002, Imaizumi et al. 2003, 2004, Ogawa et al. 2007, 2009, Ueta et al. 2009, Ueta and Saito 2014) and an interface between chromatographic separations (Abe et al. 2009, Nakane et al. 2015).

Due to the increased requirements for fibrous materials, such as good mechanical strength, chemical stability, and also the compatibility with polymeric matrix to form fiber-reinforced plastics (FRPs), various types of fibrous polymers have been synthesized in the research area of materials science. Taking advantage of the suitability of the synthetic fibrous materials, as typically illustrated in Figure 12.13, for a stationary phase and also an extraction medium, fine fibrous materials have been introduced. As the stationary phase in microscale separations, a bundle of fine fibrous materials was packed longitudinally into a relatively short capillary to prepare the column in microcolumn LC and capillary electrochromatography (CEC) (Saito and Ueta 2017). On the basis of the successful introduction of fine fibrous materials to the stationary phase in LC and CEC, further applications of the fine fibrous stationary phases have been created in gas chromatography (GC), where a polymer-coating process to the surface of the fibrous materials was carried out, if necessary (Tazawa et al. 2015, Saito and Ueta 2017). Coated with a polysiloxane-based polymeric material typically used for open-tubular GC capillary columns, the resulting selectivity could be interpreted by the selectivity of the coating and also the compatibility of the coating with the surface of the fibrous support material. The results suggested a possibility for the development of novel packed-capillary column, because a more suitable combination of the polymer-coating and the fibrous support material as the support could be expected for synthetic polymeric filaments than conventional fused-silica capillaries widely employed as the support material. As similar to the development of fibrous materials for FRPs, surface modification of the filaments was carried out to enhance the compatibility of the coating materials and the surface of the filaments. A successful surface modification of the filaments with several types of functional groups has been demonstrated (Saito et al. 2003a, Saito and Ueta 2017, Abe et al. 2005, Shirai et al. 2010, 2011).

Marcus et al. introduced polypropylene and polyester fibers with a specially-designed cross-sectional shape, called "capillary-channeled fibers" (Marcus et al. 2003). Bundles consisting of 1,000 to 3,000 filaments were packed collinearly into a conventional stainless steel LC tubing of

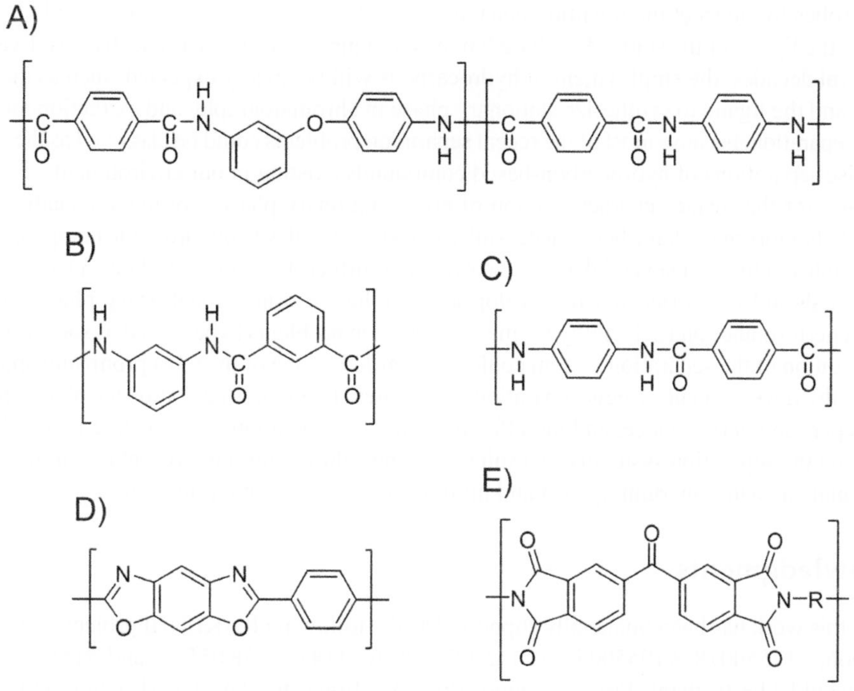

A)

B)

C)

D)

E)

Figure 12.13: Chemical structures of various fibrous polymers introduced as the stationary phases in chromatography (Saito et al. 2003b, Saito and Ueta 2017). (A) Technola, (B) Nomex, (C) Kevlar, (D) Zylon, and (E) P84-Type Polyimide.

4.6 mm i.d. On these fiber-packed columns, the separations of various mixtures, such as PAHs, organic/inorganic lead compounds, triglycerides, and proteins, were carried out typically using gradient elution (Marcus et al. 2003, Nelson and Marcus 2003). From the results of a careful theoretical consideration of hydrodynamic flow in capillary-channeled fiber-packed columns (Stanelle et al. 2005), Marcus et al. also developed a miniaturized LC column packed with capillary-channeled polymer fibers (Stanelle et al. 2006). Introducing radial compression process of the fiber-packed column from 0.8 mm i.d. to 0.7 mm i.d., the void volume of the resulting column was significantly reduced and the peak symmetry was improved.

As a general trend, synthetic fibrous materials showing a high resistance to organic solvents also have a good stability at high temperatures, and in fact, the heat resistance of Zylon and Kevlar is quite satisfactory for the use as a stationary phase in typical temperature-programmed separations in GC (Saito et al. 2001, 2005a, b, Ogawa et al. 2006). With a bundle of filaments, the resulting miniaturized column as a GC packed-capillary enables a faster temperature-programmed separation. Taking advantage of a large sample loading of the packed-capillary column and a faster temperature program, it has been demonstrated that efficient separations of complex sample mixtures consisted of a large number of components, such as homologous alkane sample mixtures (Saito et al. 2005a, b). High temperature GC separations could be realized using a metal capillary instead of fused-silica capillary having a limited heat-resistance due to the polyimide-film outside the capillary. Longitudinal miniaturization of a GC packed-capillary was also studied (Ogawa et al. 2006).

4. Conclusions

Modern stationary phases for chromatographic separation have been developed on the basis of the systematic analysis of the retention behaviors of groups of well-designed analytes. Due to the unique structure and the availability of many critical pairs, fullerenes and PAHs are widely employed as the sample probes to interpret the retention behavior. In addition, those hydrocarbons could be further applied to the ligand of the stationary phase having a unique shape and size selectivities. Even in the next several decades, the employment of hydrocarbons will be greatly expected, such as the sample analytes, and the ligand to synthesize stationary phase in chromatography and extraction medium in sample preparation, because most of the recent separation problems could be classified as the effective and precise separations of hydrocarbon-based compounds existing in our environment.

In fact, for the surface characterization of novel stationary phases, systematic analyses of the retention behaviors in LC have been successfully introduced with various aromatic compounds as the sample solutes in the past several decades. However, a further development of new characterization techniques should be carried out to develop and synthesize a novel stationary phase having an improved performance and selectivity for many separation problems to be solved as soon as possible. Miniaturization of the separation system could be regarded as one of the most promising approaches to realize the development of new separation techniques, including the stationary phases having a unique separation performance, and the effective sample preparation media with a high selectivity. Downsizing the separation system is also suitable as the solution for other recent requirements, such as rapid analysis with low running cost and minimized environmental pollution.

Acknowledgments

A part of this work has been financially supported as the following KAKENHI projects: #13740421, #15750066, #17550078, #19550086, #21550078, #24651187, #15K05537, and #18K05169. The authors would like to thank Prof. K. Jinno, Prof. Y. Hirata, the late Dr. H. Ohta, and all other collaborators in Microscale Separation Science Laboratory, Toyohashi University of Technology for their valuable suggestions and assistance during this project.

The authors also express their deep gratitude to the following individuals: Prof. H. Nagashima (Kyushu University), Prof. M. Okamoto (Meijyo University), Prof. Prof. A. Matsumoto, Prof. T. Takeichi, Prof. Y. Kiso and Prof. K. Takashima (Toyohashi University of Technology), Prof. J.J. Pesek (San Jose State University), Dr. Y.-L. Chen (J & W Scientific), Dr. J.C. Fetzer (Chevron Research and Technology Company), Prof. P.K. Zarzycki (Koszalin University of Technology), Dr. H. Wada, and Mr. K. Kotera, Mr. F. Fujimura (Shinwa Chemical Industries, Ltd.) and Mr. Y. Nomura (Toyobo Co., Ltd.).

References

Abe, A., Y. Saito, M. Imaizumi, M. Ogawa, T. Takeichi and K. Jinno. 2005. Surface derivatization of poly(*p*-phenylene terephthalamide) fiber designed for novel separation and extraction media. J. Sep. Sci. 28: 2413–2418.

Abe, A., Y. Saito, I. Ueta, K. Nakane, T. Takeichi and K. Jinno. 2009. Development of novel fiber-packed needle interface for off-line reversed-phase liquid chromatography-capillary gas chromatography. J. Chromatogr. A 1216: 7456–7560.

Catabay, A., Y. Saito, C. Okumura, K. Jinno, J.J. Pesek and E. Williamsen. 1997. Retention behavior of small polycyclic aromatic hydrocarbons with cholesteryl 10-undecenoate bonded phase in microcolumn liquid chromatography. J. Microcol. Sep. 9: 81–85.

Catabay, A., M. Taniguchi, K. Jinno, J.J. Pesek and E. Williamsen. 1998a. Separation of 1,4-benzodiazepines and analogues using cholesteryl-10-undecenoate bonded phase in microcolumn liquid chromatography. J. Chromatogr. Sci. 36: 111–118.

Catabay, A., C. Okumura, K. Jinno, J.J. Pesek, E. Williamsen, J.C. Fetzer et al. 1998b. Retention behavior of large polycyclic aromatic hydrocarbons on cholesteryl 10-undecenoate bonded phase in microcolumn liquid chromatography. Chromatographia 47: 13–20.

Catabay, A.P., H. Sawada, K. Jinno, J.J. Pesek and M.T. Matyska. 1998c. Separation of benzodiazepines using cholesterol-modified fused-silica capillaries in capillary electrochromatography. J. Capillary Electrophor. 5: 89–95.

Catabay, A.P., J.J. Pesek, M.T. Matyska and K. Jinno. 1999. Pharmaceutical applications using cholesteryl-10-undecenoate bonded phase in microcolumn liquid chromatography. J. Liq. Chromatogr. Relat. Technol. 22: 953–967.

Colin, H., J.C. Diez-Mesa, G. Guiochon, T. Czajkowska and I. Miedziak. 1978. The role of the temperature in reversed-phase high-performance liquid chromatography using pyrocarbon-containing adsorbents. J. Chromatogr. 167: 41–65.

Gehrke, C.W., R.L. Wixom and E. Bayer (eds.). 2001. Chromatography—A Century of Discovery 1900–2000, Elsevier, Amsterdam, The Netherlands.

Imaizumi, M., Y. Saito, M. Hayashida, T. Takeichi, H. Wada and K. Jinno. 2003. Polymer-coated fibrous extraction medium for sample preparation coupled to microcolumn liquid-phase separations. J. Pharm. Biomed. Anal. 30: 1801–1808.

Imaizumi, M., Y. Saito, K. Ban, H. Wada, M. Hayashida and K. Jinno. 2004. In-valve sample preparation cartridge designed for microcolumn liquid chromatography. Chromatographia 60: 619–623.

Ishii, D. (ed.). 1988. Introduction to Microscale High-Performance Liquid Chromatography. VCH, New York, NY. USA.

Jinno, K. and Y. Hirata. 1982. Investigation of the low temperature effect in normal phase micro high performance liquid chromatography. J. High Resolut. Chromatogr. Chromatogr. Commun. 5: 85–90.

Jinno, K. and K. Kawasaki. 1983. Correlation between the retention data of polycyclic aromatic hydrocarbons and several descriptors in reversed-phase HPLC. Chromatographia 17: 445–448.

Jinno, K. and K. Kawasaki. 1984. The correlation between molecular polarizability of PAHs and their retention data on various stationary phases in revered-phase HPLC. Chromatographia 18: 103–105.

Jinno, K., T. Ibuki, N. Tanaka, M. Okamoto, J.C. Fetzer, W.R. Biggs et al. 1987. Retention behavior of large polycyclic aromatic hydrocarbons in reversed-phase liquid chromatography on a polymeric octadecylsilica stationary phase. J. Chromatogr. 461: 209–227.

Jinno, K., S. Shimura, J.C. Fetzer and W.R. Biggs. 1988. Separation of polycyclic aromatic hydrocarbons in microcolumn liquid chromatography using dicoronylene as stationary phase. J. High Resolut. Chromatogr. Chromatogr. Commun. 11: 673–675.

Jinno, K., S. Shimura, N. Tanaka, K. Kimata, J.C. Fetzer and W.R. Biggs. 1989. Planarity recognition of large polycyclic aromatic hydrocarbons by various octadecylsilica stationary phases in non-aqueous reversed-phase liquid chromatography. Chromatographia 27: 285–291.

Jinno, K. 1989. CP-MAS ^{13}C nuclear magnetic resonance spectra for identification of functionality of octadecylsilica boded phases. J. Chromatogr. Sci. 27: 729–734.

Jinno, K. and H. Mae. 1990. Molecular planarity recognition of polycyclic aromatic hydrocarbons in supercritical fluid chromatography. J. High Resolut. Chromatogr. 13: 512–515.

Jinno, K., K. Yamamoto, H. Nagashima, T. Ueda and K. Itoh. 1990a. Silicas chemically bonded with multidentate phenyl groups as stationary phases in reversed-phase liquid chromatography used for non-polarity recognition of polycyclic aromatic hydrocarbons. J. Chromatogr. 517: 193–207.

Jinno, K., H. Shimura, J.C. Fetzer and W.R. Biggs. 1990b. A comparison of dicoronylene and octadecylsilica stationary phase for separation of polycyclic aromatic hydrocarbons by microcolumn liquid chromatography. J. Polycyclic Aromat. Compd. 1: 151–159.

Jinno, K., Y. Saito, R. Malhan née Chopra, J.J. Pesek, J.C. Fetzer et al. 1991. Retention behavior of polycyclic aromatic hydrocarbons on a liquid-crystal bonded phase in reversed-phase liquid chromatography. J. Chromatogr. 557: 459–468.

Jinno, K., T. Uemura, H. Nagashima and K. Itoh. 1992a. An LC-MS approach to the on-line identification of fullerenes. J. High Resolut. Chromatogr. 15: 627–628.

Jinno, K., K. Yamamoto, T. Ueda, H. Nagashima, K. Itoh, J.C. Fetzer et al. 1992b. Liquid chromatographic separation of all-carbon molecules C_{60} and C_{70} with multi-legged phenyl group bonded silica phases. J. Chromatogr. 594: 105–109.

Jinno, K., K. Yamamoto, T. Kuwamoto, H. Nagashima, T. Ueda, S. Tajima et al. 1992c. Multidenate phenyl-bonded phases in high-performance liquid chromatography. Chromatographia 34: 381–385.

Jinno, K., T. Uemura, J. Haginaka and Y. Saito. 1992d. Cu-Phthalocyanine stationary phase in microcolumn liquid chromatography. J. Microcol. Sep. 4: 325–329.

Jinno, K., K. Yamamoto, J.C. Fetzer and W.R. Biggs. 1992e. C_{60} as a stationary phase for microcolumn liquid chromatographic separation of polycyclic aromatic hydrocarbons. J. Microcol. Sep. 4: 187–190.

Jinno, K., J. Wu, M. Ichikawa and I. Takata. 1993a. Characterization of octadecylsilica stationary phase by spectroscopic methods. Chromatographia 37: 627–634.

Jinno, K., T. Uemura, H. Ohta, H. Nagashima and K. Itoh. 1993b. Separation and identification of higher molecular weight fullerenes by high-performance liquid chromatography with monomeric and polymeric octadecylsilica bonded phases. Anal. Chem. 65: 2650–2654.

Jinno, K., H. Ohta, Y. Saito, T. Uemura, H. Nagashima, K. Itoh et al. 1993c. Dimethoxyphenylpropyl bonded silica phase for higher fullerenes separation by high-performance liquid chromatography. J. Chromatogr. 648: 71–77.

Jinno, K., Y. Saito, Y.-L. Chen, G. Luehr, J. Archer, J.C. Fetzer et al. 1993d. Separation of C_{60} and C_{70} fullerenes on methoxyphenylpropyl bonded stationary phases in microcolumn liquid chromatography. J. Microcol. Sep. 5: 135–140.

Jinno, K., K. Fukuoka, J.C. Fetzer and W.R. Biggs. 1993e. Buckminsterfullerene as a stationary phase in liquid chromatography. J. Microcol. Sep. 5: 517–523.

Jinno, K., H. Matsui, H. Ohta, Y. Saito, K. Nakagawa, H. Nagashima et al. 1995a. Separation and identification of higher fullerenes in soot extract by liquid chromatography-mass spectrometry. Chromatographia 41: 353–360.

Jinno, K., K. Nakagawa, Y. Saito, H. Ohta, H. Nagashima, K. Itoh et al. 1995b. Nano-scale design of novel stationary phases to enhance selectivity for molecular shape and size in liquid chromatography. J. Chromatogr. A 691: 91–99.

Jinno, K. (ed.). 1996. Chromatographic Separations Based on Molecular Recognition. Wiley-VCH, New York, NY, USA.

Jinno, K., C. Kohrikawa, Y. Saito, H. Haginaka, Y. Saito and M. Mifune. 1996. Copper-phthalocyanine stationary phase (Cu-PCs) for fullerenes separation in microcolumn liquid chromatography. J. Microcol. Sep. 8: 13–20.

Jinno, K., K. Tanabe, Y. Saito and H. Nagashima. 1997a. Separation of polycyclic aromatic hydrocarbons with various C_{60} fullerene bonded silica phases in microcolumn liquid chromatography. Analyst 122: 787–791.

Jinno, K., K. Tanabe, Y. Saito, H. Nagashima and R.D. Trengove. 1997b. Retention behavior of calixarenes with various C_{60} bonded silica phases in microcolumn liquid chromatography. Anal. Commun. 34: 175–177.

Jinno, K. (ed.). 1999. Separation of Fullerenes by Liquid Chromatography, RSC Chromatography Monographs, The Royal Society of Chemistry, Cambridge, UK.

Kaliszan, R. and H. Lamparczyk. 1978. A relationship between the connectivity indices and retention indices of polycyclic aromatic hydrocarbons. J. Chromatogr. Sci. 16: 246–248.

Kanao, E., T. Naito, T. Kubo and K. Otsuka. 2017. Development of a C_{70}-fullerene bonded silica-monolithic capillary and its retention characteristics in liquid chromatography. Chromatography 38: 45–51.

Kanao, E., T. Kubo, T. Naito, T. Matsumoto, T. Sano, M. Yan et al. 2019. Differentiating π interactions by constructing concave/convex surfaces using a bucky bowl molecule, corannulene in liquid chromatography. Anal. Chem. 91: 2439–2446.

Kimata, K., K. Hosoya, T. Araki and N. Tanaka. 1993. [2-(1-Pyrenyl)ethyl]silyl silica packing material for liquid chromatographic separation of fullerenes. J. Org. Chem. 58: 282–283.

Kirkland, J.J., J.C. Glajch and R.D. Farlee. 1989. Synthesis and characterization of highly stable bonded phases for high-performance liquid chromatography column packings. Anal. Chem. 61: 2–11.

Kiso, Y., K. Jinno and T. Nagoshi. 1986. Liquid chromatography in a capillary packed with fibrous cellulose acetate. J. High Resolut. Chromatogr. Chromatogr. Commun. 9: 763–764.

Kiso, Y., K. Takayama and K. Jinno. 1989. Cellulose acetate as stationary phase in microcolumn LC. J. High Resolut. Chromatogr. 12: 169–173.

Kroto, H.W., J.R. Heath, S.C. O'Brien, R.F. Curl and R.E. Smalley. 1985. C_{60}: Buckminsterfullerene. Nature 318: 162–163.

Kubo, T., Y. Murakami, Y. Tominaga, T. Naito, K. Sueyoshi, M. Yan et al. 2014. Development of a C_{60}-fullerene bonded open-tubular capillary using a photo/thermal active agent for liquid chromatographic separations by π-π interactions. J. Chromatogr. A 1323: 174–178.

Kubo, T., Y. Murakami, M. Tsuzuki, H. Kobayashi, T. Naito, T. Sano et al. 2015a. Unique separation behavior of a C_{60} fullerene-bonded silica monolith prepared by an effective thermal coupling agent. Chem. Eur. J. 21: 18095–18098.

Kubo, T., Y. Murakami, T. Naito and K. Otsuka. 2015b. C_{60} fullerene bonded silica monolithic capillary for specific separations of aromatic compounds. Chromatography 36: 105–113.

Marcus, R.K., W.C. Davis, B. Knippel, L. LaMotte, T.A. Hill, D. Perahia et al. 2003. Capillary-channeled polymer fibers as stationary phases in liquid chromatography separations. J. Chromatogr. A 986: 17–31.

Nagashima, H., H. Terasaki, E. Kimura, K. Nakajima and K. Itoh. 1994. Silylmethylations of C_{60} with grignard reagents: selective synthesis of $HC_{60}CH_2SiMe_2Y$ and $C_{60}(CH_2SiMe_2Y)_2$ with selection of solvents. J. Org. Chem. 59: 1246–1248.

Nagashima, H., H. Terasaki, Y. Saito, K. Jinno and K. Itoh. 1995. Chlorosilanes and silyl triflates containing C_{60} as a partial structure: a versatile synthetic entry linking the C_{60} moieties with alcohols, phenols, and silica. J. Org. Chem. 60: 4966–4967.

Nakagami, K., O. Sumiya, T. Tazawa, T. Monobe, M. Watanabe, I Ueta et al. 2018. Polyimide filaments as a novel stationary phase in packed-capillary gas chromatography. Chromatography 39: 91–96.

Nakane, K., S. Shirai, Y. Saito, Y. Moriwake, I. Ueta, M. Inoue et al. 2011. High-temperature separations on a polymer-coated fibrous stationary phase in microcolumn liquid chromatography. Anal. Sci. 27: 811–816.

Nakane, K., T. Tazawa, Y. Mori, A. Kobayashi, I. Ueta and Y. Saito. 2015. Polymer-coated fiber-packed capillary as the sample preparation medium in liquid chromatography: heart-cutting analysis of phthalates in aqueous matrices. Chromatography 36: 61–65.

Nelson, D. and R.K. Marcus. 2003. A novel stationary phase: capillary-channeled polymer (C-CP) fibers for HPLC separations of proteins. J. Chromatogr. Sci. 41: 475–479.

Ogawa, M., Y. Saito, M. Imaizumi, H. Wada and K. Jinno. 2006. Longitudinal miniaturization of fiber-packed capillary column in high temperature gas chromatography. Chromatographia 63: 459–463.

Ogawa, M., Y. Saito, I. Ueta and K. Jinno. 2007. Fiber-packed needle for dynamic extraction of aromatic compounds. Ana. Bioanal. Chem. 388: 619–625.

Ogawa, M., Y. Saito, S. Shirai, Y. Kiso and K. Jinno. 2009. Determination of bisphenol A in water using a packed needle extraction device. Chromatographia 69: 685–690.

Ohta, H., Y. Saito, K. Jinno, H. Nagashima and K. Itoh. 1994. Temperature effect in separation of fullerene by high-performance liquid chromatography. Chromatographia 39: 453–459.

Ohta, H., Y. Saito, K. Jinno, J.J. Pesek, M.T. Matyska, Y.-L. Chen et al. 1995. Retention of fullerenes by octadecyl silica: correlation with NMR spectra at low temperatures. Chromatographia 40: 507–512.

Ohta, H., K. Jinno, Y. Saito, J.C. Fetzer, W.R. Biggs, J.J. Pesek et al. 1996. Effect of temperature on the mechanism of retention of fullerenes in liquid chromatography using various alkyl bonded stationary phase. Chromatographia 42: 56–62.

Ohta, H., Y. Saito, N. Nagae, J.J. Pesek, M.T. Matyska and K. Jinno. 2000. Fullerenes separation with monomeric type C_{30} stationary phase in high-performance liquid chromatography. J. Chromatogr. A 883: 55–66.

Ohta, H., E. Wlodarczyk, K. Piaskowski, A. Kaleniecka, L. Lewandowska, M.J. Baran et al. 2017. Unexpected differences between planar and column liquid chromatographic retention of 1-acenaphthenol enantiomers controlled by supramolecular interactions involving β-cyclodextrin at subambient temperatures. Anal. Bioanal. Chem. 409: 3695–3706.

Oña-Ruales, J.O., L.C. Sander, W.B. Wilson and S.A. Wise. 2018. Revisiting shape selectivity in liquid chromatography for polycyclic aromatic hydrocarbons (PAHs): six-ring and seven-ring *Cata*-condensed PAH isomers of molecular mass 328 Da and 378 Da. Anal. Bioanal. Chem. 410: 885–896.

Pesek, J. and T. Cash. 1989. A chemically bonded liquid crystal as a stationary phase for high performance liquid chromatography. synthesis on silica *via* an organochlorosilane pathway. Chromatographia 27: 559–564.

Pirkle, W.H. and C.J. Welch. 1991. An unusual effect of temperature on the chromatographic behavior of buckminsterfullerene. J. Org. Chem. 56: 6973–6974.

Radecki, A., H. Lamparczyk and R. Kaliszan. 1979. A relationship between the retention indices on nematic and isotropic phases and the shape of polycyclic aromatic hydrocarbons. Chromatographia 12: 595–599.

Saito, Y., H. Ohta, H. Nagashima, K. Itoh, K. Jinno, M. Okamoto et al. 1994a. Separation of C_{60} and C_{70} fullerenes with a triphenyl bonded silica phase in microcolumn liquid chromatography. J. Liq. Chromatogr. 17: 2359–2372.

Saito, Y., K. Jinno, J.J. Pesek, Y.-L. Chen, G. Luehr, J. Archer et al. 1994b. Molecular shape recognition capability of liquid-crystal bonded phases in reversed-phase high performance liquid chromatography. Chromatographia 38: 295–303.

Saito, Y., H. Ohta, H. Nagashima, K. Itoh, K. Jinno, M. Okamoto et al. 1995a. Separation of fullerenes with novel stationary phases in microcolumn high performance liquid chromatography. J. Liq. Chromatogr. 18: 1897–1908.

Saito, Y., H. Ohta, H. Nagashima, K. Itoh, K. Jinno and J.J. Pesek. 1995b. Separation of fullerenes with liquid crystal bonded silica phases in microcolumn high performance liquid chromatography. J. Microcol. Sep. 7: 41–49.

Saito, Y., H. Ohta, H. Terasaki, Y. Katoh, H. Nagashima, K. Jinno et al. 1995c. Separation of polycyclic aromatic hydrocarbons with a C_{60} bonded silica phase in microcolumn liquid chromatography. J. High Resolut. Chromatogr. 18: 569–572.

Saito, Y., H. Ohta, H. Terasaki, Y. Katoh, H. Nagashima, K. Jinno et al. 1996. Separation of calixarenes with a chemically bonded C_{60} silica stationary phase in microcolumn liquid chromatography. J. High Resolut. Chromatogr. 19: 475–477.

Saito, Y., Y. Nakao, M. Imaizumi, T. Takeichi, Y. Kiso and K. Jinno. 2000a. Fiber-in-tube solid-phase microextraction: a fibrous rigid-rod heterocyclic polymer as the extraction medium. Fresenius J. Anal. Chem. 368: 641–643.

Saito, Y., Y. Nakao, M. Imaizumi, Y. Morishima, Y. Kiso and K. Jinno. 2000b. Miniaturized solid-phase extraction as a sample preparation technique for the determination of phthalates in water. Anal. Bioanal. Chem. 373: 81–86.

Saito, Y., M. Imaizumi, K. Nakata, T. Takeichi, K. Kotera, H. Wada et al. 2001. Fibrous rigid-rod heterocyclic polymer as the stationary phase in packed capillary gas chromatography. J. Microcol. Sep. 13: 259–264.

Saito, Y. and K. Jinno. 2002. On-line coupling of miniaturized solid-phase extraction and microcolumn liquid-phase separations. Anal. Bioanal. Chem. 373: 325–331.

Saito, Y., M. Imaizumi, T. Takeichi and K. Jinno. 2002a. Miniaturized fiber-in-tube solid-phase extraction as the sample preconcentration method for microcolumn liquid-phase separations. Anal. Bioanal. Chem. 372: 164–168.

Saito, Y., M. Nojiri, M. Imaizumi, Y. Nakao, Y. Morishima, H. Kanehara et al. 2002b. Polymer-coated synthetic fibrous designed for miniaturized sample preparation process. J. Chromatogr. A 975: 105–112.

Saito, Y. and K. Jinno. 2003. Miniaturized sample preparation combined with liquid-phase separations. J. Chromatogr. A 1000: 53–67.

Saito, Y., H. Ohta and K. Jinno. 2003a. Design and characterization of novel stationary phases based on retention behavior studies with various aromatic compounds. J. Sep. Sci. 26: 225–241.

Saito, Y., A. Tahara, M. Imaizumi, T. Takeichi, H. Wada and K. Jinno. 2003b. Polymer-coated fibrous materials as the stationary phase in packed capillary gas chromatography. Anal. Chem. 75: 5525–5531.

Saito, Y., H. Ohta and K. Jinno. 2004a. Chromatographic separation of fullerenes. Anal. Chem. 76: 266A–272A.

Saito, Y., A. Tahara, M. Ogawa, M. Imaizumi, K. Ban, H. Wada et al. 2004b. Polymer-coated fibrous stationary phases in packed-capillary gas chromatography. Anal. Sci. 20: 335–339.

Saito, Y., K. Jinno and T. Greibrokk. 2004c. Capillary columns in liquid chromatography: between conventional columns and microchips. J. Sep. Sci. 27: 1379–1390.

Saito, Y., M. Imaizumi, K. Ban, A. Tahara, H. Wada and K. Jinno. 2004d. Development of miniaturized sample preparation with fibrous extraction media. J. Chromatogr. A 1025: 27–32.

Saito, Y., M. Ogawa, M. Imaizumi, K. Ban, A. Abe, T. Takeichi et al. 2005a. High-temperature separation with polymer-coated fiber in packed capillary gas chromatography. Anal. Bioanal. Chem. 382: 825–829.

Saito, Y., M. Ogawa, M. Imaizumi, K. Ban, A. Abe, T. Takeichi et al. 2005b. Short metal capillary columns packed with polymer-coated fibrous materials in high-temperature gas chromatography. J. Chromatogr. Sci. 43: 536–541.

Saito, Y. and I. Ueta. 2017. Miniaturization for the development of high performance separation systems. Chromatography. 38: 85–94.

Sander, L.C., J.B. Callis and L.R. Field. 1983. Fourier transform infrared spectrometric determination of alkyl chain conformation on chemically bonded reversed-phase liquid chromatography packings. Anal. Chem. 55: 1068–1075.

Sander, L.C. and S.A. Wise. 1984. Synthesis and characterization of polymeric C_{18} stationary phases for liquid chromatography. Anal. Chem. 56: 504–510.

Sander, L.C. and S.A. Wise. 1986. Investigations of selectivity in RPLC of polycyclic aromatic hydrocarbons. pp. 139–218. In: J.C. Giddings, E. Grushka, J. Cazes and P.R. Brown (eds.). Advances in Chromatography. Vol. 25, Marcel Dekker, New York, NY, USA.

Sander, L.C. and S.A. Wise. 1987. Effect of phase length on column selectivity for the separation of polycyclic aromatic hydrocarbons by reversed-phase liquid chromatography. Anal. Chem. 59: 2309–2313.

Sander, L.C. and S.A. Wise. 2001. The influence of column temperature on selectivity in reversed-phase liquid chromatography for shape-constrained solutes. J. Sep. Sci. 24: 910–920.

Schbron, J.F., R.J. Hurtubise and H.F. Silver. 1977. Separation of hydroaromatics and polycyclic aromatic hydrocarbons and determination of tetralin and haphthalene in coal-derived solvents. Anal. Chem. 49: 2253–2260.

Shirai, S., Y. Saito, Y. Sakurai, I. Ueta and K. Jinno. 2010. Retention behavior on aminoethyl-modified poly(p-phenylene terephthalamide) fiber stationary phases in gas chromatography. Anal. Sci. 26: 1011–1014.

Shirai, S., K. Nakane, I. Ueta and Y. Saito. 2011. Retention behavior of fluorobenzenes on fluoro-derivatized poly(p-phenylene terephthalamide) fibrous stationary phases in microcolumn liquid chromatography. Chromatography 32: 127–133.

Stanelle, R.D., L.C. Sander and R.K. Marcus. 2005. Hydrodynamic flow in capillary-channel fiber columns for liquid chromatography. J. Chromatogr. A 1100: 68–75.

Stanelle, R.D., M. Mignanelli, P. Brown and R.K. Marcus. 2006. Capillary-channeled polymer (C-CP) fibers as a stationary phase in microbore high-performance liquid chromatography columns. Anal. Bioanal. Chem. 384: 250–258.

Tanaka, N., Y. Tokuda, K. Iwaguchi and M. Araki. 1982. Effect of stationary phase structure on retention and selectivity in reversed-phase liquid chromatography. J. Chromatogr. 239: 761–771.

Tazawa, T., Y. Mori, A. Kobayashi, K. Nakane, T. Monobe, I. Ueta et al. 2015. Preconcentration of aromatic compounds in aqueous samples with a polymer-coated fiber-packed capillary and the subsequent temperature-programmed elution with water for Psudo-2D LC separations. Anal. Sci. 31: 1137–1141.

Ueta, I., Y. Saito, N.B.A. Ghani, M. Ogawa, K. Yogo, A. Abe et al. 2009. Rapid determination of ethylene oxide with fiber-packed sample preparation needle. J. Chromatogr. A 1216: 2848–2853.

Ueta, I. and Y. Saito. 2014. Needle-type extraction device designed for rapid and sensitive analysis in gas chromatography. Anal. Sci. 30: 105–110.

Welch, C.J. and W.H. Pirkle. 1992. Progress in the design of selectors for buckminsterfullerene. J. Chromatogr. 609: 89–101.

Wise, S.A., W.J. Bonnett, F.R. Guenther and W.E. May. 1981. A relationship between reversed-phase C_{18} liquid chromatographic retention and the shape of polycyclic aromatic hydrocarbons. J. Chromatogr. Sci. 19: 457–465.

Wise, S.A. and L.C. Sander. 1996. Molecular shape recognition for polycyclic aromatic hydrocarbons in reversed-phase liquid chromatography. pp. 1–64. *In*: K. Jinno (ed.). Chromatographic Separations Based on Molecular Recognition. Wiley-VCH, New York, NY. USA.

Yang, F.J. (ed.). 1989. Microbore Column Chromatography (Chromatographic Science Series, Vol. 45), Marcel Dekker, New York, NY. USA.

Zarzycki, P.K., H. Ohta, Y. Saito and K. Jinno. 2006. Chromatographic behavior of C_{60} and C_{70} fullerenes at subambient temperature with *n*-alkanes mobile phases. Chromatographia 64: 79–82.

CHAPTER 13

Granular Activated Carbon in Water Treatment

Jack G. Churchill and *Kathryn A. Mumford**

1. Introduction

Activated carbon is a widely used adsorbent across a range of process industries, including water filtration and gas adsorption. It originates from different carbonaceous sources, including coal, wood, and coconut husk. The carbonaceous material is activated by a variety of processes, with the intention of creating a material with a high surface area and pore volume with typically hydrophobic surface chemistry. These properties have encouraged the use of activated carbon to adsorb hydrophobic (often organic) compounds from aqueous systems, utilizing the high selectivity and adsorption capacity, especially for low concentration solutions. Granular Activated Carbon (GAC) is activated carbon in particle sizes typically between 0.5–4 mm, with the most commonly available commercial sizes being 8 × 30 and 12 × 40 mesh (Clark and Lykins Jr 1989). GAC is generally used in adsorber beds, with the particle size allowing for low pressure drops in large scale applications. The particle size of GAC is controlled by the particle size of the unmodified source material, or by a combination of particle size reduction and extrusion processes to generate different particle morphologies.

Drinking water, wastewater, and oil, and gas process water treatment systems have utilized the material for both bulk removal and polishing steps for decades. Treatment of environmental pollutants is a typical application, with the material performing well in removing organic compounds at even low concentrations. The initial use of activated carbon in wastewater treatment was for the removal of taste and odor compounds from drinking water. The use of GAC adsorbers in wastewater treatment became common in the 1970s as a tertiary treatment of organic contaminants known as Disinfection By-Products (DBPs), which result from chlorination treatments. GAC was found to be an effective treatment of Natural Organic Matter (NOM), the presence of which led to the formation of DBPs. By filtering out the NOM, the formation of DBPs could be avoided. GAC filters were often placed upstream of the chlorination process and downstream of sand filters. The pairing of GAC adsorbers with other tertiary treatments, including ozonation, UV disinfection and other advanced oxidation processes (AOPs), is still common in advanced wastewater treatment.

Department of Chemical Engineering, The University of Melbourne, Parkville, VIC 3010.
Email: jackgc@student.unimelb.edu.au
* Corresponding author: mumfordk@unimelb.edu.au

Since the 1970s, work has also been performed identifying the growth of biofilms in GAC adsorbers. Identifying the role of biodegradation of contaminants in the performance of GAC adsorbers in removing NOM, an integrated biophysical treatment was developed, and called Biological Activated Carbon (BAC) filtration. Combining design considerations from biofilters (using inert support materials) with the adsorptive properties of GAC further allowed for adsorbers with extended bed lifespans. In addition, the use of an attached growth process could allow for BAC adsorbers to receive secondary wastewater effluents, with biodegradation of bulk organic matter, retention and biodegradation of slowly biodegradable organic matter, and adsorption of nonbiodegradable organic micropollutants (Çeçen and Aktaş 2012).

More recently, GAC adsorbers have been identified as high performing for the removal of many organic compounds, ranging from phenols, pesticides, herbicides, and hydrocarbons to their chlorinated counter parts, dyes, surfactants, and more. Regulatory bodies have identified many new compounds as contaminants of emerging concern (CECs) ending up in important water bodies, including surface water, groundwater, and as far as drinking water, at concentrations generally from a few ng L^{-1} to several µg L^{-1}. The majority of existing wastewater treatment plants (WWTPs) are not designed to completely remove organic compounds. Instead, they are designed for the bulk removal of dissolved organic compounds and nutrient removal, such as nitrogen and phosphorus, leading to the appearance of these organic micropollutants in the aquatic environment. Several reviews have been written about the thousands of publications on CECs over the last decade, demonstrating an increasing awareness of the occurrence of CECs. An example (Barbosa et al. 2016) reviews the development of a watch list of substances for the European Union-wide monitoring efforts established in Decision 2015/495/EU.

The use of GAC and BAC processes in advanced WWTP has been identified as a strong candidate for the removal of these compounds in both individual and combined treatments. This chapter focuses on the manufacture and properties of different GAC materials, their use, and performance in WWTP applications for the removal of organic micropollutants (OMPs), and the development of combined biophysical treatment.

2. Granular Activated Carbon

2.1 Manufacturing of granular activated carbon

Granular Activated Carbon is generated from a carbonaceous source, such as coconut husk, peat, wood, lignite, or coal via carbonization/pyrolysis to generate the chemical structure of the material, and activation to generate the porosity (Boehm 1994). This is achieved by either chemical or physical activation. Physical activation involves carbonizing the material in an inert atmosphere, often steam, at high temperatures, ranging from 600–900°C. This is followed by partial gasification, resulting in the creation of pores by expanding the small carbon layer stacks that comprise the carbonized material. Further gasification increases the porosity, and encourages the formation of mesopores from the initial micropore formation (Dąbrowski et al. 2005). Chemical activation involves the impregnation of the material with a compound (for example, H_3PO_4), followed by carbonization (at significantly lower temperatures than physical activation), and then washing of the material to remove the activation chemical. The activation with the impregnated chemical causes dehydration and aromatization of the carbon structure, leading to the formation of porous structures.

The initial source of the material has some effect on the properties of the final carbon. Carbonization generally involves the dehydration of the carbohydrates of the materials, which by various methods (hydrothermal, Fe^{2+} catalyzed) results in similar chemical structures (Titirici and Antonietti 2010). However, the presence of other materials—such as proteins and lipids, or inorganic impurities—can influence the formation of pores, while also impacting the surface chemistry of the final material. Common surface impurities are the presence of carboxyl and hydroxyl groups (Dąbrowski et al. 2005). While these groups aid in adsorption of metals via cation exchange (Yin et al. 2007), they may limit the adsorption of phenolic compounds which bond to the carbon surface via π-π dispersion interactions between the basal plane of the carbon and the aromatic ring of the phenols.

The activation process of the carbon has an equally, if not bigger, influence on the adsorption properties of the activated carbon. Physical activation, done in the absence of oxygen, is thought to result in a lack of oxidized surfaces, and hence less surface groups. The surface is predominantly an aromatic carbon in the form of graphite-like layers, giving high affinity for hydrophobic adsorbents, as well as a high capacity for π-π dispersion interactions. However, oxidation of the surface can occur *in situ*, leading to (a) decline in adsorption capacity as the surface oxidizes, and (b) covalent bonding of adsorbed compounds to the surface via oxidation. In contrast, it is thought that chemical activation leads to fully oxidized surfaces. These surfaces may induce issues, such as pore blocking via coordination of water clusters to functional groups (Li et al. 2005), but the polar nature of the groups provide some explanation for the lower adsorption capacity of chemically activated carbons for hydrocarbon and aromatic compounds. The influences of the carbonaceous source and the activation method are discussed and compared in great depth in a treatise on activated carbon (Marsh and Rodríguez-Reinoso 2006).

2.2 Key properties for adsorption performance

A range of factors affect the performance of activated carbons in removing contaminants in environmental applications. Prior to describing the models for adsorption, a description of the key properties will be given.

2.2.1 Chemical surface characteristics

Activated carbon is comprised predominantly of microcrystalline planes of aromatic carbon with several impurities related to the initial material composition and activation process. As highlighted previously, the aromatic surfaces of the carbon are often oxidized due to activation, forming a number of functional groups, including ketones, carboxyl, and hydroxyl functionals groups, and some phenolic functional groups. As such, the surface characterisation is often based on the degree of oxidation, or the acidity of the surface (Karanfil and Kilduff 1999). The characterisation of a carbon surface is important in informing the potential performance for contaminant removal. A typical method is a Boehm titration, though some studies have shown it has an insufficient level of sensitivity for studying changes in surface acidity of as-received activated carbons (Karanfil et al. 1999). Oxidized surfaces may improve uptake of polar adsorbates, but overall reduce the hydrophobicity of the surface. Additional physical bonding mechanisms (such as π-π bonding with aromatic compounds and multisite bonding) may be reduced by sterically hindering the contaminant transport to the carbon surface. This has been theorized with the formation of water clusters around polar functional groups, potentially blocking micropores for lower molecular weight compounds.

2.2.2 Porosity

The number of pores, their size, and shape play a large role in the adsorption capacity and kinetics of GAC. Porosity, classified based on pore width into micropores (< 2 nm), mesopores (2–50 nm), and macropores (> 50 nm), is important, as it provides a higher specific surface area for adsorption sites. Macropore adsorption operates the same way as flat surfaces, providing sites for surface adsorption, but with low specific surface areas, and therefore low overall capacity. Mesopores allow for mono- and multilayer adsorption via the mechanism of capillary adsorbate condensation (Aktaş and Çeçen 2007). Both mesopores and macropores allow for the transport of adsorbate species towards micropores, potentially providing hierarchical transport structures to improve kinetics. Micropores-sized similarly to adsorbate species-operate via a pore-filling mechanism, with contaminant particles experiencing multi-site bonding. As such, the overall pore volume of a material is important in determining the overall capacity, while the microporosity allows for higher sorption energy. Activated carbons are often highly microporous materials with large specific pore volumes relative to other hydrophobic adsorptive materials, including other carbon-based adsorbents, allowing for high uptake of low molecular weight organic compounds. The pore size distribution can have a negative effect on adsorption via size exclusion effects for higher molecular weight organic compounds. If the molecules have higher kinetic diameters than a portion of the pores, they will be unable to access the pore adsorption sites, and the activated carbon will show a reduced adsorption capacity. This is most notably a concern for the removal of dissolved organic material (DOM), which may require more highly mesoporous materials (Karanfil et al. 1999). Additionally, when targeting a specific compound for removal, the presence of size-excluded DOM can cause pore blocking (Corwin and Summers 2010).

2.2.3 Temperature and pH

In addition to the physical properties of the carbon solid phase, process parameters also influence the adsorption performance. Temperature has an effect, as the extent of adsorption should increase with decreasing temperature due to the exothermic nature of adsorption reactions. However, temperature also increases the rate of diffusion through the liquid while also affecting the solubility of the contaminant in the liquid phase.

Adsorption of organic molecules is generally higher at pH neutral conditions, as organic molecules form positive ions at low pH and negative ions at high pH, with adsorption to the activated carbon surface being largely a result of non-polar, hydrophobic interactions. However, this relationship depends on the surface acidity of the activated carbon. Activated carbons with an oxidized surface (and therefore increased surface acidity) showed an inverse relationship between surface acidity and synthetic organic compound (SOC) adsorption capacity (Karanfil and Kilduff 1999). A proposed mechanism suggests neutralization of negative charges at the surface of the carbon at low pH values, reducing the hindrance to diffusion via pore blocking, allowing for more active adsorption sites (Aktaş and Çeçen 2007). These effects are dependent on the type and activation technique of the carbon, with differences reported between chemical and physical activation.

2.3 Adsorption Kinetics

In an adsorption system where a liquid phase contacts a solid phase, with a chemical species exhibiting affinity for both phases (solubility in the liquid phase, adsorption in the solid phase), partitioning of the species to each phase occurs. An equilibrium is reached, but requires transport of the chemical species between the phases. Adsorption kinetics is the rate of approach to equilibrium, and that rate is limited by the mass transfer mechanisms. The main transport mechanisms of an adsorbed chemical species to a surface are described in the following sections. Figure 13.1 shows the basic mass transport model for liquid adsorption. The rate of mass transfer is controlled by rate limiting steps, whereby the mass transfer mechanism that has a higher resistance to mass transfer compared to the others controls the overall kinetics. Theoretically, four steps are used to model mass transport: (1) transport of sorbate in the bulk liquid phase; (2) diffusion across a liquid film surrounding the sorbent particles; (3) migration of sorbate within the pores of the sorbent, known as intraparticle diffusion, and (4) adsorption/desorption on the solid surface.

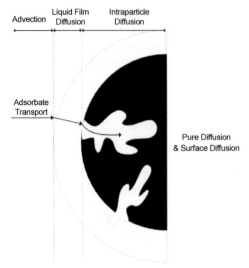

Figure 13.1: Pore and Surface Diffusion Model describing transport from a bulk liquid phase to internal micropore bonding sites.

2.3.1 Bulk liquid phase transport

In non-quiescent systems, advective forces cause mass transport of the adsorbate species to give a bulk liquid phase concentration. This gives the boundary condition of the liquid film for the diffusion model for adsorption. The model for obtaining the advection is dependent on the system, and column adsorbers generally utilize an axial dispersion model.

2.3.2 Liquid film phase transport

Mass transport from the bulk liquid to the carbon surface is determined by the diffusion through a liquid film, also known as the mass transfer boundary layer between the solid and liquid phases. It is defined as the distance from a solid to where the concentration of the chemical species is 99% of the bulk liquid or boundary condition. The mass transfer through this phase occurs via diffusion and depends on the hydrodynamic properties of the system. This is mathematically represented as

$$\text{(Mass flux)} = \frac{D}{\delta}(C_b - C_s) \tag{1}$$

where D is the diffusivity through the liquid film, δ is the boundary layer thickness, C_b is the bulk concentration of the chemical species, and C_s is the concentration of the chemical species on the adsorptive surface. The mass transfer coefficient k_b is given as

$$k_b = \frac{D}{\delta} \tag{2}$$

which gives the widely used equation for mass transfer rate through a liquid film:

$$\text{(Mass flux)} = k_b(C_b - C_s) \tag{3}$$

2.3.3 Intraparticle diffusion

Intraparticle diffusion is the transport of the adsorbate from the surface of the particle to adsorption sites within the particle. It is independent of the hydrodynamics of the system, but depends on the pore profile of the material. The two main mechanisms for intraparticle diffusion in activated carbon are pore diffusion and surface diffusion. In pore diffusion, the mean free path of the chemical species is larger than the pore width. The chemical species collide with the wall before diffusing through the length of the pore. In surface diffusion, the chemical species are more closely associated with the pore wall, and diffuse across the surface subject to a concentration gradient and surface diffusivity. Significant and ongoing research has been done in developing models to best describe the kinetics of intraparticle diffusion, with most models using a combined pore surface diffusion model (PSDM). A basic expression that can be used is as follows:

$$\text{(Mass flux)} = -D_s \rho_a \frac{\partial q}{\partial r} - \frac{D_l \varepsilon_p}{\tau_p} \frac{\partial C_p}{\partial r} \tag{4}$$

where D_s is the surface diffusion coefficient, ρ_a is the adsorbent particle density, D_l is the liquid phase diffusion coefficient, ε_p is the porosity of the particle, τ_p is the tortuosity of the path that the adsorbate takes compared to the radius, C_p is the liquid phase concentration of the adsorbate, q is the adsorbate phase concentration, and r is the radial coordinate.

This equation can be simplified, with pore diffusion being ignored in single component systems, as the pore concentration is small compared to the adsorbate surface concentration, giving the homogenous surface diffusion model (HSDM). However, the occurrence of multisolute solutions (such as the presence of dissolved organic carbon, DOC) causes a large impact on contaminant removal. DOC can block surface diffusion and limit intraparticle diffusion to pore diffusion.

2.3.4 Overall sorption rates

The kinetic models described in equations (1) to (4) have been incorporated into numerical advective-diffusive-reactive reactor models to estimate the performance of GAC beds. Simplifications of overall adsorption kinetics exist, aiming to describe the change in solid-phase adsorbate concentration with time. Common examples include the pseudo-second order equation, Linear Driving Force (LDF), and more (Plazinski et al. 2009). These models do not consider the physical theory of sorption kinetics, but approximate the kinetic models well and remain popular. The pseudo second order equation can be written in differential form as

$$\frac{dq}{dt} = k_2 (q_e - q(t))^2 \tag{5}$$

where $q(t)$ is the mass adsorbed to the carbon at time t, q_e is the mass adsorbed to the activated carbon at equilibrium at the corresponding liquid phase concentration, and k_2 is a constant. k_2 has been interpreted as a time-scaling factor, whereby a high k_2 corresponds to a short time to equilibrium, and vice versa. This equation is based on the concept that the rate of adsorption processes on the surface, described as chemical reactions, controls the kinetics of the overall system (Plazinski et al. 2013).

The LDF model is commonly used to approximate particle diffusion controlled rate of mass transfer from the liquid to the solid phase, and can be expressed as

$$\frac{dq}{dt} = \frac{60 D_p}{4 d_p^2} (q_e - q(t)) \tag{6}$$

where D_p is the particle diffusion coefficient and d_p is the mean particle diameter. Notably, both approximations are derived from zeolite-based systems which function via ion exchange processes, and the adsorbent structure is more uniform in both pore distribution and adsorption site bonding energies and mechanism. Activated carbons tend to have far more heterogeneity in their pore structure and the bonding sites, with limiting factors often being related to accessing the large adsorption capacities in micropores rather than the adsorption reactions themselves.

2.4 *Adsorption equilbrium and isotherms*

The choice of carbon type has a strong influence on the performance of the bed, with different carbons having different interactions, with target compounds depending on the carbon materials—pore size and surface chemistry—and the target compound—molecular size (often controlled as molecular weight) and hydrophobicity. The driving force for adsorption is the affinity of the target compound to the carbon surface. These forces can include Van Der Waals, dipole-dipole, induced dipole, hydrophobic, and π-π interactions.

Predicting the performance of an activated carbon system first relies on developing a thermodynamic relationship between the adsorbate and adsorbent in the form of equilibrium isotherms. Laboratory trials can be used to determine the relationships which can be modeled in a large variety of ways. Basic adsorption models have been fitted and are widely restricted to simple Freundlich and Langmuir isotherms (Limousin et al. 2007). Freundlich isotherms are the most commonly used for aqueous activated carbon systems. They describe the non-ideal and reversible adsorption and are not restricted to the formation of a monolayer. It can be applied to non-uniform distributions of heterogenous surfaces. It is typically expressed as

$$q_e = K_f C_e^{1/n} \qquad (7)$$

where q_e is the concentration of adsorbate on the solid phase, expressed as mass of adsorbate per mass of adsorbent. C_e is the equilibrium liquid phase concentration, K_f is the Freundlich adsorption capacity parameter, and $1/n$ is the Freundlich adsorption intensity parameter. Initially, the Freundlich equation was an empirical equation, but further model development highlighted the equation's consistency with the thermodynamics of heterogenous adsorption (Pikaar et al. 2006). Activated carbons are typically heterogenous in adsorption site energies and exhibit some multilayer adsorption, due to the differences in mutual orientations of adsorbing AC graphene basal planes.

In contrast, the Langmuir model assumes homogenous adsorption sites and monolayer adsorption, and is a widely used model in other environmental contaminant adsorbent materials. This is shown graphically as displaying a plateau representing a maximum adsorption capacity. The Langmuir model is presented as:

$$q_e = \frac{K_l M C_e}{1 + K_l C_e} \qquad (8)$$

where K_l is the Langmuir adsorption equilibrium constant and M is the maximum adsorption capacity of the adsorbate. Many alternative, multiparameter isotherms exist with different rationales for their use. A statistical analysis of isotherm models was performed for the sorption of organic compounds onto a wide range of activated carbons (mostly thermally activated), and it was found that dual Langmuir model showed the best fit (Pikaar et al. 2006). This highlights the potential presence of high energy and low energy sites. Another study using a thermal activated carbon used a statistical physics approach, and similarly found the presence of high and low energy sites (Sellaoui et al. 2007). A double layer model was found to provide the best fit to the data. Interestingly, the model found a spread between the number of molecules adsorbed to each site between 0.5 and 1.5 molecules per site. This might indicate the presence of multi-docking between different sites, as well as the partial adsorption of other sites. These results suggest that activated carbon may exhibit a bimodal distribution of Gibbs free energy for adsorption sites, rather than a broad distribution suggested in a Freundlich isotherm. The dual Langmuir isotherm is represented as

$$q_e = \frac{K_{l,1} M_1 C_e}{1 + K_{l,1} C_e} + \frac{K_{l,2} M_2 C_e}{1 + K_{l,2} C_e} \qquad (9)$$

where $K_{l,1}$ is the Langmuir adsorption equilibrium constant and M_1 is the maximum adsorption capacity of the adsorbate for the high energy bonding site, while $K_{l,2}$ is the Langmuir adsorption equilibrium constant and M_2 is the maximum adsorption capacity of the adsorbate for the low energy bonding site.

Accurate modelling of adsorbent behavior is critical in determining the performance of an adsorbent to a given environmental contaminant, as it yields the ability to optimize the design of adsorbent systems. Further isotherms have been proposed in the literature, including the Sips and Toth isotherm, as well as the Brunauer-Emmett-Teller isotherm (Foo and Hameed 2010). These models can be empirically based, to ascertain a best fit rather than describe the mechanism of adsorption in the activated carbons. Indeed, theoretical multiparameter isotherms can produce better fits simply due to the presence of more fitting parameters.

An important factor in developing accurate isotherm models that can accurately predict an adsorbent performance is the use of error functions to describe the goodness of the fit. Linear regression has become one of the most viable tools. However, this relies on linearization of the models, potentially introducing an inherent bias resulting from the transformation, and has led to the development of several mathematically rigorous error functions that can perform nonlinear regressions. These errors are considered to originate from linearization of the error distribution of the scatter vertical points, which would otherwise be uniform across the values of the liquid-phase residual concentration. The most common nonlinear regression method is the use of the coefficient of determination (R^2), which represents the percentage of variability in the dependent variable. Comparisons between isotherms attempt to maximise the coefficient of determination, though definitive statements on models based on slight differences in maximizing the value should be made warily (Foo and Hameed 2010).

2.4.1 Predicting contaminant adsorption

Predicting adsorption equilibria is useful due to the cost of laboratory experiments and the number of potential organic pollutants. With over 70,000 organic compounds currently in use and a wide variety of activated carbons with a large variance in adsorption performance, conducting adsorption equilibrium tests for each newly identified contaminant is costly and potentially infeasible. A recent area of research in activated carbon usage in water treatment has been the development of predictive models to generate isotherms for new compounds for a given activated carbon. Using some model compound adsorption isotherms to determine fitting parameters, the chemical and activated carbon properties are considered in pure water systems to predict the adsorption capacity. One of the first and most common models to be used is the Polanyi-Dubinin equation, combing adsorption potential theory with pore filling theory to account for the microporous nature of activated carbons (Crittenden et al. 1999). The equation is given as:

$$W = W_0 \exp\left[-\left(\frac{\varepsilon}{\beta E_0}\right)^n\right] \tag{10}$$

where W is the volume of solute absorbed, W_0 is the limiting volume of the adsorption space (i.e., the micropore volume), E_0 is the characteristic adsorption energy of the reference adsorbate, β is the affinity coefficient of the characteristic curve, and n is the exponential coefficient. s is given as:

$$\varepsilon = RTln\left[\frac{C_s}{C}\right] \tag{11}$$

where C is the aqueous phase concentration, C_s is the aqueous solubility, T is the absolute temperature, and R is the gas constant. In this equation, W_0 and n are the two parameters that describe the nature of the adsorbent, while the combination of βE_0 describes the adsorption relationship of a given compound to the reference compounds from which the affinity coefficient was found.

W_0 is considered to be the accessible pore volume, and can be estimated by the adsorption isotherm of reference compounds (Bunmahotama et al. 2015), with the adsorption capacity being fixed so long as the kinetic diameter of a molecule does not exceed the pore width, creating a molecular sieving effect (Hung and Lin 2007). The pore size distribution can be used to predict the adsorption capacity (Urano et al. 1982), and utilize the following empirical equation:

$$W_0 = 0.055 + V_{3.2} \tag{12}$$

where $V_{3.2}$ is the micropore volume for pores with a diameter less than 3.2 nm for activated carbon. W_0 can also be used as a fitting parameter along with n, as n has been shown to vary between 1 and 3, depending on the nature of the adsorbent used. Higher homogeneity in the micropores tends n to a value of 3, while higher heterogeneity tends it towards one. n is typically used as a fitting parameter. Depending on chemical groupings (aliphatic, alcohol, aromatic, sulfonated, etc.), n has typically been used in the range of 1.0 to 1.2 for activated carbons.

Key efforts in recent research to improve the accuracy of the Polanyi-Dubinin model have focussed on the inclusion of a normalizing factor to account for non-Van de Waals interactions and expand the scope of the equation to allow for the prediction of a wider range of compounds. This normalizing factor gives a modified equation represented as such (Crittenden 1999):

$$W = W_0 \exp\left[-\left(\frac{\varepsilon}{100N}\right)^n\right] \tag{13}$$

where N is the normalizing factor and 100 is a scaling factor. N accounts for the total interaction forces in a system, and has been modelled using different methods. Crittenden et al. used linear solvation energy relationships, which utilize the intrinsic molar volume, the polarity parameter, the hydrogen bonding-accepting parameter, and the hydrogen-bonding donor parameter. Other researchers have used a range of methods, such as molecular connectivity indices (Bunmahotama 2015). These utilize a three-dimensional molecular structure to create either simple or valence indices, including sigma, pi, and lone electrons. To relate N to the indices, empirical factors are attached to each index and fitted using a known isotherm.

Using a minimisation of sum of errors between the predicted volume adsorbed and experimental volume adsorbed for a range of chemicals with similar properties on a chosen activated carbon, the fitting parameters can be found, and a model produced for predicting contaminant adsorption for untested compounds. Using the modified Polanyi-Dubinin model, the volume of contaminant adsorbed compared to solution concentration can easily be converted to mass of contaminant adsorbed with a density relationship, and a data set can be created for predicting the desired isotherm model. This can limit the number of adsorption tests required when investigating the performance of chosen GAC, with potentially new contaminants being identified or designed for. However, this gives rise to the issue of multicomponent adsorption and the difficulties of competitive adsorption.

3. GAC Filtration: Adsorbent Column Design & Modelling

3.1 Types of adsorbers

Continuous-flow GAC systems generally utilize adsorbent columns, which can be fixed, expanded, fluidized, or pulsed. Fixed bed adsorbent columns are often designed in either cylindrical or rectangular geometries. Open top steel or concrete adsorbent columns with gravity flow, or closed top pressure flow adsorbent columns, are usually operated in parallel or series configurations to allow for ongoing operations during the removal of exhausted GAC, or during backwashing cycles. The exceptions are pulsed bed configurations, in which a single adsorbent column receives continuous counter current flow of fresh GAC into the bed while exhausted GAC is removed. Design considerations, including bed volumes, bed lengths, and cross-sectional areas, are based on process parameters, such as hydraulic loading (flowrate/cross sectional area) and empty bed contact time, which in turn is determined by the breakthrough curves of different GAC types and their performances in contaminant removal (Clark and Lykins Jr 1989).

3.2 Design of adsorbent columns

The two most important variables in GAC adsorbent columns systems are the contact time and breakthrough characteristics. Alongside flow quantities, these process variables determine the carbon type and bed volume.

Empty bed contact time (EBCT) is the common variable used to describe contact times. It is the volume of the empty bed divided by the superficial flow rate of the fluid phase. The EBCT represents the theoretical residence time in the filter and determines the time allowed for mass transfer of species to be removed to the adsorbent. However, as the sizing of the GAC changes, the effective contact time is not truly represented by the EBCT. An alternative is the effective contact time, which is the EBCT multiplied by the void fraction of the carbon bed. The desired EBCT determines the volume of the GAC bed, with the adsorbent column unit often including a further 20–50% volume of headspace for backwashing or expanded beds (Clark and Lykins Jr 1989).

The breakthrough characteristics of a bed are determined predominantly by the kinetics of adsorption and the chosen contact time. Adsorption of the contaminants occurs as the fluid contacts unsaturated (or non-exhausted) GAC. As the GAC begins adsorption, the kinetics are not quick enough to achieve complete saturation of the GAC, causing a front to develop in the bed. The front consists of partially-loaded GAC, and decreased concentrations of contaminant in the fluid phase. This front is called the mass transfer zone (MTZ), and is shaped as a wave in the bed. When this front reaches the end of the GAC bed, breakthrough is said to occur. The shape of MTZ is determined by the kinetics of the adsorption and the contact time (or EBCT) of the fluid. The kinetics are determined by the interactions between the activated carbon and the specified contaminant, and choice of activated carbon will greatly affect the MTZ. The breakthrough curve is determined by plotting the effluent concentration divided by the inlet concentration as a function of elapsed time. Bed Volumes are also used in describing the volume treated, as it normalizes the volumetric flowrate with the volume of the bed, allowing for better comparison between systems or processed volume (Crittenden and Borchardt 2012). Figure 13.2 shows an example breakthrough curve.

Typically, the bed is considered exhausted and changed out when a predetermined breakthrough concentration is reached. This is often defined as when 10% of the initial contaminant concentration is detected at the effluent, or when a maximum concentration is reached for a target contaminant. This depends on the overall water treatment process design beyond the GAC adsorber unit operation, but may be informed by regulatory guidelines. At this point, the bed is changed out with the length of time passed before reaching this outlet concentration, referred to as the operation time of the bed, t_b.

The Carbon Usage Rate (CUR) can be determined from the breakthrough characteristics, and is a measure of the mass of carbon divided by the cumulative volume treated at breakthrough. This quantification of the adsorbent column performance directly relates the amount of carbon required to treat a certain volume of contaminated water. The length of the MTZ has a large impact on the CUR. A larger MTZ means that at breakthrough, a larger volume of activated carbon has not reached saturation. The key process design parameter affecting the length of the MTZ is the EBCT. Increasing the EBCT allows for longer timeframes and reduces kinetic limitations of mass transfer, increasing the CUR. Additionally, increasing the length of the bed relative to the MTZ increases the CUR. Considering these design principles, large beds with large EBCTs and high CURs are ideal, but come at increasing capital cost and physical size (Crittenden and Borchardt 2012).

Hydraulic Loading Rate (HLR), also described as the superficial filter velocity, is used to determine the cross-sectional area of the bed. Paired with the EBCT, it specifies the dimensions of

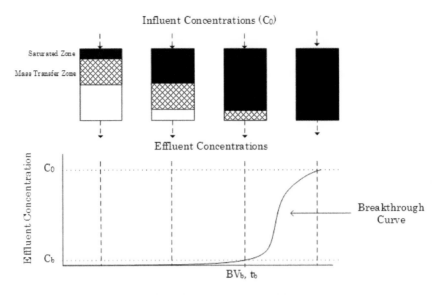

Figure 13.2: Mass transfer zone in a fixed bed GAC adsorber. When the effluent reaches a predetermined concentration (C_b), the number of Bed Volumes (BV) or elapsed time (t) is noted as the breakthrough point.

the adsorber beds. Defined as the volumetric flowrate divided by the cross-sectional area of the bed, typical ranges are 5–25 m/h. Design ranges for HLR are based predominantly on head loss in the bed, with high loadings given high pressure drops, particularly in sediment dense influents (Clark and Lykins Jr 1989).

To design a full-scale column for water treatment, determining the breakthrough curve is critical. This can be done by a pilot scale study, which will predict full scale GAC performance very accurately, but at a large expense, and needs to be conducted onsite with a typical feedwater over long time periods. Two alternatives to pilot scale studies are presented here, both using kinetic models to predict breakthrough characteristics, and therefore process parameters, such as EBCT, t_b, and HLR. Rapid Small-Scale Column Tests utilize scaling factors to allow for laboratory scale tests, while a general mathematical model uses fundamental principles and limited laboratory tests to predict the entire adsorbent column.

3.3 *Rapid Small-Scale Column Tests (RSSCTs)*

Rapid Small-Scale Column Tests allow for good accuracy in predicting breakthrough of a large scale GAC adsorbent column if appropriate scaling factors have been used. These tests utilize mathematical models to scale down a GAC adsorbent columns to produce breakthrough curves that are representative of the full scale adsorbent columns. This is done by setting the dimensionless groups that describe chemical species transport, such as the surface diffusion modulus, as equal between the small scale and large scale columns.

These tests can be used to determine key design parameters, such as EBCT or the choice of activated carbon. Refined over several papers (Crittenden et al. 1986, 1987, Hand et al. 1997), they rely on the dispersed-flow pore and surface diffusion model for the kinetics of adsorption. The procedure outlined here is taken from Crittenden and Borchardt (2012) with some modifications.

3.3.1 Scaling EBCT

The empty bed contact time of the small-scale column can be related to the EBCT of the large scale column by equating the surface or pore diffusion modulus of the two columns. Using the surface diffusion, the following relationship is found:

$$Ed_{s,SC} = Ed_{s,LC} \tag{14}$$

$$\frac{D_{s,SC} \tau_{SC} D_g}{R_{SC}^2} = \frac{D_{s,LC} \tau_{LC} D_g}{R_{LC}^2} \tag{15}$$

where $Ed_{s,SC}$ is the surface diffusion modulus of the small-scale column, τ_{SC} is the small scale packed bed contact time (found from the EBCT multiplied by ε, the bed porosity), and R_{SC} is the radius of the particle adsorbent of the small-scale column. $Ed_{s,LC}$ is the surface diffusion modulus of the large scale column, τ_{LC} is the small scale packed bed contact time (found from the EBCT multiplied by the bed porosity):

$$\tau = (EBCT)\varepsilon \tag{16}$$

and R_{LC} is the radius of the particle adsorbent of the small-scale column. D_g is the solute distribution parameter defined as the contaminant species on the adsorbent per contaminant species in bed voids at equilibrium, found from the equation

$$D_g = \frac{\rho_a q_{e,0}(1-\varepsilon)}{\varepsilon C_0} \tag{17}$$

where ρ_a is the apparent particle density of the activated carbon, $q_{e,0}$ is the equilibrium solid phase concentration of the adsorbate on the adsorbent at the initial liquid phase inlet concentration C_0, and ε is the bed porosity. Solving this equation for ratio of the bed contact times gives the equation

$$\frac{\tau_{SC}}{\tau_{LC}} = \left(\frac{R_{SC}}{R_{LC}}\right)^2 \left(\frac{D_{s,LC}}{D_{s,SC}}\right) \tag{18}$$

The same relationship can be derived from the pore diffusion, where the surface diffusion terms are replaced with pore diffusion terms. To get this point, it is assumed that the adsorption capacity and physical properties of the activated carbon and packed bed do not depend on particle size. Further to this, if the intraparticle diffusivities of the small and large GAC are identical, the relationship can be reduced to:

$$\frac{\tau_{SC}}{\tau_{LC}} = \left(\frac{R_{SC}}{R_{LC}}\right)^2 \tag{19}$$

However, if the controlling intraparticle diffusion has been observed to be dependent on the particle size (usually modelled as a power relationship), the following relationship is used:

$$\left(\frac{D_{s,LC}}{D_{s,SC}}\right) = \left(\frac{R_{LC}}{R_{SC}}\right)^x \tag{20}$$

where x is the power dependency of the diffusivity. If the dependency is linear, $x = 1$. This is the most common version used. Using these relationships, we can find that EBCT of the small column is:

$$EBCT_{SC} = EBCT_{LC} \left(\frac{\varepsilon_{LC}}{\varepsilon_{SC}} \right) \left(\frac{R_{SC}}{R_{LC}} \right)^{2-x} \tag{21}$$

where ε_{LC} and ε_{SC} are the bed porosities of the large scale and small-scale column, respectively. These can be found from the bulk density.

3.3.2 Scaling operating time

The operating time of a small column compared to a large column can be related to the two columns exhibiting the same mass throughput, defined as the ratio of the mass of contaminant species fed to the system to the mass adsorbed by the system. As a dimensionless number, this can be represented as

$$T = \frac{QC_0 t}{V \rho_a q_{e,0}(1-\varepsilon)} = \frac{\varepsilon C_0 t}{\tau \rho_a q_{e,0}(1-\varepsilon)} \tag{22}$$

where T is the mass throughput, Q is the liquid flowrate, t is the elapsed time since the contaminated feed was supplied to the filter bed, and V is the bed volume. By equating the mass throughput ratio of the large and small columns, we find the following:

$$\frac{\varepsilon_{LC} C_0 t_{LC}}{\tau_{LC} \rho_a q_{e,0}(1-\varepsilon_{LC})} = \frac{\varepsilon_{SC} C_0 t_{SC}}{\tau_{SC} \rho_a q_{e,0}(1-\varepsilon_{SC})} \tag{23}$$

which, solving and rearranging for t_{LC}, can be transformed to give the relationship:

$$t_{LC} = \left(\frac{EBCT_{SC}}{EBCT_{LC}} \right) \left(\frac{1-\varepsilon_{LC}}{1-\varepsilon_{SC}} \right) t_{SC} \tag{24}$$

By determining the EBCT for the small-scale column tests based on the particle sizes and proposed EBCT for the large scale column, adsorption tests can be performed to determine the breakthrough time t_b for the small scale column. This can then be scaled up using the relationship proposed.

It is worth noting that if the same GAC particle sizing is used in the small-scale column tests, the bed porosity values can be considered the same, and the EBCTs and breakthrough times are equivalent. However, RSSCTs generally utilize smaller adsorbent particles and have been shown to have poorer predictions of performance regardless of the diffusivity relationship. When attempting to develop breakthrough curves for dissolved organic matter (DOM), it was found that the RSSCTs overpredicted the performance (Corwin and Summers 2010). It was suggested that this was due to the adsorption capacity of the activated carbon not being constant with particle size. This was theorised to be a result of the DOM causing pore blockages in the micropores preventing adsorption through micropore channels. With larger particles, more pore volume was being blocked relative to the mass of carbon, requiring the use of a scaling factor between RSSCTs and pilot or full-scale columns.

This DOM fouling was the basis for developing a fouling factor (independent of the scaling relationship for EBCT) that can be applied to the RSSCT throughput in terms of bed volumes (BV, ratio of volume of water treated to volume of GAC bed), outlined as:

$$BV_{LC} = \frac{BV_{SC}}{FI} \tag{25}$$

where *FI* is the fouling index, found as:

$$FI = \left(\frac{R_{LC}}{R_{SC}} \right)^{Y} \tag{26}$$

where Y is the fouling factor, found from fitting the equation to data. A study (Corwin and Summers 2010) fitted for a range between 0.6 and 0.8 for Y. A more recent study (Kennedy et al. 2017) investigated the dependence of Y based on several parameters, including the initial OMP concentration, the background DOC concentration, and the bed volume at which the OMP was at 10% of the initial concentration for the RSSCT. The Y dependency also used a linear dependency of diffusivity on particle diameter and the pH-dependent octanol-water partition coefficient. Using the relationship found from the dependencies, an entire breakthrough curve can be adjusted from the RSSCT with Y. For simplicity, an empirical relationship was developed to predict the 10% breakthrough bed volume for a full-scale BV based on the RSSCT 10% breakthrough bed volume. In this equation, the requirement of deriving an entire breakthrough profile is eliminated and minimises the tests needed to predict a minimum breakthrough point of a new OMP identified at an existing GAC adsorber. The equation was given as:

$$\ln BV_{10\%}^{LC} = (0.57 \pm 0.32) + (0.855 \pm 0.029)\, BV_{10\%}^{SC} \tag{27}$$

where $BV_{10\%}^{LC}$ and $BV_{10\%}^{SC}$ are the number of bed volumes passed when 10% of the initial concentration of the contaminant is detected for the large scale and small scale columns, respectively.

3.4 Full scale GAC filtration model

The theory of adsorption kinetics and equilibrium has been discussed with some basic equations used to describe the phenomena outlined. Methods to predict adsorption equilibria have also been proposed, which is important, given the rising number of OMPs identified as environmental contaminants of concern. While rapid small scale column tests have been proposed as methods to screen the performance of GACs with new OMPs, they still require many laboratory experiments to develop a successful and accurate prediction of a full scale adsorbent column performance. Development of mathematical models to describe the performance of a GAC adsorbent column is useful as a method of determining performance in changing conditions without requiring any laboratory experiments, especially combined with predictive adsorption equilibrium relationships.

To begin the development of the equation, the transport of the contaminant species through a fixed reactive bed must be considered. Typically, the starting point is the use of the axial dispersion model of flow through a fixed bed utilizing an advection-diffusion relationship:

$$D \frac{\partial^2 C}{\partial x^2} - \bar{v} \frac{\partial C}{\partial x} = \frac{\partial C}{\partial t} \tag{28}$$

where D is the axial dispersion coefficient term, C is the contaminant species concentration in the bulk solution, x is the distance along the column length, and $\bar{v}$ is the average pore velocity. This relies on the assumptions that liquid phase concentration gradients exist only in the axial direction (the direction of flow). This equation can be modified to include a reactive term:

$$D \frac{\partial^2 C}{\partial x^2} - \bar{v} \frac{\partial C}{\partial x} - J \frac{A}{\varepsilon} = \frac{\partial C}{\partial t} \tag{29}$$

where J is a sink term describing flux to the adsorbent surface per adsorbent surface area, and A is the adsorbent surface area per volume of bed. The inclusion of the bed void fraction accounts for the liquid, and is included from the derivation of this term (Crittenden and Borchardt 2012). Some further assumptions required for this model include: (1) this gradient is small enough such that the concentration gradient across a single adsorbent particle is negligible, so that the bulk solution concentration surrounding a particle is identical; (2) the adsorbent is in a fixed position in the bed; (3) the contaminant species contained in the liquid phase of the carbon pores is negligible, and (4) the hydraulic loading is constant.

The flux term J can be expressed as the liquid film transfer rate from the bulk solution, described in equation (3). The term A can alternatively be expressed as (Crittenden and Borchardt 2012)

$$A = \frac{3}{\phi R}(1-\varepsilon) \tag{30}$$

where ϕ is the sphericity of the GAC particle. Sphericity can be derived from optical image analysis, with a range of carbons in literature typically showing a sphericity of 1.2–1.4. Substituting these equations into the Advective-Diffusive-Reactive Equation (ADRE) gives the following liquid phase mass balance partial differential equation:

$$D\frac{\partial^2 C}{\partial x^2} - \bar{v}\frac{\partial C}{\partial x} - \frac{3(1-\varepsilon)k_b}{\phi R\varepsilon}(C-C_s) = \frac{\partial C}{\partial t} \tag{31}$$

This equation is solved by the application of one initial condition and two boundary conditions. The initial condition, when no adsorbate has entered the system at the start, can be expressed as

$$C(0 \leq x \leq L, t = 0) = 0 \tag{32}$$

where L is the length of the bed.

The first boundary condition is given as the mass balance of the entire bed, which gives the equation:

$$\bar{v}C_0(t) - \bar{v}C_0(x = L,t) = \int_0^L \frac{\partial C}{\partial t}\,dx + \frac{\rho_a(1-\varepsilon)}{\varepsilon}\int_0^L \frac{\partial q_{ave}}{\partial t}\,dx \tag{33}$$

where q_{ave} is the average adsorbate solid phase concentration at bed section dx. This can be found as

$$q_{ave} = \frac{3}{4\pi R^3}\int_0^R q4\pi r^2\,dr \tag{34}$$

where r is the radial coordinate within the particle and q is the adsorbate solid phase concentration at the radial coordinate within the particle and axial coordinate within the fixed bed.

The second boundary condition is the condition at the exit boundary where $x = L$, is given as:

$$\frac{\partial^2 C(x = L,t)}{\partial x\partial t} = 0 \tag{35}$$

$$\frac{\partial C(x,t = 0)}{\partial x} = 0 \tag{36}$$

However, to solve the ADRE expression, the concentration on the adsorbent surface C_s needs to be found. Conducting a mass balance of the adsorbent phase and using the pore surface diffusion model gives the equation

$$\frac{\partial q}{\partial t} = \frac{D_s}{r^2}\frac{\partial}{\partial r}\left(r^2\frac{\partial q}{\partial r}\right) \tag{37}$$

The boundary and initial conditions are given by the expressions:

$$k_b(C-C_s) = D_s\rho_a\frac{\partial q}{\partial r} \tag{38}$$

$$\frac{\partial q(r = 0, 0 \leq x \leq L, t \geq 0)}{\partial r} = 0 \tag{39}$$

$$q(0 \leq r \leq R, 0 \leq x \leq L, t = 0) = 0 \tag{40}$$

Several solutions to these equations exist—both numerical and analytical. By making the equations dimensionless and applicable as general models, empirical relationships can be used to estimate some of the dimensionless numbers that are generated. Alternatively, numerical solutions can be found that require a partial differential equation solver.

4. Biological Activated Carbon Filtration

Biological Activated Carbon (BAC) filters have been used for the treatment of wastewater for several decades, making use of the high adsorption capacity for removal of organic compounds and biological processes for degradation of the same compounds. The growth of biological communities on activated carbon was first investigated in the late 1970s. The growth of biofilms on activated carbon systems in water and wastewater treatment was demonstrated using scanning electron microscopy (SEM) (Weber et al. 1978), with later suggestions that growth may be advantageous to the removal of organic compounds (Ying and Weber 1979). A key suggestion was that a fixed bed performed better using a combination of bio-physical processes over purely adsorption processes (straight activated carbon column) or biodegradation processes (biofilters). This was speculated to be due to the biological system being able to access substrates from both the bulk fluid phase, and the adsorption surface. However, the proof of regeneration of the activated carbon remains somewhat elusive, remaining an ongoing area of interest (Aktaş and Çeçen 2007, Abromaitis et al. 2016). This simple and advantageous system has since been investigated in an attempt to improve the performance and cost-effectiveness of activated carbon in the wastewater industry.

4.1 Adsorption and biological process integration

Several researchers have progressed towards developing conceptual and mathematical models to describe the bio-physical process of adsorption, desorption, and biodegradation. Recently, it has been shown (Aktaş and Çeçen 2007) that the major limitation in bioregeneration is the apparent hysteresis of organic compounds adsorbed to activated carbons. While biodegradation does occur in BAC processes, it is mostly from the bulk fluid phase rather than from the adsorbed phase. Further research (Abromaitis et al. 2016) showed the presence of apparent hysteresis, yet also that the biological component of the process is able to overcome the hysteresis during low bulk fluid phase loadings of the substrate.

The first models for biological activated carbon were relatively simple adsorption-diffusion-reaction equations, as described in multiple studies (Chang and Rittman 1987a, b, Speitel et al. 1987) in their biofilm on activated carbon (BFAC) models. Using spherical coordinates, a chemical species from a bulk liquid concentration diffused through a liquid boundary layer into a homogenous biofilm layer. In the biofilm layer, the species diffused through via Fickian diffusion, undergoing reaction according to Monod kinetics. The remaining species then diffused into the activated carbon, described as a homogenous solid with a Freundlich adsorption isotherm describing the transport to the adsorbed phase. Figure 13.3 shows a schematic of the model. This follows very closely with the models for abiotic granular activated carbon outlined in the previous sections.

Since Chang and Rittman's initial work, improvements on the model have been made over the last two decades. The first improvement on the BFAC model used dimensionless numbers to scale the equation from laboratory experiments to full scale water treatment plants, with the goal of eliminating the necessity for costly pilot scale studies (Badriyha et al. 2003). A similar model was used to describe the removal of azo dyes using the BAC process (Lin and Leu 2008). The model

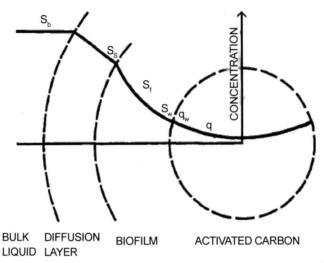

BULK   DIFFUSION   BIOFILM   ACTIVATED CARBON
LIQUID LAYER

Figure 13.3: Conceptual combined biofilm-activated carbon model. Diffusion of a chemical species from the bulk (S_b) passes through a mass transfer boundary layer and into the biofilm layer. Through the homogenous biofilm layer, Fickican diffusion controls species transport (S_f) transport while reaction occurs. Finally, the species transfers to an adsorbed phase (q_w) and goes through diffusion of the homogenous surface. Reprinted with permission from Chang, H.T. and B.E. Rittman. 1987a. Mathematical modeling of biofilm on activated carbon. Environmental Science & Technology 21: 273–280. Copyright 1987 American Chemical Society.

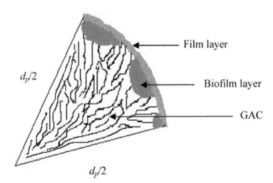

Figure 13.4: It has been proposed that complete coverage of biofilms was not always obtained (Herzberg et al. 2003). Instead, fractional coverage of the GAC by biofilms led to improved performance of BGAC over other reactors. SEM results from other studies (Freidman et al. 2017a) indicate that fractional coverage is observed for SAB degradation by biofilm processes. Reprinted with permission from Herzberg, M., C.G. Dosoretz, S. Tarre and M. Green. 2003. Patchy biofilm coverage can explain the potential advantage of BGAC reactors. Environmental Science & Technology 37: 4274–4280. Copyright 2003 American Chemical Society.

was expanded to include two chemical compounds, both of which are biodegradable and adsorb to GAC, in response to the role of dissolved organic matter competing with target compounds for adsorption sites on GAC (Liang and Chiang 2007, Liang et al. 2007). As such, a Langmuir isotherm was included instead of Freundlich isotherms to represent a maximum capacity on the GAC surface, and to allow for competitive adsorption. One study (Herzberg et al. 2003) noted a significant factor that could describe the over-performance of BGAC and a flaw in most models–partial, rather than complete coverage of GAC with biofilms, as shown in Figure 13.4.

Two studies reviewed current BAC models (Nath and Bhakhar 2011, Shen et al. 2012) and found that several factors were lacking in existing models. The main deficiencies of existing models relate to the disconnection between the advances in describing GAC adsorption and desorption, and the heterogeneous nature of biofilms. The following provides a summary of these equations.

4.1.1 Mathematical description of GAC phase

Generally, the GAC has been described as a homogenous spherical particle, with surface diffusion being used to describe the transport of chemical species through the particle. One study (Lin and Leu 2007) described the surface diffusion equation as:

$$\frac{\partial q}{\partial t} = \frac{D_s}{r^2}\frac{\partial}{\partial r}\left(r^2\frac{\partial q}{\partial r}\right) \tag{41}$$

where r is the particle radial coordinate, q is the loading of the species on the GAC particle, and D_s is the average pore diffusion coefficient. Other models have used linear driving force (LDF) approximation, which can significantly reduce the complexity of the equation, but also the accuracy of the model fit.

The initial and boundary conditions for the surface diffusion equation were defined from the following:

(1) The concentration profile is symmetric with respect to the center of the particle, therefore

$$\frac{\partial q}{\partial r}(0,t) = 0 \tag{42}$$

(2) The net flux from the biofilm/activated carbon interface and the liquid film/activated carbon interface, assuming fractional coverage of the biofilm (Herzberg et al. 2003), must equal the total flux at the activated carbon surface, therefore:

$$D_f\frac{\partial C_f}{\partial z}(0,t) + (1 - f_c)k_b(C_b - C_s) = \frac{\rho_p}{R^2}\frac{\partial}{\partial t}\int_0^R qr^2\,dr \tag{43}$$

where D_f is the diffusivity through the biofilm, C_f is the concentration in the biofilm, z is the thickness of the biofilm. R is the radius of the activated carbon particle and ρ_p is the apparent activated carbon particle density. f_c is the fractional coverage of the biofilm on the activated carbon, defined as:

$$f_c = \frac{A_{covered\ GAC}}{A_{Total}} \tag{44}$$

k_b is the liquid film mass transfer coefficient, C_b is the concentration in the bulk solution, and C_s is the concentration at the activated carbon/liquid film interface. This is found from the third boundary condition:

(3) The relationship between the adsorbed phase and the solution concentration is usually defined by adsorption equilibrium, such as the Freundlich or Langmuir isotherm.

(4) The final initial condition is described as follows:

At $t = 0$, $q = 0$. $\tag{45}$

4.1.2 Mathematical description of the biofilm phase

In existing BFAC models, the biofilm is usually described by a Fickian diffusion/reaction equation:

$$\frac{\partial C_f}{\partial t} = D_f\frac{\partial C_f}{\partial z} + \omega X_f \tag{46}$$

where X_f is the density of the biofilm and ω is the reaction term, and is often defined by Monod kinetics.

$$\omega = \frac{k_f C_f}{K_f + C_f} \tag{47}$$

where k_f is the maximum utilization rate in the biofilm, and K_f is the Monod half-velocity constant. The boundary conditions for the biofilm use the activated carbon and biofilm interface previously described, and the interface with the liquid film is described as:

$$D_f \frac{\partial C_f}{\partial z}(L_f, t) = f_c k_b (C_b - C_f) \tag{48}$$

where L_f is the thickness of the biofilm in the radial direction. A major consideration is the growth of the biofilm, with regards to the thickness and the fractional coverage. Conceptually, the biofilm will grow to a maximum thickness, at which point the decay of cells weakens the attachment to the particle surface, and shear forces are able to detach (or slough) a section of the biofilm. In BFAC models, the following equation is used:

$$\frac{\partial L_f}{\partial t} = \int_0^{L_f} (Y\omega - b_d - b_s) X_f dz \tag{49}$$

where Y is the biomass yield coefficient, b_d is the cell decay coefficient, and b_s is the cell shear coefficient. The fractional coverage is assumed to be known *a priori*, and in some models this is assumed for the biofilm thickness as well. Initial conditions are described as follows:

At $t = 0$, $L_f = L_{f0}$

At $t = 0$, $C_f = 0$

where L_{f0} is known *a priori* and is estimated as the thickness of a single cell just after attachment.

For the bulk phase, a mass balance is performed around a column reactor, and employs the previously described axial dispersion model with sink terms into the biofilm.

$$D \frac{\partial^2 C_b}{\partial x^2} - \bar{v} \frac{\partial C_b}{\partial x} - J_b A_b = \frac{\partial C_b}{\partial t} \tag{50}$$

where D is the dispersion coefficient, x is the axis along the length of the reactor, $\bar{v}$ is the average pore velocity in the reactor, J_b is a sink term for the flux into the particles, and A_b is the particle surface area.

Now,

$$J_b = f_c k_b (C_b - C_f) + (1 - f_c) k_b (C_b - C_s) \tag{51}$$

$$A_b = 4\pi (R + L_f)^2 \tag{52}$$

At $t = 0$, $C_b = 0$.

At $x = 0$, $C_b =$ inlet concentration.

At $x = L$, $\frac{\partial C_b}{\partial t} = 0$,

where L is the length of the bioreactor.

4.2 Current deficiencies

Several major deficiencies exist in these models. The first is the interaction between the biofilm and the activated carbon at time periods when the concentrations on the activated carbon is higher than that on the solution or in other words, desorption. Desorption is predominantly thought to occur via the creation of a chemical gradient, such that the loading of the chemical species is higher than the equilibrium loading given the concentration at the interface. However, this process is thought to be restricted by irreversible (or high energy) adsorption due to chemisorption of the species to the surface, either via functional groups or by oxidation with the banal plane of the activated carbon. In addition, pseudo-hysteresis is often mistaken for irreversibility, when rather the kinetic limitations of desorption are limiting the regeneration of an adsorbent (Limousin et al. 2007). Suggestions of biodegradation overcoming apparent hysteresis via increased chemical gradients (Abromaitis et al. 2016) may have more to do with kinetics rather than thermodynamics.

Indeed, many recent studies on activated carbon differ from BFAC, whereby kinetic models are widely implemented to describe processes rather than equilibrium relationships. In the BFAC model described here, the kinetics of adsorption are broadly described by a solid diffusion model. A better description is the sorbate diffusion through liquid pores, reacting to the surface. In addition, the inclusion of kinetic term to describe the transfer from the adsorbed phase to the dissolved phase may be used to better represent the data. Several kinetic models have been used for adsorption to activated carbon, including the pseudo first order, pseudo second order, and many others. The pseudo second order is shown here, as it generally gives a better fit to data and has a stronger theoretical basis.

$$\frac{\partial C_s}{\partial t} = D_s \left(\frac{\partial^2 C_s}{\partial r^2} + \frac{2}{r} \frac{\partial C_s}{\partial r} \right) - \sigma \frac{\partial q}{\partial t} \tag{53}$$

with the kinetic relationship shown as:

$$\frac{\partial q}{\partial t} = k_2 (q_e - q)^2 \tag{54}$$

where σ is a constant representing a physical parameter, q is the loading at time, t, q_e is the equilibrium loading determined from a thermodynamic relationship (such as a Langmuir isotherm), and k_2 is an empirical constant.

There have been further developments in describing chemical transport through biofilms in porous bodies than is described here. These models have been extended beyond homogenous one-dimensional biofilm descriptions to three-dimensional models describing transport through porous networks, going as far as describing the intracellular transport (Boltz et al. 2010, 2011). Figure 13.5 shows the development of these models from 1975 onwards.

However, for engineering level design, the complexity of these models prevents their use and uptake in designing field-scale equipment and processes. In developing an updated model for a BAC system that can include desorption and bioregeneration, the complexity must be enough to adequately describe and predict the performance of a BAC system, but accessible enough to be adapted to a range of processes. In the case of developing a BAC process for hydrocarbon removal in Antarctica, previous laboratory-scale results have shown relatively thin and uncomplicated biofilm structures (Figure 13.6), making multi-dimensional models of the biofilm unnecessary. In order to develop an appropriate model, an extension of the existing BFAC models to account for hysteresis is the next step in improving the accuracy of the models.

5. Applications of GAC: Case Study of Casey Station, Antarctica

The growing presence of human activity in Antarctica and the sub-Antarctic over the last sixty years has resulted in significant environmental impact, thereby challenging the perception of the continent as a pristine environment. With the adoption of the Environmental Protection Protocol in Madrid, 1998, the development of remediation practices to minimize human impacts has become an important part of Antarctic science programs (Gomez and Sartaj 2014). The historic use of tip sites for disposal of damaging environmental contaminants, alongside ongoing environmental incidents,

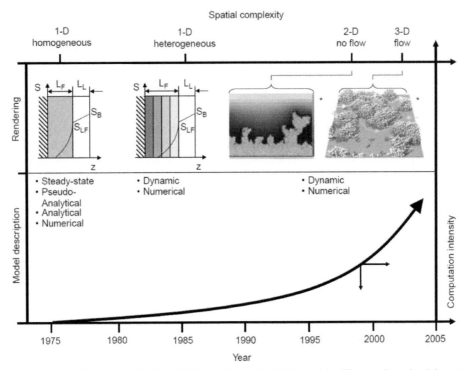

Figure 13.5: Progress of biofilm models since 1975 Generally, existing BFAC models utilize one-dimensional, homogenous descriptions of the biofilm. Extensions may include the stratification of the biofilm and incorporate growth models. Reproduced from Boltz, J.P., E. Morgenroth and D. Sen. 2010. Water Science & Technology 62(8): 1821–1836, with permission from the copyright holders, IWA Publishing.

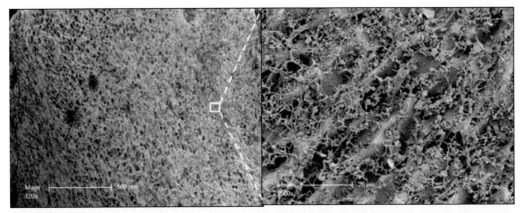

Figure 13.6: Field Emission Scanning Electron Microscopy image of biofilm formation on GC1200 Activated Carbon. Taken from a flow cell experiment designed to replicate the conditions in a Permeable Reactive Barrier. Adapted from Freidman, B.L., S.L. Gras, I. Snape, G.W. Stevens and K.A. Mumford. 2017a. A bio-reactive barrier sequence for petroleum hydrocarbon capture and degradation in low nutrient environments. International Biodeterioration & Biodegradation 116: 26–37. Copyright with permission from Elsevier.

has formed a major legacy of permanent settlements in the ice-free areas of Antarctica. One example of this is the spread of hydrocarbons through soil and meltwater as a result of numerous fuel spills (Camenzuli et al. 2015a). At Casey Station, a major Australian research facility, a number of fuel spill incidents have been recorded over many years. A significant spill occurred in 1999 from the main powerhouse (MPH) on the station (Snape et al. 2001b), while more recently, a 2015 spill was discovered in association with the activation of the emergency powerhouse (McWatters et al. 2016). This has led to the development of hydrocarbon remediation technologies for cold regions, forming an important stream within the Antarctic science program in Australia, and aiding the growth of the capability of all nations to protect the Antarctic environment (Snape et al. 2001).

After the fuel spill event in 1999 at Casey Station, initial research focused on the characterization of the hydrocarbon contamination of the soil. These studies provided information as to the degradation and evaporation potential of the contaminants, revealing that the majority of contaminants at the MPH site were predominantly a combination of Aviation Turbine Kerosene (ATK), Bergen distillate, and Special Antarctic Blend diesel (SAB) (Snape et al. 2005). These fuels are typically comprised of aliphatic hydrocarbons in the n-C9 to n-C18 range (typically ≥ 75%) with limited aromatic and polyaromatic hydrocarbons (PAHs). Approximately 30% of the compounds are unresolved by the GC-FID techniques used. In particular, the presence of naphthalene, methylnaphthalenes, and methylbenzenes in the fuel mixtures represent contaminants of concern.

Significant hurdles in remediating the contaminated soil exist, with *ex situ* biostimulation methods shown to be the most cost-effective (McWatters et al. 2016). However, a major issue is the mobilization of contaminants in meltwater as a result of excavation (Northcott et al. 2003). Excavating can influence the active layer (the soil zone experiencing seasonal freezing) and expose the permafrost, which may contain non-aqueous phase liquids (NAPL) distributed through disconnected zones with relatively high pore saturations. This exposure leads to the melting of permafrost layers and the release of NAPL, creating a high mobilization risk of the contaminants. Containment of the contaminated meltwater is an important aspect in site management to prevent off-site migration to sensitive environmental receptors. The large volumes require treatment prior to discharge, making contaminated water remediation a key issue in excavation operations in Antarctica.

During excavation of an old tip site at Thala Valley (located near Casey Station), it was determined that heavy contamination of meltwater had occurred as a result of the earthworks performed (Snape et al. 2001). Heavy metal ions were mobilized as both dissolved compounds and sorbed onto the surface on suspended solids (Northcott et al. 2003). The first technology designed was an *in situ* Permeable Reactive Barrier (PRB), using ion exchange zeolites to remove heavy metal ions from the contaminated meltwater (Woinarski et al. 2006), prior to discharge into the marine environment. However, recognition of the ecotoxicity of known fuel spills led to the concept of using GAC in the PRB to treat migrating hydrocarbons (Snape et al. 2001a).

5.1 Permeable reactive barriers

Permeable Reactive Barriers are used as a passive, *in situ* technology for the treatment of groundwater. They operate on the principle of placing reactive material in the natural flowpath of a contaminant plume. The reactive material treats the contaminant, while also displaying a hydraulic conductivity much greater than the surrounding area to ensure flow does not bypass the barrier. Treated water is then discharged from the reactive barrier along normal flow paths, minimizing the impact on the hydrogeology of the environment. PRBs have been used to treat a wide range of contaminants, ranging from heavy metals and radionuclides, to halogenated organics, to petroleum hydrocarbons. The effectiveness of treatment depends on the ability of the barrier to capture the water flow and treat the contamination in the given residence time. The choice of reactive materials relies on a range of factors, starting with the type of contaminant and including parameters, such as mechanical stability and the biogeochemical conditions of the flow path (Obiri-Nyarko et al. 2014).

5.2 PRBs for hydrocarbon capture and degradation

The initial application of GAC in permeable reactive barriers in Antarctica focused on the removal of the highly mobile phases of diesel spills—BTEX and PAH compounds. The model contaminants used were naphthalene, o-xylene, and toluene. Initial selection of material was a wood-based GAC activated using phosphoric acid, called Picabiol (Snape et al. 2001a). It was found that the hardness rating was too low when subject to freeze-thaw cycling, and material breakdown occurred. Later experiments using PicaCarb, a thermally activated coconut husk derived GAC, were found to have a significantly better freeze-thaw performance (Gore et al. 2006), with little reduction in grain size in the > 250 μm fraction. Analysis of the new material showed excellent sorption characteristics for BTEX and PAH compounds, with less than a 10% reduction in adsorption capacity at 4°C from temperate conditions (Hornig et al. 2008). The empirical Freundlich isotherm equation was used to describe the material. This work was followed by further experiments using a coconut-husk, thermally activated carbon from the supplier Pica (Arora et al. 2011a). Toluene was used as a model contaminant, and adsorption isotherms were derived in batch and column experiments. For the batch experiments, a modified Freundlich equation was fitted. Fitting of kinetic equations indicated that the carbon was particle diffusion controlled rather than film diffusion controlled. The subsequent fixed-bed column experiments (Arora et al. 2011b) used a diffusion/advection equation to describe transport through the column. This was then adapted to include a reactive term to account for the removal of toluene via adsorption, making an advective/diffusive/reactive equation. The most successful model for describing the kinetics was found to be to use the equilibrium driving force described by sorption isotherms, fitted with a Langmuir kinetic equation.

However, adsorption-based reactive barriers require changeout of material as saturation and bed breakthrough can occur. Given the relatively high cost to transport materials to Antarctic regions and high labor costs (Snape et al. 2001), this was not ideal. As such, the concept of a biotic barrier or permeable bioreactive barrier was developed and installed (Mumford et al. 2013), whereby microbial communities utilize the sorbed fuel in biodegradation process alongside the provision of additional nutrients. These communities are able to mineralize a wide range of contaminants, but require specific conditions to maximize their growth. To achieve this, five side-by-side barriers were installed and zoned into a nutrient release zone at the front, a hydrocarbon sorption/degradation zone in the middle, and an ion-exchange zone for removal of excess nutrients at the rear. Each barrier consisted of different materials or ratio of materials to compare performances in each conceptual zone. Upon subsequent analysis, it was found that degradation appeared to be occurring in the nutrient release zone, and in one case in the control material (Mumford et al. 2014). It appeared that degradation was slowed by the presence of the hydrocarbon adsorptive material (in all cases, GAC) rather than accelerated, suggesting that the adsorbed fuel was not readily available to the fuel degrading microbes. Rather, the biofilm communities established on the nutrient release materials were able to access the fuel from the bulk water flow, displaying higher numbers of heterotrophs and better degradation indices than the sorption materials (Mumford et al. 2015). This is an important result, as it significantly affects the choice of materials in future bioreactive barriers, and underlines a differentiation of the PRBs to analogous GAC filters in wastewater treatment, which can display biofilm growth as BAC. Notably, the presence of heterotrophs does not indicate the regeneration of the activated carbon-only that some of the contaminants were being degraded prior to adsorption.

These results were then expanded to the development of controlled release nutrient (CRN) materials for use as bioactive materials to degrade mobile hydrocarbons in PRBs (Freidman et al. 2016). The natural zeolite previously used in removing heavy metals and ammonium was modified by loading the surface with ammonium cations. These materials then acted as a support for thin film biofilms while providing a steady supply of ammonium via ion exchange. Compared to controls with unamended natural zeolites, biofilm formation was shown to be accelerated via a reduction in the lag phase of the biomass growth, while removal of hydrocabons from long pulse influent flow was shown to be due to biofilm formation, and not adsorption to the zeolite surfaces. Combining these materials with GAC in a sequenced media PRB showed biofilm development on the CRN, but little on the GAC (Freidman et al. 2017a). Field emission scanning electron microscopy (FESEM) showed no biofilm development, and biodegradation indices for diesel (Snape et al. 2006) showed no degradation in the pores. However, some desorption of naphthalene was observed. This has promise for the promotion of biofilm degradation of similar contaminants of concern, with desorption being the key process. It also raises the issue of contaminant remobilization with pulsed contaminant concentrations due to the diurnal freeze-thaw cycles of cold regions. Most notably from these studies, it was found that PRBs at Casey Station developed endemic microbial communities capable of degrading hydrocarbons (Friedman et al. 2017b)—namely *Burkholderiales* (*Betaproteobacteria*). This matches with some previous research, indicating that this order is dominant in BAC processes (Niemi et al. 2009). This order has been found to be capable of degrading aromatic compounds (Pérez-Pantoja et al. 2012) and efficient at mineralizing dissolved organic matter.

The issue of the lag phase of biofilm development in the bioactive PRB requires the presence of an adsorptive media as a safety net to ensure no discharge of contaminants through the PRB. The development of a biological activated carbon, which utilizes adsorption followed by bioregeneration is an ideal material, but has only been indicated and not directly observed and modeled. The current gap in knowledge for bioactive PRB design is the lack of a mathematical model capable of describing the performance of materials in sustaining biomass, and the expected degradation rates that can be achieved. Further work for both PRB development at Casey Station, Antarctica is the use of general mathematical models to describe GAC performance for broad multicomponent contamination of SAB diesel (analogous to DOM adsorption on GAC), and specific micropollutants, such as naphthalene and 1,3,5-trimethylbenzene. Incorporating biofilm degradation into the model and using constant influent contaminant concentrations in fundamental experiments will help to refine the performance of GAC in environmental remediation applications in cold regions.

6. Conclusion

This chapter has focused on and described the principles of granular activated carbon in water treatment. Granular activated carbon adsorbents are capable of removing a wide range of chemical compounds, with excellent performance in removing hydrophobic, low concentration chemical species. As such, they are a strong candidate for the removal of organic micropollutants in a range of applications, from drinking water treatment in potential potable reuse to environmental remediation applications. The basic principles of adsorption kinetics and equilibrium have been discussed, alongside predictive modeling. These principles have then been used in outlining two models for determining process parameters for the design of full-scale adsorption columns, as well as predict existing adsorption column performance for newly identified contaminants of concern.

The role of biological activity and development of biological activated carbon filters has also been discussed, with a mathematical model proposed for describing the performance of these filters. Current deficiencies in the models have been identified. Finally, a case study of granular activated carbon used in permeable reactive barriers in Antarctica was presented as a unique application in environmental remediation.

Acknowledgments

The authors acknowledge the research financial support provided from the Australian Antarctic Science Project 4029 and 4036 and The Particulate Fluids Processing Centre.

Abbreviations

AOPs	:	Advanced oxidation processes
ATK	:	Aviation Turbine Kerosene
BV	:	Bed volumes
BFAC	:	Biofilm on activated carbon

BAC : Biological Activated Carbon
CUR : Carbon Usage Rate
CECs : Contaminants of emerging concern
CRN : Controlled release nutrient
DBPs : Disinfection By-Products
EBCT : Empty bed contact time
FESEM : Field emission scanning electron microscopy
GAC : Granular Activated Carbon
HSDM : Homogenous surface diffusion model
HLR : Hydraulic Loading Rate
LDF : Linear Driving Force
MPH : Main powerhouse
MTZ : Mass transfer zone
NOM : Natural Organic Matter
NAPL : Non-aqueous phase liquids
OMPs : Organic micropollutants
PRB : Permeable Reactive Barrier
PAHs : Polyaromatic hydrocarbons
PSDM : Pore surface diffusion model
RSSCTs : Rapid Small-Scale Column Tests
SEM : Scanning electron microscopy
SAB : Special Antarctic Blend diesel
SOC : Synthetic organic compound
WWTPs : Wastewater treatment plants

References

Abromaitis, V., V. Racys, P. van der Marel and R.J.W. Meulepas. 2016. Biodegradation of persistent organics can overcome adsorption–desorption hysteresis in biological activated carbon systems. Chemosphere 149: 183–189.

Aktaş, O. and F. Çeçen. 2007. Bioregeneration of activated carbon: A review. International Biodeterioration & Biodegradation 59: 257–272.

Arora, M., I. Snape and G.W. Stevens. 2011a. The effect of temperature on toluene sorption by granular activated carbon and its use in permeable reactive barriers in cold regions. Cold Regions Science and Technology 66: 12–16.

Arora, M., I. Snape and G.W. Stevens. 2011b. Toluene sorption by granular activated carbon and its use in cold regions permeable reactive barrier: Fixed bed studies. Cold Regions Science and Technology 69: 59–63.

Badriyha, B.N., V. Ravindran, W. Den and M. Pirbazari. 2003. Bioadsorber efficiency, design, and performance forecasting for alachlor removal. Water Research 37: 4051–4072.

Barbosa, M.O., N.F.F. Moreira, A.R. Ribeiro, M.F.R. Pereora and A.M.T. Silva. 2003. Occurrence and removal of organic micropollutants: An overview of the watch list of EU Decision 2015/495. Water Research 94: 257–279.

Barbosa, M., N. Moreira, A. Riberio, M. Pereira and A. Silva. 2016. Occurence and removal of organic micropollutants: An overview of the watchlist of EU Decision 2015/495. Water Research 94: 257–279.

Boehm, H.P. 1994. Some aspects of the surface chemistry of carbon blacks and other carbons. Carbon 32: 759–796.

Boltz, J.P., E. Morgenroth, D. Brockmann and D. Sen. 2010. Mathematical modelling of biofilms and biofilm reactors for engineering design. Water Science & Technology 62: 1821–1836.

Boltz, J.P., E. Morgenroth, D. Brockmann, C. Bott, W.J. Gellner and P.A. Vanrolleghem. 2011. Systematic evaluation of biofilm models for engineering practice: components and critical assumptions. Water Science & Technology 64: 930–944.

Bunmahotama, W., W.N. Hung and T.F. Lin. 2015. Predicting the adsorption of organic pollutants from water onto activated carbons based on the pore size distribution and molecular connectivity index. Water Research 88: 521–531.

Camenzuli, D. and B.L. Freidman. 2015a. On-site and *in situ* remediation technologies applicable to petroleum hydrocarbon contaminated sites in the Antarctic and Arctic. Polar Research 34: 24492.

Camenzuli, D., K.A. Fryirs, D.B. Gore and B.L. Freidman. 2015b. Managing legacy waste in the presence of cultural heritage at Wilkes Station, East Antarctica. Polar Record 51: 151–159.

Çeçen, F. and O. Aktaş. 2012. Activated Carbon for Water and Wastewater Treatment: Integration of Adsorption and Biological Treatment. Weinheim: Wiley-VCH.

Chang, H.T. and B.E. Rittman. 1987a. Mathematical modeling of biofilm on activated carbon. Environmental Science & Technology 21: 273–280.

Chang, H.T. and B.E. Rittman. 1987b. Verification of the model of biofilm on activated carbon. Environmental Science & Technology 21: 280–288.

Clark, R.M. and B.W. Lykins Jr. 1989. Granular Activated Carbon: Design, Operation, and Cost. Chelsea, Mich.: Lewis Publishers.

Corwin, C.J. and R.S. Summers. 2010. Scaling trace organic contaminant adsorption capacity by granular activated carbon. Environmental Science & Technology 44: 5403–5408.

Crittenden, J.C., J.K. Berrigan and D.W. Hand. 1986. Design of rapid small-scale adsorption tests for a constant diffusivity. Journal (Water Pollution Control Federation) 58: 312–319.

Crittenden, J.C., D.W. Hand, H. Arora and B.W. Lykins. 1987. Design considerations for GAC treatment of organic chemicals. Journal American Water Works Association 79: 74–82.

Crittenden, J.C., S. Sanongraj, J.L. Bulloch, D.W. Hand, T.N. Rogers, T.F. Speth et al. 1999. Correlation of aqueous-phase adsorption isotherms. Environmental Science & Technology 33: 2926–2933.

Crittenden, J.C. and J.H. Borchardt. 2012. MWH's Water Treatment: Principles and Design. Hoboken, N.J.: John Wiley & Sons.

Dąbrowski, A., P. Podkościelny, Z. Hubicki and M. Barczak. 2005. Adsorption of phenolic compounds by activated carbon—a critical review. Chemosphere 58: 1049–1070.

Foo, K.Y. and B.H. Hameed. 2010. Insights into the modeling of adsorption isotherm systems. Chemical Engineering Journal 156: 2–10.

Freidman, B.L., S.L. Gras, I. Snape, G.W. Stevens and K.A. Mumford. 2016. The performance of ammonium exchanged zeolite for the biodegradation of petroleum hydrocarbons migrating in soil water. Journal of Hazardous Materials 313: 272–282.

Freidman, B.L., S.L. Gras, I. Snape, G.W. Stevens and K.A. Mumford. 2017a. A bio-reactive barrier sequence for petroleum hydrocarbon capture and degradation in low nutrient environments. International Biodeterioration & Biodegradation 116: 26–37.

Freidman, B.L., L.B.M. Speirs, J.G. Churchill, S.L. Gras, J. Tucci, I. Snape et al. 2017b. Biofilm communities and biodegradation within permeable reactive barriers at fuel spill sites in Antarctica. International Biodeterioration & Biodegradation 125: 45–53.

Gomez, F. and G. Sartaj. 2014. Optimization of field scale biopiles for bioremediation of petroleum hydrocarbon contaminated soil at low temperature conditions by response surface methodology (RSM). International Biodeterioration & Biodegradation 89: 103–109.

Gore, D.B., E.S. Heiden, I. Snape, G. Nash and G.W. Stevens. 2006. Grain size of activated carbon, and untreated and modified granular clinoptilolite under freeze-thaw: applications to permeable reactive barriers. Polar Record 42: 121–126.

Hand, D.W., J.C. Crittenden, D.R. Hokanson and J.L. Bulloch. 1997. Predicting the performance of fixed-bed granular activated carbon adsorbers. Water Science and Technology 35: 235–241.

Herzberg, M., C.G. Dosoretz, S. Tarre and M. Green. 2003. Patchy biofilm coverage can explain the potential advantage of BGAC reactors. Environmental Science & Technology 37: 4274–4280.

Hornig, G., K. Northcott, I. Snape and G.W. Stevens. 2008. Assessment of sorbent materials for treatment of hydrocarbon contaminated ground water in cold regions. Cold Regions Science and Technology 53: 83–91.

Hung, H.W. and T.F. Lin. 2007. Prediction of the adsorption capacity for volatile organic compounds onto activated carbons by the Dubinin–Radushkevich–Langmuir model. Journal of the Air & Waste Management Association 57: 497–506.

Karanfil, T. and J.E. Kilduff. 1999. Role of granular activated carbon surface chemistry on the adsorption of organic compounds. 1. Priority pollutants. Environmental Science & Technology 33: 3217–3224.

Karanfil, T., M. Kitis, J.E. Kilduff and A. Wigton. 1999. Role of granular activated carbon surface chemistry on the adsorption of organic compounds. 2. Natural organic matter. Environmental Science & Technology 33: 3225–3233.

Kennedy, A.M., A.M. Reinert, D.R.U. Knappe and R.S. Summers. 2017. Prediction of full-scale GAC adsorption of organic micropollutants. Environmental Engineering Science 34: 496–507.

Li, L., P.A. Quinlivan and D.R.U. Knappe. 2005. Predicting adsorption isotherms for aqueous organic micropollutants from activated carbon and pollutant properties. Environmental Science & Technology 39: 3393–3400.

Liang, C.H. and P.C. Chiang. 2007. Mathematical model of the non-steady-state adsorption and biodegradation capacities of BAC filters. Journal of Hazardous Materials 139: 316–322.

Liang, C.H., P.C. Chiang and E.E. Chang. 2007. Modeling the behaviors of adsorption and biodegradation in biological activated carbon filters. Water Research 41: 3241–3250.

Limousin, G., J.P. Gaudet, L. Charlet, S. Szenknect, V. Barthès and M. Krimissa. 2007. Sorption isotherms: A review on physical bases, modeling and measurement. Applied Geochemistry 22: 249–275.

Lin, Y.H. and J.Y. Leu. 2007. Kinetics of reactive azo-dye decolorization by Pseudomonas luteola in a biological activated carbon process. Biochemical Engineering Journal 39: 457–467.

Marsh, H. and F. Rodríguez-Reinoso. 2006. Activated Carbon. San Diego, CA.: Elsevier.

McWatters, R.S., D. Wilkins, T. Spedding, G. Hince, B. Raymond, G. Lagerewski et al. 2016. On site remediation of a fuel spill and soil reuse in Antarctica. Science of the Total Environment 571: 963–973.

Mumford, K.A., J.L. Rayner, I. Snape, S.C. Stark, G.W. Stevens and D.B. Gore. 2013. Design, installation and preliminary testing of a permeable reactive barrier for diesel fuel remediation at Casey Station, Antarctica. Cold Regions Science and Technology 96: 96–107.

Mumford, K.A., J.L. Rayner, I. Snape and G.W. Stevens. 2014. Hydraulic performance of a permeable reactive barrier at Casey Station, Antarctica. Chemosphere 117: 223–231.

Mumford, K.A., S.M. Powell, J.L. Rayner, G. Hince, I. Snape and G.W. Stevens. 2015. Evaluation of a permeable reactive barrier to capture and degrade hydrocarbon contaminants. Environmental Science and Pollution Research 22: 12298–12308.

Nath, K. and M.S. Bhakhar. 2011. Microbial regeneration of spent activated carbon dispersed with organic contaminants: mechanism, efficiency, and kinetic models. Environmental Science and Pollution Research International 18: 534–546.

Niemi, M.R., I. Heiskanen, R. Heine and J. Rapala. 2009. Previously uncultured b-Proteobacteria dominate in biologically active granular activated carbon (BAC) filters. Water Research 43: 5075–5086.

Northcott, K.A., I. Snape, M.A. Connor and G.W. Stevens. 2003. Water treatment design for site remediation at Casey Station, Antarctica: site characterisation and particle separation. Cold Regions Science and Technology 37: 169–185.

Obiri-Nyarko, F., S.J. Grajales-Mesa and M. Malina. 2014. An overview of permeable reactive barriers for *in situ* sustainable groundwater remediation. Chemosphere 111: 243–259.

Pérez-Pantoja, D., R. Donoso, L. Agulló, M. Córdova, M. Seeger, D.H. Pieper et al. 2012. Genomic analysis of the potential for aromatic compounds biodegradation in Burkholderiales. Environmental Microbiology 14: 1091–1117.

Pikaar, I., A.A. Koelmans and P.C.M. van Noort. 2006. Sorption of organic compounds to activated carbons. Evaluation of isotherm models. Chemosphere 65: 2343–2351.

Plazinski, W., W. Rudzinski and A. Plazinska. 2009. Theoretical models of sorption kinetics including a surface reaction mechanism: A review. Advances in Colloid and Interface Science 152: 2–13.

Plazinski, W., J. Dziuba and W. Rudzinski. 2013. Modeling of sorption kinetics: the pseudo-second order equation and the sorbate intraparticle diffusivity. Adsorption 19: 1055–1064.

Sellaoui, L., H. Guedidi, SarraWjihi, L. Reinert, S. Knani, Duclaux, Lau. et al. 2007. Experimental and theoretical studies of adsorption of ibuprofen on raw and two chemically modified activated carbons: new physicochemical interpretations. RSC Advances 6: 12363–12373.

Shen, L., Y. Lu and Y. Liu. 2012. Mathematical modeling of biofilm-covered granular activated carbon: a review. Journal of Chemical Technology & Biotechnology 87: 1513–1520.

Snape, I., C.E. Morris and C.M. Cole. 2001a. The use of permeable reactive barriers to control contaminant dispersal during site remediation in Antarctica. Cold Regions Science and Technology 32: 157–174.

Snape, I., M.J. Riddle, J.S. Stark, C.M. Cole, C.K. King, S. Duquesne et al. 2001b. Management and remediation of contaminated sites at Casey Station, Antarctica. Polar Record 37: 199–214.

Snape, I., P.M. Harvey, S.H. Ferguson, J.L. Rayner and A.T. Revill. 2005. Investigation of evaporation and biodegradation of fuel spills in Antarctica I. A chemical approach using GC–FID. Chemosphere 61: 1485–1494.

Snape, I., S.H. Ferguson, P.M. Harvey, J.L. Rayner and M.J. Riddle. 2006. Investigation of evaporation and biodegradation of fuel spills in Antarctica: II—Extent of natural attenuation at Casey Station. Chemosphere 63: 89–98.

Speitel, G., K. Dovantzis and F. DiGiano. 1987. Mathematical modeling of bioregeneration in GAC columns. Journal of Environmental Engineering 113: 32–48.

Titirici, M. and M. Antonietti. 2010. Chemistry and materials options of sustainable carbon materials made by hydrothermal carbonization. Chemical Society Reviews 39: 103–116.

Urano, K., S. Omori and E. Yamamoto. 1982. Prediction method for adsorption capacities of commercial activated carbons in removal of organic vapors. Environmental Science & Technology 16: 10–14.

Weber, W.J., P. Pirbazari and G. Melson. 1978. Biological growth on activated carbon: an investigation by scanning electron microscopy. Environmental Science & Technology 12: 817–819.

Woinarski, A.Z., G.W. Stevens and I. Snape. 2006. A natural zeolite permeable reactive barrier to treat heavy-metal contaminated waters in Antarctica: Kinetic and fixed-bed studies. Process Safety and Environmental Protection 84: 109–116.

Yin, C.Y., M.K. Aroua and W.M.A.W. Daud. 2007. Review of modifications of activated carbon for enhancing contaminant uptakes from aqueous solutions. Separation and Purification Technology 52: 403–415.

Ying, W.C. and W.J. Weber. 1979. Bio-physicochemical adsorption model systems for wastewater treatment. Journal (Water Pollution Control Federation) 51: 2661–2677.

CHAPTER 14

Carbon-based and Related Nanomaterials as Active Media for Analytical, Biomedical, and Wastewater Processing Applications

Paweł K. Zarzycki,[1,]* Renata Świderska-Dąbrowska,[1] Krzysztof Piaskowski,[1] Lucyna Lewandowska,[1] Bożena Fenert,[1] Katarzyna A. Mitura[2] and Michał J. Baran[1]

1. Introduction and Problems Overview

Pure carbon and carbon-related nanomaterials may be considered the most versatile particles that are presently studied in physics, chemistry, and bioengineering. Extensive research focusing on classical and new carbon particles is being carried out worldwide, and a number of new commercial products have been successfully implemented in different industries and technology areas, including electronic, medicine/pharmacy, cosmetic, food and agriculture, wastewater treatment or environmental protection, as well as civil engineering (Grumezescu 2017). From principles, carbon nanoparticles are easy to functionalize with various organic ligands and therefore, there are virtually an unlimited number of derivatives available, which can be adapted to the given application (Sun and Lei 2017). Particularly, carbon nanoparticles based on fullerenes, nanotubes, graphene, as well as nanodiamonds or carbon dots (CDs) are presently of great interest due to their relatively low toxicity, biocompatibility, easy synthesis, and unique physicochemical properties (Grumezescu and Holban 2018, Neethirajan et al. 2018). On the other hand, low-molecular mass carbon chemicals, such as fullerenes, are extremely difficult to analyze using classical chromatographic or electrophoretic methods. This is due to low solubility of such substances in both polar and non-polar organic solvents and the presence of higher

[1] Department of Environmental Technologies and Bioanalytics, Faculty of Civil Engineering, Environmental and Geodetic Sciences, Koszalin University of Technology, Śniadeckich 2, 75-453 Koszalin, Poland.
[2] Department of Biomedical Engineering Division, Faculty of Technology and Education, Koszalin University of Technology Śniadeckich 2, 75-453 Koszalin, Poland.
* Corresponding author: pkzarz@wp.pl

fullerenes in different isomeric forms (Jinno et al. 1993, Ohta et al. 1994, Saito et al. 1995). However, it is noteworthy to say that these problems have strongly stimulated discovery of novel stationary phases in chromatography, as well as elaboration of new determination protocols for qualitative/ quantitative analysis of such objects, also as micropollutants in environmental samples (Jinno 1999, Zarzycki et al. 2011). Most recently, we demonstrated that fullerenes can be sensitively detected using new signal processing technique involving pictures correlation methodology (voxels-volumetric pixels-correlation in time domain) (Suszynski and Zarzycki 2015).

Unique physicochemical properties of carbon-based nanoparticles and hybrid materials enable the creation of new versatile and miniaturized sensing systems. Carbon nanoparticles have, from principles, high surface area to volume ratio and a large surface area that can be functionalized with organic ligands and fluorescent molecules. On the other hand, carbon nanoparticles may act as fluorescent quantum dots. Therefore, these nanoobjects can be used for sensing applications, including low-molecular mass biomarkers and large objects such as cell structures, bacterial or fungal pathogens, as well as tissues and large biological arrangements. Particularly, a number of applications were reported using covalently bound fluorophores to core-shell nanoparticles, which were derivatives of common dyes: fluorescein, rhodamine, coumarin, and cyanine (Mitura and Włodarczyk 2018). Table 14.1 contains several examples of review and experimental papers focusing on analytical applications of carbon-based particles, enabling efficient sensing of bioanalytes, active compounds, and micropollutants in complex food, environmental, and biological samples, including cells and tissues. A literature review clearly indicated a growing interest in research focusing on the preparation of carbon nanoparticles. It was demonstrated that sensing nanoobjects, including CDs, can be efficiently and non-expensively produced from various biomass and waste materials. Particularly, carbon nanoparticles have been synthesized using raw proteins from chicken egg whites, fruit extracts, citrus peels, berries, plants, leaves, and different tissues, pollen, as well as algae biomass (Atchudan 2017, Baig and Chen 2017, Chatzimitakos et al. 2017, Kumawat et al. 2017, Liu et al. 2017a, Zhan et al. 2017, Zheng et al. 2017, Bandi et al. 2018, Ren et al. 2018, Vasimalai et al. 2018, Zhang et al. 2017a). Sensing carbon nanomaterials are primarily working as chemiluminescence, fluorescence, or photoluminescence markers, also enabling advanced cell staining and three-dimensional and/or multicolor intracellular imaging (Liu et al. 2018a, Neburkowa et al. 2017, Ye et al. 2017, Lien et al. 2012, Atchudan et al. 2017, Kumawat et al. 2017, Zhang et al. 2017a, Bandi et al. 2018). The second main detection approach is electrochemical detection working in different modes (Ju 2018, Liu et al. 2018a, Yang et al. 2018, Laurila et al. 2017). Biosensing devices involving carbon-derived nanomaterials may efficiently determine and/or quantify a wide range of analytes, including metals ions, low molecular mass inorganic and organic compounds, large proteins and intracellular structures, cells, microorganisms, and tissue patterns (Table 14.1). Low-molecular mass bioanalytes, biomarkers, and micropollutants can be detected and determined in various matrices, including complex biological and environmental samples.

The next issue is separation, determination, and quantification of low-molecular mass carbon-based nanoparticles, namely fullerenes and their derivatives, in complex mixtures. This problem was mentioned above and is extensively discussed and explained by Yoshihiro Saito and team within the chapter of this book entitled "*Fullerenes and Polycyclic Aromatic Hydrocarbons in Separation Science*". Here, we would like to highlight that there is a fundamental analytical problem with separation of fullerenes as target chemicals due to extremely low volatility and solubility of carbon nanoparticles in common organic liquids as well as the number of stereoisomeric forms of fullerenes available. The consequence of this is the extremely difficult separation process of these particles using gas chromatography or liquid phase separation systems, including high-performance liquid chromatography and/or electrophoresis. Typical chromatographic approach involves reversed-phase (RP) separation system consisting of octadecylsilane (C-18) or different *n*-alkane (e.g., C-30) bounded column as well as mobile phase that is also composed of *n*-alkanes, typically *n*-pentane, *n*-hexane, or *n*-heptane. Under such conditions, it has been found that temperature plays a critical role in achieving the efficient separation of fullerenes, and this separation effect can be particularly observed

Table 14.1: Detection of target chemicals, biomarkers, and micropollutants in various complex matrices using carbon originated nanomaterials.

No.	Active carbon nanomaterial type or different active chemicals	Target molecules, analytical matrices	Main analytical, quantification or sensing approach	Applications	Reference
1.	Nanodiamonds, fullerenes, carbon nanotubes (MWCNTs), graphene	Organic molecules, lipid memebranes, human blood serum, natural water, sea water, estrogenic pollutants, flavonoids, proteins, peptides, aromatic and non-aromatic amines, water contaminations, parabens in cosmetic products, organic dyes	Solid-phase extraction, chromatography, electrophoresis, electrochemistry, electro-oxidation	Bioanalytics, analytical chemistry, laboratory diagnostics	Review paper (Scida et al. 2011)
2.	Carbon nanotubes (MWCNTs), graphene, carbon nanoparticles, carbonaceous materials, carbon nanohorns (SWCNHs)	Pesticides, mycotoxin, antibiotics, fluoroquinolones, tetracyclines, penicillins, cephalosporins, sulfonamides, macrolides, amphenicols, benzimidazoles, B-agonists, glibenclamide, metronidazole	SPE, SPME, electrophoresis, HPLC with UV/Vis detection, electrochemical sensors, carbon nanomaterial based biosensor using enzyme, antibody or aptamer with high selectivity and sensitivity, optical detection sensors	Quantification of pharmaceutics in food products, food protection before food-borne pathogens and diseases	Review paper (Socas-Rodriguez et al. 2017)
3.	Organic polymers, organic reagents and various chromatographic stationary phases	Graphene, graphene oxide, carbon nanotubes, fullerenes, carbon nanomaterials	Target components enrichment and separation using chromatography and complementary analytical approaches including microscopic, scattering, diffraction, spectroscopic protocols	Environmental and biological matrices	Review paper (Hu et al. 2016)
4.	Carbon nanotubes (CNTs), carbon nanohorns, functionalized carbon nanofibers (CNF), three dimensional ordered graphitized mesoporous carbon (GMC-6, GMC-13), carbon dots, graphene oxide (GO), nitrogen-doped carbon nanotubes (NCNTs), carbon nanospheres, reduced graphene oxide (ERGO), single - or multiwalled CNTs (SWNTs), CNTs modified by α-N-acetylgalactosmina, graphene nanoribbon (GNR)	Various organic macromolecules and microbes, immobilized enzymes, pathogen identification, glucose oxidase (GOD), cellular markers of diseases, *Helix pomatia agglutinin* (HPA), intracellular microRNAs, (miRNAs), hemogloin, NADH, dehydrogenase substrates, glucose, ethanol, acetylthiocholine, phenol, hydrogen peroxide	Bioelectronic and biosensing devices, electrochemical biosensors, fluorescent sensing and detection, FRET, ECL, luminescent labels	Medical diagnostics and bioanalysis *in vitro* and *in vivo*, cytosensing, development of amplified biosensing strategies	Review paper (Ju 2018)

5.	Carbon dots (CDs), carbon nanotubes, graphene, ordered mesoporous carbon, carbon dots, graphene dots, silver and gold nanoparticles	Food contaminations (heterocyclic amines, macrolides) histamine, bisphenol A (in vegetable oil and mayonnaise sample), hydrazine and nitrite, organophosphorus pesticides, patogenic bacteria	Fluorescence sensing, colorimetric detection, electrochemical sensing, CDs-based biosensors, mesoporous carbon biosensors, modified nanocarbon liquid paste electrodes, plasmonic nanomaterial-based surface enhanced Raman scattering technology, chromatographic methods including SPE, HPLC, SPE, mass spectrometry detectors (MS)	Analysis of main compounds and active substances as well as micropollutants in various food products, safety of food products	Review paper (Liu et al. 2018a)
6.	Carbon nanomaterials-based nucleic acid aptamer functionalized particles, carbon dots, carbon nanohorns, carbon nanotube assemblies, graphene, nanocarbon hybrid assemblies	Cancer relevant biomarkers including nucleic acids, protein and cells	Electrochemical sensing devices	Biomedicine, detection of cancer relevant biomolecules, mobile clinic diagnostic devices in various medical fields	Review paper (Yang et al. 2018)
7.	Tailor made surfaces fabricated by combining different allotropic form of carbon, diamond-like carbon, carbon nanotubes and nanofibers, reduced graphene oxide	Various key neurotransmitters in the mammalian central nervous system including dopamine, norepinephrine, glutamate, tryptophan, seretonin, tyrosine, adenosine, glucose, cysteine, GABA, choline, uric acid, ascorbic acid and homovanilic acid	Electrochemical detection in various modes	Biomolecules detection and quantification	Review paper (Laurila et al. 2017)
8.	Coating nanodiamonds with biocompatible shells	Low-molecular mass biomarkers and proteins	Fluorescence, transmission electron microscope (TEM)	Cells imaging, Biology and medicine	Review paper (Neburkowa et al. 2016)
9.	Carbon nanotubes	Bioanalytes, biomarkers and micropollutants, proteins, phosphopeptides, pharmaceutics, naproxen, alkaloids, strychnine, pseudoephedrine, polycyclic aromatic hydrocarbons, human urine and blood	Extraction and purification techniques	Sorbents for analytical protocols, extraction techniques for bioanalysis, pharmaceutical and environmental analysis	Review paper (Ahmadi et al. 2017)
10.	Carbon dots	Various low-molecular mass nutritional compounds, food additives, aluminum (Al^{3+}), mercury (Hg^{2+}) and copper (Cu^{2+}) ions, foodborne microbes	Fluorescence sensing, chemiluminescence	Food analysis metal ion monitoring, food borne pathogens detection, organic pollutants and food nutritional components analysis	Review paper (Ye et al. 2018)

Table 14.1 contd. ...

...Table 14.1 contd.

No.	Active carbon nanomaterial type or different active chemicals	Target molecules, analytical matrices	Main analytical, quantification or sensing approach	Applications	Reference
11.	Fluorescent and magnetic nanodiamonds (FMNDs) prepared using a microwave-arcing process	Cells tissues, lung cancer cells	Flow cytometer with cells sorting functions, magnetic separation devices, fluorescence, FMND-bearing cancer cells, biomarkers detection	Cancer cell labeling and tracking, biomedical applications	Experimental paper (Lien et al. 2012)
12.	Fluorescent nanodiamonds (FNDs) in polymeric matrices/diacryloyl fluorescein (AcFl) produced through a metal-free photo-initiated RAFT process	Various metabolites in cells and tissues samples	Fluorescent probes, green fluorescence	Cell imaging and biomedical applications	Experimental paper (Chen et al. 2018a)
13.	Magnetic carboxylated nanodiamonds (MCNDs) synthesized using surface functionalization with carboxyl groups and gaining magnetic feature of carboxylated nanodiamonds by electrostatic interaction with Fe_3O_4 particles	Ziram (zincdimethyldithiocarbamate)	Separation and pre-concentration using vortex-assisted magnetic solid-phase extraction, indirect flame atomic absorption spectrometric detection	Analysis of foodstuff (rice, cracked wheat) and water samples (tap, lake) as well as various synthetic mixtures	Experimental paper (Yilmaz and Soylak 2016)
14.	Nitrogen doped carbon dots (N-CDs) produced from *Chionanthus retusus* fruit extract	Metal ions (Fe^{3+}), microbes detection (*Candida albicans* and *Cryptococcus neoformans*)	Fluorescence, fluorescence quenching, laser scanning microscopy, confocal fluorescent microscopy	Metal Ion sensing and biological applications, bioimaging, biodetection	Experimental Paper (Atchudan et al. 2017)
15.	Carbon dots generated from chicken egg whites	Bacteria (*Staphylococcus aureus* and *Escherichia coli*) and organic compound (curcumin) detection	Fluorescence, Förster resonance energy transfer	Multicolor labelling of bacteria, microbes detection, active components detection in complex turmeric powder and condensed turmeric tablets	Experimental Paper (Baig and Chen 2017)
16.	Carbon quantum dots from Citrus peels	Cell imaging, determination of iron (Fe^{3+}) and tartrazine	Fluorescence	Cell bioimaging and intracellular detection of iron	Experimental Paper (Chatzimitakos et al. 2017)

No.	Material	Application	Technique	Use	Reference
17.	Graphene quantum dots from *Mangifera indica* leaves as a carbon source	Cells analysis, cells cytoplasm	Near-infrared fluorescence, red fluorescence, nanothermometry, flow cytometry, cells staining, 3-D intracellular imaging	NIR-responsive fluorescent bioimaging, temperature sensing, biomedical nanotechnology, live intracellular temperature sensing	Experimental paper (Kumawat et al. 2017)
18.	Carbon dots from carrot juice	Cellular imaging	Fluorescence, photoluminescence	Multicolor *in vitro* imaging, biomedical applications	Experimental paper (Liu et al. 2017a)
19.	Carbon dots from water chestnut	Various structures imaging in living cells (HepG2 cells)	Fluorescence, CLSM microscopy	*In vitro* and *in vivo* bioimaging applications in cell biology	Experimental paper (Zhan et al. 2017)
20.	Nitrogen- and sulfur-co-doped carbon dots obtained from algae biomass	*Arabidopsis* guard cells and root tissues imaging	Multicolor luminescence, laser scanning confocal microscopy, fluorescence	Cells and tissues imaging, bioimaging in plant cells	Experimental paper (Zhang et al. 2017a)
21.	Pollen derived carbon dots	Bioimaging and monitoring of nitrogen, phosphorus and potassium uptake in *Brassica parachinensis* L.	Fluorescence, confocal images	*In vitro/in vivo* bioimaging, monitoring the uptake of active compounds from nutrient solutions	Experimental paper (Zheng et al. 2017)
22.	Nitrogen-doped carbon dots from *Lantana camara* berries	Detection of lead (Pb^{2+}) in live cells, water, human serum and urine samples	Multicolor fluorescence, confocal microscopy	Toxicology, bioimaging	Experimental paper (Bandi et al. 2018)
23.	Carbon Dots from natural spinach biomass	2-nitrophenol and 4-nitrophenol, various low-molecular mass nitroaromatic compounds	Fluorescence, fluorescence quenching, fluorescent ink	Biosensing and bioimaging, environmental monitoring, ecotoxicology	Experimental paper (Ren et al. 2018)
24.	Carbon dots generated from spices tissues	Cell viability measurements and cell imaging	Photoluminescence, fluorescence, confocal microscopy	*In vitro* imaging and tumour cell growth inhibition	Experimental Paper (Vasimalai et al. 2018)

Abbreviations: AcFl = Diacryloyl fluorescein; CDs = Carbon dots; CLSM = Confocal laser scanning microscopy; CNF = Carbon nanofibers; CNTs = Carbon nanotubes; ECL = Electrochemiluminescence; ERGO = reduced graphene oxide; FMNDs = Fluorescent and magnetic nanodiamonds; FRET = Fluorescence Resonance Energy Transfer; GMC = Graphitized mesoporous carbon; GNR = Graphene nanoribbon; GO = Graphene oxide; HepG2 = hepatocellular carcinoma cell; HPLC = High-performance Liquid Chromatography; MCNDs = Magnetic carboxylated Nanodiamonds; MS = Mass spectrometry; MWCNTs = Multi-walled carbon nanotubes; NADH = Nicotinamide Adenine Dinucleotide; N-CDs = Nitrogen doped carbon dots; NCNTs = Nitrogen-doped carbon nanotubes; NFDs = Fluorescent nanodiamonds; NIR = Near-infrared; RAFT = Reversible addition-fragmentation chain transfer; SPE = Solid-phase Extraction; SPME = Solid-phase Microextraction; SWCNHs = Single-walled carbon nanohorns; SWNTs = Single- or multiwalled CNTs; TEM = transmission electron microscope; UV-Vis = Ultraviolet-Visible light.

at subambient and very low temperatures. There is an important reason that separation of fullerenes should be carefully optimized. It has been found that retention profiles of C60/C70 fullerenes are nonlinear at low temperature region (subambient temperatures below –30°C) due to complex and multiple separation mechanisms based on molecular shape recognition. Several previous studies demonstrated that the solute and stationary phase interactions are primarily contributing to retention behavior of these analytes (Ohta et al. 1994, Saito et al. 1995). Additionally, changes of the geometry of alkyl chains at different temperatures may also strongly affect the retention, and consequently, baseline separation of fullerenes (Jinno 1999). Subsequently reported data indicate that liquid chromatography using mono *n*-alkane-based mobile phases and working in planar development mode (temperature-controlled micro-planar chromatography) can be efficiently applied for the separation and detection of C60/C70 fullerenes (e.g., involving fluorescence or photothermal detection and correlation images signal processing), as well as determination of such carbon chemicals in complex environmental samples (Litvinowa and Zgonnik 1997, Zarzycki 2008, Zarzycki et al. 2007, 2011, Suszynski and Zarzycki 2015).

From principles, it may be expected that selected carbon originated nanomaterials should be biocompatible and therefore, there is significant interest in seeking and application of such materials in food technology, pharmacy, and medicine. Over the last two decades, great progress has been made in case of various carbon coatings composed of nanocrystalline diamonds and amorphous carbon (Mitura et al. 1996). The main focuses of the present research, mainly involving graphene, are development of coatings and their applications for implants in surgery. Various research teams are extensively seeking versatile layers that are characterized by proper mechanical properties with emphasis on biological resistance and antibacterial properties combined with high biotolerance and biocompatibility (Liu et al. 2018b). Applications of graphene related surfaces are especially promising for controlled increasing of biotolerance of various metals, including titanium or cobalt-chromium-molybdenum-based (CoCrMo) alloys, which are commonly fabricated for orthopedic implants (Zhang et al. 2018c). Selected coating preparation modes, such as electrophoretic deposition, enable to produce almost continuous graphene films with enhanced inhibition efficiency for the corrosion of biologically and chemically driven alloys (Chen et al. 2018b). Nanodiamond coatings may also form an efficient protective barrier against metal ions emitted from metal implant to human body. Latest studies have indicated the increase of corrosion resistance for orthopedic implants protected by nanocrystalline diamonds coatings, compared to plain medical steel 316L surface (Grumezescu 2018).

Recent toxicity studies of common polymers, mainly polyethylene or polypropylene modified with nanodiamonds particles (2–5 nm), confirmed very low or no toxicity of food packaging composed of such hybrid materials (Grumezescu 2018). These materials have been found to exhibit bacteriostatic effect against virulent bacteria and fungi microorganisms. Particularly, there is a strong interest to protect the food products against Gram-negative (−) *Salmonella enteritidis* bacteria and Gram-positive (+) *Listeria monocytogenes* pathogenic microorganisms. Most recently, our research team was involved in several studies focusing on the modification of food packaging surface with nanodiamonds, primarily to change the physicochemical properties of polymeric foils (Figure 14.1) (Mitura et al. 2017, Mitura and Zarzycki 2018, Zarzycki and Mitura 2018). Such simple and non-expensive technology, especially involving detonation nanodiamond particles and plasma-chemical methods of particle applications seem to be promising ways to obtain given properties of large packaging structures. These materials can protect unstable pharmaceutical formulations and food products, particularly in terms of oxygen protective barriers and mechanical durability of packaging compartments (Grumezescu 2017, Grumezescu 2018).

Biomedical applications of carbon-related nanomaterials do not only concern the invention of effective and non-toxic protective barriers, but also include much more complex systems enabling, for example, drug delivery. Currently, there are many examples for successfully developed drug delivery systems which, as mentioned above, must be as biocompatible and safe for treated organisms as possible. Data presented in Table 14.2 listed the number of hybrid carbon forms that were currently

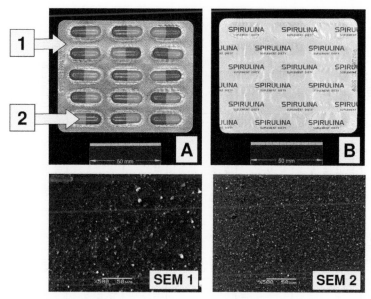

Figure 14.1: General view of food supplement packaging containing gelatin capsules filled with dehydrated Spirulina cells. (A) Optical microscope view of packaging polymer foil and (B) metal foil bottom side as well as scanning electron microscopy images (SEM; JEOL JSM-5500LV) of detonation nanodiamond particles sprayed on surface 1 (**SEM 1**) and surface 2 (**SEM 2**) areas. Patent Application No. P.424918, Patent Office of the Republic of Poland, 2018 Warsaw. Authors' unpublished results. All photos and figures copyrights belong to the authors.

investigated as the active particles for pharmaceutical and medical applications. They are mainly designed for controlled drug release and spreading active substances with high toxicity, e.g., for various cancer treatments, including leukemia, colorectal, bone, and lung cancer (Mudshinge et al. 2011, Wu et al. 2016, Chen et al. 2015, Wan et al. 2014). The main target of such systems seems to be neurological diseases and gene delivery vectors (Mengesha 2013, Luo et al. 2018). Targeted delivery systems are invented as efficient carriers of insoluble active chemicals, mainly using functionalized nanodiamond particles (Rouhani et al. 2016, Whitlow et al. 2017). These nanoparticles have also been used for treatment of urinary tract infections involving antibiotics molecules in form of nanodiamonds adducts, influencing the bioactivity of transported drug molecules (Gismondi et al. 2015).

Last year, our group published reviews and experimental papers dealing with carbon originated materials for wastewater technological processes (Świderska-Dąbrowska et al. 2018, Piaskowski et al. 2018). These works have been related to the removal of low-molecular mass compounds, namely various dyes, which may be present and considered as micropollutants in municipal and industrial wastewaters. Based on our own experimental results and data search provided by different research teams, we highlighted the issue that classical carbon adsorbents can be effective, but there is a significant problem with the selectivity of activated carbon materials to given chemical structures, high production cost, and also relatively low degree of regeneration (ŚwiderskaDąbrowska et al. 2018). On the other hand, carbon in form of nanoparticles, especially functionalized nanocarbons, can be much more selective to given micropollutants. Nevertheless, for the applications of wastewater technological processes, they must be ecologically friendly, non-expensive, and easily available at an industrial scale. For that reason, there is a particular need for the development of simple and low energy consuming technologies enabling industrial scale and green chemistry production of nanocarbons-based adsorbents using, for instance, common waste organic materials such as orange bagasse, apple pomace or fungus, and green algal biomasses (Piaskowski et al. 2018).

It is clear from the summary of the research papers presented in Tables 14.3a–d that impressive progress has been made over the last few years on the synthesis of carbon-based nanoparticles. It has been proven that raw carbon nanoparticles and/or related hybrid nanomaterials can be used

Table 14.2: Drug delivery systems involving various carbon particles and nanomaterials invented for pharmaceutical applications.

No.	Type of carbon based nanomaterials	Active molecules or drug delivery system	Expected effect, working mode	Cells, destination, and potential activity	Reference
1.	Functionalized carbon nanotubes as nanovectors	Different types of therapeutic molecules, AmB, nucleic acids	Toxicity reduction, antifungal activity	Mammalian cells, Chronic fungial infections	Review paper (Klumpp et al. 2006)
2.	Fullerenes and their derivatives	Fullerenes, fulleropyrrolidines, dendrofullerene, amino acid derivatives of fullerene, cationic, anionic and amino acid type fullerenes, carboxyfullerenes, metallofullerol	Fullerenes derived hybrid nanomaterials may act as lubricants, gene transfer agents, used in HIV treatment	Various tissues and compartments, Emerging carriers for drug delivery, treatment of leukemia and bone cancer, neurological disease including Parkinson's disease	Review paper (Mudshinge et al. 2011)
3.	Nanodiamonds	Dexamethazone, hydroxytamoxifen, paclitaxel, cisplatin, doxorubicin, folic acid	Drug controlled relase in cancer cells, enhanced drug dispersion, reduction of drug-efflux-based chemoresistance, receptor-mediated targeting of cancer cells	Chemiotherapy, drug therapy, active molecules delivery, gene delivery, biomedical applications	Review paper (Mengesha 2013)
4.	Nanodiamonds	Functionalized nanoparticles with various organic ligands (–COOH, –NH$_2$, –OH, –SH) as low-molecular mass bioactive chemicals (e.g., doxorubicin, octacylamine) as well as polymers	Influence on metabolic activity of given cells, cytotoxic effects	Targeted delivery systems for insoluble drugs, bioactive substrates for stem cells	Review paper (Whitlow et al. 2017)
5.	Multi-functionalized carbon dots as theranostic nano agent	Folate-conjugated reducible polyethylenimine passivated carbon dots (fc-rPEICdots)	Therapeutic effect, gene silencing and anti-cancer effects	Carcinomatous tissues, lung cancer targeting and treatment	Experimental paper (Wu et al. 2016)
6.	Graphene oxide nanocomposirite with Ag nanoparticles (GO@Ag)	DOX (doxorubicin)	Photothermal therapy (PTT), system used as tumor diagnostic X-ray contrast material and strong agent for photothermal ablation of tumor	Wide range of cancers	Experimental paper (Shi et al. 2014)
7.	Graphene Oxide as chemosensitizer	Irinotecan, Doxorubicin, Oxaliplatin, Cisplatin	Antitumor effects, CT26 cell killing via necrosis, chemosensitization of cancer cells, the antineoplastic effects in mice bearing the CT26 colon tumor	Colorectal and non-small cell lung cancers that have have intrinsic resistance to classical CDDP	Experimental paper (Chen et al. 2015)
8.	Nanodiamonds (ND) as anti-HIV drug delivery system	Efavirenz, unmodified and surface modified (–COOH, –NH$_2$) ND nanoparticles	Nanodrug formation with of less toxic ND than other carbon-based materials, BBB penetration, HIV-1 treatment, drug delivery and imaging	Human brain, microvascular endothelial cells	Experimental paper (Roy et al. 2018)

9.	Uniform mesoporous carbon spheres (UMCS) as drug-encapsulation reservoir	Paclitaxel (PXT), UMCS particles functionalized using hyperbranched polyethyleneimine (PEI) covalently linked with fluorescein isothiocyanate (FITC) and folic acid (FA)	Highest cytotoxicity than PXT and Taxol, antitumour effect against: KB, A549, H22 cells	Kunming mice harboring a hepatic H22 tumor	Experimental paper (Wan et al. 2014)
10.	Fluorine-doped carbon dots synthesized by RPDC method	Photoluminescent nanoparticles prepared by ring-opening reaction (RPDC)	Gene delivery, cell imaging, gene therapy, altering transfection efficiency, CDs may condense DNA into nanoparticles with proper size and surface charge	High transfection efficiency and cellular uptake than PEI 600 contrasts in various cell lines, gene vectors with high efficiency and serum tolerance	Experimental paper (Luo et al. 2018)
11.	Functionalized nanodiamonds (NDPs)	Amoxicillin	Treatment of urinary tract infections	Antibiotic drug delivery platform for UTIs	Experimental paper (Rouhani et al. 2016)
12.	Nanodiamonds adducts	Nanodiamonds particles coupled with plant bioactive metabolites, ciproten, quercetin	Reduced pro-oxidant, cytotoxic and pro-apoptotic activity of target molecules	Nanodiamond surfaces influenced the bioactivity of transported drugs, improved antiproliferative effects, on human (HeLa) and murine (B16F10) tumor cells	Experimental paper (Gismondi et al. 2015)

Abbreviations: AMB = Amphotericin B; BBB = Blood Brain Barrier; CDDP = Chemotherapy drug cisplatin; DOX = Doxorubicin; FA = Folic acid; f-CNT = Functionalized carbon nanotubes; fc-rPEICdots = Folate-conjugated reducible polyethylenimine passivated carbon dots; FITC = Fluorescein isothiocyanate; GO = Graphene oxide; HIV-1 = Human Immunodeficiency Virus Type 1; ND = Nanodiamonds; NDPs = Nanodiamond particles; PEI = Polyethyleneimine; PTT = Photothermal therapy; PXT = Paclitaxel; RPDC = Ring-opening polymerization-dehydrative carbonization; UMCS = Uniform mesoporous carbon spheres; UTIs = Urinary tract infections.

Table 14.3a: Removal of emerging pollutants with carbon-based adsorbents from water and wastewater.

No.	Adsorbent	Contaminant	Adsorption capacity/ removal efficiency (%)	Complementary data (specific surface area, initial concentration of pollutants, pH, temperature, isotherm model, kinetic model, etc.)	Reference
1.	Combustion synthesized graphene (Mg+CO$_2$)	Methyl orange	125.2 mg/g	The content of the MgO diluents (used to prevent the produced graphene from agglomeration) has an important influence on the morphology, crystallinity, surface properties, and thus adsorption performance of the prepared graphene. Monolayer adsorption.	Experimental paper (Lu et al. 2018)
2.	Co$_3$O$_4$ nanoparticles@rGO composite (reduction of GO with ascorbic acid)	Rhodamine B	97.621 mg/g	Adsorption isotherms: Freundlich and Tempkin model; Pseudo-first order model; endothermic and spontaneous process; pH-depending process: at more acidic conditions increases the number of negative charges of active sites on nanocomposite surface. The adsorption capacity decreases with the increase of ionic strength due to a strong competition between smaller ions of Na$^+$ and dye ions in the adsorption on active sites.	Experimental paper (Alwan et al. 2018)
3.	Graphene nanosheet (GNS)	Ciprofloxacin	173.40 mg/g (at 298.15 K)	t_e = 220 min; pH$_{opt}$ 6.0; kinetic Elovich model; Sips and Hill adsorption models; exothermic, spontaneous and favorable process; CP could be adsorbed onto GNS by an electrostatic interaction.	Experimental paper (Rostamian and Behnejad 2017)
4.	Porous graphene from rGO	Arsenic(III), fluoride, nitrate Oil Methylene blue and Rhodamine	313 mg/g for MB (about 95% for all tested pollutants)	SSA = 652 m^2/g; pseudo-second order kinetic model; Regeneration via heating at 600°C in air with good cycling efficiency (above 90% after 5 cycles).	Experimental paper (Tabish et al. 2018)
5.	rGO nanosheets (reduction of GO with hydrazine)	17β-estradiol 17α-ethynylestradiol	n.a.	[rGO] = 4 g/L, pH = 6.5, T = 25 ± 1°C; The adsorption capacities of estrogens increased with increasing the reduction degree of rGOs. NOM suppressed estrogens adsorption to rGOs. π-π stacking interaction could be considered as a predominant interaction to the enhanced adsorption capacities of rGOs. The adsorption via π-π stacking interaction, mainly.	Experimental paper (Jiang et al. 2018)
6.	Magnetic nanoparticles (mixture of Fe$_3$O$_4$ and FeOOH)@fluorographene	Perfluorooctanoic acid (PFOA) Perfluorooctane sulfonate (PFOS)	50.4 mg/g (PFOA) and 17.2 mg/g (PFOS)	t_e = 2 min at room temperature; Langmuir model; natural organic matter (NOM) do not affect on adsorption efficiency of PFOA and PFOS; regeneration of sorbent by methanol washing without significant decrease PFOA and PFOS removal even after 5 cycles; easily separated by a magnet.	Experimental paper (Wang et al. 2018a)
7.	Magnetic Fe$_3$O$_4$/graphene composite	Uranium(VI)	176.47 mg/g (at 298 K, pH 5.5)	SSA = 42.53 m^2/g; pH$_{IEP}$ 2.6; Langmuir model; pseudo-second-order kinetic model; U(VI) was sorbed via oxygen-containing functional groups; pH$_{opt}$ = 2.6–7.0; endothermic and favorably spontaneous process; easily separated by a magnet.	Experimental paper (Zhao et al. 2018)

8.	Graphene	Bisphenol A (BPA) Triclosan (TCS)	2.0×10^3 μg/g (BPA) 1.1×10^8 μg/g (TCS)	t_e = 240 min for BPA and 60 min for TCS, respectively; pseudo-second-order kinetic model; Langmuir model for BPA and Freundlich model for TCS; pH_{opt} = 5.0–7.0; the increase of ionic strength significantly increase of BPA sorption on graphene due to the salting-out effect. In the case of TCS, the effect of the ionic strength of the solution is negligible because of the dominant hydrophobic interaction in sorption process.	Experimental paper (Wang et al. 2017)
9.	Magnetic glucose-functionalized graphene nanosheets (rGO-N$_3$-Glc-Fe$_3$O$_4$ and rGO-diazide-Glc-Fe$_3$O$_4$)	Methylene blue	111 mg/g (rGO-N$_3$-Glc-Fe$_3$O$_4$) 105 mg/g (rGO-diazide-Glc-Fe$_3$O$_4$)	t_e = 30 min; pH_{opt} > 7 due to the ionization of glucose moieties; easily separated by a magnet.	Experimental paper (Namvari and Namazi 2016)
10.	GO nanosheets	Tetracycline (TCN) Doxycycline (DCN) Ciprofloxacin (CPN)	173.4 mg/g (CPN) $q_{max(CPN)}$ >> $q_{max(DCN)}$ > $q_{max(TCN)}$	t_e = 100 min (CPN) and 200 min (TCN and DCN), pH 6–7; Hill and Toth isotherm models; pseudo-second-order kinetic model (DCN and TCN) and Elovich (CPN) models; the mechanism for the adsorption via cation – π bonding and π-π interaction.	Experimental paper (Rostamian and Behnejad 2018)
11.	Graphene oxide@iron-aluminium oxide composite	Arsenic(III)	42.2836 mg/g (at 293 K)	SSA = 189.54 m^2/g; pH_{opt} = 6.0–8.0; pseudo-first order kinetic model; spontaneous, but unfavorable process ($\Delta S°$ < 0); the competitive effect of others ions on arsenic(III) removal efficiency: PO_4^{3-} > SO_4^{2-} > HCO_3^- :	Experimental paper (Maji et al. 2018)
12.	Graphene oxide/CuFe$_2$O$_4$ on Fe-Ni foam, synthesized via hydrothermal method	Arsenic (III) and (V)	51.64 mg/g (As(III)) 124.69 mg/g (As(V))	t_e = 180 min; pH_{PZC} 6.5; max removal efficiency (> 96%) at pH 3.7–8.8 and 2.0–7.24 for As(III) and As(V), respectively; the Langmuir isotherm model; regeneration using 0.1 M NaOH and ultrasonic for 10 min; good recyclability even after eight cycles; High adsorbent efficiency in column experiment (96% As(V) removal from water solution after 100 bed volume test).	Experimental paper (Wu et al. 2018)
13.	Superparamagnetic amino-functionalized graphene oxide composite (GO thermal annealing in ammonia and mixing with Fe(acac)$_3$)	17β-estradiol (E$_2$) Bisphenol A (BPA)	106.38 mg/g (E$_2$) 41.32 mg/g (BPA)	SSA = 251.6 m^2/g; pH_{PZC} ~ 10; equilibrium time 360 min; the pseudo-second-order kinetic model; adsorption was favorable under neutral condition; regeneration using 0.1 M NaOH; good recyclability, after five times of recycling, adsorption capacity reached 70%, easily separated by a magnet.	Experimental paper (Bai et al. 2017)
14.	Few-layered graphene oxide nanosheets	17β-estradiol (E$_2$)	149.4 mg/g (at 298 K)	SSA = 92 m^2/g; t_e = 8 h; the pseudo-second-order kinetic model; the Langmuir isotherm model; adsorption was a spontaneous process (ΔG < 0); pH-independent adsorption; mechanism of adsorption: via π-π interactions and hydrogen bonds.	Experimental paper (Jiang et al. 2016a)

Table 14.3a contd. ...

...Table 14.3a contd.

No.	Adsorbent	Contaminant	Adsorption capacity/ removal efficiency (%)	Complementary data (specific surface area, initial concentration of pollutants, pH, temperature, isotherm model, kinetic model, etc.)	Reference
15.	Graphene oxide	Metformin	50.47 mg/g (at pH 6.0 and 288 K)	SSA = 108.71 m²/g; t_e = 160 min; the pseudo-second-order kinetic model; the Langmuir isotherm model; adsorption was a spontaneous process ($\Delta G < 0$); pH-independent adsorption; mechanism of adsorption: via π-π interactions and hydrogen bonds.	Experimental paper (Zhu et al. 2017) PDF: 2017 Zhu_C
16.	Pristine MWCNTs	Acetaminophen	90 mg/g	[Acetaminophen]$_0$ = 10 mg/L; T = 25°C; the pseudo-second-order kinetic model; Freundlich isotherm model.	Experimental paper (Yanyan et al. 2018)
17.	NaOH-treated MWCNTs	Acetaminophen	130 mg/g	[Acetaminophen]$_0$ = 10 mg/L; T = 25°C; the pseudo-second-order kinetic model; Freundlich isotherm model.	Experimental paper (Yanyan et al. 2018)
18.	HNO$_3$/H$_2$SO$_4$-treated MWCNTs	Acetaminophen	160 mg/g	[Acetaminophen]$_0$ = 10 mg/L; T = 25°C; the pseudo-second-order kinetic model; Freundlich isotherm model.	Experimental paper (Yanyan et al. 2018)
19.	Chitosan-coated MWCNTs	Acetaminophen	205 mg/g	[Acetaminophen]$_0$ = 10 mg/L; T = 25°C; the pseudo-second-order kinetic model; Freundlich isotherm model.	Experimental paper (Yanyan et al. 2018)
20.	Ozone-treated MWCNTs	Acetaminophen	250 mg/g (95%)	[Acetaminophen]$_0$ = 10 mg/L; T = 25°C; the pseudo-second-order kinetic model; Freundlich isotherm model.	Experimental paper (Yanyan et al. 2018)
21.	CNTs modified with decafluorobiphenyl (CNT-DFB)	Carbamazepine (CBZ) Tetracycline (TC)	403.0 µmol/g (CBZ) 456.5 µmol/g (TC)	[CBZ/TC]$_0$ = 0.085 mg/L, [Adsorbent]$_0$ = 0.1 g/L; T = 25°C; pH 6, SSA = 128.6 m²/g; t_e = 48 h for CBZ and over 72 h for TC; the pseudo-second-order kinetic model; the Langmuir isotherm model for CBZ and Freundlich model for TC.	Experimental paper (Shan et al. 2018)
22.	CNT-based strip, mixed with polytetrafluoroethylene (CNT-PTFE)	o-cresol (OC)	20–30 mg/g (74% after 1 h and < 90% after 4 h)	Adsorption: T = 25°C, t_e = 1 h; Desorption: vacuum distillation, T = 100°C, t_e = 20 min; The removal ratio of OC remained stable at 55–60% for 10 cycles of adsorption–desorption.	Experimental paper (Yin et al. 2018)
23.	Polyimide-based carbon nanofibers (PI-CNFs)	2,4-dichlorophenol (2,4-DCP) Methylene blue (MB) Tetracycline (TC)	483.09 mg/g (2,4-DCP) 272.48 mg/g (MB) 146.63 mg/g (TC)	[Sorbate]$_0$ = 20 mg/L, [Adsorbent]$_0$ = 0.125–0.175 g/L; SSA = 715.89 m²/g; T = 25°C; The adsorption efficiency strongly depends on pH and type of sorbate: pH$_{opt}$ = 11 for MB, 4–7 for 2,4-DCP and 3–7 for TC. Pseudo-second-order kinetics model, Langmuir isotherm model; adsorption was spontaneous and endothermic.	Experimental paper (Zhang et al. 2018a)

Table 14.3b: Oxidation of emerging pollutants using various types of carbon-based nanomaterials from water and wastewater.

No.	Process	Catalyst	Irradiation/ Oxidant type	Target pollutants	Removal efficiency/ kinetic rate	Complementary data (initial concentration of pollutants, pH, temperature, etc.)	Reference
1.	Photocatalytic oxidation	Magnetic ZnFe-layered double hydroxides/rGO composite (ZnFe-CLDH/rGO)	solar light	Paracetamol, As(III) ions	95% of paracetamol after 420 min and 99% of As(III) in just 10 min	$[\text{Paracetamol}]_0 = 5$ mg/L and $[\text{As(III)}]_0 = 1$ mg/L. The photocatalytic activity of the screened ZnFe-CLDH/rGO for paracetamol degradation is about 3.5 times that of the pristine ZnFe-CLDH. The synergy of photocatalytic oxidation of As(III) to less toxic As(V) was observed.	Experimental paper (Zhu et al. 2018a)
2.	Photocatalytic oxidation	The ternary Ag-Cu$_2$O/rGO composites	visible light	Phenol	96.8% of Phenol (64.5% of TOC after 6 h)	$[\text{Phenol}]_0 = 10$ mg/L; T = 25°C; pseudo-first-order kinetics reaction ($k_1 = 0.0140$ min^{-1}), rate was 6.1 times greater than of pristine Cu$_2$O.	Experimental paper (Wei et al. 2018)
3.	Photocatalytic oxidation	3-D hollow carbon graphene-Ag$_3$PO$_4$ (HCG-Ag$_3$PO$_4$)	visible light	Phenol	N/A	$[\text{Phenol}]_0 = 8$ mg/L, $[\text{Catalyst}]_0 = 2$ g/L; Light source: 350 W Xe lamp with a an optical filter 420 nm; Z-scheme photocatalytic mechanism; performance over optimal HCG-Ag$_3$PO$_4$ is 6, 3.43, 1.92 times of pristine Ag$_3$PO$_4$, GO-Ag$_3$PO$_4$, and rGO-Ag$_3$PO$_4$, respectively.	Experimental paper (Song et al. 2018)
4.	Photocatalytic oxidation	Silver phosphate modified by MWCNTs and Cr-doped SrTiO$_3$	solar light	Malachite green (MG)	100% after 6 min	$[\text{MG}]_0 = 25$ mg/L, $[\text{Catalyst}]_0 = 0.25$ g/L; Light source: 300 W Xe lamp with a 420 nm cut filter. Z-scheme photocatalytic mechanism; inhibition under strong acid condition or at the coexistence of certain concentration of Cr(VI) and Cl.	Experimental paper (Lin et al. 2018)
5.	Photo-electrocatalytic oxidation	TiO$_2$/graphene/Cu$_2$O mesh (as catalyst/ electrode)	visible light	Bisphenol A (BPA)	max 92% at pH 4.5	$[\text{BPA}]_0 = 15$ mg/L; High conductive and interconnected three-dimensional channels inside catalyst determined its high photoelectrocatalytic activity.	Experimental paper (Yang et al. 2016a)
6.	Catalytic oxidation	rGO annealed at 900°C	Peroxy-monosulfate (PMS)	Phenol	100% after 10 min	$[\text{Phenol}]_0 = 20$ mg/L, $[\text{Catalyst}]_0 = 0.1$ g/L, $[\text{PMS}]_0 = 6.5$ mM and T = 25°C.	Experimental paper (Duan et al. 2016)

Table 14.3b contd. ...

...Table 14.3b contd.

No.	Process	Catalyst	Irradiation/ Oxidant type	Target pollutants	Removal efficiency/ kinetic rate	Complementary data (initial concentration of pollutants, pH, temperature, etc.)	Reference
7.	Catalytic oxidation	S-doped AC (synthesized by chemical activation of polythiophene with KOH under 800°C)	Persulfate (PS)	4-chlorophenol (4CP)	100% 4CP within 60 min/k = 0.083 min^{-1}	$[4CP]_0$ = 40 mg/L, $[Catalyst]_0$ = 0.05 g/L, $[PS]_0$ = 8 mM, T = 25°C.	Experimental paper (Guo et al. 2018)
8.	Photocatalytic oxidation	sulfur-doped carbon nitride	visible light/ Peroxy-monosulfate (PMS)	Bisphenol A (BPA)	N/A	$[BPA]_0$ = 50 mg/L, $[Catalyst]_0$ = 0.06 g/L; Radiation intensity 0.016 Wcm^2; T = 20–40°C, pH 3–11.	Experimental paper (Lin and Zhang 2017)
9.	Catalytic ozonation	Fe_3O_4/MWCNTs	ozone	Sulfamethazine (SMT)	39.1% at pH 4	$[SMT]_0$ = 20 mg/L; $[Catalyst]_0$ = 0.5 g/L; Acid condition was most favorable to the degradation and mineralization of SMT.	Experimental paper (Bai et al. 2018)
10.	Catalytic ozonation	Fe_3O_4/MWCNTs	ozone	Bisphenol A (BPA)	91% at catalyst concentration 1.0 g/L and at pH 7 (98% in co-existence of 25 ppm of HA)	$[BPA]_0$ = 50 ppm; $[Catalyst]_0$ = 0–1.0 g/L; $[ozone]_0$ = 3.0 mg/L; BPA removal efficiency was enhanced with increase of the initial pH, as well as concentration of humic acids (HA).	Experimental paper (Huang et al. 2017)
11.	Catalytic ozonation	GO GO/TiO$_2$ GO/Fe$_3$O$_4$ GO/TiO$_2$/Fe$_3$O$_4$	ozone	Ibuprofen (IBP)	n.a.	$[IBP]_0$ = 0.5 μM; $[ozone]_0$ = 4 mg/l; contact time 20 s; The surface modified GO suspensions (GO/Fe$_3$O$_4$ and GO/TiO$_2$/Fe$_3$O$_4$) enhance the O$_3$ exposure and •OH formation, while GO alone acts as •OH scavenger.	Experimental paper (Jothinathan and Hu 2018)

Table 14.3c: Water disinfection based on active carbon nanomaterials (CNs).

No.	CNs type	Target compound	Methods	Complementary data	Reference
1.	CHIT-CNP composite membrane	Gram-negative *Escherichia coli* (*E. coli*) ATCC 25922	Composite of carbon nano particles (CNP) and chitosan (CHIT) used to the disinfection of microorganisms in water under sunlight (SODIS) in a Compound Parabolic Collector (CPC) reactor.	System SODIS-CHIT-CNP effective exploiting the synergistic bactericidal activity of CNP and Chitosan in practical applications.	Experimental paper (Maddigpu et al. 2018)
2.	CNT$_{0-60}$/PPy/AgNPs nanocomposites	Gram-negative *Escherichia coli* (*E. coli*) Gram-positive *Staphylococcus aureus* (*S. aureus*)	The bacterial removal from water using the column filter method with nanocomposites materials containing different percentage of single wall carbon nanotubes (0–60%) obtained by oxidative polymerization of pyrrole with silver nitrate.	CNT$_{60}$/PPy/AgNPs nanocomposite columns effectively removed 100% of all analyzed bacteria, while PPy/AgNPs nanocomposite obtained the complete removal of *S. aureus*.	Experimental paper (Salam et al. 2017)
3.	Carbon-silver nano-composites: SWCNTs-Ag MWCNTs-Ag GO-Ag	Gram-negative *Escherichia coli* (*E. coli*) Gram-positive *Staphylococcus aureus* (*S. aureus*)	The synthesized carbon-silver nanocomposites with the different antibacterial properties, included single-walled carbon nanotubes-silver (SWCNTs-Ag) and multi-walled carbon nanotubes-silver (MWCNTs-Ag) and graphene oxide-silver (GO-Ag) nanoparticles, dosed such as suspensions in water with bacterial strains.	The order of *E. coli* disinfection rates: GO-Ag (99.99%) > MWCNTs-Ag (86.09%) > SWCNTs-Ag (70.24%) > SWCNTs (38.89%) > MWCNTs (38.18%) > control (4.82%) The *S. aureus* disinfection rates: GO-Ag (99.99%) > SWCNTs-Ag (95.79%) > MWCNTs-Ag (72.29%) > MWCNTs (62.42%) > control (–10.56%)	Experimental paper (Chang et al. 2016)
4.	Magnetic-graphene oxide (M-GO) nanocomposites	Gram-negative *Escherichia coli* (*E. coli*)	Magnetic-graphene oxide (M-GO) with magnetic iron oxide nanoparticles dispersed on graphene oxide (GO) nanosheets. The separation nanocomposites from aqueous solution using an external magnetic field.	M-GO synthesized by depositing magnetic iron oxide nanoparticles on the surface of GO nanosheets. The inactivation percent of *E. coli* at the M-GO concentration of 100 ug/ml was 91.49 ± 2.82%. The inactivation mechanism of *E. coli* by M-GO: the membrane and oxidation stress.	Experimental paper
5.	GO–TiO$_2$ (GOTI) nanocomposites	Gram-negative *Escherichia coli* (*E. coli*) Gram-positive *Bacillus subtilis* (*B. subtilis*)	Photocatalytic in situ synthesis of nAg particles by crumpled graphene oxide CGO–TiO$_2$ (GOTI) nanocomposites with unique open core–shell structure, for simultaneous filtration and disinfection of water.	With Ag formation, GOTI–Ag and support membranes reached enhanced antimicrobial activity ca. 3 log removal of both microorganisms.	Experimental paper (Jiang et al. 2016b)

Table 14.3c contd. ...

...Table 14.3c contd.

No.	CNs type	Target compound	Methods	Complementary data	Reference
6.	Vanadium Tetrasulfide (VS4) nanocomposites	Gram-negative *Escherichia coli* (E. coli) Gram-positive *Staphylococcus aureus* (S. aureus).	VS4 synthesized as photocatalytic disinfectants, supported by different carbon materials: VS4/CP (carbon powder), VS4/rGO (reduced graphene oxides), VS4/CF (carbon fiber), and VS4/CNT (carbon nanotube).	The best disinfection performance for removing *E. coli* under simulated visible and sunlight by VS4/CP, with a maximum inactivation rate of 9.7 log at 0.1 g L^{-1} in 30 min. Disinfection rate of S. aureus (Gram-positive) of 1.7 log inactivation in 30 min.	Experimental paper (Zhang et al. 2018b)
7.	Carbon nanotubes–silver nanoparticles (CNTs–Ag) nanocomposite Graphene oxide–silver nanoparticles (GO–Ag) nanocomposite	Gram-negative *Escherichia coli* Gram-positive *Staphylococcus aureus*	Modified photochemical preparation of CNTs–Ag and GO–Ag nanocomposites, stably dispersible in aqueous solution to eliminate infectious bacterial pathogens.	Better antibacterial activity of GO–Ag than of CNTs–Ag because of their chemical functional groups and stably high dispersion ability in aqueous medium. The antibacterial activity these nanocomposites more sensitive against *E. coli* bacteria.	Experimental paper (Dinh et al. 2015)
8.	Ag/g-C$_3$N$_4$ nanohybrids	Gram-negative (G–) *Escherichia coli (E. coli)* Gram-positive (G+) *Staphylococcus aureus (S. aureus)*	Graphitic carbon nitride (Ag-C$_3$N$_4$) nanosheets with embedded Ag nanoparticles (Ag/g-C$_3$N$_4$) and the ROS (reactive oxygen species) production under visible light irradiation for biofilm elimination.	Ag/g-C$_3$N$_4$ nanohybrids with generating of ROS has a higher activity in the oxidation and degrade the main components of biofilms (proteins, polysaccharides and nucleic acids). Efficient bactericidal agents for both gram-negative (G–) and gram-positive (G+) bacteria.	Experimental paper (Bing et al. 2015)
9.	Ag-AC hybridgranules	Gram-negative (G–) *Escherichia coli (E. coli)*	Silver nanoparticles (Ag-NP) selectively impregnated on the external surface of plasma treated activated carbon (AC) granules for continuous disinfection of water in a single continuous-flow column.	Water disinfection: zero *E. coli* count and < 100 mg/L Ag achieved in a continuous manner over a long duration, with Ag-AC used as a packing material in a continuous column, for decontaminating water.	Experimental paper (Biwas and Bandyopadhyaya 2016)
10.	1,6-diaminohexane-silver nanoparticles functionalized graphene hybridmaterial	Coliform and Fecal Coliform	Reduced graphene (G) prepared by chemically reducing graphene oxide (GO) functionalized with 1,6-diaminohexane and modified by silver nanoparticles (AgNPs) to achieved hybrid material for disinfecting wastewater.	Effective antibacterial activities. full disinfection of coliform by using concentration of the hybridmaterial in wastewater of 1 g/l without release of silver to the water.	Experimental paper (Abdelhalim et al. 2016)

Table 14.3d: Membrane removal of various contaminations using different carbon-based nanocomposites from water and wastewater.

No.	CNs type	Target compound	Membrane characterization	Complementary data	Reference
1.	TGCCM coupled with EC/EF	ECs: di-n-butyl phthalate (DnBP), di(2-ethylhexyl) phthalate (DEHP), cephalexin (CLX), sulfamethoxazole (SMX), caffeine (CAF)	Graphene-containing ceramic composite tubular membrane (TGCCM) coupled with the simultaneous electrocoagulation and electrofiltration process (EC/EF) in crossflow filtration mode to remove emerging contaminants (ECs) in model solution.	Removal efficiencies: 99% of DnBP, 32–97% of DEHP, cephalexin (CLX), sulfamethoxazole (SMX) and caffeine (CAF).	Experimental paper (Yang et al. 2016b)
2.	CNT membrane filters	PPCP: Triclosan (TCS), Acetaminophen (AAP) Ibuprofen (IBU)	SWCNT and MWCNT layers on top of a PVDF membrane to enhanced removal of pharmaceuticals and personal care products (PPCP) by nanocomposite membrane filtration.	Removals ranged 10–95%, and increased with increasing number of aromatic rings (AAP ≈ IBU < TCS), decreasing surface oxygen content (oxidized MWCNT < MWCNT), increasing specific surface area (MWCNT < SWCNT)and elimination of NOM. Greater removals obtained with neutral PPCP molecules than with ions.	Experimental paper (Wang et al. 2015)
3.	Coal-based carbon membranes	Rhodamine B (RhB)	Membranes produced by carbonization of carbonaceous precursor materials (cellulose, polyimide, poly(vinylidene chloride), phenol, etc.) under an inert atmosphere or vacuum. Removal of RhB wastewater by integrating the coal-based carbon membrane with an electric field at 0.15–0.67 V/cm.	Considerable separation performance and good electrical conductivity to combine membrane separation and electrochemical degradation for wastewater treatment. Removal of RhB from the treatment system is dominated by electrochemical degradation. With high RhB concentration (> 100 ppm) low permeability and removal efficiency because of insufficient degradation ability.	Experimental paper (Tao et al. 2018)
4.	Thin-film nanocomposite (TFN) membranes incorporated with GO	Phosphorus	TFC thin-film composite membrane modified by incorporating small quantities of GO into the microporous substrate matrix of the membrane.	Effective combination of water flux, solute rejection and phosphorus removal at different pH conditions. With increasing the phosphorus rejection in the feed solution the phosphorus rejection of the membranes reduced.	Experimental paper (Koo et al. 2018)
5.	GO/AC membrane	Tetracycline hydrochloride (TCH)	Hybrid carbon membranes composed of 2D GO sheets and AC (2:1 GO to AC mass ratio) in filtration process for the removal of TCH antibiotics.	Free-standing and flexible membrane based on an easy vacuum filtration method produce with the strong interactions between GO/AC and TCH. The high adsorption capacity (449 mg/g).	Experimental paper (Liu et al. 2017b)

Table 14.3d contd. ...

...Table 14.3d contd.

No.	CNs type	Target compound	Membrane characterization	Complementary data	Reference
6.	Graphene oxide incorporated thin film nanocomposite (TFN) membrane	Salt: Na_2SO_4, $MgSO_4$, $MgCl_2$ and NaCl	TFN membrane made by incorporating graphene oxide (GO) into polysulfone (PSf) microporous substrate.	Water permeability with rejections of TFN (0.3 wt% GO) membrane for: Na_2SO_4–95.2%, $MgSO_4$–91.1%, $MgCl_2$–62.1% and NaCl–59.5%. Applied for water softening process. Higher pure water flux, salt solution permeability and salt rejection. Overcoming the trade-off effect between water flux and salt rejection of TFC membrane.	Experimental paper (Lai et al. 2016)
7.	PS-GO membrane	Organic contaminants: ofloxacin (OFLOX), benzophenone-3 (BP-3), rhodamine b (Rh), diclofenac (DCF) and triton X-100 (TRX)	Polysulfone (PS)–graphene oxide (GO) based porous membranes (PS-GO) for removal of selected organic contaminants. Prepared by phase inversion method starting from a PS and GO mixture (5% w/w of GO).	Efficiency higher than 90% after 4 h treatments. Regeneration of PS-GO and reuse possibilities by washing with ethanol. Smaller amount of graphene-oxide to polysulfone membranes enhanced the hydrophilicity, mechanical stability and fouling resistance.	Experimental paper (Zambianchi et al. 2017)
8.	The two cation-modified graphene oxide (GO) membranes	Natural organic matter (NOM)	Trivalent cations, Al^{3+} and Fe^{3+} applied as cross linking agents to stack GO nanosheets layer by layer on a PVDF membrane supporting layer.	Compared with Al^{3+}, GO membrane cross-linked with Fe^{3+} exhibited a greater flux and NOM removal efficiency. An excellent performance in terms of flux and organic fouling control.	Experimental paper (Liu et al. 2017c)
9.	CFGO/PA nanofiltration (NF) membranes	Desalination of dye solutions	Carboxyl-functionalized graphene oxide (CFGO)/polyamide (PA) thin film nanocomposite membranes prepared via interfacial polymerization (IP) of piperazine (PIP) and trimesoyl chloride (TMC).	CFGO significantly increases the hydrophilicity and surface charge density of the membrane, which gives the CFGO/PA membranes an excellent permeability. For 0.07% CFGO/PA membranes, with 95.1% rejection of New Coccine and 25.0% retention for NaCl, a permeation flux of 110.4 L/m²/h	Experimental paper (Zhang et al. 2017b)
10.	GO-hydrogel nanocomposite membranes	Osmosis desalination	Non-swelling graphene oxide-polymer nanocomposite membrane for energy-efficient RO desalination. The crosslinked polymer networks physically bind the pristine GO laminates with remarkable water flux and salt rejection membrane properties.	Combination of mechanical robustness, remarkable water transport properties, and an anti-swelling tendency suitable for the high-pressure (RO) membrane process. Stable under a harsh oxidizing environment. High degree of NaCl rejection of 98.5% for all pressure ranges (5–35 bar).	Experimental paper (Kim et al. 2018)

No.	Membrane	Target	Description	Performance	Reference
11.	Fe(OH)$_3$/g-C$_3$N$_4$ composite membrane	Dyes: EB Evans blue MB methyl blue RB Rhodamine B MeB methylene blue	The g-C$_3$N$_4$ nanosheets deposited on anodic aluminum oxide (AAO) support and the Fe(OH)$_3$ nanoparticles deposited on the g-C$_3$N$_4$ layer. The surface zeta potential of Fe(OH)$_3$/g-C$_3$N$_4$ composite membrane: 25 mV (positively charged)	Pure water permeance of 48 Lm^{-2} h^{-1} bar^{-1}. Nearly 90% of the pure water permeance when filtrating RB and MeB solutions and only 50% when filtrating EB and MB solutions. Easily capture the negatively charged EB and MB by the positively charged Fe(OH)$_3$ layer. No adsorption of the positively charged RB and MeB dyes.	Experimental paper (Wang et al. 2018b)
12.	TFN-CQD thin-film nanocomposite membrane	Na$_2$SO$_4$	Three functionalized Carbon quantum dots (CQDs) with carboxyl, amino and sulfonic acid groups incorporated into the polyamide layer of thin-film nanocomposite (TFN) membranes via interfacial polymerization.	The highest permeate flux of TFN-SCQD membrane 42.1 Lm^{-2} h^{-1} and the Na$_2$SO$_4$ rejection of 93.6%. TFN-NCQD: better retention of divalent cations and only slightly improved antifouling performance.	Experimental paper (Sun and Wu 2018)
13.	Graphene-based membranes	Mono-/multi-valent metal ions	Graphene-based membranes with uniform 2D nanochannels (8 Å) for precise and efficient sieving of mono-/multi-valent metal ions. Built using facilely reduced graphene oxide (FRGO) nanosheets.	High selectivity, permeability and stability for sieving of mono-/multi-valent metal ions. Efficiently separate mono-valent metal ions and multi-valent metal ions, differentiated by their size.	Experimental paper (Xi et al. 2018)
14.	Ultra-wetting graphene-based membrane	Bovine serum albumin (BSA) solution	Ultrafiltration membrane for water filtration based on the amine and carboxylated graphene covalently attached to a polymer matrix composite graphene-poly(acrylonitrile-co-maleimide) (G-PANCMI).	Modified supported graphene based membrane had significantly higher hydrophilicity, with the water contact angle reduced to zero. The ultra-wetting graphene increases the water permeability of the membrane by 126% without any changes in the selectivity. Higher and more stable the TOC removal of the G-PANCMI membrane compared to PANCMA membrane.	Experimental paper (Prince et al. 2016)
15.	CNT/ZnO/TiO$_2$ nanocomposite membrane	Acid Orange 7 (AO7)	Multifunctional nanocomposite ultrafiltration (UF) membrane utilized in photo-catalytic degradation and molecular weight cut-off (MWCO) rejection applications.	High mechanical strength along with high photo-catalytic and disinfection capability to produce clean water at a high membrane flux with low fouling potential.	Experimental paper (Bai et al. 2015)

Table 14.3d contd. ...

...Table 14.3d contd.

No.	CNs type	Target compound	Membrane characterization	Complementary data	Reference
16.	Graphene oxide (GO) membranes	Water filtration	GO thin film composite (TFC) membranes prepared using polyacrylonitrile ultrafiltration membranes as porous support layers.	High water partition coefficient (0.93). The water permeability coefficient of GO membranes in the range $3.0 \times 10^{-7} - 1.0 \times 10^{-6}$ cm^2/s depending on the flake size, the oxidation level, and the drying process (similar value to that of polymeric materials used for desalination, e.g., polyamide).	Experimental paper (Cho et al. 2017)
17.	Photocatalytic membrane M-GOT	Diphenhydramine (DP)	Flat sheet ultrafiltration (UF) membranes with photocatalytic properties synthesized by simple layer-by-layer deposition of TiO$_2$ and graphene oxide GO on a polysulfone membrane in continuous filtration mode under both near-UV/Vis and visible light irradiation.	Between the cellulose membrane and the photocatalytic layer intercalated an innovative freestanding GO membrane. High photocatalytic activity, hydrophilicity and porosity of M-GOT membrane. Considerably higher contaminations removal in darkness and good photocatalytic activity under near-UV/Vis and visible light irradiation.	Experimental paper (Pastrana-Martinez et al. 2015)

Abbreviations used in Tables 3A-D: LG = (2–5 layers), Few-layer graphene; FWNT = Few-walled nanotubes; GQDs = Graphene quantum dots; GFNs = Graphene-family nanomaterials; MLG = (2–10 layers), Multilayer graphene; MWNT = Multiwall nanotubes; N/A = Non-available; pGr = Pristine graphene; ROS = Reactive oxygen species; RGO = Reduced graphene oxide; SSA = The specific surface area of sorbent; SWNT = Single-walled nanotubes; CNTs = Carbon nanotubes; SWCNTs = Single walled carbon nanotubes; MWCNTs = Multi walled carbon nanotubes; GO = Graphene oxide; rGO = Reduced graphene oxide; GNs = Graphene nanosheets.

as the active media for selective and effective removal of different classes of water contaminates. Such materials may act as the classical adsorbents (Table 14.3a), oxidation matrices/catalysts (Table 14.3b) or water disinfection media (Table 14.3c), as well as membrane components in various modes (Table 14.3d). Physicochemical properties of given carbon-based adsorbents can be accurately tuned to act as the selective matrices to remove toxic ions (Tabish et al. 2018, Zhao et al. 2018, Maji et al. 2018), dyes and colorants (Lu et al. 2018, Alwan et al. 2018), drugs and medicine residues (Rostamian and Behnejad 2017, Zhu et al. 2017, Yanyan et al. 2018), and a battery of various low-molecular mass compounds collectively described as endocrine disrupting compounds (EDCs), both natural steroids (Jiang et al. 2018, Bai et al. 2017) as well as artificial endocrine modulators such as bisphenols (Wang et al. 2017). Extensive literature review indicated that the main research interest for the study of new adsorbents is focusing on carbon nanotubes, graphene oxide, and its derivatives. There is also a strong interest in the development of new adsorbing systems involving magnetic particles with carbon nanomaterials, enabling simple, fast, and non-expensive mechanical separation by an external magnet of target chemicals and micropollutants (Wang et al. 2018a, Zhao et al. 2018, Wu et al. 2018, Bai et al. 2017).

Another process that has recently been extensively investigated and can be applied for the removal of micropollutants from water and wastewater by their partial or total mineralization is oxidation. Advanced oxidation protocols, principally involving production of nonselective hydroxyl radicals, include a number of diverse procedures, such as ozonation, Fenton reaction or Fenton-like processes, and electrochemical destruction as well as photo-catalysis. In contrast to the adsorption-based removal of target molecules described above, these processes are more complicated, and technological implementation may significantly raise the water and wastewater purification cost. The advantage of such procedures is that this methodology may result in the transformation of pollutants to simple and non-harmful mineralization products, predominantly carbon dioxide and water. Examples of several applications of carbon nanomaterials, which were used for wastewater processing by oxidation, are presented in Table 14.3b. Target micropollutants can be removed using graphene oxide or multi-wall nanotubes and their derivatives, enabling preparation of complex hybrid matrices with catalytic properties involving metal oxides (Cu_2O, TiO_2, Fe_3O_4). These hybrid materials may work under visible light conditions (Zhu et al. 2018a, Wei et al. 2018, Song et al. 2018) or using ozone (Bai et al. 2018, Huang et al. 2017, Jothinathan and Hu 2018), as well as in the presence of other primary oxidants such as peroxy-monosulfate (Duan et al. 2016, Lin and Zhang 2017). It is noteworthy to say that this type of wastewater purification was able to decrease the level of low molecular mass organic micropollutants, including dyes, phenolic compounds such as bisphenols and selected non-steroidal anti-inflammatory or pain relief drugs, including ibuprofen or paracetamol (Zhu et al. 2018a, Jothinathan and Hu 2018).

Engineered carbon nanomaterials were also broadly tested as the promising materials for alternative water disinfection technologies (Table 14.3c). These materials may exhibit strong antimicrobial properties through diverse mechanisms, also involving oxidation processes described in research papers listed within Table 14.3b. A number of carbon nanomaterials, especially functionalized nanotubes, possess antimicrobial characteristics against a wide range of microorganisms, including bacteria, such as *E. coli* and Salmonella or fungi and viruses. Graphene-related particles have been found to have a strong cytotoxic effect on fungi and both Gram-positive and Gram-negative bacteria, but a relatively low cytotoxic effect on animal and human cells (Saccucci et al. 2018). It has been found that carbon nanotubes (CNTs) possess superior microbial inactivation efficiency, however, disinfection systems based on CNTs should be well protected against leakage of carbon nanoparticles into purified water, due to potential ecotoxicological problems. Graphene and carbon nanotubes may act as strong disinfection agents via different mechanisms, including adhesion, membrane piercing and puncturing, cell wrapping, phospholipids extraction, or oxidative stress induction (Zhu et al. 2018b). Generally, water disinfection systems involving carbon nanoparticles are based on hybrid materials consisting of for instance, metals (silver, titanium, gold) or various magnetic particles (Biswas and Bandyopadhyaya 2016, Salam et al. 2017, Deng et al. 2014, Jiang et al. 2016b).

Table 14.3d lists several examples of carbon-based nanocomposites, mainly consisting of graphene and derivatives, which were recently tested and applied to enhance membrane removal of various contaminations from water and wastewater. Membrane processes, including reverse osmosis, microfiltration, or nanofiltration are considered low-cost methods, because these processes do not require additional chemical reagents. The removal rate using membrane processes is high and this technology is described as selective and time-saving (Piaskowski et al. 2018). Membrane processes are versatile and can be easily optimized using various natural or waste adsorbents, also including activated carbons. It is noteworthy to say that membrane methods can be characterized by higher removal potential with lower effective cost in comparison to common biological wastewater treatment systems. Particularly, graphene and its derivatives were extensively tested as highly selective and permeable filtration membrane components. These nanomaterials enabled high retention of low-molecular mass compounds and ions, especially for water desalination systems (Yang et al. 2016b, Wang et al. 2015, Lai et al. 2016, Zhang et al. 2017b). Recent studies, also summarized in Table 14.3d, clearly indicated that carbon-based materials can be easily embedded in polymeric membranes and usually functionalization of carbon nanoparticles improves compatibility with raw polymeric matrix. An important issue is also that these materials are characterised by high thermal resistance and mechanical stability, and possess suitable conductive and antibacterial properties. Resulting hybrid membranes used to purify water and wastewater are characterised by high flux, high rejection, and antifouling performance. Most recently, there is particular interest in development of multi-dimensional (at large scale) graphene-based nanomaterials. Porous graphene, graphene oxide (GO), and modified GO can be form new membrane material characterised by chemical robustness, unique molecular sieving properties with fast permeation rates (Cho et al. 2017).

2. Preparation of Graphene Oxide and its Preliminary Studies as Stationary Phase Component for Planar Electrochromatographic Separation

Graphene oxide (GO) is one of the most commonly investigated carbon-based nanomaterial derived from graphene, which presents an ordered honeycomb network structure. GO can be described as highly oxidative form of a graphene monolayer. The chemical structure of this substance strongly depends on the synthesis protocol. For example, sonochemical approach that is commonly used as a routine technique for graphene oxide preparation may result in unexpected and uncontrolled removal of oxygen functional groups (Le et al. 2019). Generally, GO may consist of a variety of oxygen containing ligands, including hydroxyl, carboxyl, carbonyl, and epoxy groups. These functional groups are attached to both the sides of the basal plane (e.g., hydroxyl and epoxy groups) or on the GO particle edges (e.g., carboxyl and carbonyl groups) (Nazerah et al. 2016). The consequence of the presence of functional groups in GO structure is that such particles are relatively polar and can be easily dispersed in water, forming fairly stable suspensions. This property of GO is critical for a number of industrial applications. Graphene oxide is especially attractive to the researchers because of low production costs and unique physicochemical properties, enabling various applications, as it was documented previously in Tables 14.1–14.3. It should be noted that GO can be simply synthesized using chemical exfoliation of graphite. This approach is attractive because it does not involve complex equipment or metallic catalysts and therefore, no further complex purification steps are needed. There are a number of approaches for the synthesis of GO that were described in literature, mainly based on Hummers and Offeman protocol (Hummers and Offeman 1958).

Our recent interest in graphene oxide is mainly due to separation science and water purification technological processes. Presently, we conducted preliminary experiments using planar electrochromatography to investigate the potential use of GO particles as an active stationary phase component. This may be used for further application of GO in classical chromatography techniques (liquid chromatography and electrophoresis including their planar and column forms), as well as miniaturized separation devices (e.g., paper-based separation devices as well as micro adsorption bars). Based on available data in literature, our group has elaborated on the graphene oxide synthesis procedure, starting from graphite powder (Hummers and Offeman 1958, Marcano et al. 2010, Panwar et al. 2015, Yuan et al. 2017, Zaaba et al. 2017, Saleem et al. 2018).

Detailed GO production protocol was summarized in Figure 14.2a, where all steps were described and commented on. As can be seen from the pictures included in Figure 14.2b, this approach involved as simple as possible laboratory equipment. The reaction mixture changed the colors as bulk graphite material was transformed into graphene oxide particles (Figure 14.2c). In our experiment, resulting water suspension of GO (raw graphene oxide) was split in two parts (volumes) for different drying treatment: (i) using air circulating oven (60°C, 18 hours) and (ii) lyophilization procedure (sample freezing at temperature –100°C and water evaporation at room temperature inside low pressure chamber for 18 hours). All details of lyophilization procedure and equipment that was used for dehydration of raw GO water suspension are included within Figure 14.3. Using these drying protocols, two different forms of graphene oxide were obtained, as can be seen from large scale and optical microscopy magnified pictures presented in Figures 14.4 and 14.5, respectively. Particularly, water suspension sample, which was dried under air circulating oven conditions, has resulted in a dark/brown solid material, whilst lyophilized sample of GO was transformed into light sponge material with estimated apparent density equal to 14–22 mg/mL.

Accordingly to data presented in Figure 14.6, lyophilized form of graphene oxide can be easily wetted and reconstituted as stable suspension in tap or distilled water at room temperature, in the same way as polar polymer particles, such as starch. The measured values of particle size and zeta potential parameters of lyophilized GO material reconstituted in water as well as for active carbon particles (Norit SA Super) at a concentration of 0.1 mg/mL are presented in Table 14.4. Taking into account data reported by Krishnamoorthy et al. 2013, the observed value of zeta potential at a level of –38 mV for our graphene oxide particles (dispersed in water environment characterised by pH close to 3), confirms the possibility of the existence of a stable GO suspension. This is probably due to the presence of ionized carboxylic groups on the GO surface, resulting in electrostatic repulsive forces between individual particles, and enabling the formation of a stable suspension. On the contrary, activated carbon particles cannot form stable suspension in water under similar particle size conditions (measured zeta potential value for these particles was around –20 mV).

Both forms of graphene oxide (after direct air drying and lyophilization) were inspected by scanning electron microscopy (SEM; Figure 14.7a) and analyzed using energy X-ray dispersive spectroscopy (EDS; Figures 14.7b,c). These measurements were also performed for different carbon particles, which were used as the reference materials, namely graphite (from which GO was produced) and activated carbon (Norit SA Super). SEM data presented within Figure 14.7a clearly indicate that surface of air-dried and lyophilized GO samples have a similar pattern, however, lyophilized graphene oxide is much more porous and contains large scale structures. EDS measurements have revealed that graphite sample consists of pure carbon, whilst Norit SA Super contains several metal impurities. These elements were probably included in vegetable raw material used for Norit SA Super production (ŚwiderskaDąbrowska et al. 2018). Both graphene oxide samples consisted of carbon and oxygen elements, however, in slightly different quantity. This may suggest the difference in the content of oxygen ligands for both materials. The recorded presence of sulfur atoms can be explained as an impurity that remained after GO synthesis, where sulfuric acid was used as the reagent.

In our previous studies, we used combined planar electrophoresis/chromatography system for fast screening of different natural polymers applied as stationary phases for efficient separation of low-molecular mass compounds, particularly dyes (Lewandowska et al. 2017). This simple system was very useful for fast evaluation of separation efficiency and chromatographic/electrophoretic properties of stationary phases, particularly cellulose-based strips and/or classical glass-based thin-layer chromatographic plates. Described separation equipment involved simple open-air electrophoresis box with homemade support for the positioning of active separation layers and electrode connection strips. Using this device, we were able to collect an initial set of retention/electromigration data that

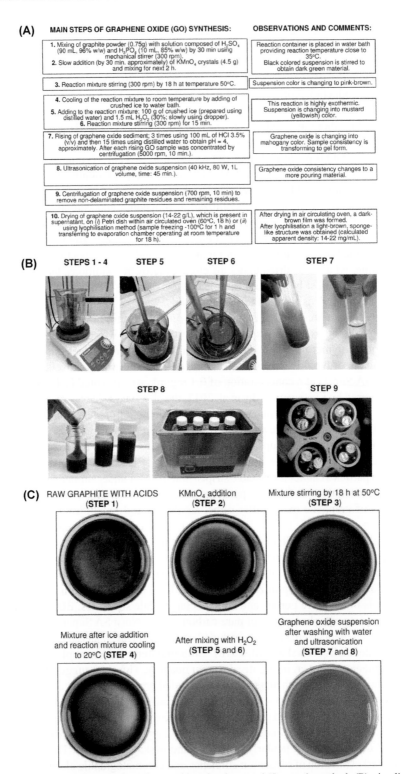

Figure 14.2: (A) Preparation protocol of graphene oxide using improved Hummer's method, (B) visualization of key experiment steps, and (C) observed color changing of the reaction mixture. Authors' unpublished results. All photos and figures copyrights belong to the authors.

Color version at the end of the book

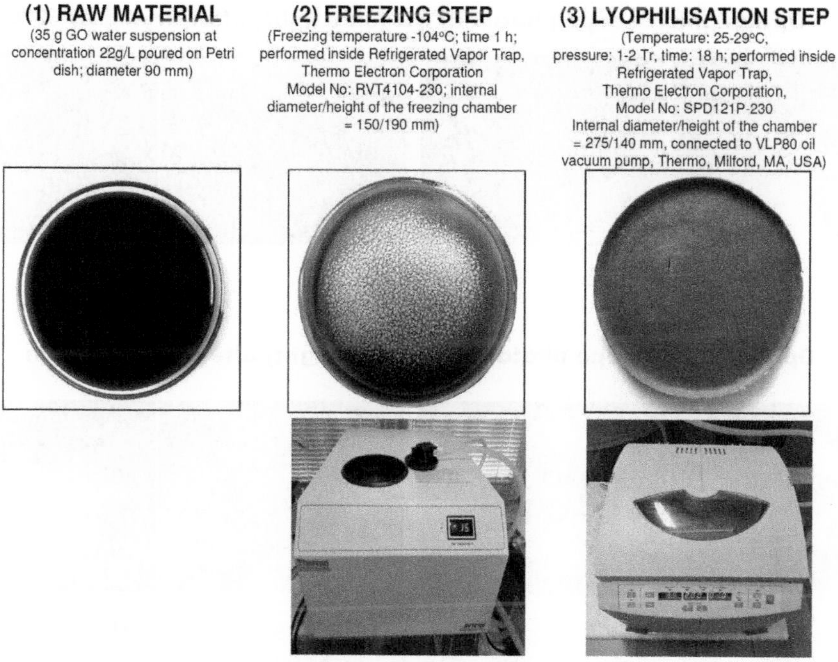

(1) RAW MATERIAL
(35 g GO water suspension at concentration 22g/L poured on Petri dish; diameter 90 mm)

(2) FREEZING STEP
(Freezing temperature -104°C; time 1 h; performed inside Refrigerated Vapor Trap, Thermo Electron Corporation Model No: RVT4104-230; internal diameter/height of the freezing chamber = 150/190 mm)

(3) LYOPHILISATION STEP
(Temperature: 25-29°C, pressure: 1-2 Tr, time: 18 h; performed inside Refrigerated Vapor Trap, Thermo Electron Corporation, Model No: SPD121P-230 Internal diameter/height of the chamber = 275/140 mm, connected to VLP80 oil vacuum pump, Thermo, Milford, MA, USA)

Figure 14.3: Lyophilization protocol of graphene oxide suspension. Authors' unpublished results. All photos and figures copyrights belong to the authors.

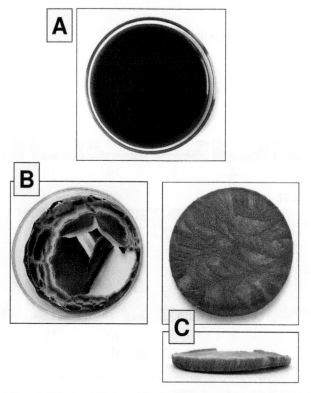

Figure 14.4: (A) Drying of graphene oxide water suspension in different conditions: (B) air circulated oven (temperature 60°C, drying time 18 hours; and (C) lyophilization (sample freezing temperature –100°C/1 h, drying time 18 hours at room temperature. Authors' unpublished results. All photos and figures copyrights belong to the authors.

Optical microscope photos of GO (side light) after drying at 60°C

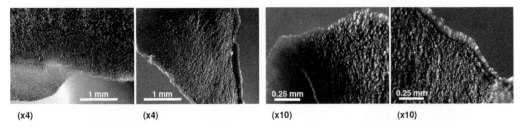

(x4)   (x4)   (x10)   (x10)

Optical microscope photos of GO (side light) after lyophilisation

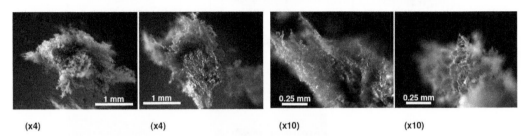

(x4)   (x4)   (x10)   (x10)

Figure 14.5: Visible light views of graphene oxide samples obtained after different drying conditions using Motic BA310 LED optical microscope (Motic China Group, Ltd., Xiamen, China) equipped with Moticam 3 (3.0 MP USB) CMOS digital camera under side light visualization mode. Digital pictures were acquired using Motic Image Plus 2.0 (Motic China Group Co., Ltd., 2007) software. Authors' unpublished results. All photos and figures copyrights belong to the authors.

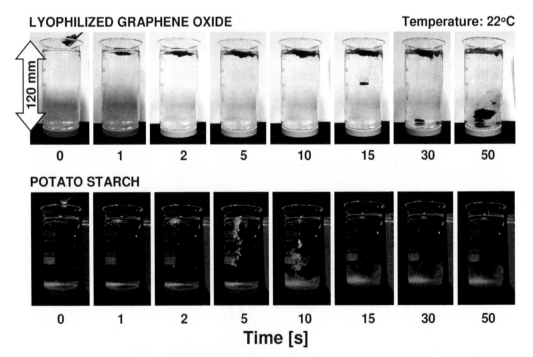

Figure 14.6: Spontaneous wetting, dispersion, and hydration of lyophilized graphene oxide sponge in distilled water (top sequence), in comparison to starch particles (bottom sequence). Authors' unpublished results. All photos and figures copyrights belong to the authors.

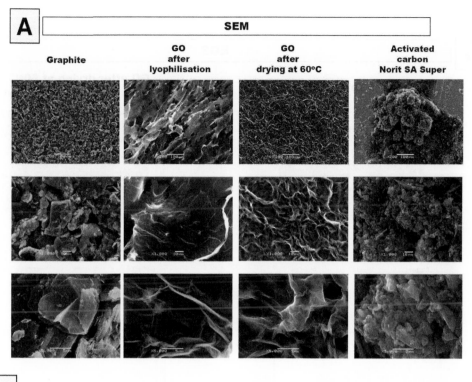

A — SEM

| Graphite | GO after lyophilisation | GO after drying at 60°C | Activated carbon Norit SA Super |

B — EDS

Graphite

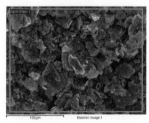

Spectrum processing :
Peak possibly omitted : 6.410 keV

Element	Weight%	Atomic%
C	100.00	100.00
Totals	100.00	

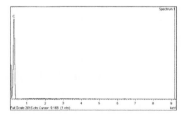

Activated carbon, Norit SA Super

Spectrum processing :
Peak possibly omitted : 2.320 keV

Element	Weight%	Atomic%
C	90.28	94.38
O	5.01	3.93
Mg	0.91	0.47
Al.	0.47	0.22
Si	0.54	0.24
Ca	1.51	0.47
Fe	1.28	0.29
Totals	100.00	

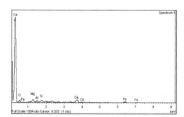

Figure 14.7 contd. ...

...Figure 14.7 contd.

EDS

GO after lyophilisation

GO after drying at 60°C inside air circulated oven

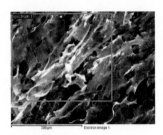

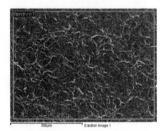

Spectrum processing :
No peaks omitted

Element	Weight%	Atomic%
C	54.01	61.51
O	44.05	37.66
S	1.94	0.83
Totals	**100.00**	

Spectrum processing :
No peaks omitted

Element	Weight%	Atomic%
C	46.07	54.10
O	50.18	44.25
S	3.75	1.65
Totals	**100.00**	

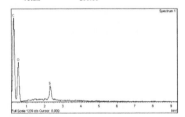

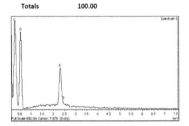

Figure 14.7: (A) SEM images of various carbon materials and (B, C) results of EDS analysis. Scanning Electron Microscopy (SEM) and Energy X-ray Dispersive Spectroscopy (EDS) analyses were performed using JEOL JSM-5500LV electron microscope (JEOL Lts, Japan). Authors' unpublished results. All photos and figures copyrights belong to the authors.

Table 14.4: Measured values of particles size and zeta potential for active carbon (Norit SA Super obtained from Cabot Concern and distributed by Brenntag, Kędzierzyn Koźle, Poland) and graphene oxide (GO was produced in our laboratory using modified Hummers method). Both carbon materials were suspended in distilled water at concentration of 0.1 mg/mL, followed by 15 minutes of ultrasonication. Graphene oxide was reconstituted in water from dry lyophilizate. The pH value of the GO suspension was 2.9 and Norit SA Super was 8.2. The analysis was carried out at a temperature of 25°C using ZetaPALS device (Zeta Potential Analyzer, Brookhaven Instruments Corporation, Holtsville, New York, USA) working with ZetaPALS particle sizing software- 9kpsdw ver. 2.31(1997) and PALS zeta potential analyzer software ver. 3.16 (1998).

Parameter measured		Norit SA Super	Graphene Oxide
Particle size [nm]	Average (*n* = 16)	1548	1368
	Standard deviation	116	214
Zeta Potential [mV]	Average (*n* = 24)	−20	−38
	Standard deviation	3	2

is necessary, for example, for fast prototyping of simple analytical systems for fractionation, and/ or separation of low-molecular mass compounds. In the present, study we used similar separation equipment to test the viability of graphene oxide material as a stationary phase component. This concept was based on considering the physicochemical properties of GO-high polarity enabling dispersion in

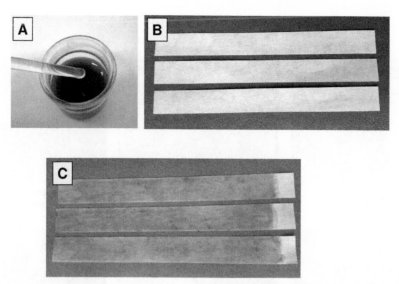

Figure 14.8: (A) Manual application of graphene oxide water suspension with GO concentration ranging between 14–22 mg/mL (B, C) on filtrating paper strips (200 mm × 20 mm, paper thickness 100 μm). Authors' unpublished results. All photos and figures copyrights belong to the authors.

water or water-based electrolytes (with expected normal phase, NP, retention mechanism) and fairly complex surface composition containing oxygen functional groups. These ligands should interact with target analytes and therefore, affect their retention/electromigration behavior. The first approach was simply to cover the cellulose strips with raw graphene oxide suspension in water manually and to dry off the resulting layer at a temperature of 60°C by 24 hours. GO sample for this experiment was the same as obtained in step 9 of the preparation protocol of graphene oxide described within Figure 14.2a. As can be seen from photos presented in Figure 14.9a, the surface of cellulose fibers are homogeneously covered by a thin layer of GO particles, and this carbon material is not filling the space between individual cellulose fibers. The same structure of GO layer on cellulose fibers was observed under SEM imaging (Figure 14.9b). Based on this observation, we have assumed that the movement of electrolyte and target analytes through GO-covered cellulose strip should be possible similarly as on initial cellulose material. Results of electromigration experiment using cellulose coated with GO are presented in Figure 14.10. As target analyte, the sunset yellow was applied. This colorant was found to have relatively high electroplanar migration measured under plain cellulose stationary phase and boric acid electrolite conditions that were tested in our previous study (Lewandowska et al. 2017). The sunset yellow spot was also characterized by no-tailing, which is an important feature due to the relatively low separation efficiency of electrochromatographic system used (Figure 14.10/1a,2b). Interestingly, we did not observe electromigration of tested dye from start line, in the case of GO-covered cellulose strips (Figure 14.10/2a,2b). To investigate this unexpected behavior of the analyte, we have prepared different cellulose strips with a thin horizontal barrier of graphene oxide (Figure 14.10/3a,3b) and also additional reference strips with horizontally positioned wall containing plain graphene powder (Figure 14.10/4a,4b). This experiment demonstrates that GO may act as an efficient barrier for sunset yellow electromigration. Under such conditions, tested dye molecules cannot penetrate the barrier as it is visible for graphite wall, but concentrate in front of the GO wall.

In subsequent experiments, we used commercially available TLC plate with cellulose stationary phase, and manually coated this layer with suspension of raw GO (as described above) and lyophilized GO reconstituted in distilled water. Cross-sections of cellulose-coated chromatographic plates presented in Figure 14.11 clearly indicate that carbon materials may penetrate 100 μm thick stationary phase, however, there are some areas that carbon particles did not penetrate. Interestingly, even such

Figure 14.9: Cellulose fibers from filtrating paper coated with graphene oxide: (A) visible light views and (B) SEM images. Authors' unpublished results. All photos and figures copyrights belong to the authors.

non-homogenous barrier was able to significantly change the migration distance of testing dyes through the plate. This is demonstrated on electrochromatographic plate scans obtained using two testing dyes: sunset yellow and Ponceau 4R. As can be seen, there is a slightly different migration effect observed for the barriers generated from raw and lyophilized GO. Importantly, the target analytes may form dense spots close to the GO barrier, indicating that separation efficiency can be significantly improved at such conditions.

At this stage of the experimental work performed, we may hypothesize that the observed phenomenon of migration of decreased dyes and the concentration of spots in the presence of GO barrier may be predominantly caused by electrostatic interactions between polar dyes consisting of negatively charged ligands and negatively charged functional groups at the surface of GO particles.

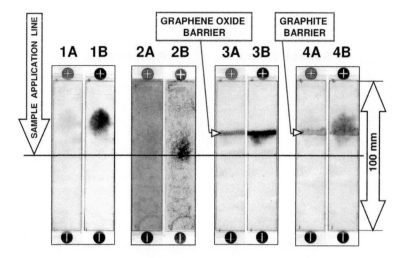

ΔV = 500 V (total length 18 cm; 28 ≈ V/cm)

Figure 14.10: Electroplanar migration of sunset yellow dye on cellulose strips (filtrating paper, active layer length 100 mm): (1A, B) blank cellulose layer, (2A, B) cellulose coated with raw graphene oxide (GO applied from the top and bottom sides), (3A, B) cellulose strips with raw graphene oxide wall (applied to the top side), (4A, B) cellulose strips with graphite wall (applied to the top side); Electrophoresis conditions: applied voltage ΔV = 500 V; run time = 20 minutes; applied dye mass: 2.5 μg; running electrolyte: boric acid 100 mM, pH = 4.9; scanned as the wet strips. Remaining experimental conditions (chamber type and system geometry) was described in the paper (Lewandowska et al. 2017); A—RGB picture with contrast auto-balance general filter applied; B—blue channel extracted from RGB picture. Authors' unpublished results. All photos and figures copyrights belong to the authors.

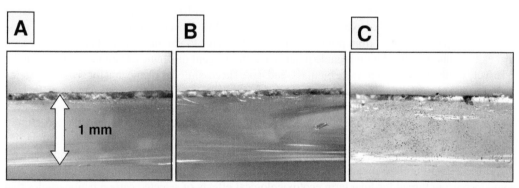

Figure 14.11: Cross-section of glass-based TLC plate with cellulose stationary phase (100 μm), manually coated with GO suspension in distilled water and dried at room temperature (A, raw GO suspension; B, lyophilized GO reconstituted in distilled water; C, activated carbon Norit Super SA). Authors' unpublished results. All photos and figures copyrights belong to the authors.

In the future studies, we will focus on detailed physicochemical analysis of graphene oxide obtained under various conditions, target analytes selection (with and without polar ligands), optimization of selectivity and separation conditions, including electrolyte or buffer composition, ionic strength and pH, separation temperature, and polymer matrix effects, as well as GO barriers geometry and its selectivity for given analytes, also under different chromatographic conditions (column and planar chromatography). It is hoped that the described phenomenon can be implemented for various microextraction bars and microfluidic paper-based analytical devices (μPADs), as well as for designing new filtrating materials that can be used for wastewater treatment and the purification of low-molecular mass micropollutants.

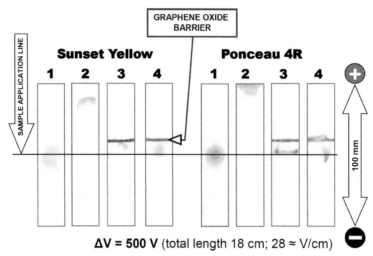

Figure 14.12: Electroplanar migration of sunset yellow and Ponceau 4R dyes on TLC cellulose plates: (1) dye sample application, (2) after 20 minutes electrophoresis, (3) after 20 minutes electrophoresis in the presence of raw graphene oxide wall, (40) after 20 minutes electrophoresis in the presence of lyophilized graphene oxide wall, Electrophoresis conditions: applied voltage ΔV = 500 V; run time = 20 minutes; applied dye mass: 2.5 µg; running electrolyte: boric acid 100 mM, pH = 4.9; scanned as the wet strips. Concentration of graphene in water suspension used for preparation of GO barriers was 22.7 mg/mL. Remaining experimental conditions (chamber type and system geometry) were described in the paper (Lewandowska et al. 2017). Authors' unpublished results. All photos and figures copyrights belong to the authors.

3. Conclusions

There is still an increasing interest in research related to carbon-based nanoparticles and particularly, development of new advanced hybrid materials with given physicochemical and mechanical properties. Carbon nanoparticles have, from principles, high surface area to volume ratio and large surface area that can be functionalized with organic ligands and fluorescent molecules. Carbon materials can be relatively easily derived from waste biomass, and were successfully implemented for sensing systems in biology and medicine. Unique physicochemical properties of carbon-based nanoparticles and hybrid materials enable the creation of new versatile and miniaturized separation systems. They are also commonly implemented as active media for large scale processes, including food, agriculture wastewater treatment, or environmental protection. On the other side, physicochemical properties of given carbon nanoparticles, such as graphene oxide, strongly depends on the production protocol, therefore, it is difficult to obtain standardized materials with predictable properties. There is also some increasing thread for long term ecotoxicological effects of selected carbon nanoparticles that can be present in soil and water ecosystems.

References

Abdelhalim, A.O.E., A. Galal, M.Z. Hussein and I.E.-T. El Sayed. 2016. Graphene functionalization by 1,6-Diaminohexane and silver nanoparticles for water disinfection. J. Nanomater. Article ID 1485280, 7 pages DOI: 10.1155/2016/1485280.

Ahmadi, M., H. Elmongy, T. Madrakian and M. Abdel-Rehim. 2017. Nanomaterials as sorbents for sample preparation in bioanalysis: A review. Anal. Chim. Acta 958: 1–21.

Alwan, S.H., H.A.H. Alshamsi and L.S. Jasim. 2018. Rhodamine B removal on A-rGO/cobalt oxide nanoparticles composite by adsorption from contaminated water. J. Mol. Struct. 1161: 356–365.

Atchudan, R. 2017. Facile green synthesis of nitrogen-doped carbon dots using Chionanthus retusus fruit extract and investigation of their suitability for metal ion sensing and biological applications. Sensor Actuator 246: 497–509.

Bai, X., C. Qin, R. Feng and Z. Ye. 2017. Binary adsorption of 17β-estradiol and bisphenol A on superparamagnetic amino-functionalized graphene oxide nanocomposites. Mater. Chem. Phys. 189: 96–104.

Bai, Z., Q. Yang and J. Wang. 2018. Catalytic ozonation of sulfamethazine antibiotics using Fe_3O_4/multiwalled carbon nanotubes. Environ. Prog. Sustain. 37(2): 678–685.

Bai, H., X. Zan, L. Zhang and D.D. Sun. 2015. Multi-functional CNT/ZnO/TiO_2 nanocomposite membrane for concurrent filtration and photocatalytic degradation. Sep. Purif. Technol. 156: 922–930.

Baig, M.M.F. and Y.C. Chen. 2017. Bright carbon dots as fluorescence sensing agents for bacteria and curcumin. J. Colloid Interf. Sci. 501: 341–349.

Bandi, R., R. Dadigala, B.R. Gangapuram and V. Guttena. 2018. Green synthesis of highly fluorescent nitrogen-doped carbon dots from *Lantana camara* berries for effective detection of lead(II) and bioimaging. J. Photoch. Photobio. B 178: 330–338.

Bing, W., Z. Chen, H. Sun, P. Shi, N. Gao, J. Ren et al. 2015. Visible-light-driven enhanced antibacterial and biofilm elimination activity of graphitic carbon nitride by embedded Ag nanoparticles. Nano Res. 8(5): 1648–1658.

Biswas, P. and R. Bandyopadhyaya. 2016. Water disinfection using silver nanoparticle impregnated activated carbon: *Escherichia coli* cell-killing in batch and continuous packed column operation over a long duration. Water Res. 100: 105–115.

Chatzimitakos, T., A. Kasouni, L. Sygellou, A. Avgeropoulos, A. Troganis and C. Stalikas. 2017. Two of a kind but different: Luminescent carbon quantum dots from Citrus peels for iron and tartrazine sensing and cell imaging. Talanta 175: 305–312.

Chang, Y.-N., J.-L. Gong, G.-M. Zeng, X.-M. Ou, B. Song, M. Guo et al. 2016. Antimicrobial behavior comparison and antimicrobial mechanism of silver coated carbon nanocomposites. Process Saf. Environ. 102: 596–605.

Chen, J., M. Liu, Q. Huang, L. Huang, H. Huang, F. Deng et al. 2018a. Facile preparation of fluorescent nanodiamond-base polymer composites through a metal-free photo-initiated RAFT process and their cellular imaging. Chem. Eng. J. 337: 82–90.

Chen, X., S. Chen, L. Liang, H. Hong, Z. Zhang and B. Shen. 2018b. Electrochemical behaviour of EPD synthesized graphene coating on titanium alloys for orthopedic implant application. Proc. CIRP. 71: 322–328.

Chen, C.-L., K.-C. Li, C.-S. Chiang and Y.-C. Hu. 2015. Graphene oxide as a chemosensitizer: Diverted autophagic flux, enhanced nuclear import, elevated necrosis and improved antitumor effects. Biomaterials 40: 12–22.

Cho, Y.H., H.W. Kim, H.D. Lee, J.E. Shin, B.M. Yoo and H.B. Park. 2017. Water and ion sorption, diffusion, and transport in graphene oxide membranes revisited. J. Memb. Sci. 544: 425–435.

Deng, C.-H., J.-L. Gong, G.-M. Zeng, C.-G. Niu, Q.-Y. Niu, W. Zhang et al. 2014. Inactivation performance and mechanism of *Escherichia coli* in aqueous system exposed to iron oxide loaded graphene nanocomposites. J. Hazard. Mater. 276: 66–76.

Dinh, N.X., D.T. Chi, N.T. Lan, H. Lan, H.V. Tuan, N.V. Quy et al. 2015. Water-dispersible silver nanoparticles-decorated carbon nanomaterials: synthesis and enhanced antibacterial activity. Appl. Phys. A 119: 85–95.

Duan, X., Z. Ao, L. Zhou, H. Sun, G. Wang and S. Wang. 2016. Occurrence of radical and nonradical pathways from carbocatalysts for aqueous and nonaqueous catalytic oxidation. Appl. Catal. B-Environ. 188: 98–105.

Gismondi, A., G. Reina, S. Orlanducci, F. Mizzoni, S. Gay, M.L. Terranova et al. 2015. Nanodiamonds coupled with bioactive metabolites: A nanotech approach for cancer therapy. Biomaterials 38: 22–35.

Grumezescu, A.M. (ed.). 2017. Food Packaging, Nanotechnology in the Agri-Food Industry. Volume 7; Academic Press/Elsevier, London.

Grumezescu, A.M. and A.M. Holban (eds.). 2018. Role of Material Science in Food Bioengineering, Handbook of Food Bioengineering. Volume 19; Elsevier, London.

Guo, Y., Z. Zeng, Y. Li, Z. Huang and Y. Cui. 2018. *In-situ* sulfur-doped carbon as a metal-free catalyst for persulfate activated oxidation of aqueous organics. Catal. Today 307: 12–19.

Hu, X., A. Sun, L. Mu and Q. Zhou. 2016. Separation and analysis of carbon nanomaterials in complex matrix. Trends in Anal. Chem. 80: 416–428.

Huang, Y., W. Xu, L. Hu, J. Zeng, C. He, X. Tan et al. 2017. Combined adsorption and catalytic ozonation for removal of endocrine disrupting compounds over MWCNTs/Fe_3O_4 composites. Catal. Today 297: 143–150.

Hummers, W.S. and R.E. Offeman. 1958. Preparation of graphitic oxide. J. Am. Chem. Soc. 80(6): 1339–1339.

Jiang, l., y. Liu, G. Zeng, S. Liu, X. Hu, L. Zhou et al. 2018. Adsorption of estrogen contaminants (17β-estradiol and 17α-ethynylestradiol) by graphene nanosheets from water: Effects of grapheme characteristics and solution chemistry. Chem. Eng. J. 339: 296–302.

Jiang, L.-H., Y.-G. Liu, G.-M. Zeng, F.-Y. Xiao, X.-J. Hu, X. Hu et al. 2016a. Removal of 17β-estradiol by few-layered graphene oxide nanosheets from aqueous solutions: External influence and adsorption mechanism. Chem. Eng. J. 284: 93–102.

Jiang, Y., D. Liu, M. Cho, S.S. Lee, F. Zhang, P. Biswas et al. 2016b. *In situ* photocatalytic synthesis of Ag nanoparticles (nAg) by crumpled graphene oxide composite membranes for filtration and disinfection applications. Environ. Sci. Technol. 50: 2514–2521.

Jinno, K., H. Ohta, Y. Saito, T. Uemura, H. Nagashima, K. Itoh et al. 1993. Dimethoxyphenylpropyl bonded silica phase for higher fullerenes separation by high-performance liquid chromatography. J. Chromatogr. 648: 71–77.

Jinno, K. (ed.). 1999. Separation of Fullerenes by Liquid Chromatography, RSC Chromatography Monographs, The Royal Society of Chemistry, Cambridge, UK.

Jothinathan, L. and J. Hu. 2018. Kinetic evaluation of graphene oxide based heterogenous catalytic ozonation for the removal of ibuprofen. Water Res. 134: 63–73.

Ju, H. 2018. Functional nanomaterials and nanoprobes for amplified biosensing. Applied Materials Today 10: 51–71.

Kim, S., R. Ou, Y. Hu, X. Li, H. Zhang, G.P. Simon et al. 2018. Non-swelling graphene oxide-polymer nanocomposite membrane for reverse osmosis desalination. J. Memb. Sci. 562: 47–55.

Klumpp, C., K. Kostarelos, M. Prato and A. Bianco. 2006. Functionalized carbon nanotubes as emerging nanovectors for the delivery of therapeutics. BBA 1758: 404–4012.

Koo, C.H., W.J. Lau, G.S. Lai, S.O. Lai, H.S. Thiam and A.F. Ismail. 2018. Thin-film nanocomposite nanofiltration membranes incorporated with graphene oxide for phosphorus removal. Chem. Eng. Technol. 41: 319–326.

Krishnamoorthy, K., M. Veerapandian, K. Yun and S.-J. Kim. 2013. The chemical and structural analysis of graphene oxide with different degrees of oxidation. Carbon 53: 38–49.

Kumawat, M.K., M. Thakur, R.B. Gurung and R. Srivastava. 2017. Graphene quantum dots from Mangifera indica: Application in near-infrared bioimaging and intracellular nanothermometry. ACS Sustainable Chem. Eng. 5: 1382–1391.

Lai, G.S., W.J. Lau, P.S. Goh, A.F. Ismail, N. Yusof and Y.H. Tan. 2016. Graphene oxide incorporated thin film nanocomposite nanofiltration membrane for enhanced salt removal performance. Desalination 387: 14–24.

Laurila, T., S. Sainio and A.M. Caro. 2017. Hybrid carbon based nanomaterials for electrochemical detection of biomolecules. Prog. Mater. Sci. 88: 499–594.

Le, G., N. Chanlek, J. Manyam, P. Opaprakasit, N. Grisdanurak and P. Sreearunothai. 2019. Insight into the ultrasonication of graphene oxide with strong changes in its properties and performance for adsorption applications. Chem. Eng. J. 375: 1212–1222.

Lewandowska, L., E. Włodarczyk, B. Fenert, A. Kaleniecka and P.K. Zarzycki. 2017. A preliminary study for the fast prototyping of simple electroplanar separation systems based on various natural polymers and planar chromatographic stationary phases. JPC-J. Planar Chromat. 30(5): 440–452.

Lien, Z.-Y., T.-Ch. Hsu, K.-K. Liu, W.-S. Liao, K.-Ch. Hwang and J.-I. Chao. 2012. Cancer cell labeling and tracking using fluorescent and magnetic nanodiamond. Biomaterials 33: 6172–6185.

Lin, Y., S. Wu, X. Li, X. Wu, C. Yang, G. Zeng et al. 2018. Microstructure and performance of Z-scheme photocatalyst of silver phosphate modified by MWCNTs and Cr-doped SrTiO₃ for malachite green degradation. Appl. Catal. B-Environ. 227: 557–570.

Lin, K.-Y.A. and Z.-Y. Zhang. 2017. Degradation of Bisphenol A using peroxymonosulfate activated by onestep prepared sulfur-doped carbon nitride as a metal-free heterogeneous catalyst. Chem. Eng. J. 313: 1320–1327.

Litvinova, L.S. and V.N. Zgonnik. 1997. Thin-layer chromatography of C-60 and C-70 fullerenes. JPC-J. Planar Chromat. 10: 38–43.

Liu, J.-M., Y. Hu, Y.-K. Yang, H. Liu, G.-Z. Fang and X. Lu. 2018a. Emerging functional nanomaterials for the detection of food contaminants. Trends in Food Science and Technology 71: 94–106.

Liu, Y., J. Wen, Y. Gao, T. Li, H. Wang, H. Yan et al. 2018b. Antibacterial graphene oxide coatings on polymer substrate. Appl. Surf. Sci. 436: 624–630.

Liu, Y., Y. Liu, M. Park, S.J. Park, Y. Zhang, M.R. Akanda et al. 2017a. Green synthesis of fluorescent carbon dots from carrot juice for *in vitro* cellular imaging. Carbon Lett. 21: 61–67.

Liu, M., Y. Liu, D. Bao, G. Zhu, G. Yang, J. Geng et al. 2017b. Effective removal of tetracycline antibiotics from water using hybrid carbon membranes. Scientific Reports 7: 43717.

Liu, T., B. Yang, N. Graham, W. Yu and K. Sun. 2017c. Trivalent metal cation cross-linked graphene oxide membranes for NOM removal in water treatment. J. Memb. Sci. 542: 31–40.

Lu, N., G. He, J. Liu, G. Liu and J. Li. 2018. Combustion synthesis of graphene for water treatment. Ceram. Int. 44: 2463–2469.

Luo, T.-Y., X. He, J. Zhang, P. Chen, J.-H. Liu, H.-J. Wang et al. 2018. Photoluminescent F-dopped carbon dots prepared by ring-opening reaction for gene delivery and cell imaging. The Royal Society of Chemistry 8: 6053–6062.

Maddigpu, P.R., B. Sawant, S. Wanjari, M.D. Goel, D. Vione, R.S. Dhodapkar et al. 2018. Carbon nanoparticles for solar disinfection of water. J. Hazard. Mater. 343: 157–165.

Maji, S., A. Ghosh, K. Gupta, A. Ghosh, U. Ghorai, A. Santra et al. 2018. Efficiency evaluation of arsenic(III) adsorption of novel graphene oxide@iron-aluminium oxide composite for the contaminated water purification. Sep. Purif. Technol. 197: 388–400.

Marcano, D.C., D.V. Kosynkin, J.M. Berlin, A. Sinitskii, Z.Z. Sun, A. Slesarev et al. 2010. Improved synthesis of graphene oxide. ACS Nano 4(8): 4806–4814.

Mengesha, A.E. 2013. Nanodiamonds for drug delivery systems. pp. 186–205. *In*: R. Narayan (ed.). Diamond Based Materials for Medical Applications. Woodhead Publishing Limited.

Mitura, A.M. and E. Włodarczyk. 2018. Fluorescent nanodiamonds in biomedical applications. J. AOAC Int. 101(5): 1297–1307.

Mitura, K.A., P.K. Zarzycki, M.M. Szczypiński, P. Louda, B. Kolesińska and S.F. Mitura. 2017. Preparation of bioactive food packaging foils; Patent Application No. P.422949, Patent Office of the Republic of Poland, 25.09.2017 Warsaw.

Mitura, K.A. and P.K. Zarzycki. 2018. Modification of packaging surfaces to obtain antioxidant properties by spraying of water dispersed nanodiamonds. Patent Application No. P.424917, Patent Office of the Republic of Poland, 16.03.2018 Warsaw.

Mitura, S., P. Miedzielski, D. Jachowicz, M. Langer, J. Marciniak, A. Stanishevsky et al. 1996. Influence of carbon coatings origin on the properties important for biomedical application. Diam. Relat. Mater. 5: 1185–1188.

Mudshinge, S.R., A.B. Deore, S. Patil and Ch.M. Bhalgat. 2011. Nanoparticles: Emerging carriers for drug delivery. Saudi Pharmaceutical Journal 19: 129–141.

Namvari, M. and H. Namazi. 2016. Magnetic sweet graphene nanosheets: preparation, characterization and application in removal of methylene blue. Int. J. Environ. Sci. Technol. 13: 599–606.

Nazerah, A., A.F. Ismail and J. Jaafar. 2016. Incorporation of bactericidal nanomaterials in development of antibacterial membrane for biofouling mitigation: A mini review. Jurnal Teknologi (Sciences & Engineering) 78:12: 53–61.

Neburkova, J., J. Vavra and P. Cigler. 2017. Coating nanodiamonds with biocompatible shells for applications in biology and medicine. Curr. Opin. Solid St. M 21: 43–53.

Neethirajan, S., V. Ragavan, X. Weng and R. Chand. 2018. Biosensors for sustainable food engineering: challenges and perspectives. Biosensors 8, 23: 1–34.

Ohta, H., Y. Saito, K. Jinno, H. Nagashima and K. Itoh. 1994. Temperature effect in separation of fullerene by high-performance liquid chromatography. Chromatographia 39: 453–459.

Panwar, V., A. Chattree and K. Pal. 2015. A new facile route for synthesizing of graphene oxide using mixture of sulfuric-nitric-phosphoric acids as intercalating agent. Physica E73: 235–241.

Pastrana-Martinez, L.M., S. Morales-Torres, J.L. Figueiredo, J.L. Faria and A.M.T. Silva. 2015. Graphene oxide based ultrafiltration membranes for photocatalytic degradation of organic pollutants in salty water. Water Res. 77: 179–190.

Piaskowski, K., R. Świderska-Dąbrowska and P.K. Zarzycki. 2018. Dyes removal from water and wastewater using various physical, chemical and biological processes. J. AOAC Int. 101(5): 1371–1384.

Prince, J.A., S. Bhuvana, V. Anbharasi, N. Ayyanar, K.V.K. Boodhoo and G. Singh. 2016. Ultra-wetting graphene-based membrane. J. Memb. Sci. 500: 76–85.

Ren, G., M. Tang, F. Chai and H. Wu. 2018. One-pot synthesis of highly fluorescent carbon dots from spinach and multipurpose applications. Eur. J. Inorg. Chem. 153–158.

Rostamian, R. and H. Behnejad. 2017. A unified platform for experimental and quantum mechanical study of antibiotic removal from water. J. Water Process Eng. 17: 207–215.

Rostamian, R. and H. Behnejad. 2018. A comprehensive adsorption study and modeling of antibiotics as a pharmaceutical waste by graphene oxide nanosheets. Ecotox. Environ. Safe. 147: 117–123.

Roy, U., V. Drozd, A. Durygin, J. Rodriguez, P. Barber, V. Atluri et al. 2018. Characterization of Nanodiamond-based anti-HIV drug delivery to the brain. Scientific Reports 8: 1603–1614.

Rouhani, P., N. Govindaraju, J.K. Iyer, R. Kaul, A. Kaul and R.N. Singh. 2016. Purification and functionalization of enanodiamond to serve as a platform for amoxicillin delivery. Mater. Sci. Eng., C 63: 323–332.

Saccucci, M., E. Bruni, D. Uccelletti, A. Bregnocchi, M.S. Sarto, M. Bossù et al. 2018. Surface disinfections: present and future. J. Nanomater. Article ID 8950143, 9 pages; DOI: 10.1155/2018/8950143.

Salam, M.A., A.Y. Obaid, R.M. El-Shishtawy and S.A. Mohamed. 2017. Synthesis of nanocomposites of polypyrrole/carbon nanotubes/silver nano particles and their application in water disinfection. RSC Adv. 7: 16878. DOI: 10.1039/c7ra01033h.

Saleem, H., M. Haneef and H.Y. Abbasi. 2018. Synthesis route of reduced graphene oxide via thermal reduction of chemically exfoliated graphene oxide. Mater. Chem. Phys. 204: 1–7. DOI: 10.1016/j.matchemphys.2017.10.020.

Saito, Y., H. Ohta, H. Nagashima, K. Itoh, K. Jinno, M. Okamoto et al. 1995. Separation of fullerenes with novel stationary phases in microcolumn high performance liquid chromatography. J. Liq. Chromatogr. 18: 1897–1908.

Scida, K., P.W. Stege, G. Haby, G.A. Messina and C.D. Garcia. 2011. Recent applications of carbon-based nanomaterials in analytical chemistry. Anal. Chim. Acta 691: 6–17.

Shan, D., S. Deng, C. He, J. Li, H. Wang, C. Jiang et al. 2018. Intercalation of rigid molecules between carbon nanotubes for adsorption enhancement of typical pharmaceutical. Chem. Eng. J. 332: 102–108.

Shi, J., L. Wang, J. Zhang, R. Ma, J. Gao, Y. Liu et al. 2014. A tumor-targeting near-infrared laser-triggered drug delivery system based on GO@Ag nanoparticles for chemo-photothermal therapy and X-ray imaging. Biomaterials 35: 5847–5861.

Socas–Rodriguez, B., J. Gonzales-Salamo, J. Hernandez-Borges and M.A. Rodriguez-Delgado. 2017. Recent applications of nanomaterials in food safety. Trends in Anal. Chem. 96: 172–200.

Song, S., A. Meng, S. Jiang and B. Cheng. 2018. Three-dimensional hollow graphene efficiently promotes electron transfer of Ag_3PO_4 for photocatalytically eliminating phenol. Appl. Surf. Sci. 442: 224–231.

Sun, H. and P. Wu. 2018. Tuning the functional groups of carbon quantum dots in thin film nanocomposite membranes for nanofiltration. J. Memb. Sci. 564: 394–403.

Sun, X. and Y. Lei. 2017. Fluorescent carbon dots and their sensing applications. TRAC-Trend. Anal. Chem. 89: 163–180.

Suszyński, Z. and P.K. Zarzycki. 2015. New approach for sensitive photothermal detection of C60 and C70 fullerenes on micro-TLC plates. Anal. Chim. Acta 863: 70–77.

Świderska-Dąbrowska, R., K. Piaskowski and P.K. Zarzycki. 2018. Preliminary studies of synthetic dyes adsorption on iron sludge and activated carbons. J. AOAC Int. 101(5): 1429–1436.

Tabish, T.A., F.A. Memon, D.E. Gomez, D.W. Horsell and S. Zhang. 2018. A facile synthesis of porous graphene for efficient water and wastewater treatment. Sci. Rep-Uk 8: 1817. DOI:10.1038/s41598-018-19978-8.

Tao, P., Y. Xu, C. Song, Y. Yin, Z. Yang, S. Wen et al. 2018. A novel strategy for the removal of rhodamine B (RhB) dye from wastewater by coal-based carbon membranes coupled with the electric field. Sep. Puri. Technol. 179: 175–183.

Vasimalai, N., V. Vilas-Boas, J. Gallo, M.F. Cerqueira, M. Menendez-Miranda, J.M. Costa-Fernandez et al. 2018. Green synthesis of fluorescent carbon dots from spices for *in vitro* imaging and tumour cell growth inhibition. Beilstein J. Nanotechnol. 9: 530–544.

Wan, L., Q. Zhao, P. Zhao, B. He, T. Jiang, Q. Zhang et al. 2014. Versatile hybrid polyethyleneimine-mesoporous carbon nanoparticles for targeted delivery. Carbon 79: 123–124.

Wang, Y., J. Zhu, H. Huang and H. Cho. 2015. Carbon nanotube composite membranes for microfiltration of pharmaceuticals and personal care products: Capabilities and potential mechanisms. J. Memb. Sci. 479: 165–174.

Wang, F., X. Lu, W. Peng, Y. Deng, T. Zhang, Y. Hu et al. 2017. Sorption behavior of bisphenol A and triclosan by graphene: comparison with activated carbon. ACS Omega 2(9): 5378–5384.

Wang, W., Z. Xu, X. Zhang, A. Wimmer, E. Shi, Y. Qin et al. 2018a. Rapid and efficient removal of organic micropollutants from environmental water using a magnetic nanoparticles-attached fluorographene-based sorbent. Chem. Eng. J. 343: 61–68.

Wang, Y., L. Liu, J. Hong, J. Cao and C. Deng. 2018b. A novel $Fe(OH)_3/g-C_3N_4$ composite membrane for high efficiency water purification. J. Memb. Sci. 564: 372–381.

Wei, Q., Y. Wang, H. Qin, J. Wu, Y. Lu, H. Chi et al. 2018. Construction of rGO wrapping octahedral Ag-Cu_2O heterostructure for enhanced visible light photocatalytic activity. Appl. Catal. B-Environ. 227: 132–144.

Whitlow, J., S. Pacelli and A. Paul. 2017. Multifunctional nanodiamonds in regenerative medicine: Recent advances and future directions. J. Control Release 261: 62–86.

Wu, Y.-F., H.-Ch. Wu, Ch.-H. Kuan, Ch.-J. Lin, L.-W. Wang, C.-W. Changa et al. 2016. Multi-functionalized carbon dots as theranostic nanoagent for gene delivery in lung cancer therapy. Scientific Reports 6: 21170. DOI: 10.1038/srep21170.

Wu, L.-K., H. Wu, H.-B. Zhang, H.-Z. Cao, G.-Y. Hou, Y.-P. Tang et al. 2018. Graphene oxide/$CuFe_2O_4$ foam as an efficient absorbent for arsenic removal from water. Chem. Eng. J. 334: 1808–1819.

Xi, Y., Z. Liu, J. Ji, Y. Wang, Y. Faraj, Y. Zhu et al. 2018. Graphene-based membranes with uniform 2D nanochannels for precise sieving of mono-/multi-valent metal ions. J. Memb. Sci. 550: 208–218.

Yanyan, L., T.A. Kurniawan, A.B. Albadarin and G. Walker. 2018. Enhanced removal of acetaminophen from synthetic wastewater using multi-walled carbon nanotubes (MWCNTs) chemically modified with NaOH, HNO_3/H_2SO_4, ozone, and/or chitosan. J. Mol. Liq. 251: 369–377.

Yang, Y., X. Yang, Y. Yang and Q. Yuan. 2018. Aptamer-functionalized carbon nanomaterials electrochemical sensor for detection cancer relevant biomolecules. Carbon 129: 380–395.

Yang, L., Z. Li, H. Jiang, W. Jiang, R. Su, S. Luo et al. 2016a. Photoelectrocatalytic oxidation of bisphenol A over mesh of TiO_2/graphene/Cu_2O. Appl. Catal. B-Environ. 183: 75–85.

Yang, G.C.C., Y. Chen, H. Yang and C. Yen. 2016b. Performance and mechanisms for the removal of phthalates and pharmaceuticals from aqueous solution by graphene-containing ceramic composite tubular membrane coupled with the simultaneous electrocoagulation and electrofiltration process. Chemosphere 155: 274–282.

Ye, S.L., J.J. Huang, L. Luo, H.J. Fu, Y.M. Sun, Y.D. Shen et al. 2017. Preparation of carbon dots and their application in food analysis as signal probe. Chinese J. Anal. Chem. 45,10: 1571–1581.

Yin, Z., H. Duoni, H. Chen, J. Wang, W. Qian, M. Han et al. 2018. Resilient, mesoporous carbon nanotube-based strips as adsorbents of dilute organics in water. Carbon 132: 329–334.

Yilmaz, E. and M. Soylak. 2016. Preparation and characterization of magnetic carboxylated nanodiamonds for vortex-assisted magnetic solid-phase extraction of ziram in food and water samples. Talanta 158: 152–158.

Yuan, R., J. Yuan, Y. Wu, L. Chen, H. Zhou and J. Chen. 2017. Efficient synthesis of graphene oxide and the mechanisms of oxidation and exfoliation. Appl. Surf. Sci. 416: 868–877.

Zaaba, N.I., K.L. Foo, U. Hashim, S.J. Tan, W.-W. Liu and C.H. Voon. 2017. Synthesis of graphene oxide using modified hummers method: solvent influence. Procedia Engineer. 184: 469–477.

Zambianchi, M., M. Durso, A. Liscio, E. Treossi, C. Bettini, M.L. Capobianco et al. 2017. Graphene oxide doped polysulfone membrane adsorbers for the removal of organic contaminants from water. Chem. Eng. J. 326: 130–140.

Zarzycki, P.K., H. Ohta, F.B. Harasimiuk and K. Jinno. 2007. Fast separation and quantification of C60 and C70 fullerenes using thermostated micro thin-layer chromatography. Anal. Sci. 23: 1391–1396.

Zarzycki, P.K. 2008. Simple horizontal chamber for thermostated micro-thin-layer chromatography. J. Chromatogr. A 1187: 250–259.

Zarzycki, P.K., M.M. Ślączka, M.B. Zarzycka, E. Włodarczyk and M.J. Baran. 2011. Application of micro-thin-layer chromatography as a simple fractionation tool for fast screening of raw extracts derived from complex biological, pharmaceutical and environmental samples. Anal. Chim. Acta 688: 168–174.

Zarzycki, P.K. and K.A. Mitura. 2018. Application of antioxidant surfaces with detonation nanodiamonds for packaging surface of food products, cosmetics, pharmaceutical formulations and food supplements. Patent Application No. P.424918, Patent Office of the Republic of Poland, 16.03.2018 Warsaw.

Zhan, Z., S. Zhao and M. Xue. 2017. Green preparation of fluorescent carbon dots from water chestnut and its application for multicolor imaging in living cells. Dig. J. Nanomater Bios. 12,2: 555–564.

Zhang, C., Y. Xiao, Y. Ma, B. Li, Z. Liu, C. Lu et al. 2017a. Algae biomass as a precursor for synthesis of nitrogen- and sulfur-co-doped carbon dots: A better probe in Arabidopsis guard cells and root tissues. J. Photoch. Photobio. B. 174: 315–322.

Zhang, H., B. Li, J. Pan, Y. Qi, J. Shen, C. Gao et al. 2017b. Carboxyl-functionalized graphene oxide polyamide nanofiltration membrane for desalination of dye solutions containing monovalent salt. J. Memb. Sci. 539: 128–137.

Zhang, Y., H. Ou, H. Liu, Y. Ke, W. Zhang, G. Liao et al. 2018a. Polyimide-based carbon nanofibers: A versatile adsorbent for highly efficient removals of chlorophenols, dyes and antibiotics. Colloid. Surface A 537: 92–101.

Zhang, B., S. Zou, R. Cai, M. Li and Z. He. 2018b. Highly-efficient photocatalytic disinfection of *Escherichia coli* under visible light using carbon supported Vanadium Tetrasulfide nanocomposites. Appl. Catal. B-Environ. 224: 383–393.

Zhang, Q., K. Li, J. Yan, Z. Wang, Q. Wu, L. Bi et al. 2018c. Graphene coating on the surface of CoCrMo alloy enhances the adhesion and proliferation of bone marrow mesenchymal stem cells. Biochem. Bioph. Res. Co. 497: 1011–1017.

Zhao, D., H. Zhu, C. Wu, S. Feng, A. Alsaedi, T. Hayat et al. 2018. Facile synthesis of magnetic Fe_3O_4/graphene composites for enhanced U(VI) sorption. Appl. Surf. Sci. 444: 691–698.

Zheng, Y., H. Zhang, W. Li, Y. Liu, X. Zhang, H. Liu et al. 2017. Pollen derived blue fluorescent carbon dots for bioimaging and monitoring of nitrogen, phosphorus and potassium uptake in Brassica parachinensis L. RSC Adv. 7: 33459–33465.

Zhu, S., Y.-g. Liu, S.-b. Liu, G.-m. Zeng, L.-h. Jiang, X.-f. Tan et al. 2017. Adsorption of emerging contaminant metformin using graphene oxide. Chemosphere 179: 20–28.

Zhu, J., Z. Zhu, H. Zhang, H. Lu, W. Zhang, Y. Qiu et al. 2018a. Calcined layered double hydroxides/reduced graphene oxide composites with improved photocatalytic degradation of paracetamol and efficient oxidation-adsorption of As(III). Appl. Catal. B-Environ. 225: 550–562.

Zhu, J., J. Hou, Y. Zhang, M. Tian, T. He, J. Liu et al. 2018b. Polymeric antimicrobial membranes enabled by nanomaterials for water treatment. J. Memb. Sci. 550: 173–197.

Index

Color Plate Section

Chapter 1

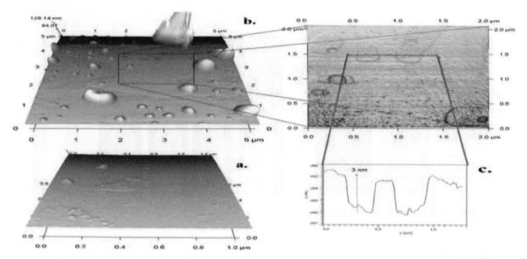

Figure 1.12: (a) and (b) AFM images after the process of chemical exfoliation, (b) and (c) AFM cross-section image and thickness measurements showing several sheets with thickness of 3 nm. Reprinted with permission taken by Elsevier (Betancur et al. 2018).

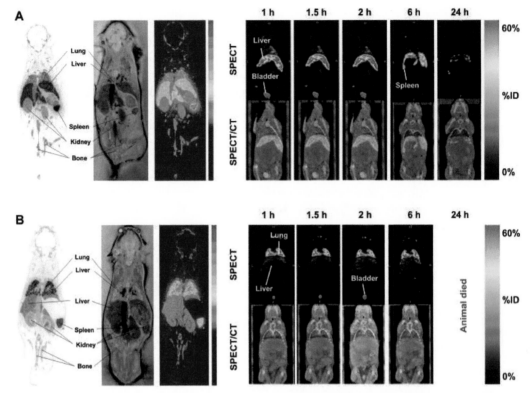

Figure 1.15: Whole body SPECT/CT tomography and volume-rendering image of ^{99m}Tc-labeled (A) PEG-NGO and (B) PEG-LGO. Reprinted with permission taken by Elsevier (Yang et al. 2013).

Chapter 3

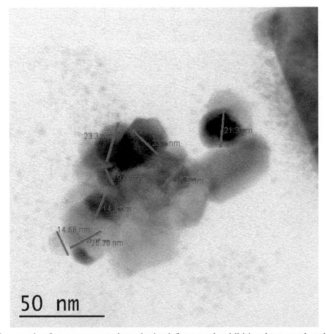

Figure 3.7: TEM micrograph of quantum nanodots obtained from coal-exhibiting hexagonal and pentagonal geometry (Manoj 2017).

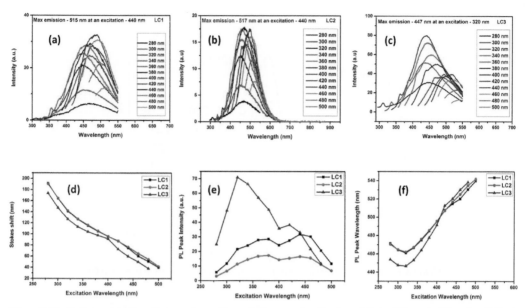

Figure 3.10: Optical properties of nanocarbon with excitation wavelengths from 280 to 500 nm in intervals of 20 nm (a) nanostructure with particle size 21 nm-LC1 (b) nanostructure with particle size 17 nm-LC2 (c) nanostructure with particle size 21 nm-LC3 (d) The stokes shift with excitation wavelength (e) PL peak intensity versus excitation wavelength (f) PL peak wavelength versus excitation wavelength (Manoj 2017).

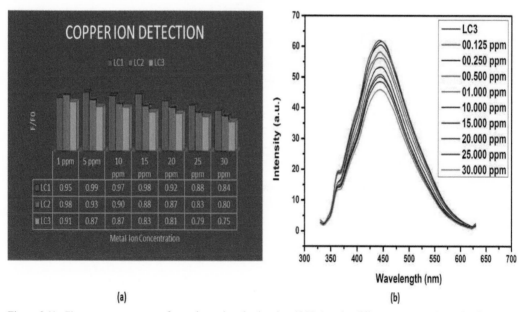

Figure 3.11: Fluorescence response of organic semiconducting dots (OSDs) under different concentrations of Cu^{2+} (Manoj et al. 2017).

Chapter 4

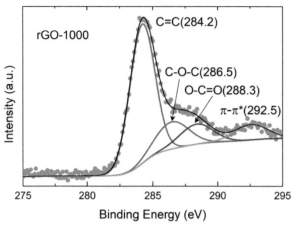

Figure 4.3: The C1s x-ray photoelectron spectroscopy (XPS) spectrum of rGO-1000.

Chapter 7

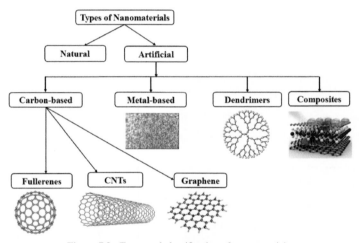

Figure 7.2: Types and classification of nanomaterials.

Chapter 9

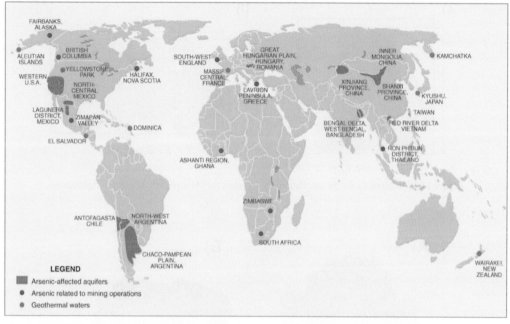

Figure 9.10: Distribution of documented world problems with As in groundwater in major aquifers, as well as water and environmental problems related to mining and geothermal sources. Reprinted from Smedley, P.L. and D.G. Kinniburgh. 2002. A review of the source, behavior, and distribution of arsenic in natural waters. Appl. Geochem. 17: 517–569. Copyright with permission from Elsevier.

Chapter 10

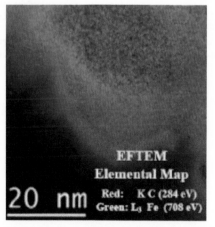

Figure 10.4: Elementary map of a carbon-coated nanoparticle, showing the spatial distribution of the carbon, coating a α-Fe$_2$O$_3$ core.

Chapter 12

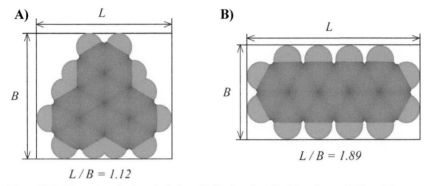

Figure 12.3: Illustrations for the calculation of L/B values for (A) triphenylene and (B) naphthacene.

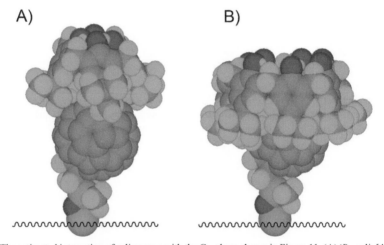

Figure 12.12: The estimated interaction of calixarenes with the C_{60} phase shown in Figure 11. (A) 'Bu-calix[4]arene and (B) 'Bu-calix[8]arene on the C_{60} bonded silica phase (Saito et al. 1996).

Chapter 14

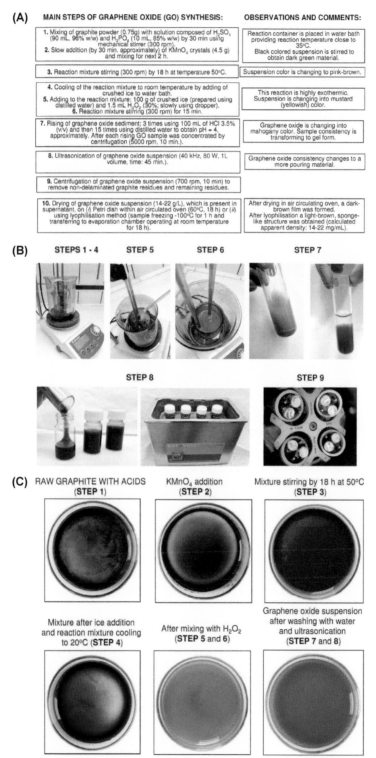

Figure 14.2: (A) Preparation protocol of graphene oxide using improved Hummer's method, (B) visualization of key experiment steps, and (C) observed color changing of the reaction mixture. Authors' unpublished results. All photos and figures copyrights belong to the authors.

Printed and bound by CPI Group (UK) Ltd, Croydon, CR0 4YY

24/10/2024

01778288-0011